Polymer Composites for Electrical Engineering

Polymer Composites for Electrical Engineering

Edited by

Xingyi Huang
Shanghai Key Laboratory of Electrical Insulation and Thermal Aging, Shanghai Jiao Tong University, Shanghai, China

Toshikatsu Tanaka
Waseda University, Tokyo, Japan

IEEE PRESS

WILEY

This edition first published 2022

Registered Offices
John Wiley & Sons, Inc., 111 River Street, Hoboken, NJ 07030, USA
John Wiley & Sons, Ltd., The Atrium, Southern Gate, Chichester, West Sussex, PO19 8SQ, UK

Editorial Office
The Atrium, Southern Gate, Chichester, West Sussex, PO19 8SQ, UK

For details of our global editorial offices, customer services, and more information about Wiley products visit us at www.wiley.com.

Wiley also publishes its books in a variety of electronic formats and by print-on-demand. Some content that appears in standard print versions of this book may not be available in other formats.

Library of Congress Cataloging-in-Publication Data Applied for:

[HB ISBN: 9781119719601]

Cover Design: Wiley
Cover Image: © Nikifor Todorov

Set in 9.5/12.5pt STIXTwoText by Straive, Pondicherry, India
Printed and bound by CPI Group (UK) Ltd, Croydon, CR0 4YY

C9781119719601_191021

Contents

List of Contributors

Lu Bai
College of Polymer Science and Engineering, Sichuan University, State Key Laboratory of Polymer Materials Engineering, Chengdu, Sichuan, P. R. China

Rui-Ying Bao
College of Polymer Science and Engineering, Sichuan University, State Key Laboratory of Polymer Materials Engineering, Chengdu, Sichuan, P. R. China

Qiannan Cai
Key Laboratory of High Performance Plastics (Jilin University), Ministry of Education. College of Chemistry, Jilin University, Changchun, P. R. China

Xue-Meng Cao
China-Spain Collaborative Research Center for Advanced Materials, School of Materials Science and Engineering, Chongqing Jiaotong University, Chongqing, China

Yu Chen
School of Electrical Engineering, Xi'an Jiaotong University, Xi'an, China

Boxue Du
Key Laboratory of Smart Grid of Education Ministry, School of Electrical and Information Engineering, Tianjin University, Tianjin, China

Chang-Ping Feng
College of Polymer Science and Engineering, Sichuan University, State Key Laboratory of Polymer Materials Engineering, Chengdu, Sichuan, P. R. China

Tao Han
Key Laboratory of Smart Grid of Education Ministry, School of Electrical and Information Engineering, Tianjin University, Tianjin, China

Takahiro Imai
Infrastructure Systems Research and Development Center, Toshiba Infrastructure Systems & Solutions Corporation, Toshiba-cho, Fuchu-shi, Tokyo, Japan

Muneaki Kurimoto
Nagoya University, Furo-cho, Chikusa-ku, Nagoya, Japan

Zhi Li
China-Spain Collaborative Research Center for Advanced Materials, School of Materials Science and Engineering, Chongqing Jiaotong University, Chongqing, China

Zhonglei Li
Key Laboratory of Smart Grid of Education Ministry, School of Electrical and Information Engineering, Tianjin University, Tianjin, China

Wang Liming
Engineering Laboratory of Power Equipment Reliability in Complicated Coastal Environments, Tsinghua Shenzhen International Graduate School, Tsinghua University, Shenzhen, Guangdong, China

Suibin Luo
Shenzhen Institute of Advanced Electronic Materials, Shenzhen Institutes of Advanced Technology, Chinese Academy of Sciences, Shenzhen, China

Takahiro Mabuchi
Advanced Technology Research and Development Center, Mitsubishi Electric Corporation, Amagasaki-shi, Tokyo, Japan

Hirotaka Muto
Advanced Technology Research and Development Center, Mitsubishi Electric Corporation, Amagasaki-shi, Tokyo, Japan

Sen Niu
Key Laboratory of High Performance Plastics (Jilin University), Ministry of Education. College of Chemistry, Jilin University, Changchun, P. R. China

Yoitsu Sekiguchi
Sumitomo Electric Industries, Ltd., Osaka, Japan

Rong Sun
Shenzhen Institute of Advanced Electronic Materials, Shenzhen Institutes of Advanced Technology, Chinese Academy of Sciences, Shenzhen, China

En Tang
China-Spain Collaborative Research Center for Advanced Materials, School of Materials Science and Engineering, Chongqing Jiaotong University, Chongqing, China

Takahiro Umemoto
Advanced Technology Research and Development Center, Mitsubishi Electric Corporation, Amagasaki-shi, Tokyo, Japan

Wang Xilin
Engineering Laboratory of Power Equipment Reliability in Complicated Coastal Environments, Tsinghua Shenzhen International Graduate School, Tsinghua University, Shenzhen, Guangdong, China

Jie Yang
College of Polymer Science and Engineering, Sichuan University, State Key Laboratory of Polymer Materials Engineering, Chengdu, Sichuan, P. R. China

Ming-Bo Yang
College of Polymer Science and Engineering, Sichuan University, State Key Laboratory of Polymer Materials Engineering, Chengdu, Sichuan, P. R. China

Wei Yang
College of Polymer Science and Engineering, Sichuan University, State Key Laboratory of Polymer Materials Engineering, Chengdu, Sichuan, P. R. China

Shuhui Yu
Shenzhen Institute of Advanced Electronic Materials, Shenzhen Institutes of Advanced Technology, Chinese Academy of Sciences, Shenzhen, China

Riming Wang
Shenzhen Institute of Advanced Electronic Materials, Shenzhen Institutes of Advanced Technology, Chinese Academy of Sciences, Shenzhen, China

Xueling Yao
School of Electrical Engineering, Xi'an Jiaotong University, Xi'an, China

Yunhe Zhang
Key Laboratory of High Performance Plastics (Jilin University), Ministry of Education. College of Chemistry, Jilin University, Changchun, P. R. China

Yutong Zhao
Key Laboratory of Smart Grid of Education Ministry, School of Electrical and Information Engineering, Tianjin University, Tianjin, China

Jia Zhidong
Engineering Laboratory of Power Equipment Reliability in Complicated Coastal Environments, Tsinghua Shenzhen International Graduate School, Tsinghua University, Shenzhen, Guangdong, China

Yao Zhou
Department of Materials Science and Engineering, The Pennsylvania State University, University Park, PA, USA

Lixue Zhu
Key Laboratory of High Performance Plastics (Jilin University), Ministry of Education. College of Chemistry, Jilin University, Changchun, P. R. China

Preface

Power grid is an indispensable infrastructure in modern society. With the increasingly severe climate change, the transmission of clean energy becomes more and more important, which puts forward higher requirements for the safe operation of power grid and power supply quality. On the other hand, the sending end of clean energy is far away from the receiving end of big cities and industrial areas, so it is necessary to develop Ultra-High Voltage (UHV), long-distance and large capacity transmission technology, as well as distributed generation technology to form modern smart micro grids.

To effectively improve the performance of power grid, especially the acceptance of clean energy by UHV power grid, advanced smart power equipment is needed. The performance of power equipment made of traditional materials is restricted by the electrical characteristics of materials, so it is difficult to meet the development needs of smart grid. Therefore, it is necessary to develop emerging high-performance power equipment materials.

Although metal materials, ceramic materials, and polymer materials are widely used in power equipment, composite materials are more widely used. This is mainly because polymer composites not only have the characteristics of easy processing, lightweight, and low cost of polymers but also have the versatility of other components, which account for an increasing proportion in power equipment. In recent years, with the wide use of clean energy and the development of UHV flexible direct current (DC) transmission technology, some emerging polymer composites (especially nanocomposites) have appeared and been used in engineering practice. Therefore, polymer composites for electrical engineering have been developed rapidly.

This book consists of 13 chapters, which are all contributed by well-recognized experts in electrical engineering or composite materials. Chapters 1 and 2 review the electrical and thermal energy storage of polymer composites, respectively, and Chapters 3 and 4 deal with the thermal properties of polymer composites in electrical engineering, including high temperature endurance and fire-retardant performance. Chapters 5, 6, 9, 10, and 12 focus on the applications of polymer composites in power cables, capacitors, motors, generators, and switchgears. The polymer composites for electric stress control and outer door insulation applications are covered in Chapters 7 and 8, respectively. Chapter 11 provides a detailed overview on design, analysis, and performance of carbon fiber reinforced polymer composites in lightning environment. Chapter 13 reviews the preparation, processing, and application of glass fiber reinforced polymer composites in electrical engineering.

As such, this book is well-structured with basics and application to transfer sufficient coordinated knowledge not only to experts in electrical engineering and material science but also to university students who intend to work in this field in near future.

We greatly thank all contributors for bringing this wonderful book to the community, which is the first comprehensive book focusing on the design, fabrication, processing, analysis, and applications of polymer composites in electrical engineering. Xingyi Huang, one of the coeditors, would like to thank Shanghai Key Laboratory of Electrical Insulation and Thermal Aging and National Natural Science Foundation of China for their long-time support.

1

Polymer Composites for Electrical Energy Storage

Yao Zhou

Department of Materials Science and Engineering, The Pennsylvania State University, University Park, PA, USA

1.1 Introduction

Electrical energy storage in dielectric materials that involves the movement and separation of opposite charges between the two electrodes under the applied electric field is ubiquitous in electrical engineering. The electrostatic capacitors are widely employed in the power electronic devices in various applications, including grid-connected wind farm and photovoltaic, hybrid electric vehicle, medical defibrillator, electric power transmission, etc. [1–4]. Polymer dielectrics have been widely adopted in electrostatic capacitors because of their high breakdown strength, light weight, ease of processing, low cost, and excellent flexibility [5–10]. However, the relatively low dielectric constant restricts the application of polymer dielectrics because of the low discharged energy density [11]. To overcome this issue, a polymer composite approach has been proposed to combine the advantages of polymers and inorganic nanofillers to attain both high dielectric constant and high breakdown strength, as well as low energy loss [12–14].

Over the past decades, great achievements have been made in developing high performance polymer composites for electrical energy storage application. A variety of material structure designs and processing strategies have been proposed to increase the energy density of the polymer composites. This chapter provides a comprehensive understanding of the fundamentals and recent development in the field of polymer composites for electrical energy storage. The strategies of achieving high electrical energy storage performance such as controlling the nanofiller dimension and morphology, controlling the nanofiller distribution and orientation, modifying the nanofiller/polymer interface, introducing multiple nanofillers, and constructing multilayer-structured composites are particularly discussed.

1.2 General Considerations

In general, the stored electrical energy density (U_s) in the dielectric materials can be expressed as

Polymer Composites for Electrical Engineering, First Edition. Edited by Xingyi Huang and Toshikatsu Tanaka.

$$U_s = \int E\mathrm{d}D \tag{1.1}$$

where E is the applied electric field and D is the electric displacement [15]. To measure the stored energy density, discharged energy density (U_d), and charge/discharge efficiency (η) of the dielectrics, the electric displacement (D) is measured as a function of applied electric field (E) during the charging and discharging processes, which is also called as the D-E loop. The schematic of typical D-E loop of the dielectric material is shown in Figure 1.1. The stored energy density can be calculated by integrating the area bounded by the charging curve, the D-axis, and the horizontal line $y = D_{max}$. Similarly, the discharged energy density can be calculated by integrating the area bounded by the discharging curve, the D-axis, and the horizontal line $y = D_{max}$. The energy loss (U_l) during the charge/discharge circle can be represented by the area bounded by the charging and discharging curves. Apparently, the stored energy density equals to the discharged energy density plus the energy loss, that is, $U_s = U_d + U_l$. The charge/discharge efficiency is defined as $\eta = U_d/U_s \times 100\%$. For linear dielectrics, the electric displacement increases linearly with the applied electric field, i.e. $D = \varepsilon_0\varepsilon_r E$, where ε_0 is the vacuum dielectric constant (8.85×10^{-12} F/m) and ε_r is the relative dielectric constant of the dielectric material [16–18]. As a result, the stored energy density of linear dielectrics can be simplified as

$$U_s = \int E\mathrm{d}D = \int \varepsilon_0\varepsilon_r E\mathrm{d}E = \frac{1}{2}\varepsilon_0\varepsilon_r E^2 \tag{1.2}$$

It is suggested from Eqs. (1.1) and (1.2) that the energy density of dielectric materials can be improved by increasing at least one of the two parameters, i.e. the dielectric constant (ε_r) and the electrical breakdown strength (E_b). This is because the dielectric constant determines the electric displacement, while the electrical breakdown strength determines the maximum electric field that can be applied on the dielectric material. For polymer dielectric materials, the dielectric constant is relatively lower than their ceramic counterparts [19–21], so inorganic nanofillers with high dielectric constant are introduced to

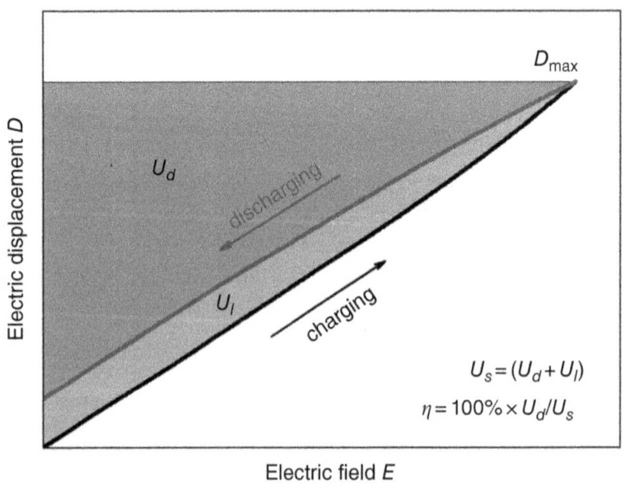

Figure 1.1 Schematic of typical D-E loop of dielectric materials.

improve the dielectric constant of polymer dielectrics. In terms of improving the breakdown strength, highly insulated nanofillers are incorporated into the polymer dielectrics.

For capacitive energy storage in dielectric materials, the stored energy density cannot be fully discharged at the removal of the applied electric field because of various energy losses, such as polarization loss, conduction loss, and hysteresis loss [22]. In most cases, the energy loss would be converted into waste heat, which accelerates the dielectric aging and even causes thermal runaway of the dielectric materials [23, 24]. Therefore, the charge/discharge efficiency is another key parameter for dielectric materials. Since the energy loss is converted into waste heat, which is harmful to the operation of dielectric materials, it is more meaningful to concurrently improve the energy density and the charge/discharge efficiency of the dielectric materials, rather than simply increasing the energy density. In order to increase the charge/discharge efficiency, the energy loss, especially conduction loss under high electric field, should be suppressed. To achieve the inhibition of conduction loss, some highly insulated nanofillers have been introduced into the polymer dielectrics.

Simultaneous enhancement of dielectric constant and breakdown strength is highly desired in the development of high performance polymer composites for electrical energy storage. However, in most cases, the increase in dielectric constant would result in the decrease of breakdown strength, which restricts the improvement of the energy density. So achieving high dielectric constant while maintaining high breakdown strength is still a great challenge. Because of the large mismatch of the electrical parameters between the inorganic nanofillers and polymer matrix, the electric field at the nanofiller/polymer interface is greatly enhanced, which results in the reduced breakdown strength. Moreover, the dispersion of inorganic nanofillers into polymer matrix is always a challenge in the development of polymer composites because of the high surface energy of nanofillers and the different surface physical and chemical properties between the nanofillers and polymer matrix. As a result, nanofiller aggregation is observed in polymer composites. The aggregation of nanofillers would introduce numerous defects in the polymer matrix, which would decrease the electrical breakdown strength and increase the conduction loss. To address these issues, various strategies have been proposed to develop high performance polymer composites for electrical energy storage, including controlling the nanofiller dimension and morphology, controlling the nanofiller distribution and orientation, modifying the nanofiller/polymer interface, introducing multiple nanofillers, and constructing multilayered composites.

1.3 Effect of Nanofiller Dimension

To increase the capacitive energy storage performance of the polymer composites, various inorganic nanofillers have been used to utilize the interfacial polarization and the high dielectric constant of the nanofillers, as well as the inhibition of conduction loss and electrical breakdown. Nanofillers with high dielectric constant, such as titanium oxide (TiO_2), barium titanate ($BaTiO_3$, BT), strontium titanate ($SrTiO_3$), barium strontium titanate ($Ba_xSr_{1-x}TiO_3$, BST), lead zirconate titanate ($PbZr_xTi_{1-x}O_3$, PZT), and calcium copper titanate ($CaCu_3Ti_4O_{12}$, CCTO), have been used to improve the dielectric constant and energy density of the polymer composites [25–30]. Moreover, nanofillers with moderate dielectric constant but highly insulating performance, such as magnesium oxide (MgO), aluminum oxide (Al_2O_3), and silicon oxide (SiO_2), have been used to enhance the breakdown strength and suppress the conduction loss [31–33].

The dimension of the nanofillers can significantly influence the performance of the polymer composites. There are mainly three categories of the nanofillers: 0D fillers, 1D fillers, and 2D fillers. 0D fillers have nanoscale dimensions in all three directions, such as spherical nanoparticles. 1D fillers have nanoscale dimensions in two directions, such as nanowires and nanofibers. 2D fillers have nanoscale dimension in only one direction, such as nanoplates and nanosheets.

Compared with 0D nanofillers, 1D nanofillers are more efficient to increase the dielectric constant of polymer composites at relatively low nanoparticle content because of the lower percolation threshold of the high-aspect-ratio 1D nanofillers compared with 0D nanofillers [34]. For example, Tang et al. showed that the use of high-aspect-ratio BT nanowires can enhance the dielectric constant of poly(vinylidene fluoride-trifluoroethylene-chlorofluoroethylene) [P(VDF-TrFE-CFE)] terpolymer more efficiently than the BT nanoparticles [35]. The dielectric constant of P(VDF-TrFE-CFE)/BT nanowire composite can reach up to 69.5 at 17.5 vol% of BT nanowires, while the dielectric constant of P(VDF-TrFE-CFE)/BT nanoparticle composite is only around 52 at 30 vol% of BT nanoparticles. As the result of enhanced dielectric constant, the P(VDF-TrFE-CFE)/BT nanowire composites exhibit a high discharged energy density of 10.48 J/cm^3, which is 45.3% higher compared with that of the P(VDF-TrFE-CFE) terpolymer of 7.21 J/cm^3.

Shen et al. compared the electrical breakdown propagation in poly(vinylidene fluoride) (PVDF)/BT nanoparticle and PVDF/BT nanofiber composites based on the phase-field simulation [36]. As shown in Figure 1.2, the breakdown phase grows from the nucleation within the composites when the electric field increases beyond the threshold (i.e. 165 kV/mm). In the PVDF/BT nanoparticle composite, the breakdown phase tends to grow at the vulnerable filler/polymer interface and then passes through the fillers near the breakdown path. While in the PVDF/BT nanofiber composite, the breakdown phase propagation is rather different. It is found that the breakdown phase in the nanofiber-based composite tends to penetrate through the nanofibers until the electric field reaches a higher threshold. As a result, the nanofiber-based composite exhibits higher electrical breakdown strength compared with the nanoparticle-based composite. The quantitative breakdown phase growth behavior indicates that the breakdown phase starts to increase when the electric field reaches 140 kV/mm. Compared with nanofiber-based composite, nanoparticle-based composite shows a higher increase rate of the breakdown phase. The breakdown phase in the nanoparticle-based composite gets saturated when the electric field approaches 200 kV/mm, indicating that the nanoparticle-based composite is totally breakdown. In stark contrast, the nanofiber-based composite gets totally breakdown when the electric field reaches 225 kV/mm. The difference in the electrical breakdown behavior comes from the different electric field distribution. The electric field in the nanoparticle-based composite concentrates at the two shoulders along the electric field direction. In the nanofiber-based composite, the electric field concentrates at the vertices of the nanofibers. So the breakdown phase is easier to get through the nanoparticles. Therefore, the nanoparticle-based composite exhibits lower breakdown strength.

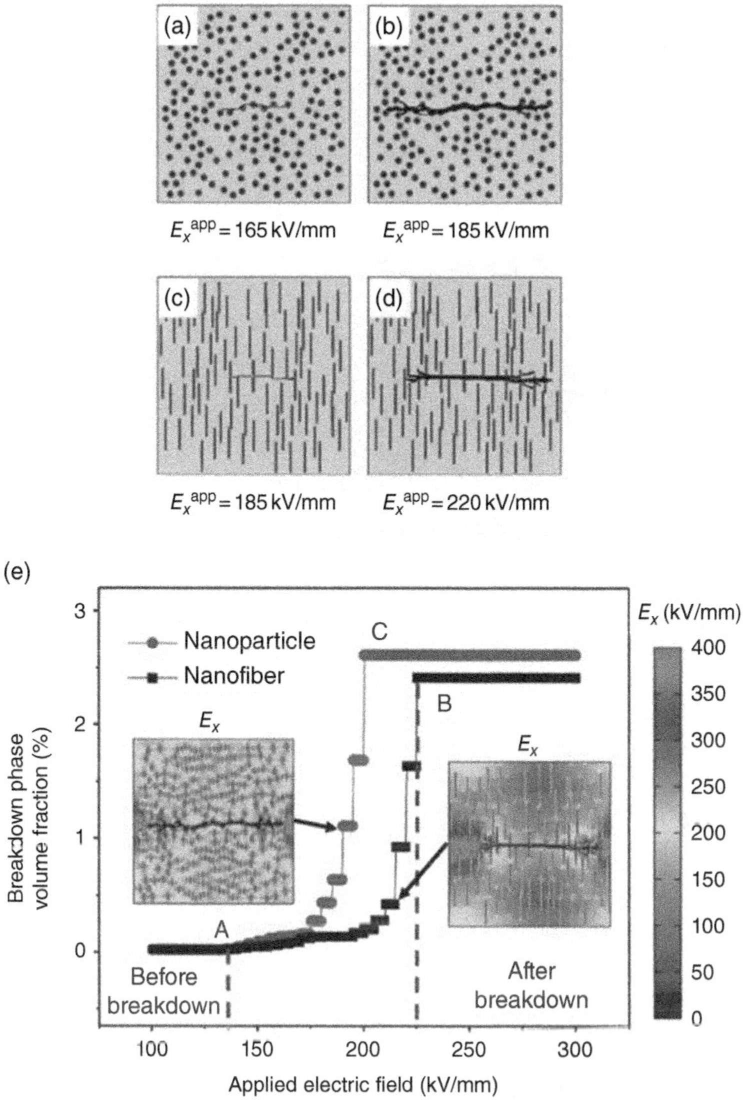

Figure 1.2 The breakdown phase propagation simulation based on phase field model in composites filled with 10 vol% of (a, b) BT nanoparticles and (c, d) BT nanofibers, (e) the evolution of breakdown phase under applied electric field. Insets in (e) show the electric field distribution in corresponding polymer composites. *Source:* Shen et al. [36]. Reproduced with permission of John Wiley & Sons.

Compared with 0D and 1D nanofillers, 2D nanofillers can impede the charge carrier transport and thus suppress the conduction loss because of the special lamellar structure [37–40]. Moreover, the 2D nanofillers can substantially improve the electrical breakdown strength of the polymer composites. Zhu et al. compared the electrical and capacitive energy storage performance of PVDF/TiO_2 composites filled with 5 wt% of 0D, 1D, and 2D TiO_2 nanofillers [41]. It is found that the 2D TiO_2 nanofillers can not only increase the dielectric constant but also suppress the dielectric loss, which is desirable for electrical energy storage application (Figure 1.3). Because the 2D TiO_2 nanofillers can effectively suppress the charge carrier transport, the leakage current of the composite with 2D TiO_2 nanofillers is only a half of that of the composite with 0D TiO_2 nanofillers. The composite with 2D TiO_2 nanofillers also exhibits higher electrical breakdown strength (i.e. 566 MV/m) in comparison to that of the composite with 0D (i.e. 295 MV/m) and 1D (i.e. 382 MV/m)

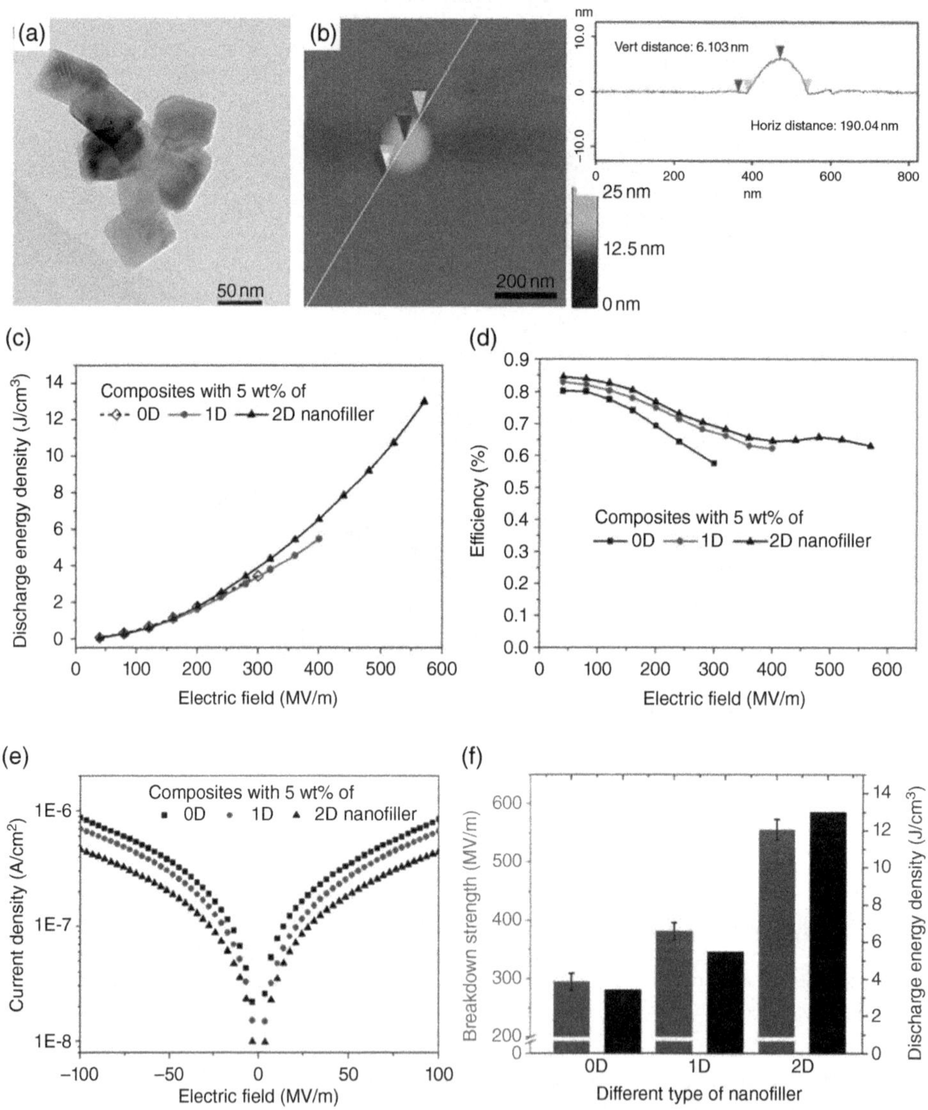

Figure 1.3 (a) TEM image of 2D TiO_2 nanofillers, (b) atomic force microscopy (AFM) image of 2D TiO_2 nanofillers and corresponding surface morphology of the selected line, (c) discharged energy density, (d) charge/discharge efficiency, (e) leakage current density, and (f) electrical breakdown strength of the PVDF/TiO_2 composites with 5 wt% of 0D, 1D, and 2D TiO_2 nanofillers. *Source:* Zhu et al. [41]. Reproduced with permission of American Chemical Society.

TiO_2 nanofillers, representing an increase of 92 and 48%, respectively. As the result of increased breakdown strength and dielectric constant, as well as the suppressed conduction loss, the composite with 2D TiO_2 nanofillers shows the highest discharged energy density of 13 J/cm^3 at 570 MV/m, which is about 236 and 382% that of the composite with 0D (i.e. 5.5 J/cm^3 at 400 MV/m) and 1D (i.e. 3.4 J/cm^3 at 300 MV/m) TiO_2 nanofillers, respectively. Moreover, the charge/discharge efficiency of the composite with 2D TiO_2 nanofillers is also the highest among the three kinds of composites because of the suppressed conduction loss.

Apart from the 2D nanofillers with high dielectric constant, such as TiO_2 nanoplates, 2D nanofillers with high insulating performance, such as hexagonal boron nitride nanosheets (BNNSs), aluminum oxide (Al_2O_3) nanoplates, and montmorillonite (MMT) nanosheets, have been widely used in polymer composites for electrical energy storage because of their unique ability to improve the breakdown strength of the polymer composites [42–44]. For example, owing to the wide bandgap (~6 eV) and excellent breakdown strength (~800 MV/m) of BNNS, the P(VDF-TrFE-CFE)/BNNS composite can reach up to a high electrical breakdown strength of 610 MV/m at the BNNS filler content of 12 wt% (Figure 1.4), which represents an enhancement of 70% as compared with that of the pristine P(VDF-TrFE-CFE) polymer of 362 MV/m [23]. Moreover, the P(VDF-TrFE-CFE)/BNNS composite shows an order of magnitude improvement in the high-field electrical resistivity over the pristine polymer. These results indicate that the highly insulated 2D nanofillers can function as an efficient barrier layer to impede the electrical conduction and breakdown. Because of the increased breakdown strength and electrical resistivity, the P(VDF-TrFE-CFE)/BNNS composite shows a maximum discharged energy density of 20.3 J/cm^3, which is 121% over that of the P(VDF-TrFE-CFE) polymer, i.e. 9.3 J/cm^3. In addition to the enhanced discharged energy density, the charge/discharge efficiency of P(VDF-TrFE-CFE)/BNNS composite is also significantly enhanced, i.e. 83% at 300 MV/m and 80% at 600 MV/m.

1.4 Orientation of Nanofillers

Considering the superiority of high-aspect-ratio nanofillers over 0D nanofillers, constructing anisotropic dielectric polymer composites with aligned nanofillers can fully utilize the high-aspect-ratio nanofillers to tailor the dielectric properties and electrical energy storage performance of the polymer composites. For example, it is shown that when the 1D nanofillers are aligned in parallel to the electric field direction, the enhancement of the dielectric constant is more obvious. While for increasing the electrical breakdown strength, nanofillers aligned perpendicular to the electric field direction are more effective [3]. These phenomena are closely related to the morphology of the nanofillers and the electric field distribution in the polymer composites. Considering that 2D nanofillers are usually used to increase the electrical breakdown strength of the polymer composites, it is obvious that the perpendicularly aligned 2D nanofillers are more effective to impede the charge transport and electrical breakdown propagation.

Using the high-throughput phase-field simulation, Shen et al. demonstrated the effect of 1D and 2D nanofiller alignment on the electrical breakdown strength of the polymer composites filled with 10 vol% nanofillers [36]. The breakdown phase growth is substantially

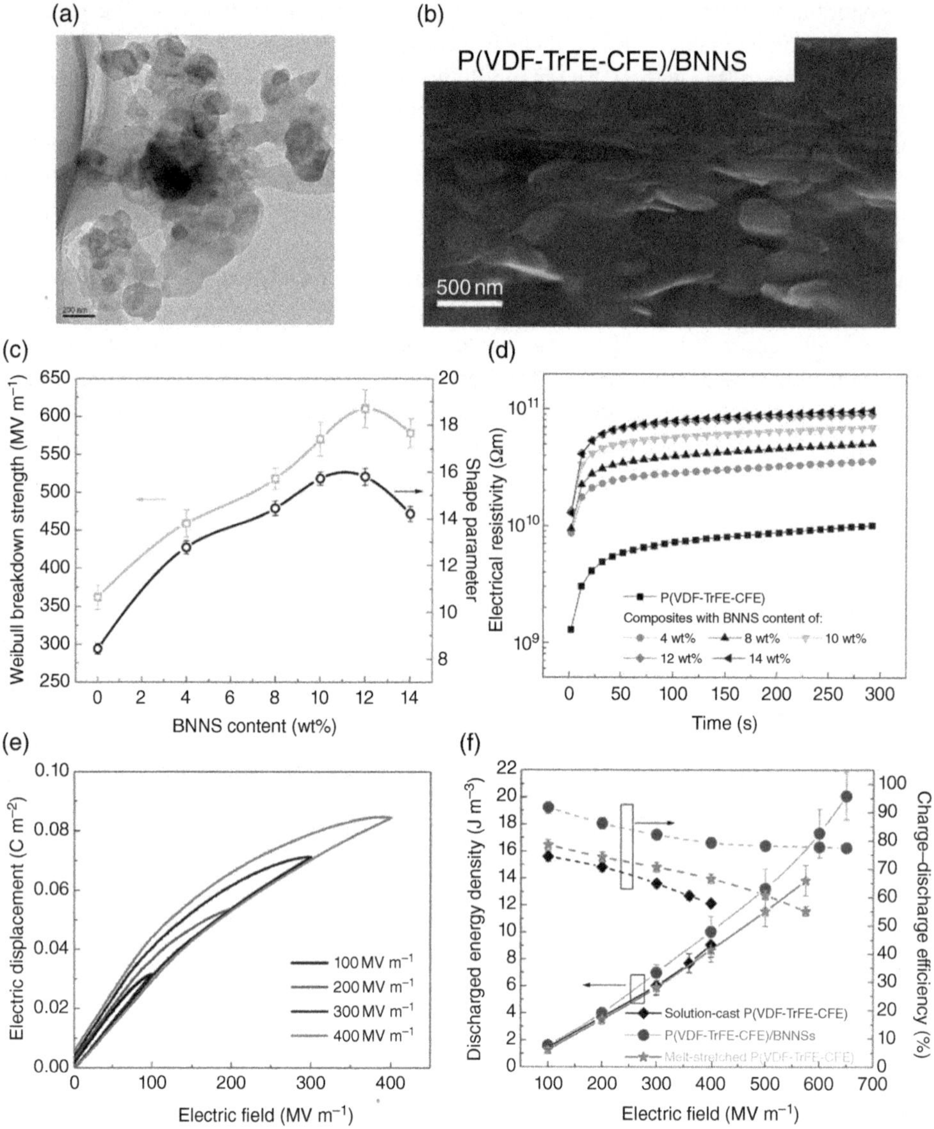

Figure 1.4 (a) TEM image of BNNSs, (b) cross-section scanning electron microscopy (SEM) image of the P(VDF-TrFE-CFE)/BNNS composite, (c) Weibull breakdown strength and shape parameter as functions of BNNS fraction, (d) electrical resistivity of P(VDF-TrFE-CFE)/BNNS composite with different contents of BNNSs under 10 MV/m, (e) *D-E* loops of P(VDF-TrFE-CFE)/BNNS composite with 12 wt% of BNNSs, and (f) discharged energy density and charge/discharge efficiency of pristine P(VDF-TrFE-CFE) and P(VDF-TrFE-CFE)/BNNS composite with 12 wt% of BNNSs. *Source:* Li et al. [23]. Reproduced with permission of The Royal Society of Chemistry.

different in the polymer composites with different nanofiller morphologies and their alignment (Figure 1.5). The polymer composite with vertically aligned nanofibers shows the lowest breakdown strength owing to the condensed electric field distribution that promotes the growth of the breakdown phase. Among the five simulated structures, only the polymer

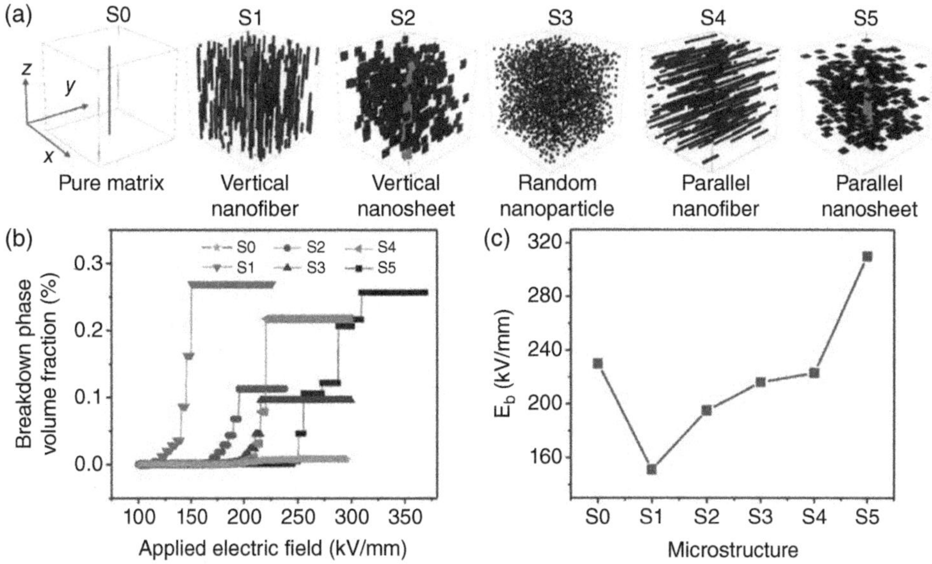

Figure 1.5 3D simulations of microstructure effects on breakdown, (a) breakdown phase morphology in the nanocomposites with different microstructures; (b) evolutions of the breakdown phase volume fraction under applied electric fields; (c) extracted breakdown strengths for corresponding nanocomposites. *Source*: Shen et al. [36]. Reproduced with permission of John Wiley & Sons.

composite with parallel nanosheets exhibits the higher electrical breakdown strength than that of the pure polymer because of the dispersed electric field. These results indicate that the volume fraction of the nanofillers should not be higher than 10 vol% where high electrical breakdown strength is desired.

Various material processing methods have been proposed to align the nanofillers in polymer composites, including dielectrophoresis, film stretching assembly, electrospinning, template-assisted fabrication, using nanorod array, and freeze casting [45–50]. Li et al. prepared polymer composites with 2D MMT nanofillers aligned in the polyethylene (PE) matrix by film blowing [51]. The film blowing process can effectively align the 2D MMT nanofillers parallel to the film surface, and the orientation degree of the nanofillers can be readily controlled by changing the shear rate of the film blowing (Figure 1.6). By changing the shear rate, three PE/MMT composites with different MMT nanofiller orientation factors are obtained. The orientation of the MMT nanofillers is observed by transmission electron microscopy (TEM) and further quantified by two-dimensional wide angle X-ray diffraction (WAXD). The orientation factors for the three composites are 0.70, 0.78, and 0.83, respectively. It is shown that the electrical breakdown strength of the PE/MMT composites significantly increases with the orientation of the 2D MMT nanofillers. The electrical breakdown strength for the composites with low, medium, and high filler orientation factors are 440, 490, and 560 MV/m, respectively, which is substantially higher than that of the pristine PE polymer of 300 MV/m. The oriented 2D MMT nanofillers can act as barriers to add the tortuosity of the electrical breakdown pathway and thus impede the propagation of electrical breakdown. The orientation of the 2D MMT

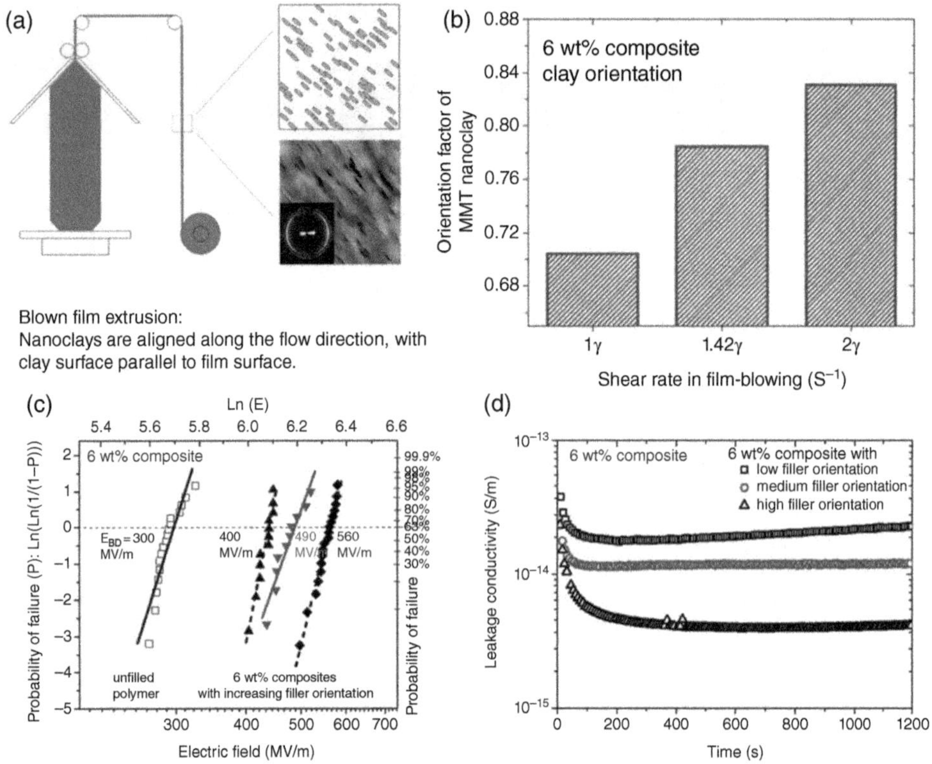

Figure 1.6 (a) Schematic of film blowing process to align the MMT nanofillers along the flow direction, with MMT surface parallel to film surface; (b) orientation factor of the MMT nanofillers in the composite films prepared by different shear rate in film blowing; (c) Weibull distribution and characteristic dielectric breakdown strength; and (d) electrical conductivity under 55 MV/m for the composite films with different MMT nanofiller orientation factor. *Source:* Li et al. [51]. Reproduced with permission of American Chemical Society.

nanofillers is also beneficial to suppress the conduction current and increase the electrical resistivity of the polymer composites. The electrical resistivity of the composite with the highly oriented MMT is four times higher than that of the composite with the lowest filler orientation factor.

Xie et al. proposed a physical-assisted casting method to fabricate the P(VDF-CTFE)/BT nanowire composites with the BT nanowires aligned in the Z direction and X–Y direction [52]. The Z-aligned nanowires are more efficient in improving the dielectric constant of the polymer composites. However, the composite with Z-aligned nanowires shows the lowest electrical breakdown strength. In the Z-aligned composite, the nanowires are aligned parallel to the direction of the applied electric field, so the inter-filler distance is relatively small, which makes the breakdown channels traveling along the nanowires at relatively low electric field. Whereas, the X–Y aligned nanowires can act as barriers for the high-energy charge carriers, which increase the electrical breakdown strength of the polymer composites.

1.5 Surface Modification of Nanofillers

Well dispersion of the nanofillers is critical in achieving high performance polymer composites [53–56]. The inorganic nanofillers are not well compatible with the polymer matrix because of their high tendency to aggregate due to the high surface energy. So the nanofillers are usually surface modified in order to improve the compatibility with the polymer matrix and achieve homogenous nanofiller dispersion. The surface modification of the nanofillers is beneficial in three aspects: (i) removal of the impurities such as water molecules and ions absorbed on the surface of the nanofillers, (ii) elimination of the filler aggregation and getting rid of the defects at the nanofiller/polymer interface such as voids and pores, and (iii) achieving uniform dispersion of the nanofiller and minimizing the local electric field distortion at the nanofiller/polymer interface.

The surface modification of the nanofillers can be accomplished by the reaction between the surface groups on the nanofillers and the small molecular coupling agent, such as the silane coupling agent, aluminate coupling agent, zirconate coupling agent, phosphonic acid, and dopamine. In order to improve the electrical breakdown strength of the high-dielectric-constant PVDF/BT composites, Hu et al. adopted the titanate coupling agent to modify the surface of BT nanoparticles [57]. It is shown that the surface modification of the BT nanoparticles can efficiently improve the compatibility between the BT nanoparticles and PVDF matrix, which ensures the uniform distribution of the BT nanoparticles. Benefited from the uniformly distributed BT nanoparticles, the electrical breakdown strength of the PVDF/BT composite can reach to 517 MV/m, which is even higher than that of the pristine PVDF polymer. Because of the concurrently increased dielectric constant and breakdown strength in the PVDF composite with surface-modified BT nanoparticles, the discharged energy density reaches 11.27 J/cm^3, which is nearly the double of that of the composite with non-surface-modified BT nanoparticles.

To further improve the compatibility between the inorganic nanofillers and polymer matrix, core-shell structured nanoparticles can be prepared by various "grafting from" and "grafting to" methods [58–62]. The core-shell structured nanoparticles can provide unique merits and flexibility in optimizing the dielectric properties of the shell layer. The mismatch of the electrical parameters (i.e. dielectric constant and electrical resistivity) of the nanofillers and polymer matrix would result in local electric field concentration at the nanofiller/polymer interface, which largely decreases the electrical breakdown strength of the polymer composites. In order to mitigate the electric field concentration, the shell layer of the core-shell structured nanofillers can act as a buffer layer by carefully controlling the electrical parameters of the shell layer. For example, the shell layer with medium dielectric constant between the high-dielectric-constant nanofillers and the low-dielectric-constant polymer matrix can obviously mitigate the local electric field distortion.

Zhu et al. prepared a series of BT-based core-shell structured nanoparticles with various polymer shells, i.e. poly(methyl methacrylate) (PMMA), poly(hydroxyethyl methacrylate) (PHEMA), and poly(glycidyl methacrylate) (PGMA), by surface-initiated reversible-addition-fragmentation chain transfer (RAFT) polymerization [63]. The different polymer shells have different electrical parameters, for example, the order of the dielectric constants of the shell layers is PHEMA>PGMA>PMMA and the order of the electrical conductivities of the shell layers is PHEMA>PGMA>PMMA. The experimental results show that the core-shell structured nanoparticles with high-dielectric-constant shell layer (i.e. PHEMA) can result in high dielectric constant and high discharged energy density of the polymer composite (Figure 1.7). However, due to the high electrical conductivity of PHEMA, the corresponding composite also shows the lowest breakdown strength and charge/discharge efficiency because of the high electrical conduction in the polymer composite. In contrast to PHEMA shell layer, the PMMA shell layer with low dielectric constant but high electrical resistivity can effectively improve the breakdown strength and charge/discharge efficiency of the polymer composite. So the optimized shell layer should simultaneously possess high dielectric constant and high electrical resistivity. However, it is sometimes

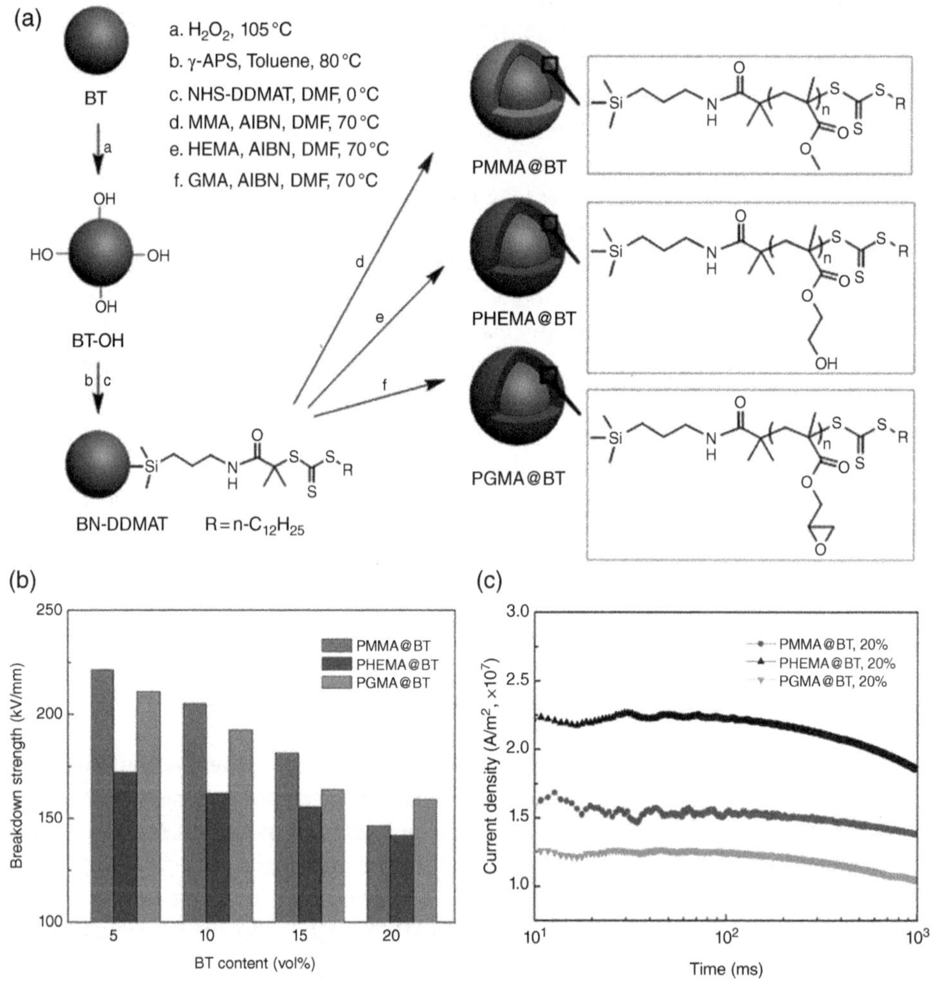

Figure 1.7 (a) Schematic illustrating the preparation process of the core-shell structured polymer@BT nanoparticles by surface-initiated RAFT, (b) breakdown strength, and (c) leakage current density at 100 kV/mm of the polymer@BT-based PVDF composites. *Source:* Zhu et al. [63]. Reproduced with permission of American Chemical Society.

difficult to find one material with both high dielectric constant and high electrical resistivity. To solve this problem, core-double-shell structured nanoparticles have been proposed to combine the advantages of the two shell layers [64].

In addition to improving the compatibility of the nanofillers, the shell layer of the core-shell structured nanoparticles can also act as a functional layer to tailor the performance of the polymer composites. It is well documented that the deep traps introduced by chemical modification of grafting functional groups onto polymer chains can slow down the charge carrier transport, thus suppressing the conduction current in polymers [65–68]. So, the functional group grafted polymers can be used to act as the shell layer of the core-shell structured nanoparticles. Zhou et al. modified the surface of MgO nanoparticles with polypropylene-graft-maleic anhydride (PP-g-mah) and blended the PP-mah-MgO core-shell structured nanoparticles with polypropylene (PP) matrix [69]. PP-g-mah is used because the functional groups in PP-g-mah can offer deep traps to suppress the conduction current. Moreover, the similar physical and chemical characteristics of PP-g-mah and PP matrix make the PP-mah-MgO nanofillers highly miscible with the PP matrix (Figure 1.8). By using the proposed nano-isothermal surface potential decay (nano-ISPD) method, the local deep charge traps introduced by the shell layer is directly probed with nanoscale spatial resolution at the nanofiller/polymer interface. It is found that the surface potential decay at the interfaces in the PP-mah-MgO/PP composite is much slower than that of the un-MgO/PP composite, indicating that the interfacial charge traps in PP-mah-MgO/PP composite are much higher than that of the un-MgO/PP composite, which verifies the design of the shell layer. Benefited from the interfacial deep charge traps introduced by the shell layer, the PP-mah-MgO/PP composite exhibits superior electrical energy storage performance with excellent temperature stability.

In addition to the polymer shell layers, inorganic materials including SiO_2, TiO_2, and Al_2O_3 have been employed as the shell layers of the core-shell structured nanofillers [70–74]. The highly insulated shell layers, such as SiO_2 and Al_2O_3, can serve as a barrier layer to suppress the electrical conduction and breakdown, yielding increased electrical breakdown strength and charge/discharge efficiency. The shell layers with medium dielectric constant, such as TiO_2, can act as a buffer layer to mitigate the local electric field distortion in the high-dielectric-constant nanofiller/polymer composites. Moreover, the core-shell structured strategy is also applicable to the high-aspect-ratio 1D and 2D nanofillers.

1.6 Polymer Composites with Multiple Nanofillers

As mentioned before, the discharged energy density is related to both the dielectric constant and electrical breakdown strength of the dielectrics. So both high dielectric constant and high electrical breakdown strength are required. However, high-dielectric-constant nanofillers can only increase the dielectric constant of the polymer composites but decrease the electrical breakdown strength because of the local electric field distortion. While the high-insulating nanofillers usually exhibit low dielectric constant. Then, a strategy of intro ducing multiple nanofillers with different functionalities has been proposed to simultaneously increase the dielectric constant and the electrical breakdown strength. The benefit of the multiple nanofillers doping comes from the combined advantages of every single

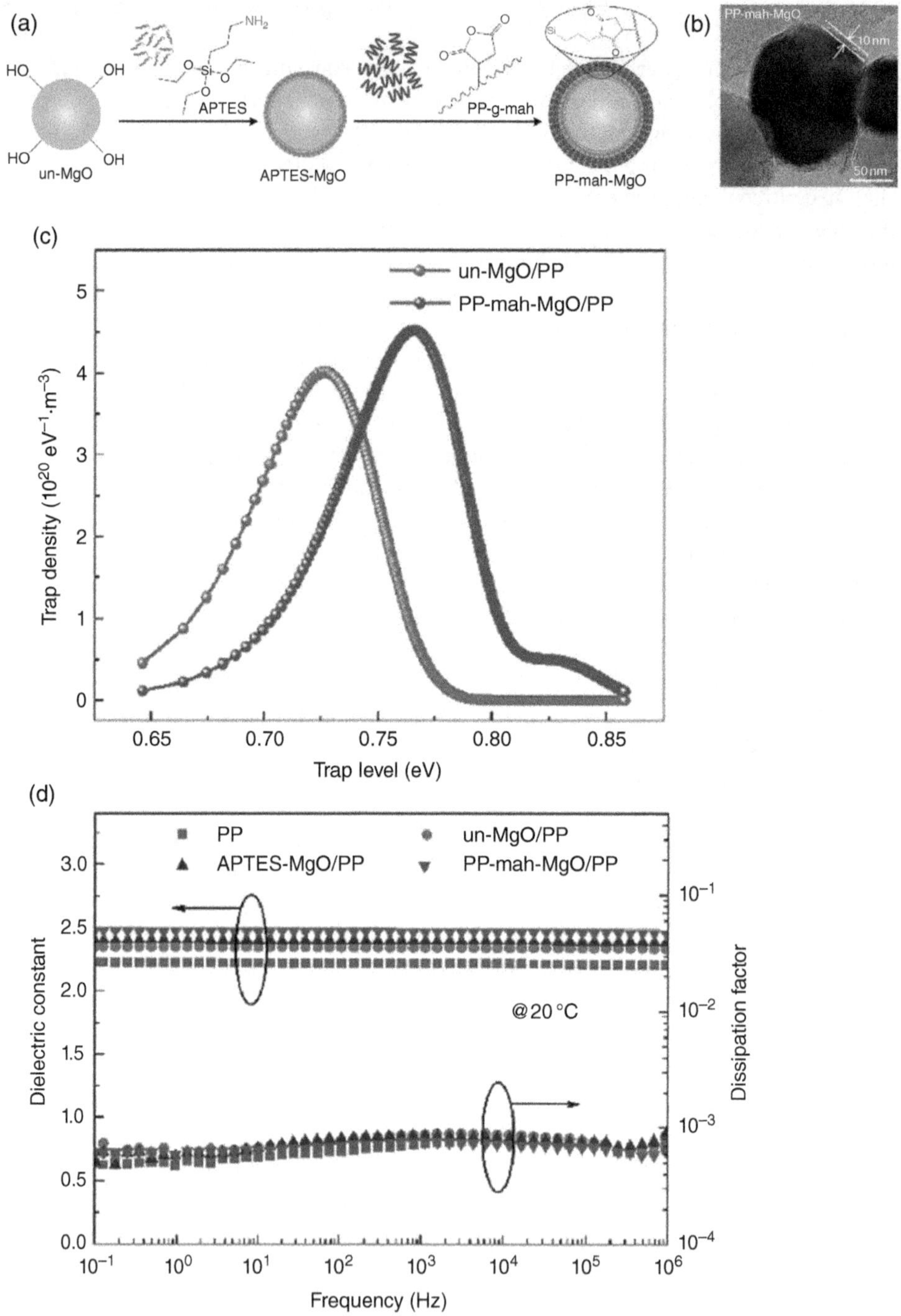

Figure 1.8 (a) schematic of the preparation of the core-shell structured pp-mah-mgo nanoparticles, (b) tem image of the pp-mah-mgo nanoparticle, (c) local charge trap level distribution at the interfacial region obtained from nano-ispd measurement, (d) frequency-dependent dielectric constant and dissipation factor at room temperature, (e) temperature-dependent breakdown strength, and (f) discharged energy density and charge/discharge efficiency at 120 °C. *Source:* zhou et al. [69]. Reproduced with permission of elsevier.

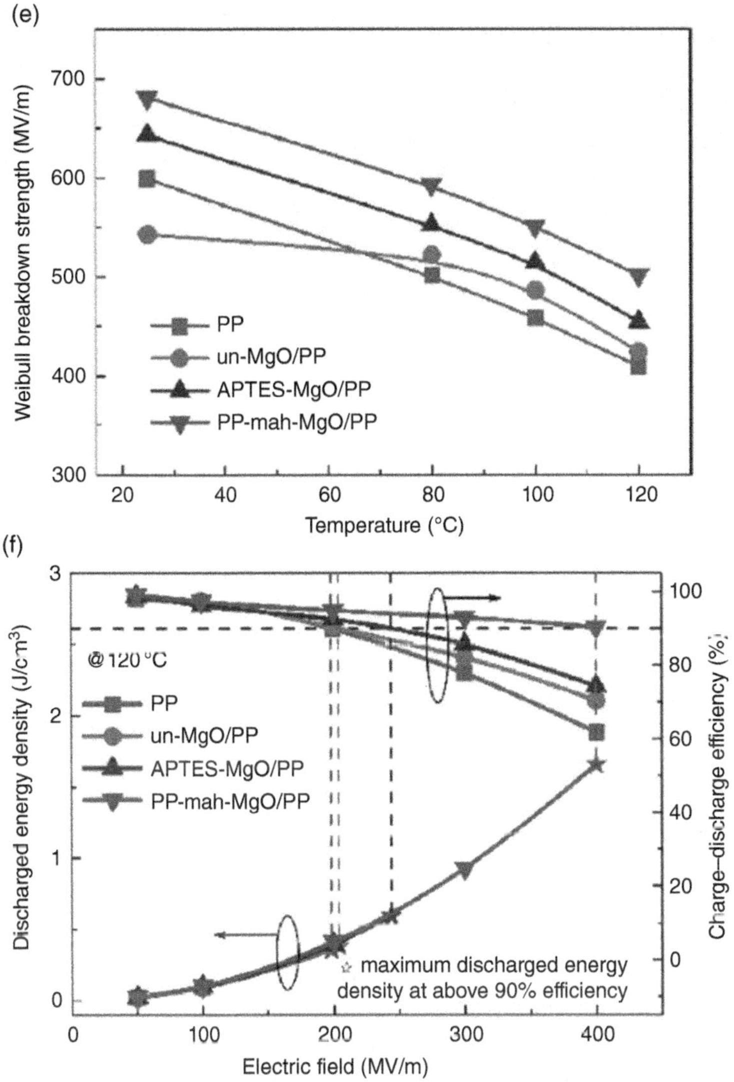

Figure 1.8 (Continued)

nanofiller, i.e. highly insulated nanofillers to increase the electrical breakdown strength and high-dielectric-constant nanofillers to promote the dielectric constant [75–78].

Liu et al. introduced the highly insulated BNNS and high-dielectric-constant BST nanowires into P(VDF-TrFE-CFE) terpolymer [79]. With the inclusion of BST nanowires, the dielectric constant of the ternary P(VDF-TrFE-CFE)/BNNS/BST nanowires composite reaches 52.7, which is 15% improvement over the pristine P(VDF-TrFE-CFE) polymer (Figure 1.9). The electrical breakdown strength of the ternary composite with 5 wt% BST nanowires and 12 wt% BNNS is 589 kV/mm, which is 62% higher than that of the pristine

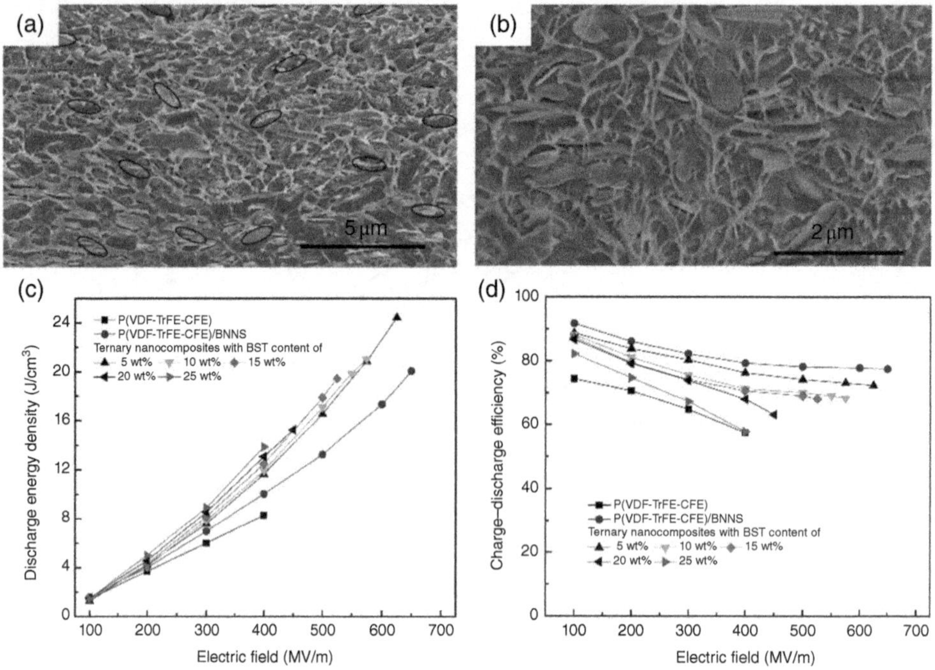

Figure 1.9 (a) Large-scale cross-section SEM image of the ternary nanocomposite, the dashed lines point out the location of BST nanowires while the ellipses indicate the existence of the BNNS. (b) Cross-section SEM image of the ternary nanocomposite, (c) discharged energy density, and (d) charge/discharge efficiency of the ternary nanocomposites. *Source:* Liu et al. [79]. Reproduced with permission of Elsevier.

P(VDF-TrFE-CFE) of 363 kV/mm. Owing to the concurrently increased dielectric constant and breakdown strength, the ternary composite with 5 wt% BST nanowires and 12 wt% BNNS exhibits a super discharged energy density of 24.4 J/cm^3 at an electric field of 625 MV/m, which is 295% that of the pristine polymer matrix.

In addition to directly adding multiple nanofillers in the polymer matrix, Luo et al. prepared a hybrid nanofiller with BT nanoparticles tightly embedded in BN by calcining the BT nanoparticles and BN at high temperature [80]. By employing the BN as the host of the BT nanoparticles, not only the dispersion of the BT nanoparticles is promoted but also the local electric field distortion caused by the BT nanoparticles is mitigated. Introducing the hybrid nanofillers into PVDF can achieve a high electrical breakdown strength of 580 kV/mm and a high discharged energy density of 17.6 J/cm^3 (Figure 1.10). Some other methods such as covalent bonding can also be used to prepare the hybrid nanofillers [81].

1.7 Multilayer-structured Polymer Composites

In addition to incorporating multiple nanofillers strategy, one can also combine the advantages of each kind of nanofiller by constructing multilayer-structured polymer composites with different functional layers. For example, polymer composite with

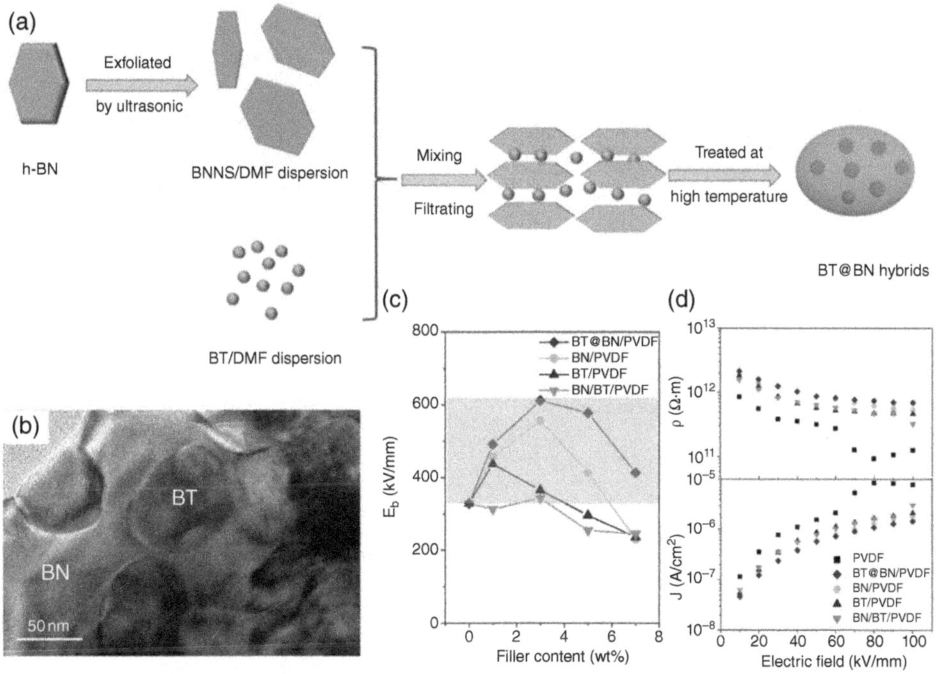

Figure 1.10 (a) Schematic of the preparation process of the BT@BN hybrid nanofillers, (b) TEM image of the prepared BT@BN hybrid nanofillers, (c) electrical breakdown strength, and (d) electrical resistivity and leakage current density of PVDF-based composites with different type of nanofillers. *Source:* Luo et al. [80]. Reproduced with permission of John Wiley & Sons.

high-dielectric-constant nanofillers can act as the high electric displacement layer, which promotes the polarization and achieves high dielectric constant. While the polymer composite with highly insulating nanofillers can act as the charge barrier layer, which impedes the charge carrier transport and electrical breakdown propagation. The multilayer-structured polymer composites can integrate the advantages of each functional layer by rationally designing every single layer, as well as the spatial arrangement of the different layers. With the rationally designed multilayer structure, the polymer composite can simultaneously achieve the high discharged energy density and high charge/discharge efficiency [82–86]. Moreover, the multilayer-structured strategy is also compatible with other strategies, such as nanofiller surface modification and nanofiller orientation, which can further improve the electrical energy storage performance of the polymer composites.

Liu et al. prepared a trilayer-structured polymer composite composed of two outer layers of PVDF/BNNS composite and the middle layer of PVDF/BST nanowire composite [87]. The outer PVDF/BNNS layers functionalized with highly insulating BNNS provide the trilayer-structured film with high electrical breakdown strength, while the middle PVDF/BST nanowire layer improves the polarization of the trilayer-structured film. By carefully controlling the nanofiller content in each layer, a high breakdown strength of 588 MV/m can be obtained by the redistribution of the electric field and the suppression of electrical tree development by the outer PVDF/BNNS layers (Figure 1.11). With the high dielectric

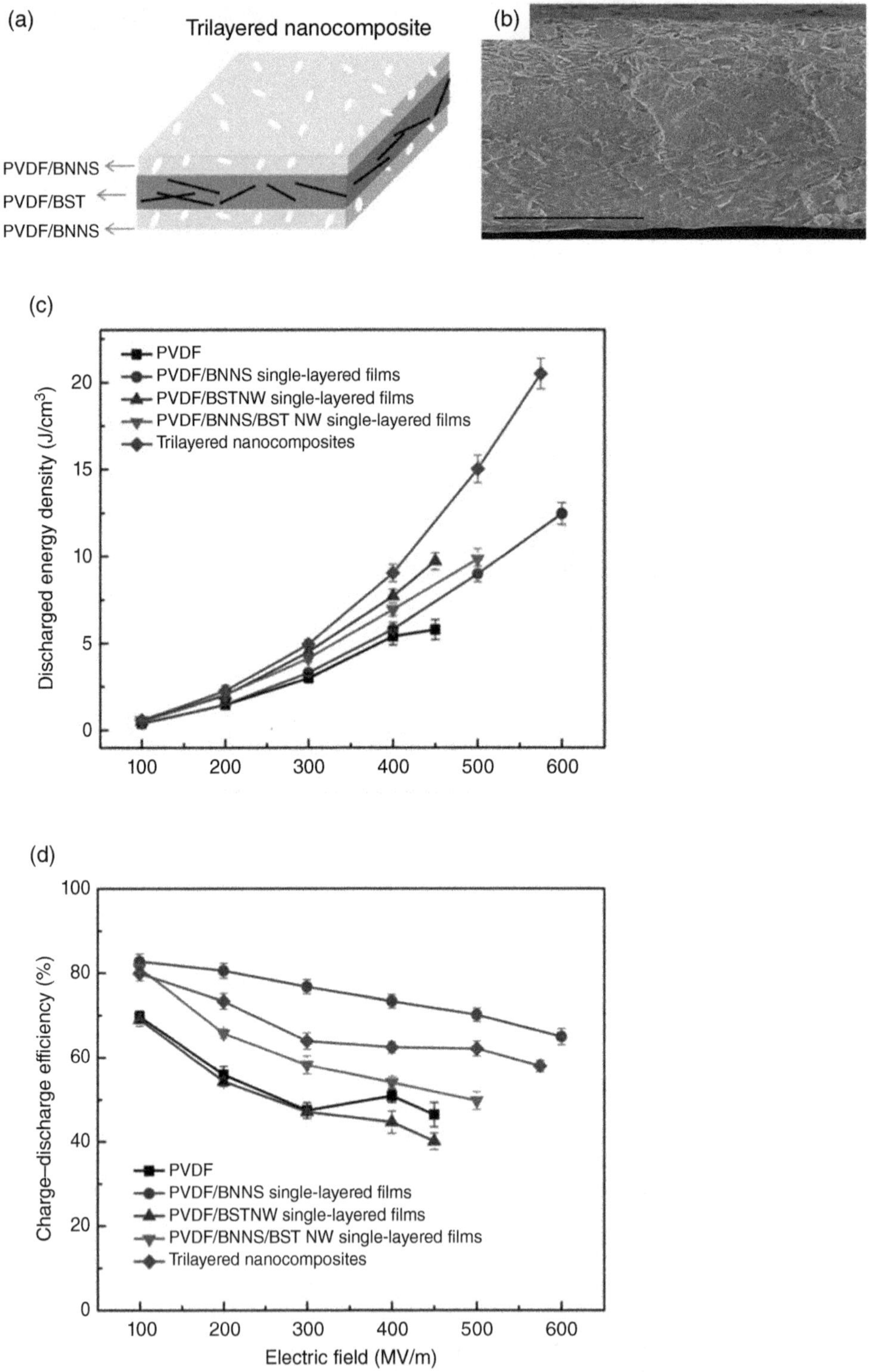

Figure 1.11 (a) Schematic of the trilayer-structured film composed of PVDF/BNNS as outer layers and PVDF/BST as the middle layer, (b) cross-sectional SEM image of trilayer-structured polymer composites, (c) discharged energy density, and (d) charge/discharge efficiency of PVDF-based composites with various compositions and structures. *Source:* Liu et al. [87]. Reproduced with permission of John Wiley & Sons.

constant and high breakdown strength of the trilayer structure, the polymer composite exhibits a discharged energy density of 20.5 J/cm^3. In addition to multilayer-structured composites with different nanofillers in different layers, polymer composites with same nanofiller but different filler contents in different layers can also improve the performance of the multilayer-structured composites [88–90]. Furthermore, the multilayer-structured composites with the gradient distribution of different nanofillers can be designed [91–93].

The multilayer-structured polymer composites can also employ all-inorganic layers. Zhu et al. fabricated a series of PVDF-based multilayer-structured films interlayered by assembled BNNS layer [94]. The BNNS in the middle layer is aligned in the in-place direction by the layer-by-layer solution casting method (Figure 1.12). By controlling the BNNS concentration, the thickness of the middle BNNS layer can be adjusted from 100 to 700 nm. The simulations of the electric field distribution and electrical tree propagation indicate that the assembled BNNS interlayer can effectively mitigate the electric field distortion and impede the electrical breakdown. The composite film interlayered with BNNS exhibits remarkably suppressed leakage current, improved breakdown strength of 612 kV/mm, and enhanced discharged energy density of 14.3 J/cm^3. Compared with the composite with randomly distributed BNNS nanofillers, the proposed interlayered BNNS can significantly reduce the content of the BNNS in the film, i.e. the BNNS content in the layer-structured film is only 0.16 vol%, while high loading of 5–10 vol% BNNS nanofiller is needed for the conventional composite to achieve desired performance. Also, all-inorganic layers can be coated onto the surface of the polymer films to construct multilayer-structured polymer composites [95]. Zhou et al. proposed a roll-to-roll plasma-enhanced chemical vapor deposition (roll-to-roll PECVD) method to deposit SiO_2 layer onto various polymer films [96]. The proposed roll-to-roll PECVD method enables the scalable, high-throughput, and environmentally benign deposition of uniform and compact SiO_2 layer on the surface of polymer films. The deposited SiO_2 layer would act as the barrier at the electrode/dielectric interface, which impedes the charge injection and enables the substantial improvement in the electrical energy storage performance of the coated films.

1.8 Conclusion

Advances in the nanomaterials and nanotechnology have promoted the development of polymer composites for electrical energy storage application. The field of dielectric polymer composites has witnessed great progress, and the progress is still accelerating continuously. The discharged energy density achieved in the polymer composites has already exceeded 20 J/cm^3, which is comparable to other energy storage methods, such as electrochemical capacitors. Various innovative material structure designs and processing methods, such as nanofiller morphology control, nanofiller surface modification, nanofiller alignment, and multilayer-structured composites, have been proposed to improve the performance of the dielectric polymer composites.

To further improve the performance of the dielectric polymer composites to meet the increasing demand from application, the following points should be addressed in the future studies. The majority of the high discharged energy density dielectrics are based on PVDF-based polymers with high dielectric constant. However, these polymer composites suffer

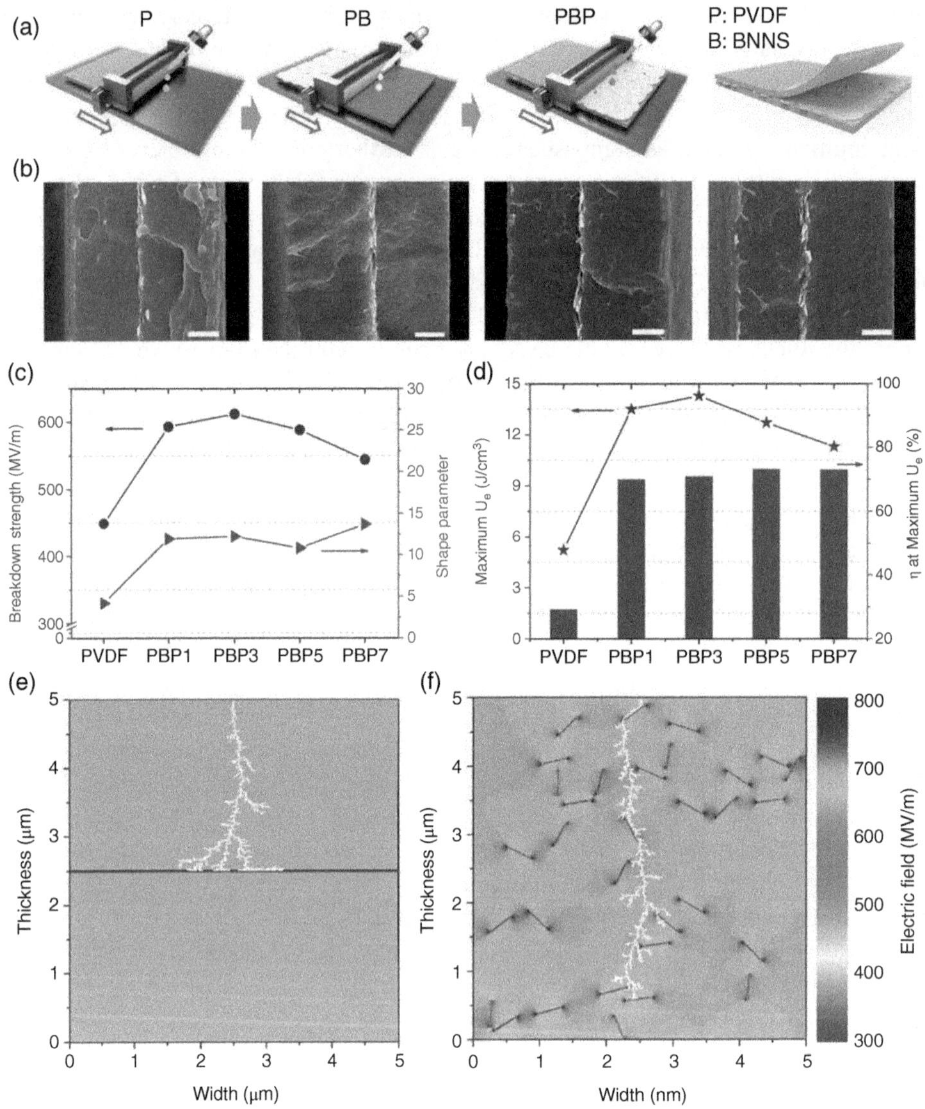

Figure 1.12 (a) Illustration of the fabrication process of sandwich-structured films (P stands for the PVDF layer, and B denotes the BNNSs layer); (b) cross-sectional SEM pictures of PBP1, PBP3, PBP5, and PBP7, respectively (all scale bars are 2 μm); (c) Weibull breakdown strength; and (d) maximum discharged energy density and charge/discharge efficiency of the series films, comparison of electric field distribution, and electrical tree propagation path of (e) PBP3 and (f) PVDF/BNNS composites. *Source:* Zhu et al. [94]. Reproduced with permission of John Wiley & Sons.

from the high energy loss, i.e. >20% in most PVDF-based polymer composites at high electric field, which is not acceptable in many applications. So efforts should be made to improve the charge/discharge efficiency of the high-energy-density polymer composites. The charge/discharge efficiency should be comparable to the benchmark of biaxially oriented polypropylene (BOPP) film. It should be noted that the current high-energy-density polymer composites are mostly designed for room temperature application because of the relatively low thermal stability of the polymer matrix. However, the emerging applications under extreme conditions require polymer dielectrics capable of high temperature application. Although great achievements have been made in high temperature polymer dielectrics, the discharged energy density is still relatively low, i.e. $<5\,J/cm^3$ at 150 °C. To further improve the performance of the polymer composites, fundamental understanding of the interfacial properties should be advanced, which would assist the rational design of the material structure. Moreover, the application of big data and machine learning technologies should be explored in the material structure design and optimization to guide the development of dielectric polymer composites. Last but not least, the low-cost, scale-up film processing method should be developed to produce the high-quality, large-scale polymer composite films. All these efforts together would push the field of dielectric polymer composites to a new stage to meet the emerging demand for electrical energy storage in various industrial applications.

References

1 Li, Q., Yao, F., Liu, Y. et al. (2018). High-temperature dielectric materials for electrical energy storage. *Annual Review of Materials Research* 48: 219–243.

2 Chen, Q., Shen, Y., Zhang, S. et al. (2015). Polymer-based dielectrics with high energy storage density. *Annual Review of Materials Research* 45: 433–458.

3 Huang, X., Sun, B., Zhu, Y. et al. (2019). High-k polymer nanocomposites with 1D filler for dielectric and energy storage applications. *Progress in Materials Science* 100: 187–225.

4 Dang, Z.M., Zheng, M.S., and Zha, J.W. (2016). 1D/2D carbon nanomaterial-polymer dielectric composites with high permittivity for power energy storage applications. *Small* 12 (13): 1688–1701.

5 Zhou, Y. and Wang, Q. (2020). Advanced polymer dielectrics for high temperature capacitive energy storage. *Journal of Applied Physics* 127 (24): 240902.

6 Pan, J., Li, K., Chuayprakong, S. et al. (2010). High-temperature poly (phthalazinone ether ketone) thin films for dielectric energy storage. *ACS Applied Materials & Interfaces* 2 (5): 1286–1289.

7 Dang, Z.M., Yuan, J.K., Yao, S.H. et al. (2013). Flexible nanodielectric materials with high permittivity for power energy storage. *Advanced Materials* 25 (44): 6334–6365.

8 Li, Q. and Wang, Q. (2016). Ferroelectric polymers and their energy-related applications. *Macromolecular Chemistry and Physics* 217 (11): 1228–1244.

9 Thakur, V.K. and Gupta, R.K. (2016). Recent progress on ferroelectric polymer-based nanocomposites for high energy density capacitors: synthesis, dielectric properties, and future aspects. *Chemical Reviews* 116 (7): 4260–4317.

10 Li, H., Liu, F., Fan, B. et al. (2018). Nanostructured ferroelectric-polymer composites for capacitive energy storage. *Small Methods* 2 (6): 1700399.

11 Rabuffi, M. and Picci, G. (2002). Status quo and future prospects for metallized polypropylene energy storage capacitors. *IEEE Transactions on Plasma Science* 30 (5): 1939–1942.

12 Shen, Y., Zhang, X., Li, M. et al. (2017). Polymer nanocomposite dielectrics for electrical energy storage. *National Science Review* 4 (1): 23–25.

13 Wang, Q. and Zhu, L. (2011). Polymer nanocomposites for electrical energy storage. *Journal of Polymer Science Part B: Polymer Physics* 49 (20): 1421–1429.

14 Li, Q. and Cheng, S. (2020). Polymer nanocomposites for high-energy-density capacitor dielectrics: fundamentals and recent progress. *IEEE Electrical Insulation Magazine* 36 (2): 7–28.

15 Cao, Y., Irwin, P.C., and Younsi, K. (2004). The future of nanodielectrics in the electrical power industry. *IEEE Transactions on Dielectrics and Electrical Insulation* 11 (5): 797–807.

16 Yao, K., Chen, S., Rahimabady, M. et al. (2011). Nonlinear dielectric thin films for high-power electric storage with energy density comparable with electrochemical supercapacitors. *IEEE Transactions on Ultrasonics, Ferroelectrics, and Frequency Control* 58 (9): 1968–1974.

17 Fan, B., Liu, F., Yang, G. et al. (2018). Dielectric materials for high-temperature capacitors. *IET Nanodielectrics* 1 (1): 32–40.

18 Sebastian, M.T. and Jantunen, H. (2008). Low loss dielectric materials for LTCC applications: a review. *International Materials Reviews* 53 (2): 57–90.

19 Sarjeant, W.J., Zirnheld, J.F., and MacDougall, W. (1998). Capacitors. *IEEE Transactions on Plasma Science* 26 (5): 1368–1392.

20 Sarjeant, W.J., Clelland, I.W., and Price, R.A. (2001). Capacitive components for power electronics. *Proceedings of the IEEE* 89 (6): 846–855.

21 Yao, Z., Song, Z., Hao, H. et al. (2017). Homogeneous/inhomogeneous-structured dielectrics and their energy-storage performances. *Advanced Materials* 29 (20): 1601727.

22 Li, Q., Chen, L., Gadinski, M.R. et al. (2015). Flexible high-temperature dielectric materials from polymer nanocomposites. *Nature* 523 (7562): 576–579.

23 Li, Q., Zhang, G., Liu, F. et al. (2015). Solution-processed ferroelectric terpolymer nanocomposites with high breakdown strength and energy density utilizing boron nitride nanosheets. *Energy & Environmental Science* 8 (3): 922–931.

24 Yuan, C., Zhou, Y., Zhu, Y. et al. (2020). Polymer/molecular semiconductor all-organic composites for high-temperature dielectric energy storage. *Nature Communications* 11: 3919.

25 Li, J., Seok, S.I., Chu, B. et al. (2009). Nanocomposites of ferroelectric polymers with TiO_2 nanoparticles exhibiting significantly enhanced electrical energy density. *Advanced Materials* 21 (2): 217–221.

26 Li, J., Claude, J., Norena-Franco, L.E. et al. (2008). Electrical energy storage in ferroelectric polymer nanocomposites containing surface-functionalized $BaTiO_3$ nanoparticles. *Chemistry of Materials* 20 (20): 6304–6306.

27 Yao, L., Pan, Z., Zhai, J. et al. (2018). High-energy-density with polymer nanocomposites containing of $SrTiO_3$ nanofibers for capacitor application. *Composites Part A: Applied Science and Manufacturing* 109: 48–54.

28 Lu, X., Zhang, L., Tong, Y. et al. (2019). BST-P(VDF-CTFE) nanocomposite films with high dielectric constant, low dielectric loss, and high energy-storage density. *Composites Part B: Engineering* 168: 34–43.

29 Jain, A., Prashanth, K.J., Sharma, A.K. et al. (2015). Dielectric and piezoelectric properties of PVDF/PZT composites: a review. *Polymer Engineering & Science* 55 (7): 1589–1616.

30 Shen, Y., Gu, A., Liang, G. et al. (2010). High performance $CaCu_3Ti_4O_{12}$/cyanate ester composites with excellent dielectric properties and thermal resistance. *Composites Part A: Applied Science and Manufacturing* 41 (11): 1668–1676.

31 Chen, S.S., Hu, J., Gao, L. et al. (2016). Enhanced breakdown strength and energy density in PVDF nanocomposites with functionalized MgO nanoparticles. *RSC Advances* 6 (40): 33599–33605.

32 Kim, S.H., Kang, H.S., Sohn, E.H. et al. (2020). High discharge energy density and efficiency in newly designed PVDF@SiO_2-PVDF composites for energy capacitors. *ACS Applied Energy Materials* 3 (9): 8937–8945.

33 Li, H., Ai, D., Ren, L. et al. (2019). Scalable polymer nanocomposites with record high-temperature capacitive performance enabled by rationally designed nanostructured inorganic fillers. *Advanced Materials* 31 (23): 1900875.

34 Zhang, H., Marwat, M.A., Xie, B. et al. (2019). Polymer matrix nanocomposites with 1D ceramic nanofillers for energy storage capacitor applications. *ACS Applied Materials & Interfaces* 12 (1): 1–37.

35 Tang, H., Lin, Y., and Sodano, H.A. (2013). Synthesis of high aspect ratio $BaTiO_3$ nanowires for high energy density nanocomposite capacitors. *Advanced Energy Materials* 3 (4): 451–456.

36 Shen, Z.H., Wang, J.J., Lin, Y. et al. (2018). High-throughput phase-field design of high-energy-density polymer nanocomposites. *Advanced Materials* 30 (2): 1704380.

37 Shen, X., Zheng, Q., and Kim, J.K. (2021). Rational design of two-dimensional nanofillers for polymer nanocomposites toward multifunctional applications. *Progress in Materials Science* 115: 100708.

38 Pan, Z., Liu, B., Zhai, J. et al. (2017). $NaNbO_3$ two-dimensional platelets induced highly energy storage density in trilayered architecture composites. *Nano Energy* 40: 587–595.

39 Pan, Z., Ding, Q., Yao, L. et al. (2019). Simultaneously enhanced discharge energy density and efficiency in nanocomposite film capacitors utilizing two-dimensional $NaNbO_3$@Al_2O_3 platelets. *Nanoscale* 11 (21): 10546–10554.

40 Bao, Z., Hou, C., Shen, Z. et al. (2020). Negatively charged nanosheets significantly enhance the energy-storage capability of polymer-based nanocomposites. *Advanced Materials* 32: 1907227.

41 Zhu, Y., Yao, H., Jiang, P. et al. (2018). Two-dimensional high-k nanosheets for dielectric polymer nanocomposites with ultrahigh discharged energy density. *The Journal of Physical Chemistry C* 122 (32): 18282–18293.

42 Ma, Y., Tong, W., Wang, W. et al. (2018). Montmorillonite/PVDF-HFP-based energy conversion and storage films with enhanced piezoelectric and dielectric properties. *Composites Science and Technology* 168: 397–403.

43 Xie, Y., Wang, J., Yu, Y. et al. (2018). Enhancing breakdown strength and energy storage performance of PVDF-based nanocomposites by adding exfoliated boron nitride. *Applied Surface Science* 440: 1150–1158.

44 Wang, Y., Li, Z., Wu, C. et al. (2020). High-temperature dielectric polymer nanocomposites with interposed montmorillonite nanosheets. *Chemical Engineering Journal* 401: 126093.

45 Pohl, H.A. and Crane, J.S. (1972). Dielectrophoretic force. *Journal of Theoretical Biology* 37 (1): 1–13.

46 Yao, S.H., Yuan, J.K., Zhou, T. et al. (2011). Stretch-modulated carbon nanotube alignment in ferroelectric polymer composites: characterization of the orientation state and its influence on the dielectric properties. *The Journal of Physical Chemistry C* 115 (40): 20011–20017.

47 Agarwal, S., Greiner, A., and Wendorff, J.H. (2013). Functional materials by electrospinning of polymers. *Progress in Polymer Science* 38 (6): 963–991.

48 Banerjee, P., Perez, I., Henn-Lecordier, L. et al. (2009). Nanotubular metal-insulator-metal capacitor arrays for energy storage. *Nature Nanotechnology* 4 (5): 292–296.

49 Liao, S., Shen, Z., Pan, H. et al. (2017). A surface-modified TiO_2 nanorod array/P (VDF-HFP) dielectric capacitor with ultra high energy density and efficiency. *Journal of Materials Chemistry C* 5 (48): 12777–12784.

50 Zeng, X., Ye, L., Yu, S. et al. (2015). Facile preparation of superelastic and ultralow dielectric boron nitride nanosheet aerogels via freeze-casting process. *Chemistry of Materials* 27 (17): 5849–5855.

51 Li, B., Xidas, P.I., and Manias, E. (2018). High breakdown strength polymer nanocomposites based on the synergy of nanofiller orientation and crystal orientation for insulation and dielectric applications. *ACS Applied Nano Materials* 1 (7): 3520–3530.

52 Xie, B., Zhang, H., Zhang, Q. et al. (2017). Enhanced energy density of polymer nanocomposites at a low electric field through aligned $BaTiO_3$ nanowires. *Journal of Materials Chemistry A* 5 (13): 6070–6078.

53 Luo, H., Zhou, X., Ellingford, C. et al. (2019). Interface design for high energy density polymer nanocomposites. *Chemical Society Reviews* 48 (16): 4424–4465.

54 Zhou, Y., He, J., Hu, J. et al. (2016). Surface-modified MgO nanoparticle enhances the mechanical and direct-current electrical characteristics of polypropylene/polyolefin elastomer nanodielectrics. *Journal of Applied Polymer Science* 133 (1): 42863.

55 Zhang, X., Li, B.W., Dong, L. et al. (2018). Superior energy storage performances of polymer nanocomposites via modification of filler/polymer interfaces. *Advanced Materials Interfaces* 5 (11): 1800096.

56 Zhou, Y., Dang, B., Wang, H. et al. (2018). Polypropylene-based ternary nanocomposites for recyclable high-voltage direct-current cable insulation. *Composites Science and Technology* 165: 168–174.

57 Hu, P., Gao, S., Zhang, Y. et al. (2018). Surface modified $BaTiO_3$ nanoparticles by titanate coupling agent induce significantly enhanced breakdown strength and larger energy density in PVDF nanocomposite. *Composites Science and Technology* 156: 109–116.

58 Huang, X. and Jiang, P. (2015). Core-shell structured high-k polymer nanocomposites for energy storage and dielectric applications. *Advanced Materials* 27 (3): 546–554.

59 Paniagua, S.A., Kim, Y., Henry, K. et al. (2014). Surface-initiated polymerization from barium titanate nanoparticles for hybrid dielectric capacitors. *ACS Applied Materials & Interfaces* 6 (5): 3477–3482.

60 Yang, K., Huang, X., Xie, L. et al. (2012). Core-shell structured polystyrene/$BaTiO_3$ hybrid nanodielectrics prepared by in situ RAFT polymerization: a route to high dielectric constant and low loss materials with weak frequency dependence. *Macromolecular Rapid Communications* 33 (22): 1921–1926.

61 Tchoul, M.N., Fillery, S.P., Koerner, H. et al. (2010). Assemblies of titanium dioxide-polystyrene hybrid nanoparticles for dielectric applications. *Chemistry of Materials* 22 (5): 1749–1759.

62 Li, Z., Fredin, L.A., Tewari, P. et al. (2010). In situ catalytic encapsulation of core-shell nanoparticles having variable shell thickness: dielectric and energy storage properties of high-permittivity metal oxide nanocomposites. *Chemistry of Materials* 22 (18): 5154–5164.

63 Zhu, M., Huang, X., Yang, K. et al. (2014). Energy storage in ferroelectric polymer nanocomposites filled with core-shell structured polymer@$BaTiO_3$ nanoparticles: understanding the role of polymer shells in the interfacial regions. *ACS Applied Materials & Interfaces* 6 (22): 19644–19654.

64 Xie, L., Huang, X., Huang, Y. et al. (2013). Core@double-shell structured $BaTiO_3$-polymer nanocomposites with high dielectric constant and low dielectric loss for energy storage application. *Journal of Physical Chemistry C* 117 (44): 22525–22537.

65 Zhou, Y., Hu, J., Dang, B. et al. (2016). Mechanism of highly improved electrical properties in polypropylene by chemical modification of grafting maleic anhydride. *Journal of Physics D: Applied Physics* 49 (41): 415301.

66 Meunier, M., Quirke, N., and Aslanides, A. (2001). Molecular modeling of electron traps in polymer insulators: chemical defects and impurities. *The Journal of Chemical Physics* 115 (6): 2876–2881.

67 Yuan, H., Zhou, Y., Zhu, Y. et al. (2020). Origins and effects of deep traps in functional group grafted polymeric dielectric materials. *Journal of Physics D: Applied Physics* 53 (47): 475301.

68 Chen, L., Batra, R., Ranganathan, R. et al. (2018). Electronic structure of polymer dielectrics: the role of chemical and morphological complexity. *Chemistry of Materials* 30 (21): 7699–7706.

69 Zhou, Y., Yuan, C., Wang, S. et al. (2020). Interface-modulated nanocomposites based on polypropylene for high-temperature energy storage. *Energy Storage Materials* 28: 255–263.

70 He, D., Wang, Y., Chen, X. et al. (2017). Core-shell structured $BaTiO_3$@ Al_2O_3 nanoparticles in polymer composites for dielectric loss suppression and breakdown strength enhancement. *Composites Part A: Applied Science and Manufacturing* 93: 137–143.

71 Bi, K., Bi, M., Hao, Y. et al. (2018). Ultrafine core-shell $BaTiO_3$@SiO_2 structures for nanocomposite capacitors with high energy density. *Nano Energy* 51: 513–523.

72 He, D., Wang, Y., Song, S. et al. (2017). Significantly enhanced dielectric performances and high thermal conductivity in poly(vinylidene fluoride)-based composites enabled by SiC@ SiO_2 core-shell whiskers alignment. *ACS Applied Materials & Interfaces* 9 (51): 44839–44846.

73 Zhang, X., Shen, Y., Zhang, Q. et al. (2015). Ultrahigh energy density of polymer nanocomposites containing $BaTiO_3$@TiO_2 nanofibers by atomic-scale interface engineering. *Advanced Materials* 27 (5): 819–824.

74 Zhang, X., Shen, Y., Xu, B. et al. (2016). Giant energy density and improved discharge efficiency of solution-processed polymer nanocomposites for dielectric energy storage. *Advanced Materials* 28 (10): 2055–2061.

75 Li, Q., Han, K., Gadinski, M.R. et al. (2014). High energy and power density capacitors from solution-processed ternary ferroelectric polymer nanocomposites. *Advanced Materials* 26 (36): 6244–6249.

76 Li, H., Ren, L., Ai, D. et al. (2020). Ternary polymer nanocomposites with concurrently enhanced dielectric constant and breakdown strength for high-temperature electrostatic capacitors. *InfoMat* 2 (2): 389–400.

77 He, D., Wang, Y., Zhang, L. et al. (2018). Poly(vinylidene fluoride)-based composites modulated via multiscale two-dimensional fillers for high dielectric performances. *Composites Science and Technology* 159: 162–170.

78 Zhao, M., Fu, Q., Hou, Y. et al. (2019). $BaTiO_3$/MWNTs/polyvinylidene fluoride ternary dielectric composites with excellent dielectric property, high breakdown strength, and high-energy storage density. *ACS Omega* 4 (1): 1000–1006.

79 Liu, F., Li, Q., Li, Z. et al. (2018). Ternary PVDF-based terpolymer nanocomposites with enhanced energy density and high power density. *Composites Part A: Applied Science and Manufacturing* 109: 597–603.

80 Luo, S., Yu, J., Yu, S. et al. (2019). Significantly enhanced electrostatic energy storage performance of flexible polymer composites by introducing highly insulating-ferroelectric microhybrids as fillers. *Advanced Energy Materials* 9 (5): 1803204.

81 Li, Y., Zhou, Y., Zhu, Y. et al. (2020). Polymer nanocomposites with high energy density and improved charge-discharge efficiency utilizing hierarchically-structured nanofillers. *Journal of Materials Chemistry A* 8 (14): 6576–6585.

82 Baer, E. and Zhu, L. (2017). 50th anniversary perspective: dielectric phenomena in polymers and multilayered dielectric films. *Macromolecules* 50 (6): 2239–2256.

83 Yin, K., Zhou, Z., Schuele, D.E. et al. (2016). Effects of interphase modification and biaxial orientation on dielectric properties of poly (ethylene terephthalate)/poly (vinylidene fluoride-co-hexafluoropropylene) multilayer films. *ACS Applied Materials & Interfaces* 8 (21): 13555–13566.

84 Yao, L., Wang, D., Hu, P. et al. (2016). Synergetic enhancement of permittivity and breakdown strength in all-polymeric dielectrics toward flexible energy storage devices. *Advanced Materials Interfaces* 3 (13): 1600016.

85 Mackey, M., Schuele, D.E., Zhu, L. et al. (2012). Reduction of dielectric hysteresis in multilayered films via nanoconfinement. *Macromolecules* 45 (4): 1954–1962.

86 Li, Q., Liu, F., Yang, T. et al. (2016). Sandwich-structured polymer nanocomposites with high energy density and great charge-discharge efficiency at elevated temperatures. *Proceedings of the National Academy of Sciences* 113 (36): 9995–10000.

87 Liu, F., Li, Q., Cui, J. et al. (2017). High-energy-density dielectric polymer nanocomposites with trilayered architecture. *Advanced Functional Materials* 27 (20): 1606292.

88 Wang, Y., Cui, J., Yuan, Q. et al. (2015). Significantly enhanced breakdown strength and energy density in sandwich-structured barium titanate/poly (vinylidene fluoride) nanocomposites. *Advanced Materials* 27 (42): 6658–6663.

89 Jiang, J., Shen, Z., Qian, J. et al. (2019). Synergy of micro-/mesoscopic interfaces in multilayered polymer nanocomposites induces ultrahigh energy density for capacitive energy storage. *Nano Energy* 62: 220–229.

90 Hu, P., Wang, J., Shen, Y. et al. (2013). Highly enhanced energy density induced by hetero-interface in sandwich-structured polymer nanocomposites. *Journal of Materials Chemistry A* 1 (39): 12321–12326.

91 Jiang, J., Shen, Z., Cai, X. et al. (2019). Polymer nanocomposites with interpenetrating gradient structure exhibiting ultrahigh discharge efficiency and energy density. *Advanced Energy Materials* 9 (15): 1803411.

92 Zhang, X., Jiang, J., Shen, Z. et al. (2018). Polymer nanocomposites with ultrahigh energy density and high discharge efficiency by modulating their nanostructures in three dimensions. *Advanced Materials* 30 (16): 1707269.

93 Zhang, Y., Chi, Q., Liu, L. et al. (2018). PVDF-based dielectric composite films with excellent energy storage performances by design of nanofibers composition gradient structure. *ACS Applied Energy Materials* 1 (11): 6320–6329.

94 Zhu, Y., Zhu, Y., Huang, X. et al. (2019). High energy density polymer dielectrics interlayered by assembled boron nitride nanosheets. *Advanced Energy Materials* 9 (36): 1901826.

95 Azizi, A., Gadinski, M.R., Li, Q. et al. (2017). High-performance polymers sandwiched with chemical vapor deposited hexagonal boron nitrides as scalable high-temperature dielectric materials. *Advanced Materials* 29 (35): 1701864.

96 Zhou, Y., Li, Q., Dang, B. et al. (2018). A scalable, high-throughput, and environmentally benign approach to polymer dielectrics exhibiting significantly improved capacitive performance at high temperatures. *Advanced Materials* 30 (49): 1805672.

2

Polymer Composites for Thermal Energy Storage

Jie Yang, Chang-Ping Feng, Lu Bai, Rui-Ying Bao, Ming-Bo Yang, and Wei Yang

College of Polymer Science and Engineering, Sichuan University, State Key Laboratory of Polymer Materials Engineering, Chengdu, Sichuan, P. R. China
Jie Yang and Chang-Ping Feng contribute equally.

2.1 Introduction

It is predicted that there will be a one-third increase in energy demand by 2035 from 2011.[1] Although fossil fuels, as limited and nonrenewable resources, can provide vast quantities of energy, the irreversible burning releases large amounts of pollutants, resulting in the deterioration of environment and global warming. The focus of the energy battlefield is also gradually shifting to green and renewable energy. However, the utilization efficiency of renewable energy sources, including solar energy, wind energy, tidal energy, geothermal energy, etc., only accounts for approximately 20% of today's energy consumption owing to the limitations of modern technologies and geographical environments.[2] The discontinuity and regionalism seriously affect their utilization efficiency. More importantly, during the process of using various energy resources, a large amount of energy is used in the form of thermal energy or converted from thermal energy for reuse. Therefore, the development of thermal energy storage (TES) technologies and systems through sensible heat, latent heat, and thermochemical reaction is of great significance to solve the issue of spatiotemporal mismatch between energy supply and demand, improve energy utilization efficiency, alleviate energy crisis, and manage environmental pollution.[3–5] TES not only serves as an integrated buffer in energy conversion systems but also can be incorporated into thermal management technologies for electronics, human body, and buildings.

Phase change materials (PCMs), as an advanced energy storage technique, can absorb or release a lot of energy in the form of latent heat during phase transition for efficient storage and utilization of energy, while maintaining constant temperature (Figure 2.1).[6–8] The selection criteria and characteristics of ideal PCMs feasible for TES are listed in Table 2.1.[7–11] However, it is troublesome for the majorities of PCMs to fully satisfy these criteria. Recent advancements in the innovations of nanomaterials and nanotechnologies for energy-related applications provide possibilities for PCMs with enhanced comprehensive performance.

Polymer Composites for Electrical Engineering, First Edition. Edited by Xingyi Huang and Toshikatsu Tanaka.

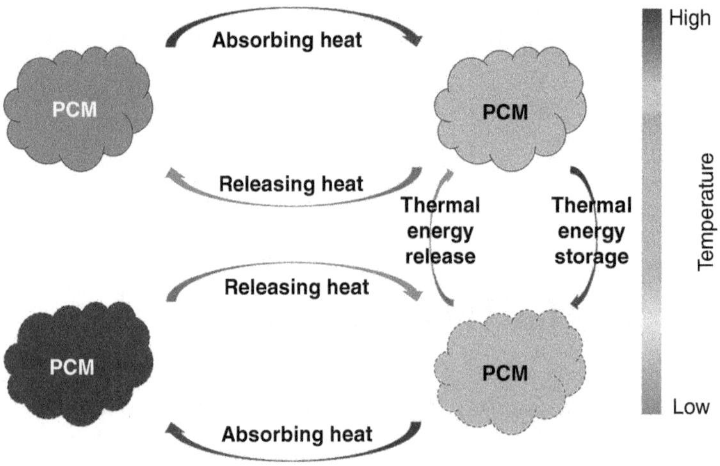

Figure 2.1 Working principle of PCMs.

Table 2.1 The features of ideal PCMs feasible for TES.

Aspects	Details
Thermal properties	• Appropriate phase transition temperature to satisfy practical applications • Large phase change enthalpy to provide high latent heat storage capacity • High thermal conductivity to achieve fast charging or discharging rate • Excellent thermal stability to prolong service life
Kinetic properties	• No/negligible supercooling • No phase separation • High nucleation rate • Adequate crystallization rate
Physical properties	• Subtle volume variation during phase transformation • Low vapor pressure at the operating temperature
Chemical properties	• Long-term chemical stability • Noncorrosive, nontoxic, nonpolluting, nonflammable, and nonexplosive to ensure safety and harmless to the surroundings
Economics	• Abundance • Easy availability • Low cost • Good recyclability

In the light of the phase transition mode, PCMs can be classified into solid–solid, solid–liquid, solid–gas, and liquid–gas systems (Figure 2.2). Although solid–gas and liquid–gas PCMs possess a very high phase change latent heat, the large volume change during phase transition restricts their applications in TES systems.[7] Therefore, solid–solid and solid–liquid PCMs have received extensive attention in recent decades owing to their small volume change. Generally, solid–liquid PCMs exhibit a higher energy density than

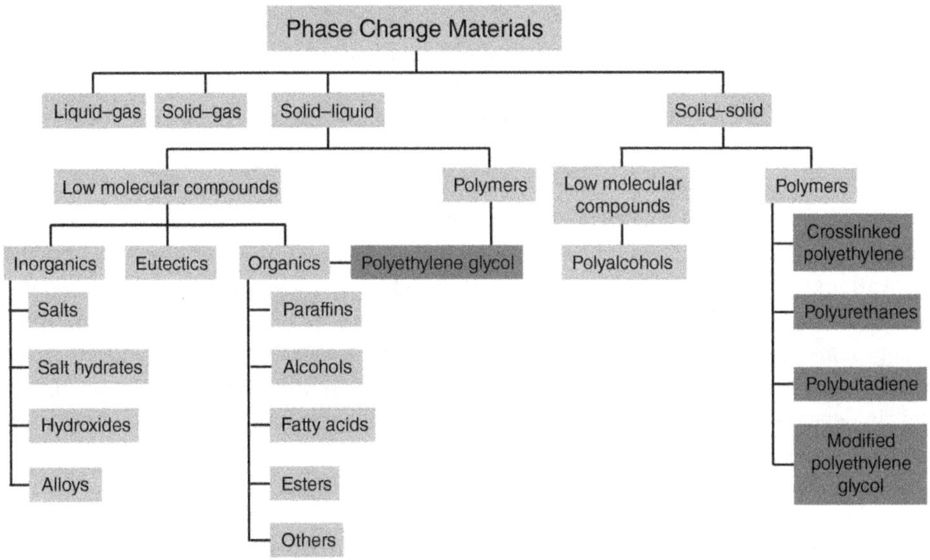

Figure 2.2 Classification of PCMs. *Source:* Based on [7,12].

Table 2.2 Advantages and disadvantages of organic PCMs and inorganic PCMs.

Solid–liquid PCMs	Advantages	Disadvantages
Inorganic PCMs	• High energy storage densities • High thermal conductivities • Low costs	• Supercooling • Phase separation
Organic PCMs	• High energy storage densities • Wide range of phase change temperature for convenient use • Isothermal characteristics • No phase separation • Low or negligible supercooling • Good compatibility • Nontoxicity and noncorrosion • Desirable thermal and chemical stability for long-term use • Abundant natural resources	• Poor shape stability (leakage during phase transition) • Low thermal conductivity • Weak energy conversion ability

solid–solid PCMs. Solid–liquid PCMs can be broadly classified into organic PCMs and inorganic PCMs.[6] Compared with inorganic PCMs not discussed in this chapter, organic PCMs represented by low-molecular paraffin wax (PW) or alkane and polyethylene glycol (PEG) have attracted ever-growing interest owing to their high energy storage capacity, low or negligible supercooling, and desirable stability. However, the majorities of organic PCMs possess fatal disadvantages of poor formability and low thermal conductivity, which seriously limit their practical applications. More details are presented in Table 2.2. Although solid–solid PCMs possess excellent shape stability, they also exhibit inherently low thermal conductivity and limited energy conversion ability. Incorporating functional components (supporting or conductive materials) into these PCMs is regarded as a promising route to fabricate thermally conductive or leakage-proof phase change composites for high-efficiency TES systems.

Polymeric phase change composites, as TES units, can be roughly divided into two categories: phase change polymers presented in Figure 2.2 as working substance and common polymers as supporting component. In this chapter, we put our emphases on the innovations of shape-stabilized and thermally conductive polymeric phase change composites, covering processing method, structural design, and materials selection. Also, energy conversion routes and potential applications associated with polymeric phase change composites are highlighted. Finally, a brief perspective on future opportunities and challenges for high-performance and multifunctional polymeric phase change composites is proposed.

2.2 Shape-stabilized Polymeric Phase Change Composites

Organic solid–liquid PCMs suffer from liquid phase leakage in the process of phase transition, resulting in decrease of energy storage density, pollution of related devices, and security risks. Shape-stabilized PCMs are composed of working substance and supporting material, and the latter predominately includes polymers with good compatibility and thermal stability, nanomaterials with desirable structure, and porous scaffolds with excellent mechanical properties.[13] Micro/nanoencapsulated techniques, physical blending with supporting materials, incorporating porous scaffolds, and crosslinking-solidity strategy have been adopted to improve the shape stability of PCMs, which are magnified in this chapter.

To the best of our knowledge, the methods used to evaluate the shape stability of PCMs have not been unified and standardized. Thermal mechanical analyzer (TMA) and dynamic rheometer are usually used to quantitatively measure the size change of materials to evaluate their shape stability. With the increase of temperature, the size or dimension of samples will automatically adjust when subjected to constant normal force, and thus the shape stability can be determined by the change degree of the size or dimension. Generally speaking, the smaller the change of the sample size, the better the shape-stabilizing effect of the material.[14] In addition to monitoring change of sample size in rheological test, dynamic temperature scanning can be used to compare the difference of storage modulus between pure sample and composites, further to judge the shape-stabilizing effect.[15] Moreover, facile and qualitative leakage experiments are conducted to intuitively reflect the shape-stabilizing effect in most cases. The samples are directly placed on a hot stage or in an oven and then heated to different temperatures for varying durations, followed by tracking real-time photography with a digital camera to record the change of shape and size and leakage status of samples. It is more reasonable to calculate the mass change of samples before and after leakage test.[16, 17] Additionally, Shi et al.[18] measured the dropping point to determine the shape-stabilizing effect of samples.

2.2.1 Micro/Nanoencapsulated Method

Micro/nanoencapsulation is one of the common techniques to fabricate shape-stabilized PCMs with core-shell structure, in which the core and the shell are, respectively, the working medium performing energy storage and functional coating layer providing complete and stable structure. In general, PCMs with phase change temperature in the range of −10–80 °C can be encapsulated.[19] Micro/nanostructures of phase change capsules can be tuned by using different preparation methods and controlling synthesis conditions, mainly forming single core-shell, multi-shell and polynuclear structures (Figure 2.3).[20] Microencapsulated phase change products present different appearances, such as sphere, tubular, and other irregular shapes. The coating materials mainly include organic polymers, inorganics (e.g. silica, metal oxides, and hydroxides), and their hybrids.[21] The choice of coating material depends on the compatibility with the core materials and the synthesis methods. Usually, monomer polymerization (e.g. emulsion polymerization, interfacial polymerization, and *in-situ* polymerization) and sol–gel method are adopted for preparing organic and inorganic shell layers, respectively.[21, 22]

Owing to their low cost, light weight, outstanding flexibility, good processability, and desirable compatibility, polymers, including polystyrene (PS), polymethylmethacrylate (PMMA), polyurethane (PU), urea–formaldehyde resin (UF), melamine–formaldehyde resin (MF), and many others, are widely used in the preparation of core-shell phase change composites.[19, 23, 24] For example, a direct miniemulsion method has been adopted to prepare spherical poly(ethyl methacrylate) (PEMA)/*n*-octadecane nanocapsules with the average size of 140 nm and PMMA/*n*-octadecane nanocapsules with the average size of 119 nm, and the obtained composite capsules exhibit a relatively high encapsulation ratio

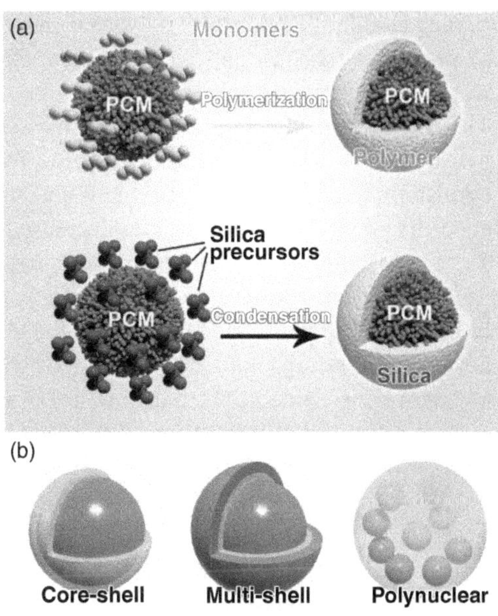

Figure 2.3 Illustrations of (a) preparation routes and (b) various architectures of micro/nanoencapsulated PCMs. *Source:* Aftab et al. [20]. Reproduced with the permission from the Royal Society of Chemistry.

(89.5%), which is the ratio of the fusion enthalpy of encapsulated phase change composites to that of bulk PCMs.[25] Docosane-loaded microcapsules with PU shell were prepared via miniemulsion interfacial polymerization. The size of the microcapsules plays a vital role in the thermal properties of the final product.[26]

Aside from supporting effect, these polymer shell materials are required to impart some important characteristics such as biocompatibility, thermal stability, and flame retardancy to micro/nanoencapsulated phase change composites. For example, the introduction of phosphorus–nitrogen containing diamine into MF/*n*-octadecane phase change composites significantly increased the limiting oxygen index (LOI) and effectively inhibited the release of heat and smoke. This flame-retardant nanoencapsulated phase change product exhibits considerable potential in TES applications such as energy saving construction and thermoregulated textile.[27]

Although polymers have many excellent properties, their applications are limited by several key problems, such as poor thermal response and flammability. To solve the dilemmas, inorganic materials (e.g. SiO_2, [28, 29] TiO_2,[30] $CaCO_3$,[31] and hybrids [32]) with excellent thermal conductivity, thermal stability, flame retardancy, and mechanical strength have been used to prepare encapsulated phase change composites with organic PCMs, especially for low-molecular compounds. Among them, SiO_2 becomes the most widely studied inorganic shell material for composite phase change micro/nanocapsules, which is attributed to its unique superiorities such as high melting point and low thermal expansion coefficient. However, they suffer from brittleness and poor compatibility. Therefore, the high thermal conductivity of inorganic shell and the good compatibility of polymer shell are integrated to develop emerging encapsulated phase change composites with organic–inorganic hybrid shell, which are expected to become the next-generation star products. The typical organic–inorganic hybrid encapsulation method includes microemulsion polymerization, pickering emulsion polymerization, seeded emulsion polymerization, and soap-free emulsion polymerization.

Yang et al.[33] fabricated a polyurea/TiO_2 hybrid shell to encapsulate *n*-octadecane by a two-step liquid phase deposition method. The *n*-octadecane/polyurea pre-microcapsules were first synthesized through interfacial polymerization, followed by the deposition of TiO_2 on the surface of pre-microcapsules. Tang et al.[34] synthesized a PMMA/SiO_2 hybrid shell to encapsulate *n*-octadecane PCMs using a novel photocurable pickering emulsion method, and the introduction of ultraviolet (UV) irradiation effectively shortened the polymerization time and decreased the processing temperature. Furthermore, TiC was introduced into the PMMA-SiO_2 hybrid shell to improve the thermal conductivity and thermal stability of the polymeric phase change composites.[35] Similarly, a PMMA/TiO_2/boron nitride (BN) hybrid shell with high thermal conductivity has been developed to encapsulate PW through a pickering emulsion method (Figure 2.4a).[36] MF, graphene oxide (GO), and PW acted as shell, extra protective screen, and core, respectively, to prepare microencapsulated phase change composites with high encapsulation ratio (up to 93.9 wt%) through *in-situ* polymerization (Figure 2.4b).[37] Along this line, carbon nanotube (CNT) as thermally conductive enhancement component was added into MF/GO shell system to facilitate thermal charging/discharging of microencapsulated phase change composites.[38]

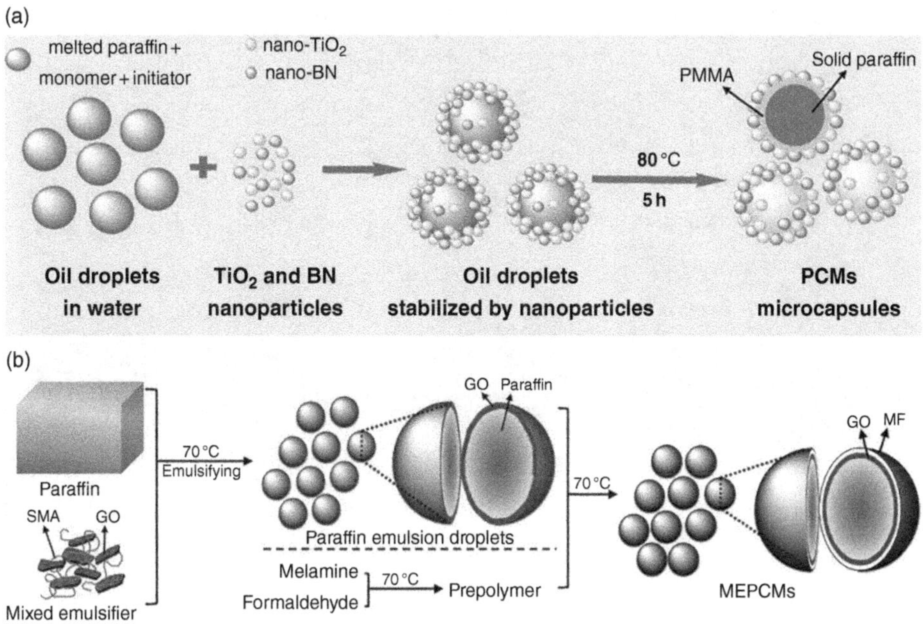

Figure 2.4 Schematic diagrams of the fabrication route of composite phase change microcapsules with paraffin core and (a) PMMA/BN/TiO_2 hybrid shell. *Source:* Sun and Xiao [36]. Reproduced with the permission from the American Chemical Society or (b) MF/GO hybrid shell. *Source:* Zhang et al. [37]. Reproduced with the permission from Elsevier Ltd.

2.2.2 Physical Blending

It is quite appealing that supporting materials are directly introduced into the phase change matrices via physical blending to produce shape-stabilized phase change composites. Compounding organic PCMs, especially PW, with robust or flexible polymers featuring high melting temperature or network structure is an effective strategy to achieve this goal. Accordingly, a facile melt blending has been conducted to fabricate polymeric phase change composites, including polyethylene (PE)/PW,[39] ethylene propylene diene monomer (EPDM)/PW,[40] styrene-b-(ethylene-co-butylene)-b-styrene (SEBS)/PW,[41] olefin block copolymer (OBC)/PW,[42] OBC/hexadecane,[43] epoxy/PW,[44] polyolefin elastomer (POE)/lauric acid,[45] POE/stearic acid,[46] PMMA/stearic acid,[47] etc. For example, Chen et al.[48] introduced high-density PE (HDPE), low-density PE (LDPE), and linear LDPE (LLDPE) into PW through melt blending to fabricate three kinds of polymeric PCMs. The results indicated that the leakage behavior could be controlled by the co-continuous structure, and the PE phase had a minor influence on the crystallinity structure and thermal transition temperatures of PW phase. For these polymeric PCMs, the organic PCMs content is maintained in the range of 30–60 wt%, and most of them can be used as shape memory devices. To impart high thermal conductivity to these polymeric PCMs, functional fillers can be added to prepare thermally conductive polymeric phase change composites,[49, 50] and more discussion will be provided in the next section.

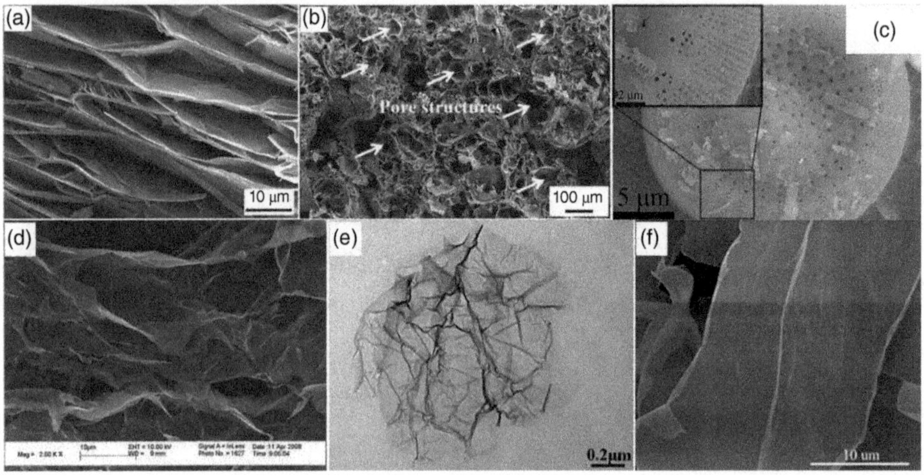

Figure 2.5 Microstructures of (a) EVM. *Source:* Deng et al. [51]. (b) EP. *Source:* Zhang et al. [52]. (c) diatomite. *Source:* Qian et al. [53]. (d) EG. *Source:* Wang et al. [54]. (e) graphene. *Source:* Shi et al. [18]. and (f) GO. *Source:* Yang et al. [55].

The innovations of nanotechnology open up new frontiers in energy conversion and storage materials and systems, and thus multifarious functional nanomaterials can act as supporting materials to prepare shape-stabilized polymeric phase change composites. Effective supporting nanomaterials shown in Figure 2.5 possess the following three main characteristics:

1) Micro/nanoporous structure can provide capillary force to accommodate PCMs. Typical candidates include expanded vermiculite (EVM),[51] expanded perlite (EP),[52] diatomite,[53] expanded graphite (EG),[54] and activated carbon[56].
2) Large specific surface area can yield strong surface tension to absorb PCMs. The typical representatives are two-dimensional (2D) graphene with a specific surface area of $2630\,m^2\,g^{-1}$[57] and MXene[58].
3) Hydrogen bonding interaction can restrict the molecular chain motion of PCMs. Typical materials are SiO_2,[59] GO,[14] and cellulose[60]. There are strong hydrogen bonding interactions between these materials and PEG with oxygen-containing functional groups. A typical example is the PEG-based shape-stabilized composites with ultra-low content of GO (4wt%).[14]

2.2.3 Porous Supporting Scaffolds

Considering the complicated technological process of micro/nanoencapsulated PCMs and the low utilization efficiency of functional materials in the processing of phase change composites through physical blending, impregnating PCMs into porous supporting scaffolds to develop form-stable PCMs with high energy storage density has attracted extensive attention in recent years. Lightweight three-dimensional (3D) porous supporting architectures not only inherit the intrinsic characteristics from individually functional components but also create new collective properties from the monoliths, including large specific surface area, high porosity, and unique network structure, and these features are beneficial to improving the comprehensive performance of PCMs through the strong interactions between PCMs and supporting components in the form of hydrogen bonding, interfacial adhesion, capillary force, and van der Waals forces. Therefore, their physicochemical characteristics, such as pore size, specific surface area, and hydrophilicity or hydrophobicity, have an important influence on the properties of the final composites. The porous supporting scaffolds for polymeric phase change composites can be divided into three categories: polymeric porous scaffolds, inorganic porous scaffolds, and organic–inorganic hybrid porous scaffolds.

Polymeric porous scaffolds with high strength/weight ratio and porosity, such as poly(vinyl alcohol) (PVA) aerogel,[61] PU foam,[62] and polypropylene (PP) aerogel,[63] have been acted as excellent supporting materials for PCMs. More interestingly, a rapid preparation of PEG-based phase change composites containing 3D cellulose network constructed by UV-induced thiol-ene click chemistry has been achieved by a solvent exchange strategy without additional freeze-drying operation.[64] In comparison to polymeric porous scaffolds, inorganic porous scaffolds can provide supporting effect and conductive network simultaneously. The above-mentioned partial supporting nanomaterials can be constructed into 3D porous scaffolds (e.g. biological porous carbon,[16] GO aerogel,[65] GO/BN scaffolds,[66] and hybrid graphene aerogels[55]) by high-temperature pyrolysis (biomass carbonization), self-assembly, and other methods, and then they can be introduced into phase change matrices to effectively prepare shape-stabilized polymeric phase change composites. Additionally, it is difficult for a portion of functional fillers to form 3D structural materials by themselves, and organic components are often required as additives. The most popular one is cellulose-based composite scaffolds, successfully developing cellulose/CNT,[67] cellulose/graphene nanoplatelets (GNPs),[68] cellulose/BN,[69] cellulose/MXene,[70] and cellulose/black phosphorus[71] hybrid aerogels for polymeric phase change composites.

2.2.4 Solid–Solid Composite PCMs

Different from solid–liquid PCMs, solid–solid PCMs can store heat via phase transition from one crystalline form to another similar form without generation/leakage of liquid or gas and additional encapsulation. The main strategy to obtain solid–solid PCMs is to construct secondary structure capable of preventing liquid noncrystalline phase from flowing through chemical bonding. In solid–solid polymeric PCMs, the phase change component as the "soft segment" is structurally incorporated into the macromolecular backbone as the "hard segment" via side-chain grafting, block-polymerization, hyperbranching, or crosslinking copolymerization approaches. Phase transition behaviors and thermophysical properties of solid–solid polymeric PCMs can be tailored by adjusting the relative length and physicochemical structure of the soft and hard segments.[72] At present, solid–solid polymeric PCMs have been designed and optimized as thermoplastics and thermosets for TES systems.[73, 74]

Owing to the active terminal groups on PEG and its derivatives, great progresses in PEG-based solid–solid PCMs have been witnessed in recent decades.[3] Commonly, it gains wide popularity in the preparation of PU-based solid–solid PCMs via the direct reaction of

PEG with crosslinker-like isocyanate.[75] Alternatively, the preparation route of PU-based PCMs involves pre-polymerization of PEG with diisocyanate and chain extending or crosslinking with the participation of the chemicals containing multifunctional groups like 1,4-butanediol,[76] tetrahydroxy compound,[77] and hyperbranched polyester[78]. For example, PU-based solid–solid PCMs with the maximum latent heat storage capacity of $136.8\,J\,g^{-1}$ were synthesized utilizing PEG and hexamethylene diisocyanate trimer through one-step and solvent-free method.[79]

Compared with the sophisticated encapsulation procedure and weak physical interaction between supporting materials and phase change parts in leakage-proof polymeric phase change composites discussed above, chemical grafting or blocking is a promising route to fabricate shape-stabilized PCMs with superior chemical stability and thermal reliability, but it is accompanied by a decrease in phase change latent heat or energy storage density. The origin of this reduction may result from several aspects: dilution of working substance, mobility or packing restrictions of the crystalline moieties, and strong interaction between phase change part and polymer backbone. Figure 2.6 summarizes the melting enthalpy and melting temperature ranges of polymeric solid–solid PCMs.[72] Besides, the majorities of these solid–solid polymeric PCMs are fabricated in solvents, which is not environmentally friendly and hinders their applications.

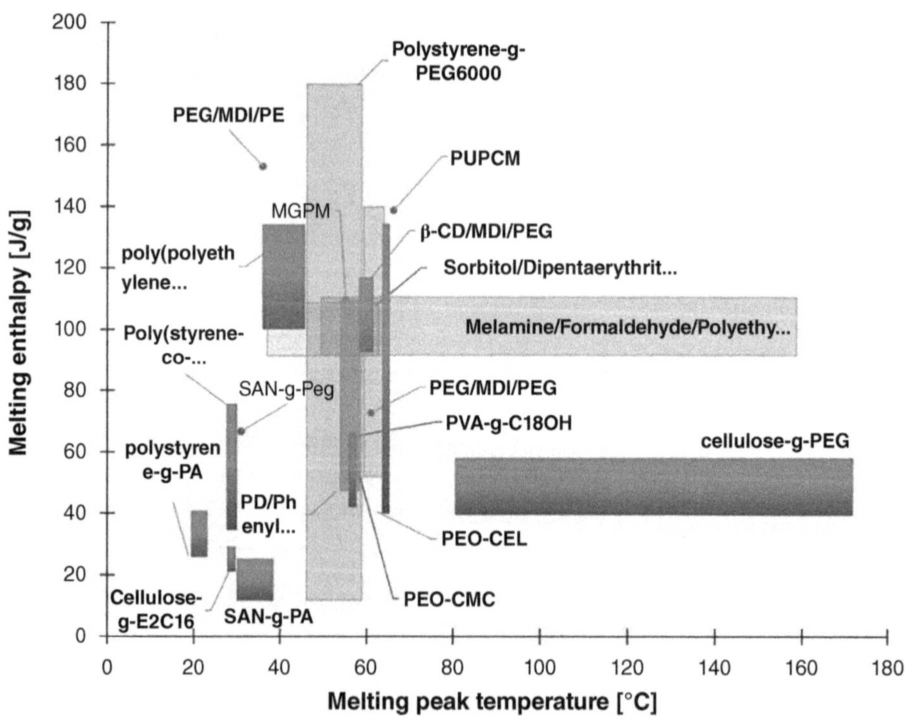

Figure 2.6 The melting enthalpy and melting temperature ranges for solid–solid polymeric PCMs. *Source:* Fallahi et al. [72]. Reproduced with permission from Elsevier Ltd.

2.3 Thermally Conductive Polymeric Phase Change Composites

Heat is transferred mainly by heat conduction in solids. The main carriers are photons, electrons, and phonons, and phonon mechanism (lattice vibration) for heat transferring predominates in polymeric phase change composites. Apart from poor formability, another fatal shortcoming for PCMs is low thermal conductivity (0.1–0.3 $W m^{-1} K^{-1}$), which affects the working efficiency during heat charge/discharge process. Thermally conductive fillers, mainly including metals, carbon materials, ceramics, and their hybrids, are usually employed as enhancement component to improve the thermal conductivity of polymeric phase change composites. The thermal conductivity of the composites is closely related to the intrinsic thermal conductivity, addition content, geometric size, and distribution state/stacking mode of the functional fillers as well as the interface interaction between the filler and the matrix. The thermal conductivities of common organic PCMs and thermally conductive fillers are shown in Table 2.3.

The thermal conductivity can be determined by either steady-state technique tracking the heat flow across a sample with a known thickness or transient technique measuring the energy dissipation through a sample when subjected to a heat pulse. Transient plane source

Table 2.3 Thermal conductivities of frequently used organic PCMs and thermally conductive fillers.

Materials		**Thermal Conductivity ($W m^{-1} K^{-1}$)**
Organic PCMs	PEG	0.30
	PW	0.25
	n-Octadecane	0.15
	Erythritol	0.72
	Octadecanoic acid	0.18
Metals	Silver	429
	Cupper	380–400
	Aluminum	204
	Nickel	158
Carbon materials	Carbon fiber (CF)	1000
	CNT	2000–6000
	Graphite	100–400
	Graphene	5300
Ceramics	BN	250–300
	Aluminum nitride (AlN)	200
	Silicon carbide (SiC)	120
	Aluminum oxide (Al_2O_3)	30–40

Sources: Based on [80–82].

(TPS) or hot disk and laser flash as transient techniques are commonly applied to measure thermal conductivity of the composites.[83] It is worth mentioning that more and more researchers use infrared thermal imagers to record the temperature distribution of samples during heating or cooling, further reflecting the thermal response rate of the sample. Generally speaking, the dispersion of thermally conductive fillers in the composites prepared by facile solution or melt blending is random, and thus forming disordered heat conduction network with large thermal resistance and limited thermal conductivity enhancement. In other words, in order to significantly improve the thermal conductivity of the composites, a large number of thermally conductive fillers, especially non-carbon materials, have to be added, which in turn affects the energy storage density of the phase change composites. To balance the thermal conductivity and inherent energy storage density and make full use of the performance characteristics of functional fillers, the introduction of 3D structural materials assembled from small building blocks into the phase change matrices has become a research hotspot.[80] The preconstructed 3D structural materials can improve the utilization rate of thermally conductive filler and endow the composites with optimized thermally conductive network. Also, the use of 3D architectures can avoid the unstable factors such as sedimentation during the long-term recycle.

2.3.1 Metals

Among metal materials, the outstanding thermal conductivity of silver and copper is favored in the preparation of high thermally conductive composites. Qian et al.[84] fabricated leakage-proof PEG/diatomite phase change composites. When a large amount of diatomite was added, the thermal conductivity improvement of the composite was very limited. Furthermore, the thermal conductivity of PEG/diatomite/silver nanoparticle ternary phase change composites was greatly improved by depositing spherical silver nanoparticles with the diameter of 3–10 nm on diatomite and then compounding with PEG. Compared with PEG/diatomite composites, the thermal conductivity of the ternary composites with silver nanoparticle loading of 7.2 wt% was increased by 127%, up to $0.82\,W\,m^{-1}\,k^{-1}$. Similarly, physical blending and impregnation were adopted to prepare PEG/EVM/silver nanowire composites, in which EVM with porous structure and silver nanowire with the length of 5–20 μm and the diameter of 50–100 nm acted as supporting skeleton and thermally conductive component to improve the shape stability and thermal conductivity of the phase change composites, respectively.[51] Apart from silver, thermal conductivity improvement of form-stable PEG/SiO_2 phase change composites has been achieved by *in-situ* cupper doping.[85]

Three-dimensional metal foams (copper foam, nickel foam, and aluminum foam) with continuous skeleton structure exhibit excellent mechanical properties, ultra-low relative density, high porosity, and high thermal conductivity, but they are mainly used for the preparation of PW-based phase change composites. In addition, it is worth noting that the pore size of commercial metal foams is generally large, which is not conducive to the improvement of shape stability of phase change composites. Meanwhile, copper nanowire aerogel[86] and silver/polyvinylpyrrolidone (PVP) nanowire aerogel[87] with tunable structures have been developed to yield phase change composites with enhanced thermal and electrical conductivities, exhibiting great potentials in metal-based polymeric phase change composites for thermal energy conversion and storage.

2.3.2 Carbon Materials

Owing to the low density, excellent thermal/chemical stability, and high thermal conductivity, carbon materials can be considered as one of the most promising candidates for production of thermally conductive phase change composites. CNT,[88–90] CF,[91] EG,[54] and GNP[89, 92, 93] have been employed to develop PEG-based phase change composites with enhanced thermal conductivity. Additionally, carbon materials [94] can be added into leakage-proof composites containing supporting polymers to improve the thermal conductivity. Noteworthy, it is essential to conduct thermal annealing treatment at high temperature to repair the defects corresponding to phonon scattering when GO is employed as precursor to yield graphitized graphene with high intrinsic thermal conductivity for thermally conductive phase change composites.

In addition to direct introduction of carbon-based fillers into phase change matrices, 3D carbon-based structural materials have been widely used to prepare phase change composites in recent years. Three-dimensional thermally conductive networks have been successfully constructed in polymeric PCMs by incorporating biological porous carbon,[16] nanofibrillated cellulose/CNT aerogel,[67] cellulose/graphene aerogel,[68] microcrystalline cellulose/GNP aerogel,[95] GO/GNP hybrid aerogel,[55] and cellulose nanofiber/GNP hybrid-coated melamine foam[96]. Owing to that the nanostructured materials are physically overlapped through π–π interaction in common carbon-based aerogel/foam prepared by self-assembly strategy, there is pronounced thermal contact resistance at the joints. Covalent-bond diamond and graphene foams prepared by chemical vapor deposition (CVD) are able to provide continuous phonon transmission pathway. Unfortunately, these hydrophobic foams are preferential for nonpolar PCMs like PW. Hydrophilic modification such as plasma treatment is considered as an effective solution to enrich the variety of phase change composites.

2.3.3 Ceramics

Compared with electrically conductive metal and carbon materials, ceramics represented by BN gain unprecedented popularity in the fabrication of thermally conductive yet electrically insulating phase change composites, expanding their application scenarios. Thermally conductive PEG/silicon nitride (Si_3N_4) nanowire phase change composites have been developed via a facile solution blending.[97] AlN[98] and Al_2O_3[99] have been used as functional fillers to improve thermal conductivity of shape-stabilized PEG/SiO_2 phase change composites. Similarly, SiC nanowires have been incorporated into leakage-proof PEG/EVM composites to improve the thermal conductivity.[100] PEG and BN have compounded by melt blending,[101] solution blending,[17] and interfacial polyelectrolyte complex spinning[102] to fabricate polymeric phase change composites with high thermal conductivity.

In order to improve the utilization efficiency of functional fillers and construct thermally conductive pathways, ceramic-based porous scaffolds have been proposed for polymeric phase change composites. Unfortunately, it is more difficult to exfoliate and functionalize BN than graphite owing to the higher electronegativity of N atoms originating from the partially ionic B-N bonds, and thus the formation of 3D-BN structural materials without the assistance of auxiliary materials remains a challenging problem. It has been proved that chitosan,[103] cellulose,[69] and GO [66, 104, 105] can be used as supporting materials or binders to help BN build 3D conductive network for polymeric phase change composites. For example, GO/BN porous scaffolds with isotropic and aligned structures have been designed through a facile yet powerful freeze-casting to fabricate thermally conductive PEG-based phase change composites,[66, 104, 105] and the final composites with a phase change enthalpy of 143.6 $J\,g^{-1}$ exhibit enhanced thermal conductivity as high as 3.18 $W\,m^{-1}\,K^{-1}$ at BN loading of c. 28.7 wt%.

Furthermore, hybrid fillers with different geometries, such as silver-graphene[57] and GNP-BN[106], are capable of forming synergistic network structure to improve the thermal conductivity of PEG-based phase change composites. Metals, carbon materials, ceramics, and their hybrids can be employed as enhanced component to produce electrically conductive or insulating polymeric phase change composites with high thermal conductivity through facile solution blending, melt blending, and vacuum impregnation methods. More details are summarized in Table 2.4, in which thermal conductivity enhancement (TCE) is defined by Eq. (2.1). Constructing 3D conductive network in phase change composites seems to be an efficient route to achieve a trade-off among shape stability, thermal conductivity, and energy storage density, attracting ever-growing interest.

$$\mathrm{TCE} = \frac{k_c - k_m}{k_m} \tag{2.1}$$

where k_c and k_m represent the thermal conductivities of composites and matrix or matrix with supporting materials, respectively.

2.4 Energy Conversion and Storage Based on Polymeric Phase Change Composites

In recent years, energy conversion, storage, and utilization based on PCMs have become the research frontiers, especially for solar energy, industrial waste heat, and surplus heat. As-prepared phase change composites with improved comprehensive performance can achieve efficient energy conversion and storage via electro-to-heat (Figure 2.7a),[95] photo-to-heat (Figure 2.7b),[88, 95, 109] magnetism-to-heat (Figure 2.7c),[107] heat-to-electricity (Figure 2.7d),[108] and photo-to-heat-to-electricity (Figure 2.7e)[105] routes, providing a strong support for the popularization of advanced energy-related devices and systems associated with functional PCMs. Also, both solid–liquid and solid–solid polymeric phase change composites can realize these energy conversion routes.

2.4.1 Electro-to-Heat Conversion

Electric thermal storage technique based on PCMs is expected to improve the utilization efficiency of power generation equipment and exhibits broad prospects in the field of off-peak electricity. Unfortunately, the electrical conductivity of PCMs is inherently low, which makes it impossible to directly employ them for electro-to-heat conversion. It is usually necessary to incorporate functional fillers to improve the electrical conductivity of phase

Table 2.4 Thermal conductivity of polymeric phase change composites.

Material systems	Processing methods	Thermally conductive filler loading	Melting enthalpy (Jg^{-1})	Thermal conductivity ($W\ m^{-1}\ K^{-1}$)	Thermal conductivity enhancement (%)
PEG/diatomite/silver nanoparticle[84]	Vacuum impregnation	7.2 wt%	111.3	0.82	127
PEG/EVM/silver nanowire[51]	Physical blending and impregnation	19.3 wt%	99.1	0.68	172
PEG/SiO_2/cupper[85]	Sol–gel and *in-situ* doping method	2.1 wt%	110.2	0.414	15
PW/silver-PVP nanowire aerogel[87]	Vacuum impregnation	5.43 wt%	~163	0.49	~133
PEG-co-*N*,*N*′-dihydroxyethyl aniline/single-walled CNT[88]	Vacuum evaporation	—	100.5	0.334	25
PEG/single-walled CNT[90]	Solution blending	10 wt%	165.4	3.43	1329
PEG/SiO_2/CF[91]	Sol–gel and *in- situ* doping method	3 wt%	142.6	0.45	73
PEG/EG[54]	Melt blending	10 wt%	161.2	1.324	344
PEG/GO/GNP[92]	Solution blending	6 wt%	167.4	1.72	493
PEG/unsaturated polyester resin/GNP[93]	Free radical copolymerization and solution blending	2 wt%	140.8	0.67	131
PEG/single-walled CNT[89]	Vacuum impregnation	8 wt%	162.1	2.73	950
PEG/GNP[89]	Vacuum impregnation	4 wt%	169.3	3.11	1096
Poly(hexadecyl acrylate)/cellulose/graphene[94]	Atom transfer radical polymerization (ATRP) and injection molding	9 wt%	78	1.32	560
PEG/biological porous carbon[16]	Vacuum impregnation	14.6 wt%	158.8	4.489	953
PEG/cellulose-graphene aerogel[68]	Vacuum impregnation	5.3 wt%	156.1	1.35	463

(*continued*)

Table 2.4 (Continued)

Material systems	Processing methods	Thermally conductive filler loading	Melting enthalpy (Jg^{-1})	Thermal conductivity ($W m^{-1} K^{-1}$)	Thermal conductivity enhancement (%)
PEG/microcrystalline cellulose-GNP aerogel[95]	Vacuum impregnation	1.51 wt%	182.6	1.03	232
PEG/GO-GNP aerogel[55]	Vacuum impregnation	2.23 wt%	181.5	1.43	361
PEG/cellulose nanofiber-GNP hybrid-coated melamine foam[96]	Vacuum impregnation	0.65 wt%	178.9	0.26	189
PEG/Si_3N_4 nanowires[97]	Solution blending	10 wt%	152.3	0.362	89
PEG/SiO_2/Al_2O_3[99]	Ultrasound-assisted sol–gel and *in-situ* doping method	12.6 wt%	123.8	0.435	21
PEG/EVM/SiC nanowires[100]	Physical blending and impregnation	3.29 wt%	64.93	0.53	96
PEG/epoxy/BN[101]	Melt blending and curing	40 wt%	60.7	2.962	~887
PEG/GO/BN[17]	Solution blending	34 wt%	107.4	3.00	900
PEG/crosslinked cellulose -chitosan/BN[102]	Interfacial polyelectrolyte complex spinning	47.4 wt%	48.3	4.005	2256
PEG/chitosan-BN scaffold[103]	Nondirectional freezing and vacuum impregnation	27 wt%	136.9	2.77	~794
PEG/cellulose-BN nanosheet scaffold[69]	Vacuum impregnation	10 vol%	136.8	4.764	~1344
PEG/GO-BN scaffold[104]	Nondirectional freezing and vacuum impregnation	19.2 wt%	145.9	1.84	479
PEG/aligned GO-BN scaffold[66]	Unidirectional freezing and vacuum impregnation	28.7 wt%	143.6	3.18	864
PEG/silver-graphene[57]	Solution blending	8 wt%	166.1	0.414	95
PEG/BN-GNP[106]	Solution blending	31 wt%	122.2	1.33	329

change composites, and carbon materials, including biomass carbon,[110] carbon nanofiber (CNF),[111] CNT,[109] and graphene[112], are preferentially considered. Of critical importance is the electro-to-heat conversion and storage efficiency (η_e), which can be defined by the ratio of stored heat energy to applied electric energy during phase change process and determined by the electro-thermal calculation Eq. (2.2).

$$\eta_e = \frac{m\Delta H}{UIt} \tag{2.2}$$

where m is the mass, ΔH is the phase change enthalpy determined by differential scanning calorimeter (DSC), U and I, respectively, represent the applied voltage and current, and t is phase change time.

The majorities of the systems used to conduct electro-to-heat conversion are organic nonpolymeric solid–liquid PCMs. Leakage-proof phase change composites composed of microcrystalline cellulose/GNPs aerogel and PEG have been reported. The conductive composites have the ability of electrical energy transition and release (Figure 2.7a).[95] When an electrical field is applied, electrical energy can be inverted into thermal energy by generating Joule heat. Once the accumulated heat reaches the phase transition temperature of working substance, phase change and heat storage behaviors occur. Likewise, shape-stabilized PW-based composites containing commercial melamine foam incorporated by GO and GNPs exhibited high electrical conductivity (2.787 $\mathrm{S\,cm^{-1}}$) at a filler loading of 4.89 wt% and efficient electro-to-heat conversion capacity with an efficiency of 62.5%.[113] In addition, Chen et al.[114] employed solid–solid PCMs to realize electro-to-thermal energy conversion. PEG was introduced into graphite foam and then *in-situ* polymerized, giving rise to the formation of PU-based solid–solid phase change composites. When a relatively low voltage of 1.2 or 1.4 V is applied, the phase change composites can complete the electro-to-heat conversion, and the estimated conversion efficiency is above 80%. Also, an efficient electro-to-heat conversion for PU-based solid–solid phase change composites has been achieved after introducing electrically conductive graphene aerogel.[115, 116]

2.4.2 Light-to-Heat Conversion

Not only can electricity be converted into thermal energy, but also solar energy can be transformed into heat and stored in PCMs. Solar energy, a renewable and clean energy source, is considered to be one of the most effective methods to solve the energy shortage issue. However, there are still technical challenges in the effective utilization of solar energy owing to the intermittence and discontinuity of solar radiation in time and space, which can be exactly solved with the assistance of PCMs storing and releasing heat during the phase transition process. The weak photoabsorption capacity of PCMs makes them unable to convert solar energy into heat directly and effectively. In recent years, dyes,[117] carbon materials (biomass carbon,[118] CNT,[88] EG,[119] graphene,[94] and GO[120]), polydopamine (PDA),[121] Fe_3O_4@graphene,[122] and MXene[70] have been employed as photothermal absorbers for PCMs to realize the efficient conversion and storage of solar energy. The solar-to-heat conversion and storage efficiency (η_s) can be calculated from the ratio of the stored heat with respect to the input solar energy according to the light-thermal calculation Eq. (2.3).

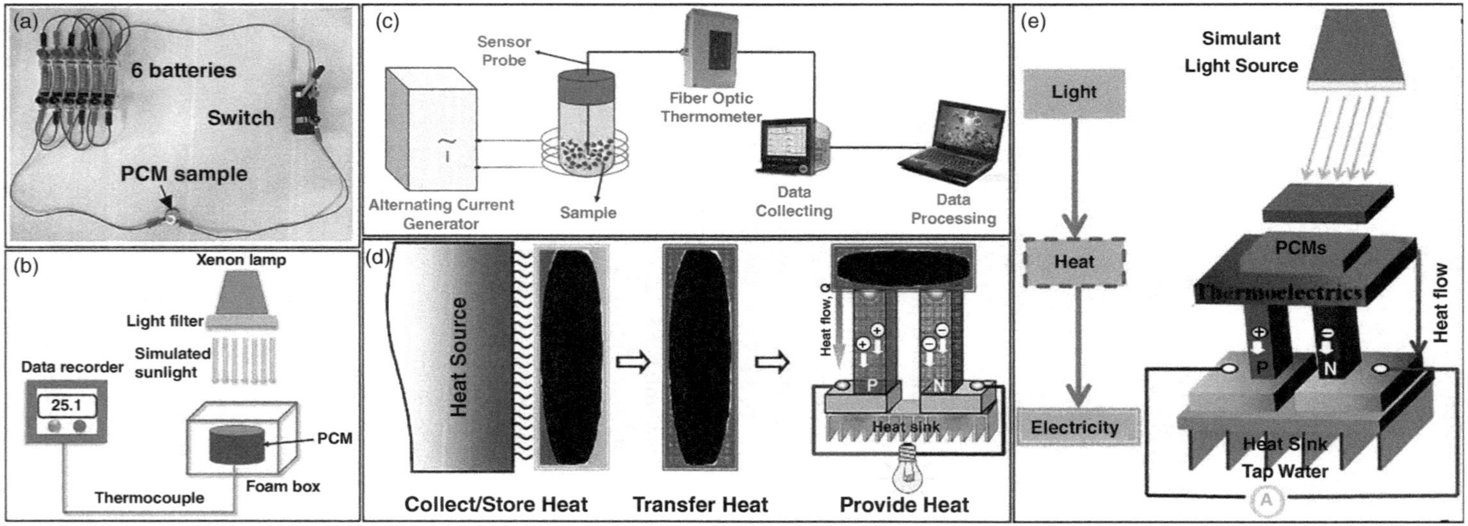

Figure 2.7 Energy conversion routes associated with polymeric phase change composites: (a) electro-to-heat. (b) photo-to-heat. *Source:* Wei et al. [95]. Reproduced with permission from Elsevier Ltd. (c) magnetism-to-heat. *Source:* Based on Wang et al. [107]. (d) heat-to-electricity. *Source:* Jiang et al. [108]. Reproduced with permission from the Royal Society of Chemistry and (e) photo-to-heat-to-electricity. *Source:* Yang et al. [105]. Reproduced with permission from the Royal Society of Chemistry.

$$\eta_s = \frac{m\Delta H}{\rho St} \tag{2.3}$$

where ρ and S are the light irradiation intensity and the surface area, respectively.

Note that carbon materials are mainly incorporated in the above-mentioned electric-driven phase change composites, and thus most of them can achieve electro-to-heat and light-to-heat conversions simultaneously. Considering that PEG with abundant oxygen-containing functional groups, dyes, GO, PDA, and MXene serving as photon harvesters and molecular heaters have been proved to be good candidates for the fabrication of photodriven polymeric phase change composites. The working mechanism and experimental setup of solar-to-heat conversion associated with polymeric phase change composites are shown in Figure 2.7b. During the solar radiation from simulated light source, phase change composites can absorb thermal energy in the form of sensible heat, leading to a rapid increase in temperature. When the temperature reaches the phase change temperature, an inflection point emerges in the temperature evolution curves, indicating that phase transition and TES in the form of latent heat take place with small temperature fluctuation. After the phase transformation, the temperature rises again, and the converted heat energy is expressed as sensible heat in this phase. Upon removing the incident light, the temperature drops sharply and another phase change platform emerges later, demonstrating that the stored heat energy is released to maintain the temperature during the solidification. After that the temperature of the system slowly decreases to room temperature.

Aside from photodriven polymeric solid–liquid phase change composites, Zhou et al.[75] developed a photothermal energy conversion system by introducing graphene into crosslinked PU solid–solid PCMs. When the content of graphene in the phase change composites increased from 0.5 to 9.1 wt%, the photothermal energy conversion efficiency of the system increased from 42.9 to 78.7% accordingly. Furthermore, multifunctional PU-based phase change composites with zinc oxide (ZnO) and graphene aerogel were fabricated.[116] Owing to the enhanced photoabsorption from ZnO and graphene aerogel and electrical conductivity from graphene aerogel, the photo-/electro-thermal energy conversion efficiencies of the system containing the resultant polymeric solid–solid phase change composites were up to 80.1 and 84.4%, respectively.

2.4.3 Magnetism-to-Heat Conversion

Tang's group has proposed a magnetism-to-heat conversion system through incorporating magnetic particles into phase change composites for the first time.[107] Magnetic Fe_3O_4 was deposited on the surface of graphene and then introduced into PEG matrix to yield phase change composites. The obtained nanocomposites can not only absorb solar energy to perform light-to-heat conversion but also realize magnetism-to-heat conversion under an alternating magnetic field. The energy conversion process and experimental apparatus are illustrated in Figure 2.7c. Together with Fe_3O_4/graphene hybrids as nanoheaters, alternating magnetic field induced by an alternating current generator can trigger magnetism-to-heat conversion of phase change composites. The magnetic thermal conversion efficiency (η_m) can be defined by the ratio of stored heat energy to energy exerted by magnetic field on the basis of the magnetism-thermal calculation Eq. (2.4), which increases with the increase of content of functional materials in the composites, up to 41.7%. Moreover, the results and conclusions have been confirmed in another magnetic thermal energy conversion system containing a novel $PEG/SiO_2/Fe_3O_4$ ternary phase change composite.[123]

$$\eta_m = \frac{m\Delta H}{Pt} \tag{2.4}$$

where P is alternating magnetic field power.

2.4.4 Heat-to-Electricity Conversion

The above three systems are capable of converting other forms of energy into heat energy for storage. How to use the thermal energy collected and stored by PCMs? In 2015, Jiang et al.[108] creatively proposed a proof-of-concept thermoelectric conversion by combining phase change energy storage materials with n/p-type semiconductors, achieving the conversion of heat into applicable electricity. As shown in Figure 2.7d, the thermal energy from heat source can be collected and stored by PCMs, and then the collected heat is transferred to a commercial thermoelectric device, forming a temperature difference at its two ends and generating electricity via Seebeck effect. Similarly, Yu et al.[124] designed a thermoelectric energy conversion system based on graphene aerogel embedded shape-stabilized PCMs. Two PCMs (PEG and *1*-tetradecanol) with different phase transition temperatures are, respectively, used as heat and cold sources at the ends of thermoelectric devices to tune the temperature difference. Based on the experimental data and numerical predictions, the heating and cooling energy conversion efficiencies are 52.81 and 30.29%, respectively. Furthermore, one-step photo-to-electricity conversion systems (Figure 2.7e)[66, 104, 105] integrating light-to-heat with heat-to-electricity have been developed. To achieve light-to-heat-to-electricity energy conversion, a prerequisite is that phase change composites ought to possess excellent photoabsorption.

Compared with electro/photo-to-heat energy conversion efficiency, magnetism-to-heat and heat-to-electricity conversion efficiencies are relatively low and need to be further improved. As the most of the above-mentioned energy conversion systems are exposed to the surrounding environment, a portion of the converted heat energy is more or less dissipated, affecting the energy conversion efficiency. One solution is to carry out these energy conversion operations in a vacuum (thermal insulation) environment. Additionally, there is a lack of evaluation index for light-to-electricity energy conversion. For the case of photodriven phase change composites, polymer-based photothermal absorbers such as artificial melanin, polyaniline (PANI), and polypyrrole (PPy) are competitive candidates. In heat/solar-to-electricity route, a commercial thermoelectric module is used directly, and thus an integrated device consisting of PCMs-based component and thermoelectric component needs to be developed. Additionally, new ways of energy conversion associated with PCMs are waiting to be discovered.

2.5 Emerging Applications of Polymeric Phase Change Composites

The foregoing energy conversion routes of PCMs are closely related to their practical applications. Polymeric phase change composites exhibit great potentials in thermal management of electronics,[61] smart textiles,[125] and shape memory devices[126].

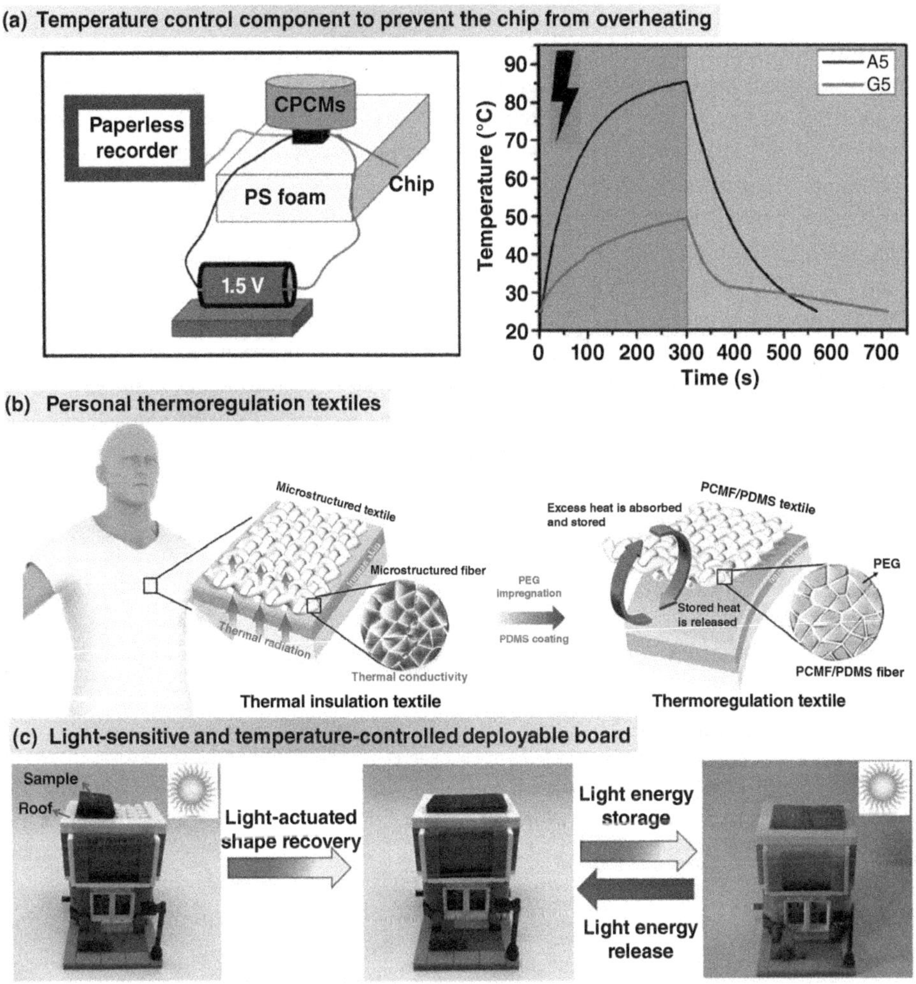

Figure 2.8 Potential applications of polymeric phase change composites. (a) Leakage-proof phase change composites as a temperature control component to prevent the chip from overheating and maintain a low working temperature. *Source:* Yang et al. [61]. Reproduced with permission from the American Chemical Society. (b) Flexible polymeric phase change composite fibers for personal thermoregulation textiles. *Source:* Wu et al. [127]. Reproduced with permission from the American Chemical Society. (c) PCMs-based shape memory composites as a light-actuated deployable roof for energy-saving buildings. *Source:* Wu et al. [126]. Reproduced with permission from Elsevier Ltd.

2.5.1 Thermal Management of Electronics

Thermal management systems are widely used in electronics, food preservation, construction, and many other fields. The addition of functional PCMs can significantly improve their performance. For example, the electrochemical performance and cycling life of the batteries depend heavily on the operating temperature, especially in high-power scenarios, and thus the thermal management of the battery is particularly important. Conventional heat dissipation methods such as forced air cooling and liquid cooling have been widely developed. Although these thermal management systems can ensure heat safety, the whole system is too bulky, sophisticated, and costly in terms of fans, pumps, pipelines, and other accessories, resulting in an increase in the weight of the whole system and the dependence on external energy input. It has been proved that the thermal management system associated with PCMs for the batteries can overcome these shortcomings to a great extent, prolonging their service life.[128]

Polymeric phase change composites have been employed as thermal management component to protect the electronic systems, including batteries,[129] chips,[61] mainboards,[130] etc. Yang et al.[61] fabricated leakage-proof phase change composites through impregnating low-temperature PEG into hierarchically porous PVA aerogel. The obtained composites can work as a temperature control component to prevent the chip from overheating (Figure 2.8a). Flexible melamine foam-supported shape-stabilized PCMs can be packed in the electronic systems to work as temperature-controlled deployable panel for the mainboard.[130] Furthermore, graphene and BN nanosheets have been deposited on the surface of melamine foam framework to improve the thermal response performance of the composites as a thermal protection component in electronic systems.[131] A solid–solid polymeric PCMs/EG composite with high cooling efficiency has been designed for the battery modules.[129] Apart from thermal management performance of polymeric phase change composites, other capabilities such as electromagnetic interference shielding and heat dissipation are crucial for the next-generation electronic devices.

2.5.2 Smart Textiles

Apart from the thermal management of electronics, PCMs can be utilized to tune the on-body temperature, named as smart clothing, which exhibits great potentials in the clothing for extreme conditions such as spacesuits and fire protective garments. Comfort related with temperature is an important function of clothing, and one of the basic functions of smart clothing is to maintain relatively stable temperature for human body. Integrating PCMs featuring large heat storage capacity and almost constant phase change temperature with personal thermal management textiles can greatly improve their heat preservation performance and temperature regulation function. At present, the main approach to make smart clothing with PCMs is to infiltrate microencapsulated phase change products into fibers, fabrics, and foams. In addition, PCMs with the working temperature range of 18–35 °C are most suitable for the production of smart clothing.[125]

Phase change component can be directly integrated into the fabrics through printing, coating, and padding routes.[132, 133] Another plausible alternative is to construct core-shell structure, in which phase change matrix and polymer fiber serve as core and shell, respectively. Apart from typical coaxial electrospinning,[134] melt spinning as a clean and high-efficiency spinning strategy is available for the preparation of polymeric phase change fibers.[135] Very recently, a novel freeze-spinning strategy integrating freeze-casting with spinning has been proposed to yield polymeric phase change composite fibers for personal thermoregulation textiles (Figure 2.8b).[127, 136] Porous aerogel fibers are likely to be competitive candidates for the carriers of phase change clothing. Wu et al.[137] developed PU/graphite foam solid–solid phase change composites with the ability of electro-to-thermal energy conversion. The obtained composites are able to be assembled into the fabric as an active protective layer (or a thermal buffering layer) to fight the cold.

2.5.3 Shape Memory Devices

Shape memory polymers (SMPs), as a class of intelligent materials, are capable of fixing the deformed (temporary) shape and recovering to the original (permanent) shape upon exposure to external stimuli, such as heat, light, electricity, moisture, chemicals, etc.[138] Common SMPs consist of two structural components, namely, the reversible switching phase determining the shape memory transition and the permanent phase responsible for memorizing the original state. Enlightened by this dual-phase structure, the polymeric phase change composites prepared by compounding PCMs with elastic supporting materials are able to exhibit shape memory function. With this in mind, the supporting component plays a significant role in the macroscopic physical properties of shape memory polymeric phase change composites. Elastomers represented by SEBS and OBC as well as melamine foam with 3D porous structure have been proved to be good supporting candidates for PCMs-based shape memory composites.[41, 42, 130]

PCMs-based shape memory composites present a rigid state when the temperature is below the phase change temperature (melting point) of PCMs. Once the temperature is higher than their melting temperature, PCMs melt into a liquid. Meanwhile, the composites soften and can be deformed into a targeted shape. Therefore, accompanying the phase transition of PCMs, the transformation of the composites from rigidity to flexibility is reversible. PCMs and elastic supporting materials can be facilely blended to produce binary thermally induced phase change composites.[130] Additional functional component such as EG can facilitate the performance owing to the improvement of thermal conductivity.[139] For light-actuated shape memory composites (Figure 2.8c), photothermal absorbers such as PDA,[140] CNT,[141] and graphene[126] have been added to improve their photoadsorption ability.

2.6 Conclusions and Outlook

Great advances in PCMs with excellent shape-stability and high thermal conductivity have been witnessed in past few decades, largely extending their applications. In this chapter, polymeric phase change composites with enhanced comprehensive performance have been highlighted. Micro/nanoencapsulation, facile blending, porous structural scaffolds and solid–solid transition routes have been adopted to fabricate leakage-proof phase change composites. Further to improve the thermal conductivity, conductive metals and carbon materials as well as insulating ceramics have been employed as thermally conductive components. Polymeric phase change composites exhibit immense potential applications in energy conversion, thermal management, smart clothing, and shape memory device. High-performance and multifunctional phase change composites as an advanced TES technique are bound to play an increasingly important role in energy storage-related applications.

Although high-performance polymeric phase change composites have been greatly developed, many technologies are not mature enough to achieve mass production. There is a long way to go from the laboratory to the factory, in particular high-efficiency 3D porous macroarchitectures. Of note, the introduction of functionality is at the expense of energy storage density, but high energy storage density is the intrinsic essence of PCMs. It is unwise to ignore the original intention when developing new polymeric phase change composites. Therefore, it is of significance to optimize the structural design and processing technology to improve the utilization efficiency of functional materials. Moreover, the effect of nanomaterials and nano/microstructures on the phase change behaviors needs to be further clarified. Accordingly, green, safety, efficiency, and cost, these timeless pursuits, are still the concerns. More attention is well-advised to be paid to sustainable, flexible, and intelligent polymeric phase change composites in the future. (i) For the perspective of polymeric supporting component, biodegradable materials are considered as plausible alternatives, including nanomaterials represented by nanocellulose and biomasses like porous wood. (ii) With the emergence of Internet of Things (IoT), flexible devices for wearable applications are becoming research priorities, and flexible phase change energy storage composites are an indispensable part. (iii) Breaking through existing application scenarios, polymeric phase change composites are likely to be more intelligent in the future, such as multiresponsive materials, self-adaptive devices, infrared stealth systems, advanced thermotherapy, smart drug delivery, and so on.

Acknowledgments

This work is financially supported by the National Natural Science Foundation of China (51873126 and 51721091).

References

1 Xu, B., Li, P., and Chan, C. (2015). Application of phase change materials for thermal energy storage in concentrated solar thermal power plants: a review to recent developments. *Applied Energy* 160: 286–307.

2 Chu, S. and Majumdar, A. (2012). Opportunities and challenges for a sustainable energy future. *Nature* 488: 294–303.

3 Sundararajan, S., Samui, A.B., and Kulkarni, P.S. (2017). Versatility of polyethylene glycol (PEG) in designing solid–solid phase change materials (PCMs) for thermal management and their application to innovative technologies. *Journal of Materials Chemistry A* 5: 18379–18396.

4 Amaral, C., Vicente, R., Marques, P.A.A.P., and Barros-Timmons, A. (2017). Phase change materials and carbon nanostructures for thermal energy storage: a literature review. *Renewable and Sustainable Energy Reviews* 79: 1212–1228.

5 Sharma, S.D., Kitano, H., and Sagara, K. (2004). Phase change materials for low temperature solar thermal applications. *Research Reports of the Faculty of Engineering, Mie University* 29: 31–64.

6 Shchukina, E.M., Graham, M., Zheng, Z., and Shchukin, D.G. (2018). Nanoencapsulation of phase change materials for advanced thermal energy storage systems. *Chemical Society Reviews* 47: 4156–4175.

7 Pielichowska, K. and Pielichowski, K. (2014). Phase change materials for thermal energy storage. *Progress in Materials Science* 65: 67–123.

8 Wu, S., Yan, T., Kuai, Z., and Pan, W. (2020). Thermal conductivity enhancement on phase change materials for thermal energy storage: a review. *Energy Storage Materials* 25: 251–295.

9 Sharma, A., Tyagi, V.V., Chen, C.R., and Buddhi, D. (2009). Review on thermal energy storage with phase change materials and applications. *Renewable and Sustainable Energy Reviews* 13: 318–345.

10 Zhang, P., Xiao, X., and Ma, Z.W. (2016). A review of the composite phase change materials: fabrication, characterization, mathematical modeling and application to performance enhancement. *Applied Energy* 165: 472–510.

11 Hu, H. (2020). Recent advances of polymeric phase change composites for flexible electronics and thermal energy storage system. *Composites Part B: Engineering* 195: 108094.

12 Sharma, R.K., Ganesan, P., Tyagi, V.V. et al. (2015). Developments in organic solid–liquid phase change materials and their applications in thermal energy storage. *Energy Conversion and Management* 95: 193–228.

13 Umair, M.M., Zhang, Y., Iqbal, K. et al. (2019). Novel strategies and supporting materials applied to shape-stabilize organic phase change materials for thermal energy storage – a review. *Applied Energy* 235: 846–873.

14 Qi, G.-Q., Liang, C.-L., Bao, R.-Y. et al. (2014). Polyethylene glycol based shape-stabilized phase change material for thermal energy storage with ultra-low content of graphene oxide. *Solar Energy Materials and Solar Cells* 123: 171–177.

15 Ye, S., Zhang, Q., Hu, D., and Feng, J. (2015). Core–shell-like structured graphene aerogel encapsulating paraffin: shape-stable phase change material for thermal energy storage. *Journal of Materials Chemistry A* 3: 4018–4025.

16 Zhao, Y., Min, X., Huang, Z. et al. (2018). Honeycomb-like structured biological porous carbon encapsulating PEG: a shape-stable phase change material with enhanced thermal conductivity for thermal energy storage. *Energy and Buildings* 158: 1049–1062.

17 Yang, J., Tang, L.-S., Bao, R.-Y. et al. (2018). Hybrid network structure of boron nitride and graphene oxide in shape-stabilized composite phase change materials with enhanced thermal conductivity and light-to-electric energy conversion capability. *Solar Energy Materials and Solar Cells* 174: 56–64.

18 Shi, J.-N., Ger, M.-D., Liu, Y.-M. et al. (2013). Improving the thermal conductivity and shape-stabilization of phase change materials using nanographite additives. *Carbon* 51: 365–372.

19 Jamekhorshid, A., Sadrameli, S.M., and Farid, M. (2014). A review of microencapsulation methods of phase change materials (PCMs) as a thermal energy storage (TES) medium. *Renewable and Sustainable Energy Reviews* 31: 531–542.

20 Aftab, W., Huang, X., Wu, W. et al. (2018). Nanoconfined phase change materials for thermal energy applications. *Energy & Environmental Science* 11: 1392–1424.

21 Liu, C., Rao, Z., Zhao, J. et al. (2015). Review on nanoencapsulated phase change materials: preparation, characterization and heat transfer enhancement. *Nano Energy* 13: 814–826.

22 Milián, Y.E., Gutiérrez, A., Grágeda, M., and Ushak, S. (2017). A review on encapsulation techniques for inorganic phase change materials and the influence on their thermophysical properties. *Renewable and Sustainable Energy Reviews* 73: 983–999.

23 Giro-Paloma, J., Martínez, M., Cabeza, L.F., and Fernández, A.I. (2016). Types, methods, techniques, and applications for microencapsulated phase change materials (MPCM): a review. *Renewable and Sustainable Energy Reviews* 53: 1059–1075.

24 Jacob, R. and Bruno, F. (2015). Review on shell materials used in the encapsulation of phase change materials for high temperature thermal energy storage. *Renewable and Sustainable Energy Reviews* 48: 79–87.

25 Zhang, G.H., Bon, S.A.F., and Zhao, C.Y. (2012). Synthesis, characterization and thermal properties of novel nanoencapsulated phase change materials for thermal energy storage. *Solar Energy* 86: 1149–1154.

26 De Castro, P.F., Ahmed, A., and Shchukin, D.G. (2016). Confined-volume effect on the thermal properties of encapsulated phase change materials for thermal energy storage. *Chemistry* 22: 4389–4394.

27 Du, X., Fang, Y., Cheng, X. et al. (2018). Fabrication and characterization of flame-retardant nanoencapsulated n-octadecane with melamine–formaldehyde shell for thermal energy storage. *ACS Sustainable Chemistry & Engineering* 6: 15541–15549.

28 He, F., Wang, X., and Wu, D. (2014). New approach for sol–gel synthesis of microencapsulated n-octadecane phase change material with silica wall using sodium silicate precursor. *Energy* 67: 223–233.

29 Geng, L., Wang, S., Wang, T., and Luo, R. (2016). Facile synthesis and thermal properties of nanoencapsulated n-dodecanol with SiO_2 shell as shape-formed thermal energy storage material. *Energy & Fuels* 30: 6153–6160.

30 Liu, H., Wang, X., and Wu, D. (2017). Fabrication of graphene/TiO_2/paraffin composite phase change materials for enhancement of solar energy efficiency in photocatalysis and latent heat storage. *ACS Sustainable Chemistry & Engineering* 5: 4906–4915.

31 Wang, T., Wang, S., Luo, R. et al. (2016). Microencapsulation of phase change materials with binary cores and calcium carbonate shell for thermal energy storage. *Applied Energy* 171: 113–119.

32 Li, M., Liu, J., and Shi, J. (2018). Synthesis and properties of phase change microcapsule with SiO_2-TiO_2 hybrid shell. *Solar Energy* 167: 158–164.

33 Zhao, A., An, J., Yang, J., and Yang, E.-H. (2018). Microencapsulated phase change materials with composite titania-polyurea (TiO_2-PUA) shell. *Applied Energy* 215: 468–478.

34 Wang, H., Zhao, L., Chen, L. et al. (2017). Facile and low energy consumption synthesis of microencapsulated phase change materials with hybrid shell for thermal energy storage. *Journal of Physics and Chemistry of Solids* 111: 207–213.

35 Wang, H., Zhao, L., Song, G. et al. (2018). Organic-inorganic hybrid shell microencapsulated phase change materials prepared from SiO_2/TiC-stabilized pickering emulsion polymerization. *Solar Energy Materials and Solar Cells* 175: 102–110.

36 Sun, N. and Xiao, Z. (2017). Synthesis and performances of phase change materials microcapsules with a polymer/BN/TiO_2 hybrid shell for thermal energy storage. *Energy & Fuels* 31: 10186–10195.

37 Zhang, L., Yang, W., Jiang, Z. et al. (2017). Graphene oxide-modified microencapsulated phase change materials with high encapsulation capacity and enhanced leakage-prevention performance. *Applied Energy* 197: 354–363.

38 Liu, Z., Chen, Z., and Yu, F. (2019). Enhanced thermal conductivity of microencapsulated phase change materials based on graphene oxide and carbon nanotube hybrid filler. *Solar Energy Materials and Solar Cells* 192: 72–80.

39 Trigui, A., Karkri, M., and Krupa, I. (2014). Thermal conductivity and latent heat thermal energy storage properties of LDPE/wax as a shape-stabilized composite phase change material. *Energy Conversion and Management* 77: 586–596.

40 Lu, X., Yu, H., Zhang, L. et al. (2020). Flexible ethylene propylene diene monomer/paraffin wax vulcanizate with simultaneously increased mechanical strength, thermal-energy storage, and shape-memory behavior. *Energy & Fuels* 34: 9020–9029.

41 Zhang, Q., Song, S., Feng, J., and Wu, P. (2012). A new strategy to prepare polymer composites with versatile shape memory properties. *Journal of Materials Chemistry* 22: 24776–24782.

42 Zhang, Q. and Feng, J. (2013). Difunctional olefin block copolymer/paraffin form-stable phase change materials with simultaneous shape memory property. *Solar Energy Materials and Solar Cells* 117: 259–266.

43 Zhang, Q., Cui, K., Feng, J. et al. (2015). Investigation on the recovery performance of olefin block copolymer/hexadecane form stable phase change materials with shape memory properties. *Solar Energy Materials and Solar Cells* 132: 632–639.

44 Ma, T., Li, L., Wang, Q., and Guo, C. (2018). High-performance flame retarded paraffin/epoxy resin form-stable phase change material. *Journal of Materials Science* 54: 875–885.

45 Du, J., Zhang, Z., Liu, D. et al. (2019). Triple-stimuli responsive shape memory effect of novel polyolefin elastomer/lauric acid/carbon black nanocomposites. *Composites Science and Technology* 169: 45–51.

46 Du, J., Liu, D., Zhang, Z. et al. (2017). Dual-responsive triple-shape memory polyolefin elastomer/stearic acid composite. *Polymer* 126: 206–210.

47 Wang, Y., Xia, T.D., Feng, H.X., and Zhang, H. (2011). Stearic acid/polymethylmethacrylate composite as form-stable phase change materials for latent heat thermal energy storage. *Renewable Energy* 36: 1814–1820.

48 Chen, F. and Wolcott, M. (2015). Polyethylene/paraffin binary composites for phase change material energy storage in building: a morphology, thermal properties, and paraffin leakage study. *Solar Energy Materials and Solar Cells* 137: 79–85.

49 Ding, Z., Yang, W., He, F. et al. (2020). GO modified EPDM/paraffin shape-stabilized phase change materials with high elasticity and low leakage rate. *Polymer* 204: 122824.

50 Li, W.-W., Cheng, W.-L., Xie, B. et al. (2017). Thermal sensitive flexible phase change materials with high thermal conductivity for thermal energy storage. *Energy Conversion and Management* 149: 1–12.

51 Deng, Y., Li, J., Qian, T. et al. (2016). Thermal conductivity enhancement of polyethylene glycol/expanded vermiculite shape-stabilized composite phase change materials with silver nanowire for thermal energy storage. *Chemical Engineering Journal* 295: 427–435.

52 Zhang, X., Wen, R., Tang, C. et al. (2016). Thermal conductivity enhancement of polyethylene glycol/expanded perlite with carbon layer for heat storage application. *Energy and Buildings* 130: 113–121.

53 Qian, T., Li, J., Min, X. et al. (2015). Diatomite: a promising natural candidate as carrier material for low, middle and high temperature phase change material. *Energy Conversion and Management* 98: 34–45.

54 Wang, W., Yang, X., Fang, Y. et al. (2009). Preparation and thermal properties of polyethylene glycol/expanded graphite blends for energy storage. *Applied Energy* 86: 1479–1483.

55 Yang, J., Qi, G.-Q., Liu, Y. et al. (2016). Hybrid graphene aerogels/phase change material composites: thermal conductivity, shape-stabilization and light-to-thermal energy storage. *Carbon* 100: 693–702.

56 Feng, L., Zheng, J., Yang, H. et al. (2011). Preparation and characterization of polyethylene glycol/active carbon composites as shape-stabilized phase change materials. *Solar Energy Materials and Solar Cells* 95: 644–650.

57 Zhang, Y., Wang, J., Qiu, J. et al. (2019). Ag-graphene/PEG composite phase change materials for enhancing solar-thermal energy conversion and storage capacity. *Applied Energy* 237: 83–90.

58 Lu, X., Huang, H., Zhang, X. et al. (2019). Novel light-driven and electro-driven polyethylene glycol/two-dimensional MXene form-stable phase change material with enhanced thermal conductivity and electrical conductivity for thermal energy storage. *Composites Part B: Engineering* 177: 107372.

59 Feng, L., Zhao, W., Zheng, J. et al. (2011). The shape-stabilized phase change materials composed of polyethylene glycol and various mesoporous matrices (AC, SBA-15 and MCM-41). *Solar Energy Materials and Solar Cells* 95: 3550–3556.

60 Şentürk, S.B., Kahraman, D., Alkan, C., and Gökçe, İ. (2011). Biodegradable PEG/cellulose, PEG/agarose and PEG/chitosan blends as shape stabilized phase change materials for latent heat energy storage. *Carbohydrate Polymers* 84: 141–144.

61 Yang, L., Yang, J., Tang, L.-S. et al. (2020). Hierarchically porous PVA aerogel for leakage-proof phase change materials with superior energy storage capacity. *Energy & Fuels* 34: 2471–2479.

62 Qu, L., Li, A., Gu, J., and Zhang, C. (2018). Thermal energy storage capability of polyurethane foams incorporated with microencapsulated phase change material. *ChemistrySelect* 3: 3180–3186.

63 Hong, H., Pan, Y., Sun, H. et al. (2018). Superwetting polypropylene aerogel supported form-stable phase change materials with extremely high organics loading and enhanced thermal conductivity. *Solar Energy Materials and Solar Cells* 174: 307–313.

64 Zhou, L., Tang, L.-S., Tao, X.-F. et al. (2020). Facile fabrication of shape-stabilized polyethylene glycol/cellulose nanocrystal phase change materials based on thiol-ene click chemistry and solvent exchange. *Chemical Engineering Journal* 396: 125206.

65 Tang, L.-S., Yang, J., Bao, R.-Y. et al. (2017). Polyethylene glycol/graphene oxide aerogel shape-stabilized phase change materials for photo-to-thermal energy conversion and storage via tuning the oxidation degree of graphene oxide. *Energy Conversion and Management* 146: 253–264.

66 Yang, J., Tang, L.-S., Bai, L. et al. (2018). Photodriven shape-stabilized phase change materials with optimized thermal conductivity by tailoring the microstructure of hierarchically ordered hybrid porous scaffolds. *ACS Sustainable Chemistry & Engineering* 6: 6761–6770.

67 Du, X., Qiu, J., Deng, S. et al. (2020). Alkylated nanofibrillated cellulose/carbon nanotubes aerogels supported form-stable phase change composites with improved n-alkanes loading capacity and thermal conductivity. *ACS Applied Materials & Interfaces* 12: 5695–5703.

68 Yang, J., Zhang, E., Li, X. et al. (2016). Cellulose/graphene aerogel supported phase change composites with high thermal conductivity and good shape stability for thermal energy storage. *Carbon* 98: 50–57.

69 Lei, C., Wu, K., Wu, L. et al. (2019). Phase change material with anisotropically high thermal conductivity and excellent shape stability due to its robust cellulose/BNNSs skeleton. *Journal of Materials Chemistry A* 7: 19364–19373.

70 Tang, L., Zhao, X., Feng, C. et al. (2019). Bacterial cellulose/MXene hybrid aerogels for photodriven shape-stabilized composite phase change materials. *Solar Energy Materials and Solar Cells* 203: 110174.

71 Du, X., Qiu, J., Deng, S. et al. (2020). Flame-retardant and form-stable phase change composites based on black phosphorus nanosheets/cellulose nanofiber aerogels with extremely high energy storage density and superior solar-thermal conversion efficiency. *Journal of Materials Chemistry A* 8: 14126–14134.

72 Fallahi, A., Guldentops, G., Tao, M. et al. (2017). Review on solid–solid phase change materials for thermal energy storage: molecular structure and thermal properties. *Applied Thermal Engineering* 127: 1427–1441.

73 Xi, P., Xia, L., Fei, P. et al. (2012). Preparation and performance of a novel thermoplastics polyurethane solid–solid phase change materials for energy storage. *Solar Energy Materials and Solar Cells* 102: 36–43.

74 Fu, X., Xiao, Y., Hu, K. et al. (2016). Thermosetting solid–solid phase change materials composed of poly(ethylene glycol)-based two components: flexible application for thermal energy storage. *Chemical Engineering Journal* 291: 138–148.

75 Zhou, Y., Liu, X., Sheng, D. et al. (2018). Polyurethane-based solid–solid phase change materials with in situ reduced graphene oxide for light-thermal energy conversion and storage. *Chemical Engineering Journal* 338: 117–125.

76 Meng, Q. and Hu, J. (2008). A poly(ethylene glycol)-based smart phase change material. *Solar Energy Materials and Solar Cells* 92: 1260–1268.

77 Xi, P., Zhao, F., Fu, P. et al. (2014). Synthesis, characterization, and thermal energy storage properties of a novel thermoplastic polyurethane phase change material. *Materials Letters* 121: 15–18.

78 Du, X., Wang, H., Wu, Y. et al. (2017). Solid–solid phase-change materials based on hyperbranched polyurethane for thermal energy storage. *Journal of Applied Polymer Science* 134: 45014.

79 Lu, X., Fang, C., Sheng, X. et al. (2019). One-step and solvent-free synthesis of polyethylene glycol-based polyurethane as solid–solid phase change materials for solar thermal energy storage. *Industrial & Engineering Chemistry Research* 58: 3024–3032.

80 Yang, J., Tang, L.-S., Bai, L. et al. (2019). High-performance composite phase change materials for energy conversion based on macroscopically three-dimensional structural materials. *Materials Horizons* 6: 250–273.

81 Han, Z. and Fina, A. (2011). Thermal conductivity of carbon nanotubes and their polymer nanocomposites: a review. *Progress in Polymer Science* 36: 914–944.

82 Burger, N., Laachachi, A., Ferriol, M. et al. (2016). Review of thermal conductivity in composites: mechanisms, parameters and theory. *Progress in Polymer Science* 61: 1–28.

83 Apostolopoulou-Kalkavoura, V., Munier, P., and Bergstrom, L. (2021). Thermally insulating nanocellulose-based materials. *Advanced Materials* 33: 2001839.

84 Qian, T., Li, J., Min, X. et al. (2015). Enhanced thermal conductivity of PEG/diatomite shape-stabilized phase change materials with Ag nanoparticles for thermal energy storage. *Journal of Materials Chemistry A* 3: 8526–8536.

85 Tang, B., Qiu, M., and Zhang, S. (2012). Thermal conductivity enhancement of PEG/SiO_2 composite PCM by in situ Cu doping. *Solar Energy Materials and Solar Cells* 105: 242–248.

86 Zhang, L., An, L., Wang, Y. et al. (2019). Thermal enhancement and shape stabilization of a phase-change energy-storage material via copper nanowire aerogel. *Chemical Engineering Journal* 373: 857–869.

87 Zhang, L. and Feng, G. (2020). A one-step-assembled three-dimensional network of silver/polyvinylpyrrolidone (PVP) nanowires and its application in energy storage. *Nanoscale* 12: 10573–10583.

88 Wang, Y., Tang, B., and Zhang, S. (2013). Single-walled carbon nanotube/phase change material composites: sunlight-driven, reversible, form-stable phase transitions for solar thermal energy storage. *Advanced Functional Materials* 23: 4354–4360.

89 Qian, T., Zhu, S., Wang, H. et al. (2019). Comparative study of single-walled carbon nanotubes and graphene nanoplatelets for improving the thermal conductivity and solar-to-light conversion of PEG-infiltrated phase-change material composites. *ACS Sustainable Chemistry & Engineering* 7: 2446–2458.

90 Qian, T., Li, J., Feng, W., and Nian, H.E. (2017). Single-walled carbon nanotube for shape stabilization and enhanced phase change heat transfer of polyethylene glycol phase change material. *Energy Conversion and Management* 143: 96–108.

91 Liu, Z., Wei, H., Tang, B. et al. (2018). Novel light–driven CF/PEG/SiO_2 composite phase change materials with high thermal conductivity. *Solar Energy Materials and Solar Cells* 174: 538–544.

92 Qi, G.-Q., Yang, J., Bao, R.-Y. et al. (2015). Enhanced comprehensive performance of polyethylene glycol based phase change material with hybrid graphene nanomaterials for thermal energy storage. *Carbon* 88: 196–205.

93 He, L., Wang, H., Yang, F., and Zhu, H. (2018). Preparation and properties of polyethylene glycol/unsaturated polyester resin/graphene nanoplates composites as form-stable phase change materials. *Thermochimica Acta* 665: 43–52.

94 Qian, Y., Han, N., Zhang, Z. et al. (2019). Enhanced thermal-to-flexible phase change materials based on cellulose/modified graphene composites for thermal management of solar energy. *ACS Applied Materials & Interfaces* 11: 45832–45843.

95 Wei, X., Xue, F., Qi, X. et al. (2019). Photo- and electro-responsive phase change materials based on highly anisotropic microcrystalline cellulose/graphene nanoplatelet structure. *Applied Energy* 236: 70–80.

96 Wu, H., Deng, S., Shao, Y. et al. (2019). Multiresponsive shape-adaptable phase change materials with cellulose nanofiber/graphene nanoplatelet hybrid-coated melamine foam for light/electro-to-thermal energy storage and utilization. *ACS Applied Materials & Interfaces* 11: 46851–46863.

97 Zhang, X., Liu, H., Huang, Z. et al. (2016). Preparation and characterization of the properties of polyethylene glycol @ Si_3 N_4 nanowires as phase-change materials. *Chemical Engineering Journal* 301: 229–237.

98 Wang, W., Yang, X., Fang, Y. et al. (2009). Enhanced thermal conductivity and thermal performance of form-stable composite phase change materials by using β-Aluminum nitride. *Applied Energy* 86: 1196–1200.

99 Tang, B., Wu, C., Qiu, M. et al. (2014). PEG/SiO_2–$Al2O_3$ hybrid form-stable phase change materials with enhanced thermal conductivity. *Materials Chemistry and Physics* 144: 162–167.

100 Deng, Y., Li, J., and Nian, H. (2018). Polyethylene glycol-enwrapped silicon carbide nanowires network/expanded vermiculite composite phase change materials: form-stabilization, thermal energy storage behavior and thermal conductivity enhancement. *Solar Energy Materials and Solar Cells* 174: 283–291.

101 Luo, F., Yan, P., Qian, Q. et al. (2020). Highly thermally conductive phase change composites for thermal energy storage featuring shape memory. *Composites Part A: Applied Science and Manufacturing* 129: 105706.

102 Fang, H., Lin, J., Zhang, L. et al. (2020). Fibrous form-stable phase change materials with high thermal conductivity fabricated by interfacial polyelectrolyte complex spinning. *Carbohydrate Polymers* 249: 116836.

103 Jia, X., Li, Q., Ao, C. et al. (2020). High thermal conductive shape-stabilized phase change materials of polyethylene glycol/boron nitride@chitosan composites for thermal energy storage. *Composites Part A: Applied Science and Manufacturing* 129: 105710.

104 Yang, J., Tang, L.-S., Bao, R.-Y. et al. (2016). An ice-templated assembly strategy to construct graphene oxide/boron nitride hybrid porous scaffolds in phase change materials with enhanced thermal conductivity and shape stability for light–thermal–electric energy conversion. *Journal of Materials Chemistry A* 4: 18841–18851.

105 Yang, J., Yu, P., Tang, L.S. et al. (2017). Hierarchically interconnected porous scaffolds for phase change materials with improved thermal conductivity and efficient solar-to-electric energy conversion. *Nanoscale* 9: 17704–17709.

106 Yang, J., Tang, L.-S., Bao, R.-Y. et al. (2017). Largely enhanced thermal conductivity of poly (ethylene glycol)/boron nitride composite phase change materials for solar-thermal-electric energy conversion and storage with very low content of graphene nanoplatelets. *Chemical Engineering Journal* 315: 481–490.

107 Wang, W., Tang, B., Ju, B. et al. (2017). Fe_3O_4-functionalized graphene nanosheet embedded phase change material composites: efficient magnetic- and sunlight-driven energy conversion and storage. *Journal of Materials Chemistry A* 5: 958–968.

108 Jiang, Y., Wang, Z., Shang, M. et al. (2015). Heat collection and supply of interconnected netlike graphene/polyethyleneglycol composites for thermoelectric devices. *Nanoscale* 7: 10950–10953.

109 Chen, L., Zou, R., Xia, W. et al. (2012). Electro- and photodriven phase change composites based on wax-infiltrated carbon nanotube sponges. *ACS Nano* 6: 10884–10892.

110 Li, Y., Samad, Y.A., Polychronopoulou, K. et al. (2014). From biomass to high performance solar–thermal and electric–thermal energy conversion and storage materials. *Journal of Materials Chemistry A* 2: 7759–7765.

111 Zhang, K., Han, B., and Yu, X. (2012). Electrically conductive carbon nanofiber/paraffin wax composites for electric thermal storage. *Energy Conversion and Management* 64: 62–67.

112 Li, G., Zhang, X., Wang, J., and Fang, J. (2016). From anisotropic graphene aerogels to electron- and photo-driven phase change composites. *Journal of Materials Chemistry A* 4: 17042–17049.

113 Xue, F., Lu, Y., Qi, X. et al. (2019). Melamine foam-templated graphene nanoplatelet framework toward phase change materials with multiple energy conversion abilities. *Chemical Engineering Journal* 365: 20–29.

114 Chen, R., Yao, R., Xia, W., and Zou, R. (2015). Electro/photo to heat conversion system based on polyurethane embedded graphite foam. *Applied Energy* 152: 183–188.

115 Zhou, Y., Wang, X., Liu, X. et al. (2019). Polyurethane-based solid–solid phase change materials with halloysite nanotubes-hybrid graphene aerogels for efficient light- and electro-thermal conversion and storage. *Carbon* 142: 558–566.

116 Zhou, Y., Wang, X., Liu, X. et al. (2019). Multifunctional ZnO/polyurethane-based solid–solid phase change materials with graphene aerogel. *Solar Energy Materials and Solar Cells* 193: 13–21.

117 Wang, Y., Tang, B., and Zhang, S. (2012). Novel organic solar thermal energy storage materials: efficient visible light-driven reversible solid–liquid phase transition. *Journal of Materials Chemistry* 22: 18145–18150.

118 Huang, X., Xia, W., and Zou, R. (2014). Nanoconfinement of phase change materials within carbon aerogels: phase transition behaviours and photo-to-thermal energy storage. *Journal of Materials Chemistry A* 2: 19963–19968.

119 Zhang, Q., Wang, H., Ling, Z. et al. (2015). RT100/expand graphite composite phase change material with excellent structure stability, photo-thermal performance and good thermal reliability. *Solar Energy Materials and Solar Cells* 140: 158–166.

120 Xiong, W., Chen, Y., Hao, M. et al. (2015). Facile synthesis of PEG based shape-stabilized phase change materials and their photo-thermal energy conversion. *Applied Thermal Engineering* 91: 630–637.

121 Yang, J., Qi, G.-Q., Tang, L.-S. et al. (2016). Novel photodriven composite phase change materials with bioinspired modification of BN for solar-thermal energy conversion and storage. *Journal of Materials Chemistry A* 4: 9625–9634.

122 Wang, Z., Tong, Z., Ye, Q. et al. (2017). Dynamic tuning of optical absorbers for accelerated solar-thermal energy storage. *Nature Communications* 8: 1478.

123 Fan, X., Xiao, J., Wang, W. et al. (2018). Novel magnetic-to-thermal conversion and thermal energy management composite phase change material. *Polymers* 10: 585.

124 Yu, C., Yang, S.H., Pak, S.Y. et al. (2018). Graphene embedded form stable phase change materials for drawing the thermo-electric energy harvesting. *Energy Conversion and Management* 169: 88–96.

125 Sarier, N. and Onder, E. (2012). Organic phase change materials and their textile applications: an overview. *Thermochimica Acta* 540: 7–60.

126 Wu, H., Li, S., Shao, Y. et al. (2020). Melamine foam/reduced graphene oxide supported form-stable phase change materials with simultaneous shape memory property and light-to-thermal energy storage capability. *Chemical Engineering Journal* 379: 122373.

127 Wu, J., Hu, R., Zeng, S. et al. (2020). Flexible and robust biomaterial microstructured coloured textiles for personal thermoregulation. *ACS Applied Materials & Interfaces* 12: 19015–19022.

128 Wang, Z., Zhang, Z., Jia, L., and Yang, L. (2015). Paraffin and paraffin/aluminum foam composite phase change material heat storage experimental study based on thermal management of Li-ion battery. *Applied Thermal Engineering* 78: 428–436.

129 Xiao, C., Zhang, G., Li, Z., and Yang, X. (2020). Custom design of solid–solid phase change material with ultra-high thermal stability for battery thermal management. *Journal of Materials Chemistry A* 8: 14624–14633.

130 Jing, J., Wu, H., Shao, Y. et al. (2019). Melamine foam-supported form-stable phase change materials with simultaneous thermal energy storage and shape memory properties for thermal management of electronic devices. *ACS Applied Materials & Interfaces* 11: 19252–19259.

131 Xue, F., Jin, X.Z., Xie, X. et al. (2019). Constructing reduced graphene oxide/boron nitride frameworks in melamine foam towards synthesizing phase change materials applied in thermal management of microelectronic devices. *Nanoscale* 11: 18691–18701.

132 Nejman, A., Cieślak, M., Gajdzicki, B. et al. (2014). Methods of PCM microcapsules application and the thermal properties of modified knitted fabric. *Thermochimica Acta* 589: 158–163.

133 Geng, X., Li, W., Wang, Y. et al. (2018). Reversible thermochromic microencapsulated phase change materials for thermal energy storage application in thermal protective clothing. *Applied Energy* 217: 281–294.

134 Babapoor, A., Karimi, G., Golestaneh, S.I., and Mezjin, M.A. (2017). Coaxial electro-spun PEG/PA6 composite fibers: fabrication and characterization. *Applied Thermal Engineering* 118: 398–407.

135 Xia, W., Fei, X., Wang, Q. et al. (2021). Nano-hybridized form-stable ester@F-SiO_2 phase change materials for melt-spun PA6 fibers engineered towards smart thermal management fabrics. *Chemical Engineering Journal* 403: 126369.

136 Wang, Y., Cui, Y., Shao, Z. et al. (2020). Multifunctional polyimide aerogel textile inspired by polar bear hair for thermoregulation in extreme environments. *Chemical Engineering Journal* 390: 124623.

137 Wu, W., Huang, X., Li, K. et al. (2017). A functional form-stable phase change composite with high efficiency electro-to-thermal energy conversion. *Applied Energy* 190: 474–480.

138 Leng, J., Lan, X., Liu, Y., and Du, S. (2011). Shape-memory polymers and their composites: stimulus methods and applications. *Progress in Materials Science* 56: 1077–1135.

139 Wu, W., Wu, W., and Wang, S. (2019). Form-stable and thermally induced flexible composite phase change material for thermal energy storage and thermal management applications. *Applied Energy* 236: 10–21.

140 Wu, H., Chen, R., Shao, Y. et al. (2019). Novel flexible phase change materials with mussel-inspired modification of melamine foam for simultaneous light-actuated shape memory and light-to-thermal energy storage capability. *ACS Sustainable Chemistry & Engineering* 7: 13532–13542.

141 Qi, X., Shao, Y., Wu, H. et al. (2019). Flexible phase change composite materials with simultaneous light energy storage and light-actuated shape memory capability. *Composites Science and Technology* 181: 107714.

3

Polymer Composites for High-Temperature Applications

Sen Niu, Lixue Zhu, Qiannan Cai, and Yunhe Zhang

Key Laboratory of High Performance Plastics (Jilin University), Ministry of Education, College of Chemistry, Jilin University, Changchun, P. R. China

3.1 Applications of Polymer Composite Materials in High-Temperature Electrical Insulation

Electrical insulation materials are one of the important fundamental materials in the field of electrical engineering, which provide insulation protection for electrical components. Based on differences in the chemical composition, the electrical insulation materials can be generally divided into three categories: inorganic, organic, and polymer-based composite. Also, according to the different states of existence, they can be divided into gas, liquid, and solid electrical insulation materials. The commercialized polymer-based composite electrical insulation materials did not have any significant requirements in terms of temperature resistance before the 1960s. However, with the further development of high-tech fields such as aerospace, military, atomic energy, etc., higher requirements are set on the heat resistance of electrical insulation materials, and their heat resistance plays a decisive role in the technical and economic indicators as well as the operating life of electrical equipment. Meanwhile, improvements in the heat resistance of electrical insulation materials contribute a critical role in miniaturization and large capacity electrical appliances and motors. Therefore, investigations on high-temperature-resistant polymer-based electrical insulation materials are a topic of major focus these days [1–4].

3.1.1 High-Temperature-Resistant Electrical Insulating Resin Matrix

High-temperature-resistant electrical insulating polymer matrix often has strong conjugation between molecular structures or easily forms hydrogen bonds within the molecular structure, resulting in strong interaction forces within and between the molecular chains. It directly leads to possessing outstanding high-temperature resistance, dielectric properties, environmental stability, and mechanical properties [5, 6]. At present, the widely

Polymer Composites for Electrical Engineering, First Edition. Edited by Xingyi Huang and Toshikatsu Tanaka.

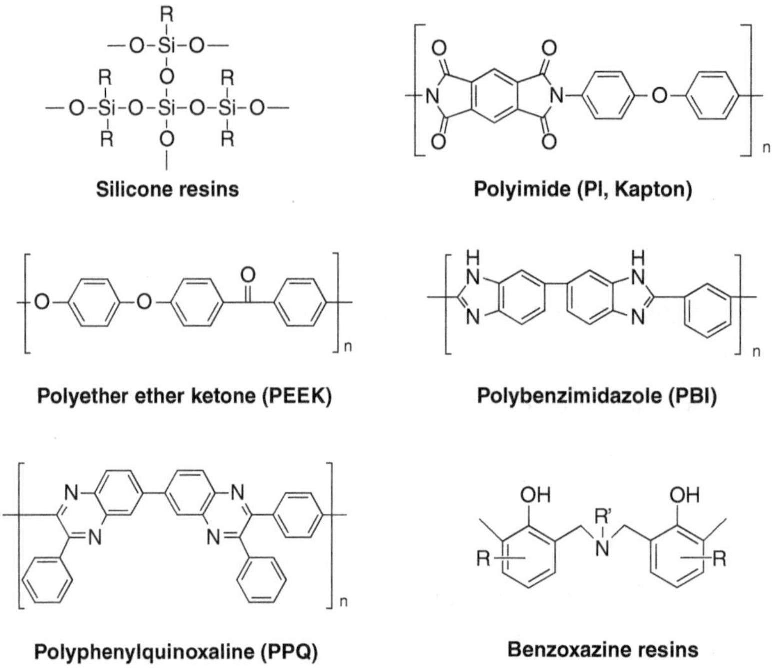

Figure 3.1 The chemical structures of heat-resistant polymer insulating materials.

used high-temperature-resistant electrical insulating polymer matrix mainly includes silicone resin, fluoro resin, polyamide, polyimide, polybenzimidazole, polyphenylquinoxaline, polyether ether ketone (Figure 3.1), etc.

3.1.1.1 Silicone Resins

Silicone resins are stable over wide ranges of temperature, humidity, and frequency. Meanwhile, silicone resins have excellent dielectric properties, electrical insulation, excellent resistance to oxidation, chemicals, and radiation [7]. Therefore, silicone impregnating paint has many unique and excellent properties, and developing this new material has draw intense attention all over the world. In the early stages of research, solvent-based silicone insulating paints were the mainstay. With the continuous advancement of science and technology, the quality and reliability of impregnating paints are increasingly in demand. In particular, the introduction of vacuum impregnation requires the impregnating varnish free from solvent. Also, people's awareness of environmental protection has increased. To reduce pollution and improve the operating conditions, the development of solvent-free silicone impregnating varnishes began in the late 1960s and has been widely used in the manufacturing of motors and other electrical devices.

3.1.1.2 Polyimide

Polyimide (PI) is one of the five-membered ring polymers [8]. It is a macromolecule that has a large number of imide groups, and hence it has low absorption of water. And its heterocyclic ring protects the carbon-nitrogen bond, which results in good stability. From the 1950s to 1960s, it rapidly developed into one of the most valuable high-temperature-resistant materials. Initially, PI was used to make paints. In the 1860s, Du Pont announced the Pyre-ML PI paint and was tested on motors. It demonstrated the usage of PI lacquer in the coil windings of high-temperature motor electrical appliances and electronic components. Not only does the paint film show significant hardness and non-softening but also has high adhesion to the conductor. Thus, it was first industrially used as the insulation coating of electromagnetic wire and in the flexible printed circuit board of the PI copper clad foil.

Zhou Sheng et al. [9, 10] obtained insulating paint from pyromellitic dianhydride and 4,4′-diaminodiphenyl ether, which was coated on a glass sheet and baked to obtain a thin PI film through different imidization processes. Using the characteristic wavenumber of $725\,cm^{-1}$ in the infrared spectrum, the imidization rate of the homophenyl PI insulating paint under different imidization processes was calculated, and the results showed that the temperature was the controlling parameter in the homopolymer PI insulating paint. The imidization was completed at 300 °C. The higher the degree of imidization, the better the electrical insulation and mechanical properties, but the electrical strength started to decline at 300 °C.

3.1.1.3 Polyether Ether Ketone

Polyether ether ketone (PEEK) is a type of semicrystalline polymer material with ether bond and ketocarbonyl group in its molecular structure [11]. Traditional PI dielectric materials require using additional adhesive layers such as fluorine-containing coatings, acrylates, epoxy resins, etc. to achieve thermal insulation of components owing to their insolubility. The additional adhesive layer tends to deteriorate the high-temperature insulation effect. The semicrystalline structural characteristics of PEEK make it hot-melt characteristics, which can be used to directly perform high-temperature hot winding with copper wires at high temperatures, and then to produce insulating products.

3.1.1.4 Polybenzimidazole

Polybenzimidazole (PBI) is the highest heat-resistant class of thermoplastic polymer materials. Commercial PBI has a glass transition temperature of up to 427 °C and can be used above 300 °C for a long time [12]. PBI resin can be dissolved in polar solvents such as *N*,*N*-dimethylacetamide (DMAc), *N*-methylpyrrolidone (NMP), and formulated into a solution with certain solid content. It then can be used as a varnish for the insulation protection of copper wires.

3.1.1.5 Polyphenylquinoxaline

Polyphenylquinoxaline (PPQ) is a type of polymer containing a phenyl-substituted quinoxaline ring in its molecular structure. The content of polar groups in its molecular structure is very low. It is currently the only cyclized aromatic heterocyclic polymer that can be prepared at room temperature [13]. In the 1980s, the United States began to evaluate the performance of commercial PPQ film (IP200®) as a high temperature electrical insulation material for the enameled wire insulation. By comparing the comprehensive characteristics of PPQ film, Kapton® film, Upilex® S film, and other materials, it was found that the weight loss of PPQ film after aging at 90 °C and 100% relative humidity for 1222 h was 0.6%.

Under the same conditions, the surface of Kapton® film became very viscous in less than 500 h and then completely hydrolyzed. This shows that PPQ film possesses excellent stability both under high-temperature and high-humidity environments, and therefore can be applied to electrical or electronic insulation in similar environments [14].

3.1.1.6 Benzoxazine

Benzoxazine is a six-membered heterocyclic compound produced by the Mannich reaction of phenol, aldehyde, and primary amine [15]. It undergoes ring-opening polymerization under high temperature or catalyst to produce a nitrogen-containing phenolic resin-like network polymer, referred to as benzoxazine resin or polybenzoxazine. Because of its flexible molecular design, low cure shrinkage, and good mechanical properties, benzoxazine resin has become an important new type of high-performance composite matrix resin. F grade and H grade high-temperature-resistant electrical insulating laminated glass cloth developed by Gu Yi, Sichuan University, [16] have been used in practical applications such as in motor insulation materials, vacuum pump rotors, and other mechanical parts.

3.1.2 Modification of Resin Matrix with Reinforcements

At present, with the development of aerospace technology, high-voltage power generation equipment, and high-voltage power transmission equipment, simple polymer materials cannot meet the performance requirements of industry for electrical insulation materials. Also, inorganic insulation materials face difficulties such as processing and shape, besides being expensive. To overcome these problems, doping of inorganic materials into a polymer matrix to form a composite material is an effective method. Polymer-based composite insulation materials possess the advantages of polymers. At the same time, due to the addition of inorganic materials, the mechanical and electrical properties of the insulation materials are improved effectively. The commonly used inorganic materials are glass fiber and inorganic nanoparticles.

3.1.2.1 Mica

Mica is a special flaky silicate mineral and has a lamellar crystal structure composed of two layers of silicon-oxygen tetrahedron sandwiching a layer of aluminum-oxygen octahedron (Figure 3.2).

The special layered structure perpendicular to the cleavage plane of mica determines its high electrical insulation; at the same time, it has excellent transparency, peelability, and chemical stability. It is a very important insulating material that can maintain excellent physical and chemical properties, as mentioned above. Mica insulation materials include mica flakes, mica paper, mica boards, mica tapes, etc. After being compounded with a polymer matrix, the insulation and heat resistance of the material can be improved effectively.

3.1.2.2 Glass Fiber

Glass fiber is an important raw material in the insulation material industry. Glass fiber is used in the following categories: impregnated fiber products, laminate products, and electrical plastic products. Glass fiber usually accounts for 30–50% of the total mass of insulating materials. It is another major raw material in addition to matrix resin and

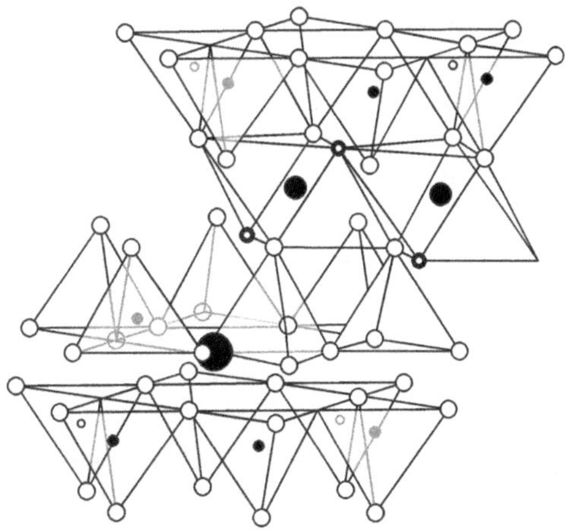

Figure 3.2 The crystal structure of mica.

occupies a pivotal position in the insulation material industry. Compared to glass fiber used in other industries, the glass fiber for insulating materials has unique requirements as the material employed must have the function of insulation. It requires all raw materials to contain substances with no or less conductivity. Generally, the insulating material industry chooses non-alkali glass fiber and its products.

3.1.2.3 Inorganic Nanoparticles

The commonly used inorganic nanoparticles are SiO_2, ZrO_2, Al_2O_3, TiO_2, BN, etc. The addition of inorganic nanoparticles not only improves the insulation performance of the material but also enhances the corona resistance of the material. Generally, nanodielectrics are known as "future insulating materials" and are composed of a nanofiller (particle size <100 nm) dispersed into the polymer matrix [17]. "Nanodielectrics" was first proposed by Michel Fréchette of the Canadian HydroQuébec Institute (IREQ) in 2001 [18]. At present, the research on nanodielectrics mainly revolves around the polymer matrix mentioned above, nanofillers and their interfacial problems, as well as composite processes.

Recently, Lei Qingquan et al. have systematically studied the synthesis of corona-resistant PI thin film dielectrics and the relationship between their structure and performance [19–21]. Through the studies on the performance of the prepared hybrid composite membrane, it has been found that introducing inorganic fillers significantly increases the corona resistance of PI membrane. At the same time, to improve the compatibility of the two inorganic–organic phases, adding an appropriate amount of coupling agent is necessary.

3.1.3 Modifications in the Thermal Conductivity of Resin Matrix

With the development of electrical and electronic equipment toward high power density, small size, lightweight, and high integration, the heat generated in the unit volume of the equipment increases sharply. The continuous accumulation of heat and the resulting temperature rise accelerates the aging of insulating materials and significantly reduces the reliability and life of the electrical and electronic equipment [22]. Increasing the thermal conductivity of insulating materials is the fundamental way to enhance the heat dissipation capacity of the electrical and electronic equipment. Thus, high thermal conductivity and high-temperature-resistant insulating polymer matrix composite materials have also become the focus of research.

3.1.3.1 Mechanism of Thermal Conductivity

Heat is a kind of energy that involves the movement, rotation, and vibration of molecules, atoms, electrons, etc. The heat conduction of matter is the result of collision and transfer of internal microscopic particles. There are three types of heat-conducting carriers in solids: electrons, phonons, and photons. Polymer insulating materials are generally the saturated systems, indicating that free electrons are not present. Also, their molecular movement is difficult, and therefore heat conduction is mainly due to lattice vibration, and phonons are the main thermal energy loaders. According to Debye theory, the thermal conductivity of polymers mainly depends on their crystallinity and orientation order. Due to the large relative molecular weight and random entanglement of the molecular chain, the polymer contains more amorphous parts; and the entire molecular chain is not completely free to move, whereby relatively only a part of atoms, groups, or links can move. Therefore, the thermal conductivity of polymers is generally very low.

The studies on high thermal conductivity insulating materials mainly focus on two aspects: (i) the molecular structure of polymers is modified to obtain intrinsic high thermal conductivity insulating materials, (ii) and the filler with high thermal conductivity is doped in the polymer matrix to improve the overall thermal conductivity of the material.

3.1.3.2 Intrinsic High Thermal Conductivity Insulating Material

The thermally conductive carrier phonons of the polymer are subject to the dynamic scattering of molecules and lattice inharmonic vibrations and scattering of interfaces and defects. Hence, the thermal conductivity is closely related to the microstructural regularity of the material. By changing the structure of the polymer molecules and chain links, the polymer can attain a physical structure with a high degree of crystallinity or orientation. Then, phonon scattering is reduced to increase the free path of phonons and increase the transmission speed of phonons, which can effectively improve the thermal conductivity of the polymer [23, 24].

For example, in recent years, investigations on intrinsically conductive epoxy resins are intensively focused on introducing mesogenic elements in the epoxy resins to form crystal-like or liquid crystal structures. They use epoxy prepolymer and curing agent with crystal-like or liquid crystal structure to increase the "order" of the molecular structure of the resin and improve the coordination of molecular and lattice vibrations. The molecular defects are then reduced in the resin material, and the probability of order structure is expanded. Furthermore, the free path of the phonon is increased, and the thermal conductivity between the molecular chains is improved.

Hitachi Group has conducted continuous and in-depth research on the technologies of preparation and controlling the structure of the epoxy resin containing mesogens as well as its application in high thermal conductivity copper clad laminate substrate materials [25–27]. They developed two new epoxy monomers with a crystal-like structure. The intermediate groups of these two epoxy monomers are two benzoic acid or one diphenyl groups, and the curing agent is an aromatic diamine. As the intermediate groups of epoxy monomers are highly ordered, this feature is conducive to the formation of a crystal-like structure, which can be interconnected with the internal amorphous structure through covalent bonds, thereby forming a microscopic highly ordered structure to suppress the scattering of phonons effectively (Figure 3.3).

The thermal conductivity of this epoxy resin can reach 0.96 W/(m·K), which is more than four times that of a standard epoxy resin.

3.1.3.3 Filled High Thermal Conductivity Insulating Material

Filled high thermal conductivity insulating materials are generally obtained by adding a large amount of high thermal conductivity inorganic fillers to the polymer matrix. This method is easy to process, low in cost, and easy to industrialize, and hence several studies are directed toward these materials. Fillers used for these high thermal conductivity polymer-based composite insulation materials are generally metal oxides (BeO, MgO, Al_2O_3, CaO, NiO, etc.), nitrides (AlN, BN, etc.), and carbides (SiC, BC, etc.). An improvement in the thermal conductivity of the material mainly depends on the content, the level of thermal conductivity, the shape, and distribution of the filler in the matrix as well as its interfacial bonding with the matrix.

Du Boxue et al. [28] enhanced the thermal conductivity and surface erosion resistance of the composite materials by filling nano-BN in a silicone rubber matrix. The experimental results show that as the amount of added nano-BN increases, the thermal conductivity of the composite material improves, and the maximum temperature in the discharge area of the sample surface is suppressed. After the addition of nano-BN, the erosion depth and erosion loss quality of the composite silicone rubber material decreased significantly compared to without its addition. Namitha et al. [29] obtained AlN/silicone rubber composite material by hot pressing technology, which showed mechanical flexibility. The results show that the dielectric loss and relative dielectric constant of the composite material change linearly with the addition of AlN, and the thermal conductivity increases with the addition of AlN. When the volume fraction of AlN is 0.4%, the thermal conductivity reaches 1.0 W/(m·K) (Figure 3.4).

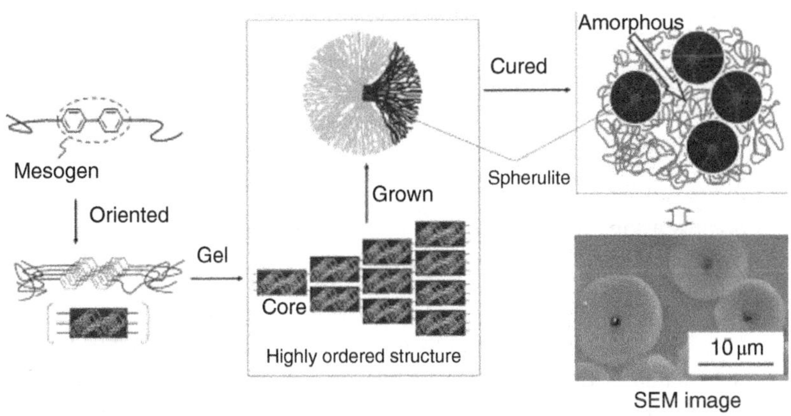

Figure 3.3 Schematic representation of the microscopic pattern of spherulites in the mesogen-containing resin system.

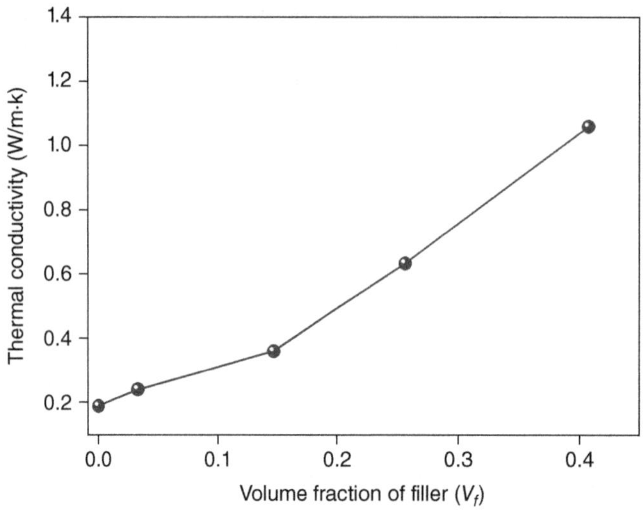

Figure 3.4 Variations in the thermal conductivity of mechanically flexible aluminum nitride–silicone rubber composites (SAN) composites with the contents of filler.

3.2 High-Temperature Applications for Electrical Energy Storage

Electrical energy storage is the most important application of polymer composite dielectrics in high-temperature electrical engineering. The rapid development of polymer composite dielectrics is due to their combined advantages of the polymer matrix and ceramic fillers. The specific characteristics such as lightweight, low cost, easy processing, reliability, or excellent electrical insulation performances make polymer composite dielectrics the ideal materials for dielectric capacitors applications, which are widely used in the power systems and electronic devices that need to operate at elevated temperatures [30]. In such cases, the demand for high temperature is necessary. The thermal stability of polymer materials is usually poor, and continuous operation under high temperature and strong electric field will result in tremendous conductivity losses. which greatly limits the applications at high temperatures. For example, biaxially oriented polypropylenes (BOPP) as the dielectric used in mainstream capacitors causes the mismatch between ambient (140 °C) and maximum operating temperature of BOPP (~105 °C), making it necessary to introduce heat-resistant fillers into the polymer matrix [30].

3.2.1 General Considerations for High-Temperature Dielectrics

Dielectric is a medium that can be polarized. Its main feature is the positive and negative charges that do not coincide with each other in the form of polarization, transfer, and storage of electricity, where its main role is binding the charge. The deformation/displacement of an isolated neutral atom's electron cloud under the influence of an electric field causes dipole moments, leading to polarization. Polarization is a frequency-dependent phenomenon and can be attributed to four main mechanisms: electron, atom/ion, dipole (orientation) and interface polarization. Each of them contributes to the dielectric constant. Dielectric constant (relative permittivity) of dielectric material is the ratio of dielectric permittivity to vacuum permittivity. Dissipation factor, or dielectric loss tangent, is a measure of the rate of energy loss during polarization and depolarization of dielectric materials. Under an external electric field, polarization occurs, and the energy is stored in the dielectric in the form of an electric field. The stored energy density can be obtained from the following Eq. (3.1).

$$U_e = \int E\mathrm{d}D \tag{3.1}$$

where U_e is the energy density, E is the electric field strength, and D is the electric displacement inside the dielectric. For linear dielectric, Eq. (3.1) can be simplified as follows (Eq. 3.2).

$$U_e = 1/2\varepsilon_0\varepsilon_r E^2 \tag{3.2}$$

Where ε_0 is the vacuum permittivity (8.85×10^{-12} F/m), and ε_r is the relative permittivity, which is a constant related to the material. Energy storage density is an important indicator for evaluating the energy storage performance of dielectric materials. Eq. (3.2) shows that both high dielectric constant and high breakdown field strength are beneficial to increase the energy storage density of dielectric materials.

3.2.2 High-Temperature-Resistant Polymer Matrix

Dielectric polymers mainly face two problems when they are applied at high temperatures. First, the thermal stability of the polymer materials is generally poor. When the temperature is above the glass transition temperature (T_g), the mobility of the polymer chain and the free volume between the molecular chains are significantly increased, and the mechanical and dielectric properties of the polymer material deteriorate simultaneously. Second, leakage current accompanied by a large amount of Joule heat eventually leads to heat collapse. Improving the thermal stability of dielectric polymers, suppressing their loss, and increasing their thermal conductivity are the key issues for high-temperature energy storage dielectric materials. To maintain the structural stability at high temperatures, dielectric polymers usually contain strong intramolecular or intermolecular forces, such as hydrogen bonds (–NH–, –CO–NH–, etc.), conjugated π bonds (imide ring, imidazole ring, etc.), aromatic or heterocyclic aromatic molecular skeleton, high-strength chemical bonds (C-F, etc.), etc. [31]. Besides, high crystallinity or crosslinked structure is also beneficial to improve the thermal stability of the dielectric polymer. Thus, high-performance engineering polymers, such as polyetherimide (PEI), polyimide (PI), polyphenylene sulfide (PPS), poly(arylene ether nitrile) (PEN), fluorene polyester (FPE) [32, 33], divinyltetramethyldisiloxane-bis(benzocyclobutene) (BCB), poly(ether ketone ketone) (PEKK), etc., are well employed in the high-temperature polymer matrix.

3.2.3 Polymer Composites for High-Temperature Energy Storage Applications

Dielectric composites with inorganic ceramics, copolymers, or hybrid polymer nanocomposite systems are widely studied in film capacitors. Researchers found that the introduction of nanoscale inorganic fillers into the high-temperature polymer matrix can affect the mechanical, thermal, and electrical properties [34–39], as well as indirectly affect the energy storage performance. The improvement of mechanical properties could avoid softening of polymer nanocomposites with an increase in temperature, thereby improving the breakdown strength of dielectric to some extent. On the other hand, the interfacial interaction and physical confinement between fillers and polymer matrix could improve the values of T_g [38, 39]. Based on Eq. (3.2), both high relative dielectric permittivity and the high breakdown strength are beneficial for high energy density. Thus, a series of high-k dopants are considered such as barium titanate ($BaTiO_3$), lithium niobate ($LiNbO_3$), silicon nitride (Si_3N_4), silicon oxide (SiO_2), zinc oxide (ZnO), lead titanate ($PbTiO_3$), and lead zirconate titanate ($Pb[Zr_xTi_{1-x}]O_3$) [34]. For example, the enhanced energy density at elevated temperatures was achieved by doping high-k $BaTiO_3$ nanoparticles with PI polymer matrix [40]. However, a large contrast of dielectric constant between the fillers and polymer matrix leads to a highly nonhomogeneous distribution of local electric fields and consequently reduces E_b [41]. A small fraction of nanofillers with a dielectric constant similar to that of the matrix can increase the dielectric constant ideally [4]. As shown in Figure 3.5(a), nanoparticles of different types and sizes are studied for this purpose. Recently, boron nitride with a high aspect ratio, wide bandgap (5.9 eV), insulating performance, and thermal conductivity is considered as effective barriers of breakdown, leakage current, and conduction loss. Besides, Li et al. [42] reported on the *c*-BCB nanocomposites with 7.5 vol% Al_2O_3 nano particles, nano wires, and nano plates at 150 °C. Among which, alumina nanosheets as nanofillers with better insulation properties and higher dielectric constant improved the discharge energy density of the nanocomposites further. These results demonstrate an effective pathway to generate high energy storage polymers with low loss and high operating temperature.

The large enhancement in the dielectric constant contributes to excellent energy storage properties in the PEI/Al_2O_3 nanocomposites at high temperatures up to 150 °C [43–45]. Extensive studies have shown that alumina may be the most potential nanofiller in high-temperature energy storage. Ferroelectric polymers with high dielectric constants and high breakdown strengths are currently one of the hot research areas in high energy storage density dielectric materials. Poly(tetrafluoroethylene/hexafluoropropylene/vinyli-dene fluoride) with insulating Al_2O_3 nanofillers exhibits low loss and high-temperature stability [46]. With the introduction of nano-sized aluminium nitride particles, a significant enhancement of ε_r (~4.4) and high thermal stability up to 250 °C were obtained [47].

3.2.4 Surface Modification of Nanocomposite for High-Temperature Applications

An interface between the filler and polymer leads to an increased loss and reduced breakdown strength inevitably. One of the effective methods to mitigate the problem is through the surface modification of nanoparticles and construction of core-shell structures. Huang et al. designed a series of core–shell structured nanoparticles and grafted with polymers to achieve high dielectric constant [48]. As shown in Figure 3.6, in the “grafting to” method, particles and polymer coatings can be combined by physical absorption and chemical reactions, such as electrostatic interaction, van der Waals interaction, and reactions of

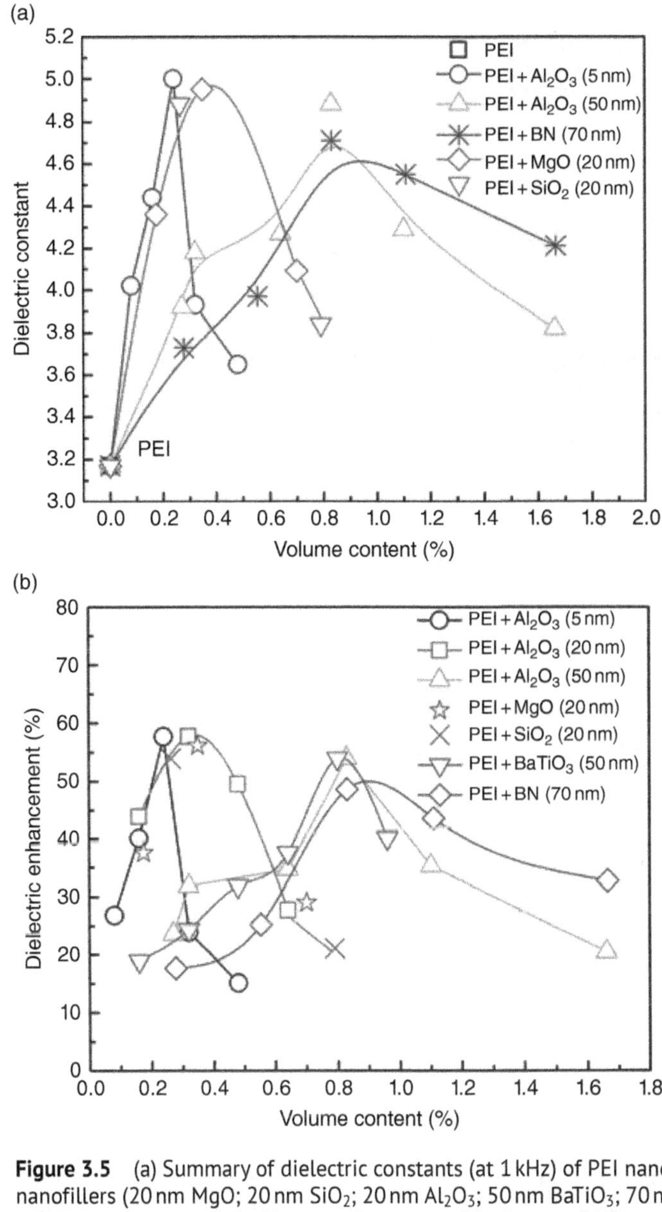

Figure 3.5 (a) Summary of dielectric constants (at 1 kHz) of PEI nanocomposites with different nanofillers (20 nm MgO; 20 nm SiO_2; 20 nm Al_2O_3; 50 nm $BaTiO_3$; 70 nm BN). Experimental data points are shown, and the lines are drawn to guide the eye. (b) Summary of dielectric constants (at 1 kHz) of various nanocomposites studied. The data are presented as the percentage increase of the dielectric constant of the nanocomposites with respect to the original polymer matrix. Experimental data points are shown, and the lines are drawn to guide the eye.

functional groups. While in the "grafting from" method, the particle surface acts as initiation sites for the polymerization of monomers [45].

Jung et al. [49] developed a polystyrene nanocomposite comprising of core-shell $BaTiO_3$ nanoparticles with diblock copolymer shielding layers, in which polystyrene-block-poly(styrene

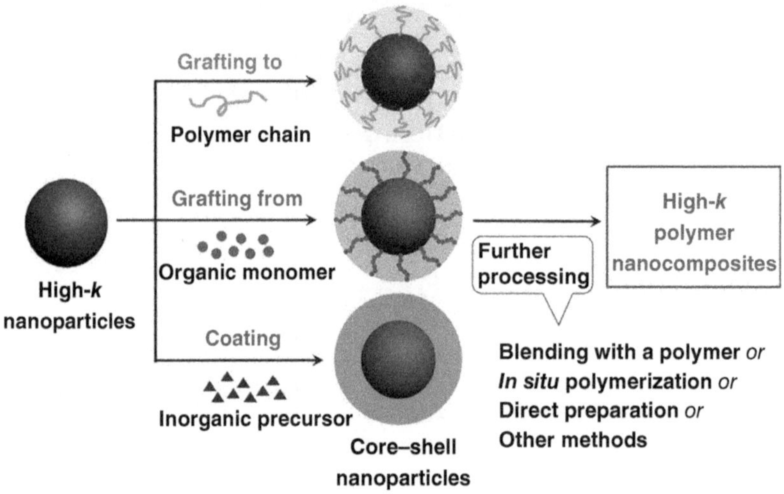

Figure 3.6 General methods associated with the design and construction of core–shell nanoparticles for high-*k* polymer nanocomposites. *Source:* Reprinted with permission from Huang and Jiang [45].

co-vinylbenzylchloride) self-assembled to form a double-layered shell on the nanoparticle. The polar ammonium-containing blocks form an inner layer that bonds to the nanoparticle, and the surface of which can be charged largely with ammonium groups, which lead to high polarization.

Crosslinking can reduce the defects at the interface, which can effectively inhibit the leakage current and reduce interfacial polarization, contributing to electric breakdown strength and energy loss reduction. Moreover, the self-crosslinking of the polymer can reduce dielectric relaxation originating from the movement of polymer chain segments at elevated temperatures and thereby improving the mechanical properties and dielectric stability in a broad temperature range. Li et al. [4] designed *c*-BCB/boron nitride nanoplates (BNNS) and obtained an increased breakdown strength of 50% with the nanocomposite containing 10 vol% boron nitride nanosheets at 250 °C. At the same time, the leakage current density and conduction loss sharply decreased. As such, *c*-BCB/BNNS outperforms all the earlier high-temperature polymer dielectrics in terms of high-temperature energy storage capability (Figure 3.7).

Xu et al. found that giant discharged energy density and high charge-discharge efficiency are simultaneously achieved at elevated temperatures by the bioinspired polymer nanocomposites, BN-BCB@ crosslinkabe poly(arylene ether sulfone)s (DPAES). The superior capacitive performance is shown in Figure 3.8. The successful improvement of interfacial structure and silk-inspired strategy via nanoconfinement effect could be essential to the development of future high-temperature dielectrics. [50].

On this basis, Liu et al. [51] proposed a novel strategy of "dual crosslinking" and achieved significant breakdown strength and discharged energy density by BT-BCB@DPAES nanocomposites at elevated temperature (Figure 3.9).

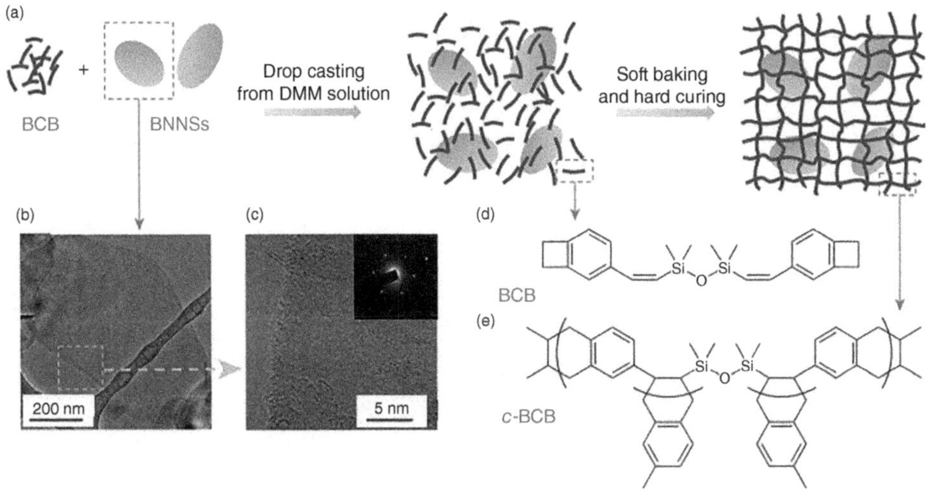

Figure 3.7 (a) Schematic presents the preparation of *c*-BCB/BNNS films. (b), (c) Transmission electron microscopy (TEM) images of BNNSs exfoliated from h-BN powders. Inset to (c) is the electron-diffraction pattern of BNNSs, exhibiting its hexagonal symmetry. (d) Chemical structure of BCB monomer. (e) The repeating unit of *c*-BCB. *Source:* Reprinted with permission from Li et al. [4].

3.2.5 Sandwich Structure of Nanoparticles for High-Temperature Applications

To mitigate the local electric-field distortion between the polymer matrix and inorganic fillers in the dielectric polymer composites and to further improve the dielectric properties and high-temperature capacitive performance, a range of novel approaches have been developed, e.g. using large-aspect-ratio fillers and modifying the interface layers [52–54]. Among which, the construction of a sandwich structure is more promising. The high-k of the central layer integrating with high E_b of the outer layers makes it an outstanding dielectric at high temperatures.

Li et al. [55] constructed sandwich-structured nanocomposites with *c*-BCB/BT as the central layer to improve the dielectric constant and *c*-BCB/BNNS as the outer layer to block charge injection, which results in stable energy storage density of 0.93 J cm^{-3} up to 150 °C at 200 MV m^{-1} (Figure 3.10).

The existence of effective and strong barrier interfaces between the adjacent layers in the sandwich structure protects the composite films from total breakdown and suppresses the formation of conductive paths in the hard layer. In contrast, high polarization induced by the barium titanate (BT) nanoparticles in the soft layers guarantees a high dielectric polarization. Further improvements in the breakdown strength and energy density have been realized for a group of three-tiered ferroelectric poly(vinylidene fluoride) films with an increase in the content of BT in a layer-by-layer gradient [56]. On this basis, a more time-saving chemical vapor deposition (CVD) method for hexagonal BN (h-BN) was adopted to develop high-temperature polymer nanocomposite dielectrics. Superior high-temperature

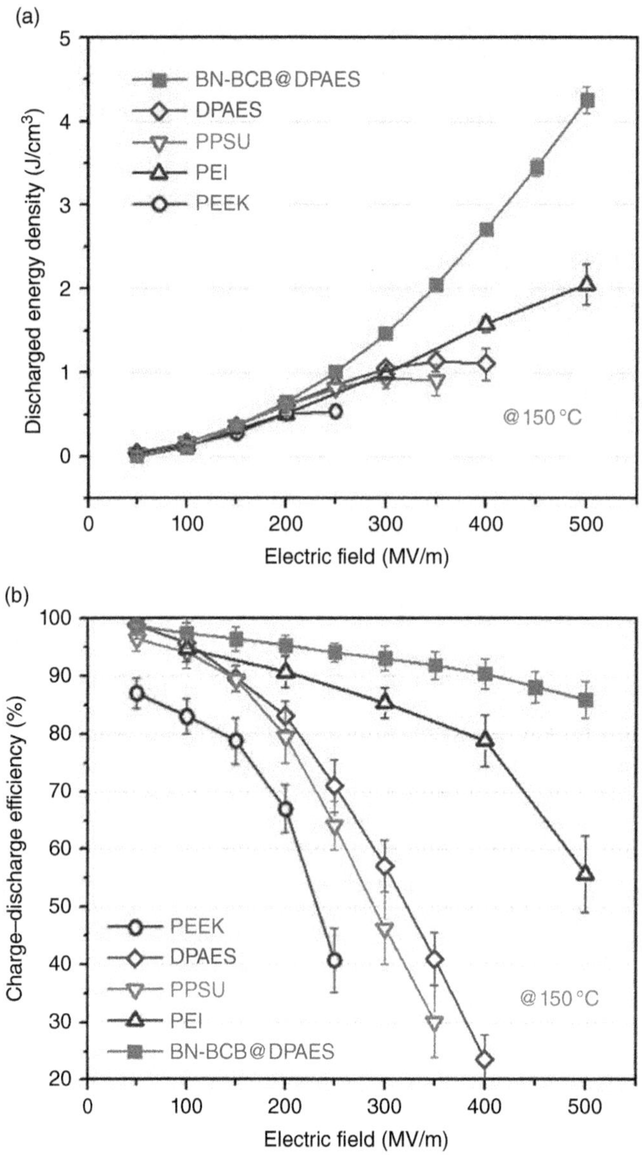

Figure 3.8 (a) Discharged energy density and (b) charge–discharge efficiency of dielectrics as a function of electric fields at 150 °C. (c) The discharged energy density of dielectrics achieved at above 85 and 90% charge–discharge efficiency. Nanocomposites exhibit significantly superior properties compared to other dielectrics at high temperature. (d) Reported discharged energy density and charge–discharge efficiency of the state-of-the-art polymer nanocomposites under 150 °C.

energy storage properties (U_e of 1.19 Jcm^3 and $\eta > 90\%$) were achieved at 200 °C in the case of h-BN-coated PEI nanocomposites [57]. It is envisioned that the excellent performance, along with the outstanding cyclability and dielectric stability at elevated temperatures, paves the way for the applicability of a wide range of engineering polymers in high-temperature electronics and energy devices.

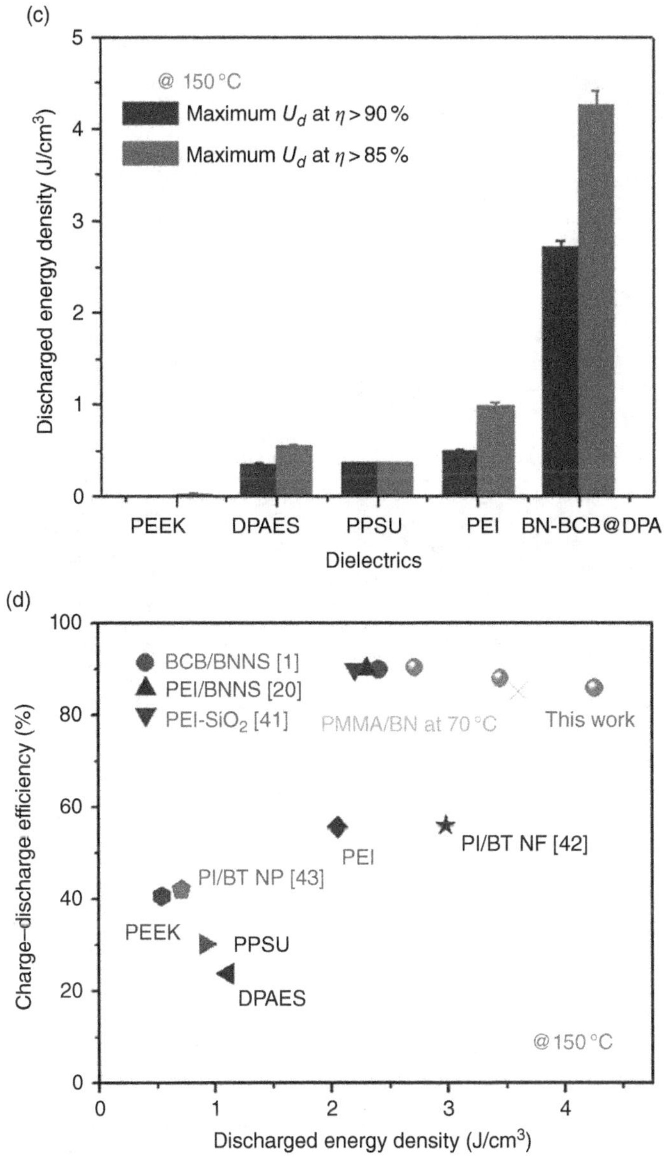

Figure 3.8 (Continued)

3.3 Applications of High-Temperature Polymer in Electronic Packaging

In the current century, electronic packaging with high-temperature resistance, high density, and high performance, as well as many new developments are being emerged. The concept of electronic packaging has also changed from traditional devices to systems, that is, in the package of signal transmission, support carrier, and based on traditional functions such as heat conduction and chip protection, it is expanding further. The thin film, thick

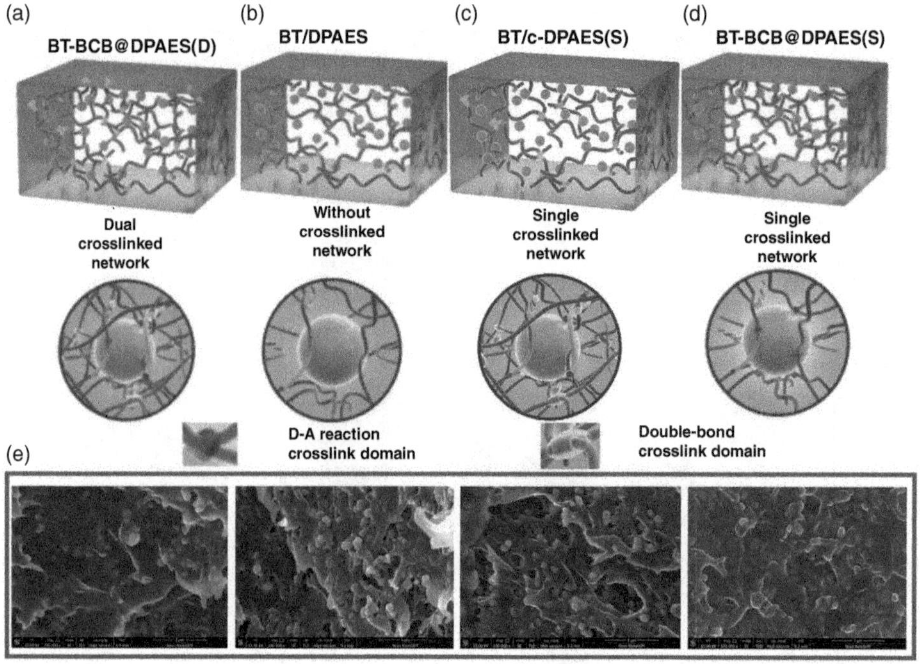

Figure 3.9 Types of crosslinked networks: schematic diagram of crosslinked network structures of (a) 10 vol% BT-BCB@DPAES (D), (b) 10 vol% BT/DPAES, (c) 10 vol% BT/c-DPAES(S), (d) 10 vol% BT-BCB@DPAES (S), and (e) the corresponding scanning electron microscopy (SEM) images.

film, and embedding processes are used to integrate the signal transmission circuit of the system and most active and passive components and are integrated with high-density packaging and the components of the chip. The external bonding process is combined to achieve the integration of the packaging system to obtain the highest packaging density [1–3, 58].

The concept of electronic packaging refers to arranging and fixing semiconductor devices, built-in chips, and other components of integrated circuits on the frame or substrate using film and fine connection technologies, connecting through lead terminals, and insulating them. The medium is the packaging material to encapsulate and fix it, thereby forming a large-scale integrated circuit with a complex multilevel structure. The electronic packaging has a very important role in large-scale integrated circuits. It can protect built-in chips of the integrated circuits, enhance the environmental adaptability of the integrated circuits, and also in electrical connection, field shielding, size transition, heat and moisture protection, standardization, and many other functions [4].

Conventional electronic packaging materials are generally high-molecular-weight organic materials [59], which are widely used due to their excellent processability and low cost. Still, their poor thermal conductivity seriously affects the normal operation of electrical appliances. To ensure the safe and reliable use of electronic equipment for a long time, heat needs to be continuously transferred from high to low temperature. Therefore, materials with low dielectric constant, low dielectric loss, and high thermal conductivity should be selected for electronic product packaging [60]. The application of polymers in electronic packaging has a history of more than 100 years. Polymer materials have been widely used

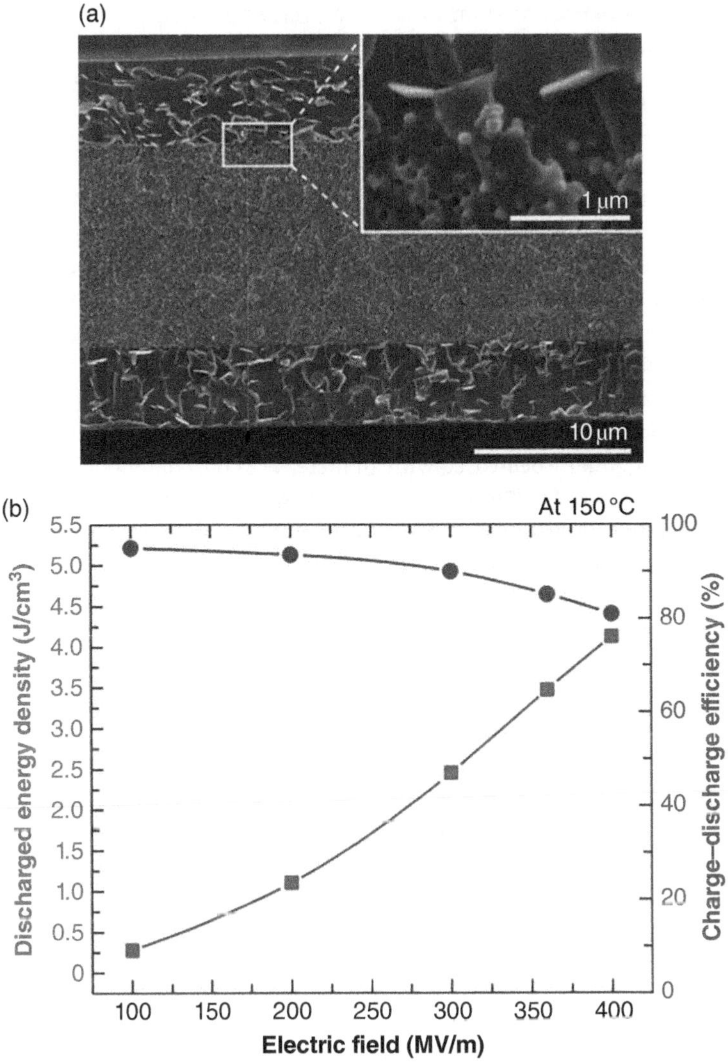

Figure 3.10 (a) Cross-sectional SEM image and (b) energy storage performance of the sandwich-structured nanocomposites measured at 150 °C. *Source:* Adapted from Yin et al. [52].

in the field of electronic packaging due to their small density, excellent electrical insulation properties, excellent dielectric properties, cheap raw materials, and easy processing.

At present, high-temperature-resistant polymers including aromatic PIs, polyaryl ether sulfones, polyaryl ether ketones, heterocyclic aromatic polymers, etc. are considered to be potentially applicable to packaging materials for advanced chips. Therefore, it is of significance to examine and develop low dielectric constant high-temperature-resistant organic polymer materials to meet the increasing complexity of the integration density of the chip circuit. The current research on this type of material mainly includes the following aspects.

3.3.1 Synthesis of Low Dielectric Constant Polymer Materials Through Molecular Structure Design

3.3.1.1 Fluorine-Containing Low Dielectric Constant Polymer

Because the nucleus of the fluorine element has a strong attraction to the electrons outside the nucleus and the interaction between the electrons and the nucleus is large, the electron cloud density is high, and the polarization rate when polarized by an external electric field is low. In addition, due to the small atomic size of the fluorine element and the high binding energy of the carbon–fluorine bond, the introduction of fluorine-containing groups can also improve the thermal properties of the polymer to a certain extent (Figure 3.11). High-performance low dielectric constant fluorine-containing PI and polyarylether materials have received extensive attention in recent years.

Ichino et al. synthesized a series of fluorine-containing PI materials with long fluorine-containing alkoxy side chains. Studies also found that the dielectric constant of Fluorine-containing polyimide (FPI) gradually decreased with an increase in the content of F in the molecular structure, from 2.9 of 6FDA/7F to 2.6 of 6FDA/15F. The glass transition temperature is 189 °C, and the temperature for 10% weight loss is only 455 °C. The structure and performance of such materials are shown in Table 3.1 [61].

Besides, Yaning Lu et al. applied thermal crosslinking on the synthesized material to obtain a fluorine-containing dense PI material with high thermal stability and low dielectric constant. As the proportion of monomers increases gradually, the dielectric constant of the final material can be reduced to 2.37 [62]. All these characteristics make the fluorine-containing PI materials to be extensively used in the microelectronics field.

One method of introducing fluorine into poly (arylene ether) materials is to replace the carbon chain with the main chain containing a perfluoro structure in the polymer. Labadie and Hedrick used polyfluoromonomers containing a perfluoro backbone structure to polymerize with a variety of bisphenols having different structures to obtain a range of polyaryl ethers with a perfluoro backbone structure (Figure 3.12). The dielectric constant

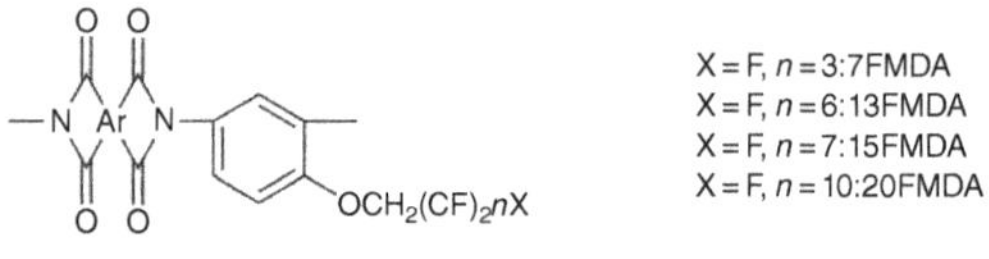

Figure 3.11 Structure of FPI.

Table 3.1 The chemical structure and properties of PI derived from molecular chains with different fluorine content (nFMDA) monomer.

PI	Fluorine content (%)	ε (1 MHz)	T_d (°C)	$T_{10\%}$ (°C)	T_g (°C)
6FDAP7F	34.6	2.9	411	487	266
6FDAP13F	41.8	2.7	398	461	228
6FDAP15F	43.6	2.6	395	459	226
6FDAP20F	47.2	2.6	392	455	189

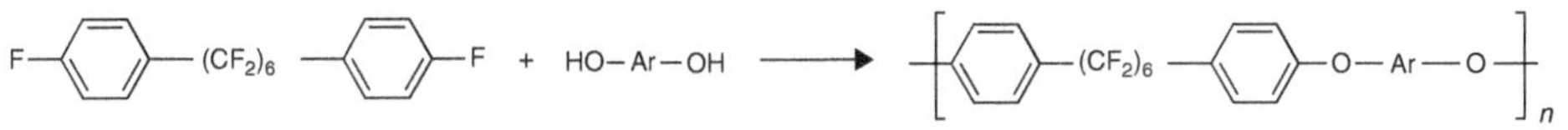

Figure 3.12 The chemical structure of polyarylether containing perfluorohexyl group.

($\varepsilon = 2.35$ at 2 MHz) of the highest fluorine content polyaryl ether material was prepared from hexafluorobisphenol A and a difluoromonomer with a perfluoro structure in the main chain [63], indicating that introducing perfluoro and main chain can reduce the dielectric constant of the material.

Also, the fluorine-containing low dielectric constant polyaryl ether ketone material can be prepared by introducing trifluoromethylphenyl side groups into the polymer. Guibin Wang et al. synthesized a bisphenol monomer (3FHQ) containing pendant trifluoromethylphenyl groups through a series of synthetic reactions and polymerized it with 4,4′-difluorobenzophenone (DFB) to obtain a soluble fluorine-containing polyaryl ether ketone (FPEEK) with lower dielectric constant. The measured dielectric constant is about 2.70 (1 MHz), and the glass transition temperature (T_g) is 140 °C. FPEEK is an amorphous polymer so that it shows excellent solubility in common organic solvents [64–66]. The preparation process of the polymer is relatively simple, the yield is high, and the comprehensive performance is excellent. It can be used as a low dielectric constant polymer matrix material for further modification of compounds.

3.3.1.2 Low Dielectric Constant Polymer Material Containing Nonpolar Rigid Bulk Group

The introduction of nonpolar rigid and bulky groups into the polymer molecular structure can increase the rigidity of the polymer molecular chain, reduce the packing density of the polymer material, and increase its free volume, thereby reducing the dielectric constant of the polymer material and increasing glass transition temperature.

K. Goto et al. studied low dielectric constant PI materials with a fluorene structure on the side chain. The study found that the introduction of a huge fluorene structure increased the solubility of PI while maintaining the excellent thermal properties of the material, and the dielectric constant has been significantly reduced. The structure and properties of this type of material are shown in Figure 3.13 and Table 3.2 [67].

Timothy ML et al. [68] envisioned the introduction of bulky triptycene units into polymer molecules, thereby destroying the stacking of polymer chains and increasing the free volume of the polymer, which is equivalent to introducing "molecular-sized pores" into the polymer. Thus, a low dielectric material is obtained. They successfully synthesized a series

Figure 3.13 The chemical structure of fluorine-containing PI.

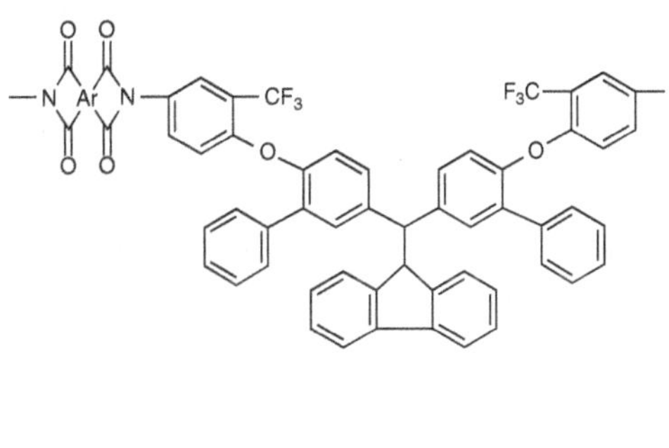

Table 3.2 Performance of fluorine-containing PI.

Ar	ε(1 MHz)	$T_{5\%}$(°C)	T_g(°C)
PMDA	2.69	500	290
6FDA	2.46	538	269

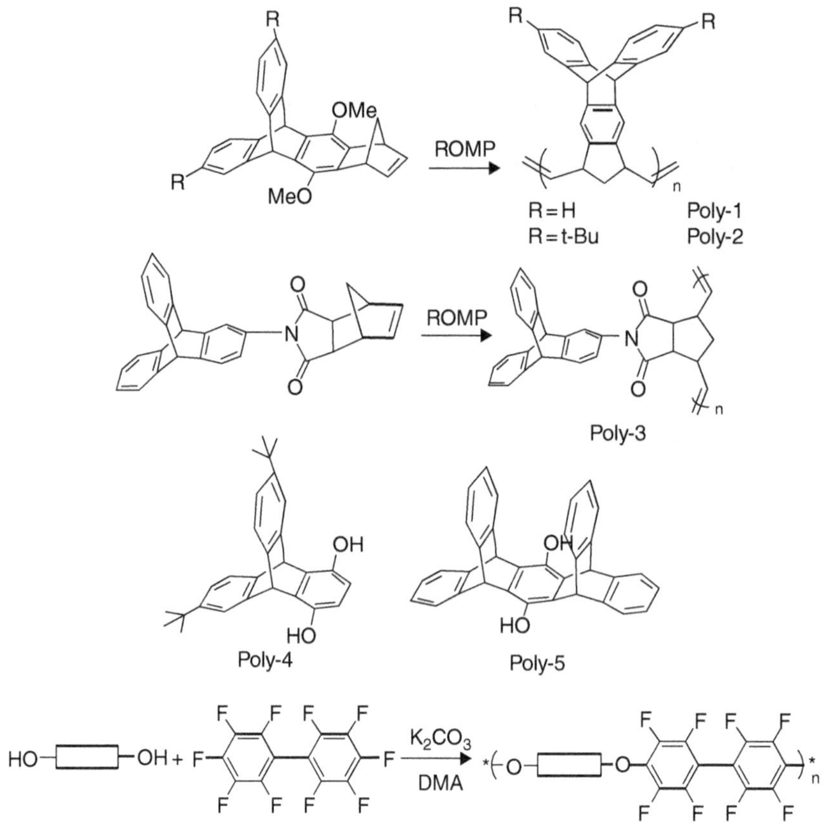

Figure 3.14 The chemical structure of triptycene-containing polymer.

of polymers containing triazene. The structure of these polymers is shown in Figure 3.14, and their properties are shown in Table 3.3. These polymers have both low dielectric constant (2.4~3.0) and high thermal decomposition temperature (greater than 400 °C).

3.3.2 High-Temperature-Resistant Low Dielectric Constant Polymer Composite Material

The introduction of functional fillers into the polymer matrix material by an appropriate method can give the polymer composite material corresponding functionality and maintain the basic properties of the polymer matrix material. Therefore, in recent years, the research

Table 3.3 Performance of triptycene-containing polymer.

Polymer	ε (10 kHz)	Water absorption (wt%)	T_g (°C)	T_{dec} (°C)
Poly-1	2.78	1.0	None	416
Poly-2	2.59	1.1	None	473
Poly-3	2.98	3.7	None	464
Poly-4	2.41	0.9	None	504
Poly-5	—	1.1	None	594

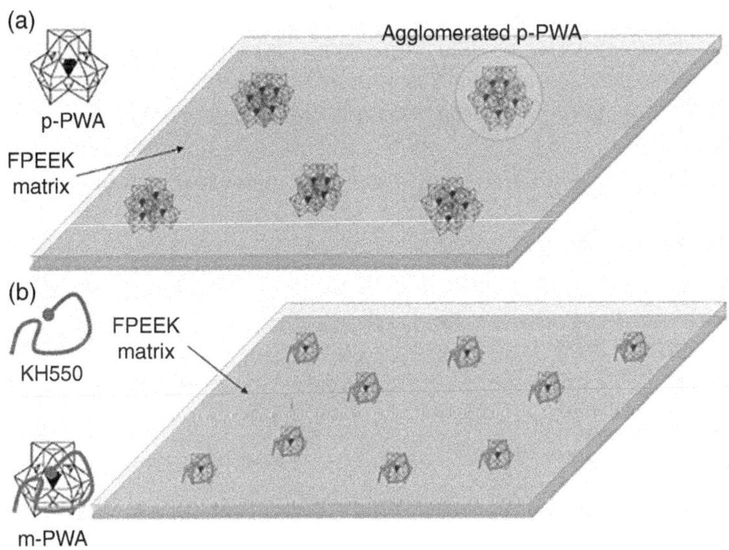

Figure 3.15 Schematic diagram of (a) p-PWA/FPEEK hybrid films, (b) mPWA/FPEEK hybrid films.

on the preparation of low dielectric constant polymer composite materials by introducing nanoparticles into polymers to utilize their dielectric confinement effect has attracted widespread attention.

3.3.2.1 Low Dielectric Constant Polyoxometalates/Polymer Composite

Zhi Geng et al. prepared a series of composite film materials composed of low dielectric constant FPEEK with excellent solubility and phosphotungstic acid (PWA) subjected to surface modification by the silane coupling agent KH-550 (m-PWA). The properties of the composite materials were investigated from different aspects and compared with the composite film composed of untreated PWA (p-PWA) and FPEEK. The effect of inorganic particle's size on the performance of the material, especially the dielectric property was discussed [69]. The schematic diagrams of the composite materials are presented in Figure 3.15. The experiment results show that the dielectric constant of the as-prepared m-PWA/FPEEK composite films vary with the range of 2.28–2.62 (1 MHz). The dielectric constant can be reduced to a lowest level of 2.28 (1 MHz) from the pure FPEEK matrix's dielectric constant of 2.74. This lowest number was recorded at a m-PWA content in the composite film of 10%. Compared with the scenario of the lowest dielectric constant, the dielectric constant of the p-PWA/FPEEK composite materials at the p-PWA contents of 5 and 10% are 2.70 and 2.63 (1 MHz), respectively. The dielectric constant of these p-PWA/FPEEK composite materials are slight reduced when they are compared with the dielectric constant of the pure FPEEK polymer matrix ($\varepsilon = 2.74$ at 1 MHz). This was caused by the fact that PWA particles agglomerated and lost the unique spontaneous polarization inductive effect of the nanoparticles.

Honggang Chen et al. selected silicotungstic acid nanoparticles with a hollow structure as the filler. First, they were surface modified with a silane coupling agent to make the surface carry amino functional groups that can be further reacted, and then the modified nanoparticles were introduced into the PI molecular structure by means of covalent bonds, and a PI composite containing silicotungstic acid nanoparticles was prepared (Figure 3.16). The nanoparticles are well dispersed in the PI. The author tested the dielectric properties of the composite material and found that as the mass fraction of the nanoparticles in the composite material increases, the dielectric constant decreases continuously. When the mass fraction reaches at 10 wt%; the dielectric constant of the composite material can be reduced to 1.22 [70].

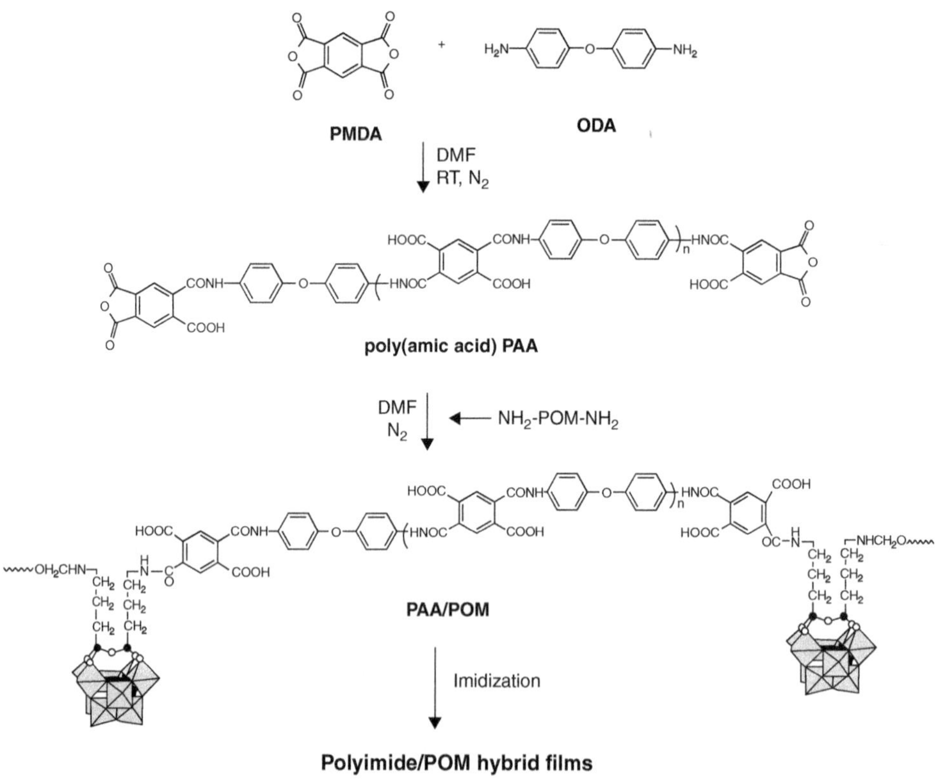

Figure 3.16 Preparation of silicotungstic acid/polyimide hybrid material.

3.3.2.2 Low Dielectric Constant POSS/Polymer Composite

Leu CM et al. selected cage-type polyhedral oligomeric silsesquioxane (POSS) with a hollow structure as the nanoparticle filler, first synthesized a PI matrix polymer with hydroxyl groups on the side chain, and then selected POSS with chlorine substituents as the reactant. With the active group of the PI side chain, POSS reacts with the group and successfully grafted to the side chain of PI to prepare a PI composite material containing POSS nanoparticles in the side chain (Figure 3.17). The test results of the dielectric constant of the composite material show that as the content of POSS in the composite material increases, the dielectric constant of the composite material continues to decrease, and the lowest can be reduced from 3.35 of the matrix material to 2.40 [71].

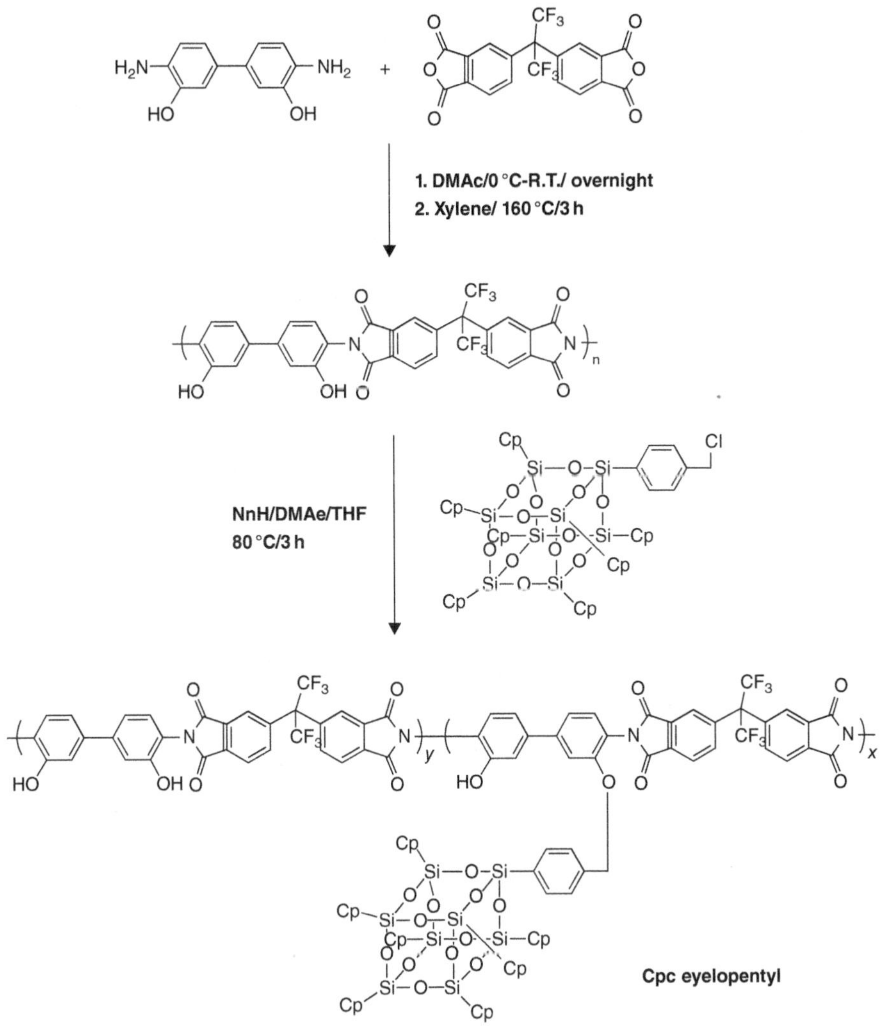

Figure 3.17 Preparation of silicotungstic acid/polyimide hybrid material.

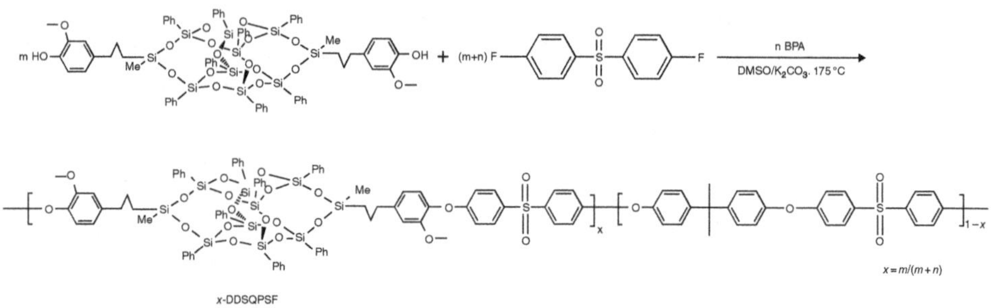

Figure 3.18 Synthesis of POSS-containing poly(aryl ether sulfone) hybrid material.

Weihai Zhang et al. synthesized a bisphenol monomer containing POSS structure through a series of organic synthesis reactions and used this monomer to copolymerize with bisphenol A and fluorosulfone to prepare a series of POSS/POSS with different main chain POSS structure content. The synthesis of the polyaryl ether sulfone hybrid material is shown in Figure 3.18. When the proportion of POSS monomers in the molecular chain structure of the hybrid material is increased to 100%, the dielectric constant of the hybrid material can be reduced to about 2.6 (1 MHz). At the same time, the water contact angle and hydrophobic properties of the hybrid materials gradually increase with the increase of the content of POSS groups in their chemical structure [72].

Zhi Geng et al. prepared a series of soluble NH_2-POSS/PEEK-CF_3-COOH nanocomposite materials featuring mildly crosslinked structures by grafting aminopropylisobutyl POSS (NH_2-POSS) to the side chain of PEEK-CF_3-COOH polymer matrix that contains reactive groups via the side chain grafting reactions [73]. The synthesis route and chemical structure of the hybrid material are presented in Figure 3.19.

It can be seen from Tables 3.4 and 3.5, respectively. The dielectric constant of the as-prepared NH_2-POSS/PEEK-CF_3-COOH nanocomposite material can be reduced down to 1.71 (1 MHz). Besides, the introduction of POSS can not only significantly reduce the dielectric constant of the composite material but also dramatically improve the thermal properties and mechanical properties of the material.

3.4 Applications of Polymer Composite Materials in the Field of High-Temperature Wave-Transmitting and Wave-Absorbing Electrical Fields

Wave-transmitting materials have a good wave-transmitting effect on electromagnetic waves within a certain frequency range. They are widely used in the antenna system of the military aerospace field and have several unique functions such as diversion, heat protection, wave penetration, and bearing. Their applications, in general, can be divided into radomes and antenna windows. These materials usually employ ceramic compounds and insulating polymer materials. The absorbing material refers to a material that can absorb or greatly reduce the electromagnetic wave energy projected on its surface, thereby reducing the interference of electromagnetic waves. In engineering applications, the

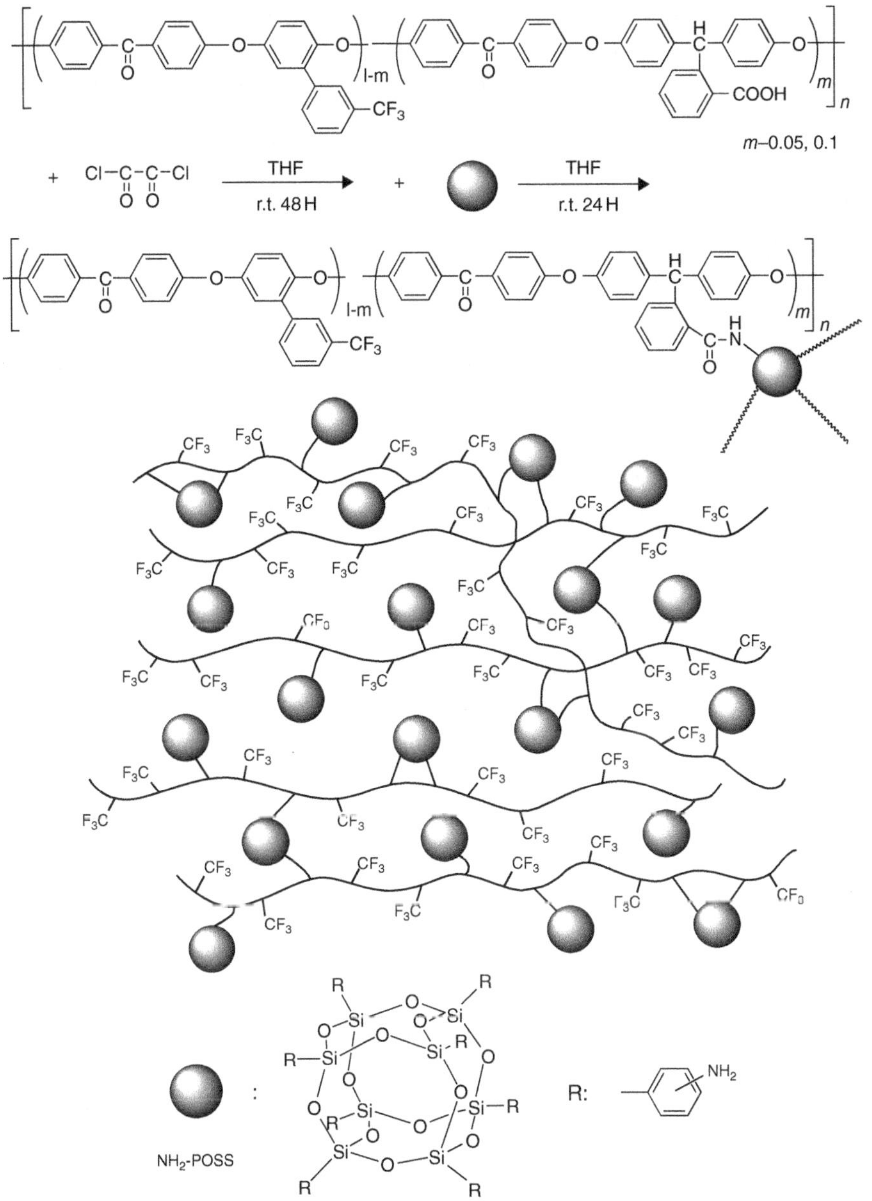

Figure 3.19 Synthesis of the NH_2-POSS/PEEK-CF_3-COOH composite material.

absorbing material is required to have a high absorption rate of electromagnetic waves in a wide frequency band, besides possessing other properties such as lightweight, temperature resistance, humidity resistance, and corrosion resistance. Since the spacecraft has higher friction and impact with the air during flight, for the wave-transmitting and absorbing materials, it also requires to have excellent heat resistance.

Table 3.4 Dielectric properties of the NH_2-POSS/PEEK-CF_3-COOH composite films.

Samples	Dielectric constant (ε) (at 1 MHz)	Dielectric constant decrease ($\Delta\varepsilon$)	Dielectric loss (at 1 MHz)
PEEK-CF_3-COOH-5%	2.84	—	0.015
PEEK-CF_3-COOH-10%	2.99	—	0.024
NH_2-POSS/PEEK-CF_3-COOH-5%	2.04	−0.80	0.0061
NH_2-POSS/PEEK-CF_3-COOH-10%	1.71	−1.28	0.0094

Table 3.5 Thermal properties of the NH_2-POSS/PEEK-CF_3-COOH composite films.

Samples	T_g (°C)[a]	DT_5 (°C)[b]	DT_{10} (°C)[c]
PEEK-CF_3-COOH-5%	140	274	504
PEEK-CF_3-COOH-10%	142	257	476
NH_2-POSS/PEEK-CF_3-COOH-5%	163	480	516
NH_2-POSS/PEEK-CF_3-COOH-10%	179	461	505

[a] Glass transition temperature.
[b] Heat distortion temperature.
[c] Heat distortion temperature.

3.4.1 Wave-Transmitting Materials

The ceramic materials are broadly exploited in the area of wave transmission and have the advantages of high-temperature resistance, good dielectric properties, and high strength, but they are brittle, poor in toughness, poor in thermal shock resistance, and complicate the molding process. In contrast, resin-based composite materials have a higher specific strength, and all the dielectric, mechanical, and thermal properties of the materials could be controlled through design. Hence, they have become the choice of materials in the case of radome manufacturers. However, the high-temperature resistance of resin-based polymer materials is only relative to other types of polymers, and they are not appropriate for hypersonic missile radomes. Subsonic aircraft radomes mostly employ low-temperature resin-based composite materials, which mainly include reinforcement materials that provide strength and modulus and the resin matrix provides adhesion [74].

3.4.1.1 The High-Temperature Resin Matrix

The resin acts as a binder and is an essential part of the composite material [75]. The interfacial strength of the composite material principally depends on the performance of the resin matrix. The resin matrix for radome includes unsaturated polyester resin (UPR), epoxy resin, etc. and is essentially a medium- and low-temperature thermosetting resin. UPR is one of the earliest thermosetting resins used in the radome and with improvements its temperature for long-term use can reach 150 °C. Epoxy resin is the most widely used resin matrix for radomes. It has a low cure shrinkage rate, excellent dielectric and adhesion properties, as well as chemical resistance. Some high-performance thermoplastic resins, such as PEEK, polytetrafluoroethylene, PI, etc. can also be used in the preparation of radar radomes.

3.4.1.2 Reinforced Materials

Reinforcement materials are the fundamental parts of resin matrix composites that provide strength and modulus (Table 3.6). They are applied for the development of the main load-bearing structure of single-layer radome and load-bearing skin of sandwich radome. The regularly utilized reinforcement materials for aircraft radomes chiefly include glass fiber, quartz fiber, and organic fiber. Among them, D glass fiber is developed for radome, and its dielectric constant and loss tangent are the lowest among glass fibers. High silica glass fiber is a type between ordinary glass fiber and quartz fiber, and its dielectric properties and strength are better than D glass fiber, second only to quartz glass fiber, and its cost is much lower than quartz fiber. Quartz fiber has the best dielectric performance owing to its SiO_2 content, which is more than 99.9% and can achieve low loss wave permeability of the radome. Organic fibers have a lower dielectric constant than their inorganic counterparts and have the characteristics of high strength and high modulus. The typical representatives in this class are aramid and polyethylene fibers [76].

3.4.2 Absorbing Material

The advanced radar detection and precision guidance capabilities of many countries around the world are increasing day by day, and few large combat weapons, such as aircraft, missiles, ships, and tanks, are currently facing the threat of being destroyed when they are noticed. The survival and defense capabilities of weapons and equipment have vital military significance. The ideal electromagnetic wave absorbing material should have the characteristics of lightweight, less thickness, strong absorption capacity and absorption frequency bandwidth, as well as its chemical composition should be stable, thereby meeting the needs of diverse environments. The conductive polymer, which originates in the middle of the

Table 3.6 Properties of the typically reinforced fibers.

Fiber	Density/ $g \cdot cm^{-3}$	Tensile strength/MPa	Elastic modulus/GPa	Dielectric constant (10 GHz)	Dielectric loss (10 GHz)
E glass fiber	2.54	3140	73.0	6.13	0.0055
S glass fiber	2.49	4020	82.9	5.21	0.0068
M glass fiber	2.77	3700	91.6	7.00	0.0039
D glass fiber	2.10	2000	48.0	4.50	0.0026
Quartz fiber	2.20	1700	72.0	3.78	0.0002
High silica glass fiber	2.30	2500	52.0	4.00	0.0048
Kevlar 49	1.46	3620	152.0	2.50	0.0015
Spectra1000	0.90	5010	192.8	3.00	0.0001

twentieth century, has the advantages of a lightweight, low density, good corrosion resistance, adjustable electrical conductivity, and a special conjugated large π system. Thus, it displays remarkable potential in the field of absorbing materials. However, relying on the conductive polymer itself concerning the absorption of electromagnetic waves is very limited. Therefore, combining it with other materials, while using the excellent electrical properties of the conductive polymer, can further adjust the electromagnetic parameters, as well as improve both the impedance matching and absorbing performance of the material [77].

3.4.2.1 The High-Temperature Resin Matrix

At present, the most commonly used polymer matrix is polyaniline (PANI) and polypyrrole (PPy). In 1862, PANI was first discovered. Until 1987, MacDiarmid proposed benzene and quinone, which became the final definite structural model of PANI (Figure 3.20) [78]. PANI can be regarded as a copolymer of phenylenediamine and quinodiimide. Undoped PANI is composed of two parts: oxidation and reduction. The difference in the proportion of these two parts directly affects the conductivity of PANI. And PANI with a special conjugated structure is the easiest to achieve a wide range of applications. At the same time, suitable dopants and doping levels as well as controlling the reaction conditions during synthesis can regulate the conductivity and water solubility of PANI, where highly doped PANI demonstrates excellent conductivity. Through the designing of molecular structures, it is compounded with other materials so that a wide frequency band and high absorption rate could be realized.

PPy is a new type of functional polymer material and is the research hotspot in recent studies, which has good conductivity and electrochemical redox reversibility (Figure 3.21). Its monomer pyrrole is a five-membered heterocyclic ring composed of C and N elements. It is slightly soluble in water at room temperature and easily soluble in organic solvents. In the presence of an electric field or oxidation environment, electrons easily transport to form cation free radicals. An intermediate product dimer is then formed by the coupling of free radical addition and then oxidized. Following this, through continuous chain coupling reaction, long-chain PPy can be obtained, which has a unique π-bonded macromolecular structure composed of single and double bonds and shows excellent electrical conductivity and electrochemical performance [79].

3.4.2.2 Inorganic Filler

To further improve the absorption of the material, the polymer matrix is doped with inorganic fillers, and the two commonly used fillers are (i) magnetic nanoparticles that include carbonyl iron, ferrite, ultrafine magnetic metal powder, etc. and (ii) carbon materials that include graphene sheet and carbon fiber.

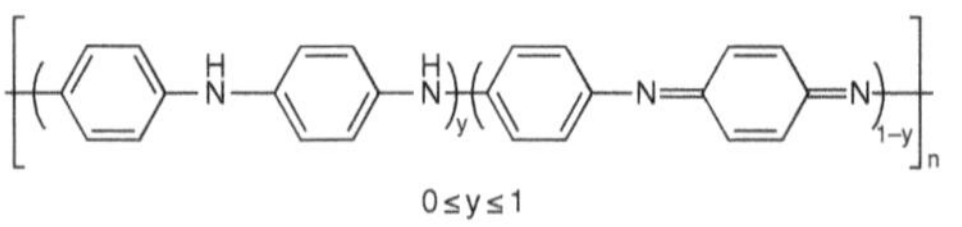

Figure 3.20 Structure of polyaniline.

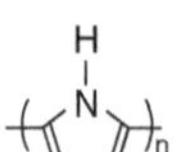

Figure 3.21 Structure of polypyrrole.

Sun et al. [80] first obtained Fe_3O_4 nanoparticles by a simple redox method and then polymerized them *in situ* to achieve Fe_3O_4/PANI nanoparticles. These nanoparticles are irregular spheres with Fe_3O_4 core and PANI polymer shell, with an average particle size of 72.4 nm. The optimal reflection loss observed at 16.7 GHz is 35.1 dB, and the thickness is 1.7 mm. Its excellent electromagnetic performance is mainly due to the dielectric resonance of its special core–shell microstructure and the natural resonance of Fe_3O_4 nanoparticles.

Chen et al. [81] prepared laminated graphene/PPy composite materials through *in situ* interlayer polymerization. It was found that during the exfoliation of graphite oxide, a chemical hybrid HN-CO was formed as an interlayer composite inside the graphene oxide layer, resulting in a new type of graphene oxide-based composite material. The obtained material exhibited stronger polarization, which significantly increased the electromagnetic loss of the broadband response. In the sample with a thickness of 96 mm, the reflection loss curve showed a minimum of −58.1 dB and a frequency of 12.4 GHz.

Ruey-Bin Yang et al. [82] prepared Fe_3O_4/PPy/carbon nanotube electromagnetic protection wave-absorbing composite material, and the effects of carbon nanotubes and the content of PPy were examined. The Fe_3O_4 conductive polymer was covered with a mixture of PPy and carbon nanotubes. The dielectric constant and permeability were analyzed in the frequency range of 2–18 GHz, and good wave absorption effect was noted. The reflection loss was less than −0. The absorbing frequency band of dB was as wide as 8–12.5 GHz, and the effective complementary nanocomposite material between this complex dielectric constant and permeability further improves the minimum reflectivity from −15.8 to −25.9 dB, paving the way for the ideal microwave absorber with reflection loss.

3.5 Summary

In recent years, a large number of exploratory studies have been conducted on the preparation, application, structure, and performance of high-temperature materials. And these materials are also widely used in various fields of electrical and electronics. However, their performance cannot still meet the needs of many high-end applications. Hence, much attention is directed toward these materials, as they hinder the manufacturing development of electrical and electronic equipment.

First, for intrinsic high-temperature-resistant insulation polymers, the basic theoretical research should be strengthened. The processing conditions and external factors are critical for the orientation of molecular chains and internal groups. In the case of filled high-temperature-resistant polymer composites, attention should be paid to the surface modification of micro- and nanofillers, as well as the control of dispersion and spatial distribution of fillers in the matrix.

Second, polymer composite dielectrics, the mostly insulating materials with the main function of storing charge, are highly desired for electrical engineering applications. However, a few problems are yet to be solved. First, due to the conflict between low dielectric loss and high polarization/dielectric constant, one can only achieve an accurate balance between them. Second, the interface problems between the polymer matrix and fillers are important factors that affect the dielectric properties of the composite materials. Third, the synthesis of new polymers and modification of fillers fundamentally affect the performance of the composite materials.

Third, today's integrated circuits are developing toward miniaturization, high-density assembly, low cost, high performance, and high reliability, which puts forward higher requirements for substrates, wiring materials, sealing materials, and interlayer dielectric materials. The emergence of electronic packaging materials with good performance and low cost is needed. This provides a huge space for the development of polymer-based electronic packaging materials. By changing the molecular structure of the polymer and the type and content of the filler in the polymer-based electronic packaging materials, it is necessary to find an electronic packaging material that not only matches the thermal performance of the substrate but also has good mechanical properties, and the manufacturing method is also economical and applicable, which is the development direction of polymer-based electronic packaging composite materials.

Fourth, the temperature of the polymer-based wave-transmitting composite material cannot be applied to ultra-high-speed aircraft. Therefore, investigations related to resin-based wave-transmitting composite materials should be directed to low dielectric constant glass fiber products as well as the modification and synthesis of high-performance matrix resins. Taking advantages of their excellent foldability, reliability, lightweight, ease of use and installation, high carrying capacity, good fatigue performance, and high portability, especially in high frequency and harsh environment with excellent dielectric performance characteristics, they are promising in the future defense antennas, aerospace, electronic appliances, military radar, and other communication antennas. The ideal absorbing material should have the following characteristics of "thin, light, wide, and strong." However, conductive polymer absorbing materials still have the following shortcomings: (i) A single conductive polymer mainly has an electrical loss, lack of magnetic loss, and hence by adding appropriate magnetic substances, composite absorbing materials could be obtained with excellent properties, and also the dispersion of magnetic substances in the composite materials is required in the development. (ii) The interaction between the composite systems is not just a simple superposition. The influence of interaction between the components on the performance of absorption needs to be further explored. (iii) In the case of improving the performance of absorbing materials, such as multi-band and low frequency, further developments are necessary. Thus, the development of wide-spectrum high-performance absorbing materials has become an important area in the research and development of wave-absorbing materials.

In summary, the current studies indicate that the polymer composites for high-temperature applications are in the key position in the field of electrical and electronics. In conclusion, groundbreaking developments are witnessed in materials science. It is to be expected that these developments should lead to the further development of polymer composites for high-temperature applications that will revolutionize electronic and electrical devices built for harsh environments.

References

1 Xiumin, Z., Xiaowen, P., Kang, L. et al. (2015). New research and application progress of nanodielectrics. *Insulating Materials* 38 (10): 1–9.

2 Xiumin, Z. and Xiaodong, Z. (2006). Recent progress of high voltage outdoor insulator materials. *Insulating Materials* 39 (1): 46–51.

3 Ahamd, Z. (2012). *Polymeric Dielectric Materials in Dielectric Material.* Croatia: Intech.

4 Li, Q., Chen, L., Gadinski, M.R. et al. (2015). Flexible high-temperature dielectric materials from polymer nanocomposites. *Nature* 523: 576–579.

5 Venkat, N., Dang, T.D., Bai, Z.W. et al. (2010). High temperature polymer dielectrics for aerospace power conditioning capacitor applications. *Materials Science and Engineering B* 168 (1–3): 16–21.

6 Qi, L.J., Petersson, L., and Liu, T.L. (2014). Review of recent activities on dielectric films for capacitor applications. *Journal of International Council on Electrical Engineering* 4 (1): 1–6.

7 Miao, Q.J., Fang, Z.-P., and Cai, G.P. (2003). Silica supported Karstedt-type catalyst for hydrosilylation reactions. *Catalysis Communication* 4: 637–639.

8 Liaw, D.J., Wang, K.L., Huang, Y.C. et al. (2012). Advanced polyimide materials: syntheses, physical properties and applications. *Progress in Polymer Science* 37 (7): 907–974.

9 Chung, H., Lee, J., Wonbong, J. et al. (2000). Stress behaviors and thermal properties of polyimide thin films depending on different curing process. *Journal of Polymer Science Part B: Polymer Physics* 38: 2879–2890.

10 Kotera, M., Nishino, T., and Nakamae, K. (2000). Imidization processes of aromatic polyimide by temperature modulated DSC. *Polymer* 41: 3615–3619.

11 Kang, P.H., Lee, C., and Kim, K.Y. (2007). Radiation and thermal effects on the dielectric relaxation properties of PEEK. *Journal of Industrial and Engineering Chemistry* 13 (2): 250–256.

12 Chen, B.S., Luan, D.C., Jiao, G.P. et al. (2009). Preparation of high temperature resistance polybenzimidazole resin. *Frontiers of Chemistry in China* 4 (2): 207–209.

13 Liu, J.G.(eds. O. Olabisi and K. Adewale). *Polyphenylquinoxaline in Handbook of Thermoplastics*, 2e. Boca Raton: CRC Press.

14 Cheng, L., Jingang, L., and Yang, S. (2010). Recent progress on research and development of poly-phenylquinoxalines. *Polymer Bulletin* 2: 1–10.

15 Gardiner, G. (2014). BMI and benzoxazine battle for future OOA aerocomposites. *High Performance Composites* 22 (1): 34–39.

16 Zhang, S., Yi, G., Fu, Q., and Ran, Q. (2019). Controlled polymerization of 3,4-dihydro-2H-1,3-benzoxazine and its properties tailored by Lewis acids. *Reactive and Functional Polymers* 139: 75–84.

17 Nelson, J.K. (2007). Overview of nanodielectrics: insulating materials of the future. Electrical Insulation Conference and Electrical Manufacturing Expo, Nashville, TN, USA.

18 Fréchette, M.F., Trudeau, M., Alamdari, H.D. et al. (2001). Introductory remarks on nanodielectrics. *Proceedings of the Conference on Electrical Insulation and Dielectric Phenomena*, Kitchener, Ontario, Canada.

19 Lei, Q.Q., Tian, F.Q., Yang, C. et al. (2010). Modified isothermal discharge current theory and its application in the determination of trap level distribution in polyimide films. *Journal of Electrostatics* 68 (3): 243–248.

20 Liu, L.Z., Liang, B., Wang, W. et al. (2006). Preparation of polyimide/inorganic nanoparticle hybrid films by sol-gel method. *Journal of Composite Materials* 40 (23): 21752183.

21 Zhang, M.Y., Zeng, S.J., Fan, Y. et al. (2008). Dielectric and conduction properties of polyimide/silica nano-hybrid films. *Polymer Composites* 29 (6): 617–622.
22 Huang, X.Y., Jiang, P.K., and Tanaka, T. (2011). A review of dielectric polymer composites with high thermal conductivity. *IEEE Electrical Insulation Magazine* 27 (4): 8–16.
23 Harada, M., Hamaura, N., Ochi, M. et al. (2013). Thermal conductivity of liquid crystalline epoxy/BN filler composites having ordered network structure. *Composites Part B Engineering* 55 (9): 306–313.
24 Chowdhury, S.R. (2016). Polymer matrices functionalized with liquid crystals for enhanced thermal conductivity. WO 2014105227A1, 28 June 2016.
25 Song, S., Katagi, H., and Takezawa, Y. (2012). Study on high thermal conductivity of mesogenic epoxy resin with spherulite structure. *Polymer* 53 (20): 4489–4492.
26 Takezawa, Y., Akatsuka, M., and Farren, C. (2010). High thermal conductive epoxy resins with controlled high order structure. *Journal of Applied Polymer Science* 89 (9): 2464–2467.
27 Koda, T., Toyoshima, T., Komatsu, T. et al. (2013). Ordering simulation of high thermal conductivity epoxy resins. *Polymer Journal* 45 (4): 444–448.
28 Du, B.X. and Xu, H. (2014). Effects of thermal conductivity on DC resistance to erosion of silicone rubber/BN nanocomposites. *IEEE Transactions on Dielectrics and Electrical Insulation* 21 (2): 511–518.
29 Namitha, L.K., Ananthakumar, S., and Sebastian, M.T. (2015). Aluminum nitride filled flexible silicone rubber compositesfor microwave substrate applications. *Journal of Materials Science: Materials in Electronics* 26 (2): 891–897.
30 Zhang, T. and Chi, Q. (2021). High temperature dielectric materials for electrical energy storage. In: Du, B. (eds) *Polymer Insulation Applied for HVDC Transmission.*
31 Rabuffi, M. and Picci, G. (2002). Status quo and future prospects for metallized polypropylene energy storage capacitors. *IEEE Transactions on Plasma Science* 30: 1939–1942.
32 Zhang, X.M, Liu, J.G., Yang, S.Y. (2016). *Review on Advanced Materials Science* 46 (1): 22–38.
33 Liaw, D.J., Wang, K.L., Huang, Y.C. et al. (2012). *Progress in Polymer Science* 37 (7): 907–974.
34 Asimakopoulos, I.A., Psarras, G.C., and Zoumpoulakis, L. (2014). Barium titanate/ polyester resin nanocomposites: development, structure-properties relationship and energy storage capability. *Express Polymer Letters* 8: 692–707.
35 An, N.L., Liu, S.L., Fang, C.Q. et al. (2015). Preparation and properties of beta-phase graphene oxide/PVDF composite films. *Journal of Applied Polymer Science* 132: n/a–n/a.
36 Fredin, L.A., Li, Z., Lanagan, M.T. et al. (2013). Sustainable high capacitance at high frequencies: metallic aluminum–polypropylene nanocomposites. *ACS Nano* 7: 396–407.
37 Yim, A., Chahal, R.S., and St. Pierre, L.E. (1973). The effect of polymer-filler interaction energy on the Tg of filled polymers. *Journal of Colloid and Interface Science* 43: 583–590.
38 Fragiadakis, D., Pissis, P., and Bokobza, L. (2005). Glass transition and molecular dynamics in poly(dimethylsiloxane)/silica nanocomposites. *Polymer* 46: 6001–6008.
39 Qi, L., Zhang, G., Zhang, X. et al. (2015). Electrocaloric effect: relaxor ferroelectric-based electrocaloric polymer nanocomposites with a broad operating temperature range and high cooling energy. *Advanced Materials* 27 (13): 2236–2241.

40 Rajib, M., Martinez, R., Shuvo, M. et al. (2016). Enhanced energy storage of dielectric nanocomposites at elevated temperatures. *International Journal of Applied Ceramic Technology* 13 (1): 125–132.

41 Chen, Q., Shen, Y., Zhang, S. et al. (2015). Polymer-based dielectrics with high energy storage density. *Annual Review of Materials Research* 45 (1): 433–458.

42 Li, H., Ai, D., Ren, L. et al. (2019). *Advanced Materials* 31 (23): 1900875.1–1900875.7.

43 Lin, X., Salari, M., Arava, L.M.R. et al. (2016). High temperature electrical energystorage: advances, challenges, and frontiers. *Chemical Society Reviews* 45 (21): 5848–5887.

44 Thakur, Y., Zhang, T., Iacob, C. et al. (2017). Enhancement of the dielectricresponse in polymer nanocomposites with low dielectric constant fillers. *Nanoscale* 9 (31): 10992–10997.

45 Sharma, V., Wang, C., Lorenzini, R.G. et al. (2014). Rational design of all organicpolymer dielectrics. *Nature Communications* 5: 4845.

46 Thakur, Y., Lean, M.H., and Zhang, Q.M. (2017). Reducing conduction losses in high energy density polymer using nanocomposites. *Applied Physics Letters* 110 (12): 122905.

47 Pan, C., Kou, K., Jia, Q. et al. (2016). Fabrication and characterization of micronano AlN co-filled PTFE composites with enhanced thermal conductivity: amorphology-promoted synergistic effect. *Journal of Materials Science: Materials in Electronics* 27 (11): 11909–11916.

48 Huang, X. and Jiang, P. (2015). Core–shell structured high-k polymer nanocomposites for energy storage and dielectric applications. *Advanced Materials* 27 (3): 546–554.

49 Jung, H.M., Kang, J.H., Yang, S.Y. et al. (2010). Barium titanate nanoparticles with diblock copolymer shielding layers for high-energy density nanocomposites. *Chemistry of Materials* 22: 450–456.

50 Xu, W.H., Liu, J., Chen, T. et al. (2019). Bioinspired polymer nanocomposites exhibit giant energy density and high efficiency at high temperature. *Small* 15 (28): 1901582.

51 Liu, J., Shen, Z., Xu, W. et al. (2020). Interface-strengthened polymer nanocomposites with reduced dielectric relaxation exhibit high energy density at elevated temperatures utilizing a facile dual crosslinked network. *Small* 16 (22): e2000714.

52 Fredin, L.A., Zhong, L., Ratner, M.A. et al. (2012). Enhanced energy storage and suppressed dielectric loss in oxide core–shell–polyolefin nanocomposites by moderating internal surface area and increasing shell thickness. *Advanced Materials* 24 (44): 5946–5953.

53 Song, Y., Shen, Y., Liu, H. et al. (2012). Improving the dielectric constants and breakdown strength of polymer composites: effects of the shape of the BaTiO3 nanoinclusions, surface modification and polymer matrix. *Journal of Materials Chemistry* 22 (32): 16491–16498.

54 Li, Z., Fredin, L.A., Tewari, P. et al. (2010). In situ catalytic encapsulation of core shell nanoparticles having variable shell thickness: dielectric and energy storage properties of high-permittivity metal oxide nanocomposites. *Chemistry of Materials* 22 (18): 5154–5164.

55 Li, Q., Liu, F., Yang, T. et al. (2016). Sandwich-structured polymer nanocomposites with high energy density and great charge-discharge efficiency at elevated temperatures. *PNAS* 113: 9995–10000.

56 Wang, Y., Wang, L., Yuan, Q. et al. (2017). Ultrahigh electric displacement and energy density in gradient layer-structured $BaTiO_3$/PVDF nanocomposites with an interfacial barrier effect. *Journal of Materials Chemistry A* 5: 10849–10855.

57 Azizi, A., Gadinski, M.R., Li, Q. et al. (2017). High-performance polymers sandwiched with chemical vapor deposited hexagonal boron nitrides as scalable high-temperature dielectric materials. *Advanced Materials* 29 (35): 1701864.
58 Tian, Z., Wang, X., Shu, L. et al. (2009). *Ceramic Society* 92: 830–833.
59 Merkel, T., Graeber, M., and Pagel, L. (1999). A new technology for fluidic microsystems based on PCB technology. *Sensors and Actuators A Physical* 77 (2): 98–105.
60 Bao, J., Jeppson, K., Edwards, M. et al. (2016). Synthesis and applications of two-dimensional hexagonal boron nitride in electronics manufacturing. *Electronic Materials Letters* 12 (1): 1–16.
61 Ichino, T., Sasaki, S., Matsuura, T., and Nishi, S. (1990). Synthesis and properties of new polyimides containing flourinated alkoxy side chains. *Journal of Polymer Science Part A: Polymer Chemistry* 28 (2): 323–331.
62 Yao, H.Y., Zhang, Y.H., You, K.Y. et al. (2014). Synthesis and properties of soluble cross-linkable fluorinated co-polyimides. *Reactive & Functional Polymers* 82: 58–65.
63 Chen, Y., Wang, W., Yu, W. et al. (2004). Nanoporous low-κ polyimide films via poly(amic acid)s with grafted poly(ethylene glycol) side chains from a reversible addition–fragmentation chain-transfer-mediated process. *Advanced Functional Materials* 14 (5): 471–478.
64 Wang, G., Chen, C., Zhou, H. et al. (2000). Synthesis and characterization of polyaryl ether ketone containing fluorine side groups. *Chemical Journal of Chinese Universities* 21 (8): 1325–1327.
65 Wang, G., Wang, D., Jiang, Z. et al. (2001). Study on the synthesis and properties of fluorine-containing polyaryl ether ketone. *Chemical Journal of Chinese Universities* 22 (6): 1053–1056.
66 Wang, G. (2000). *Study on Fluorinated Polyaryl Ether Ketone*. Jilin: College of Chemistry, Jilin University.
67 Goto, K., Kakuta, M., Inoue, Y. et al. (2000). Low dielectric and thermal stable polyimides with fluorene structure. *Journal of Photopolymer Science and Technology* 13 (2): 313–315.
68 Long, T.M. and Swager, T.M. (2003). Molecular design of free volume as a route to low-k dielectric materials. *Journal of the American Chemical Society* 125: 46.
69 Geng, Z., Zhang, S., Mu, J. et al. (2013). Design and preparation of poly (aryl ether ketone)/phosphotungstic acid hybrid films with low dielectric constant. *Journal of Applied Polymer Science* 129 (6): 3219–3225.
70 Chen, H.G., Xie, L., Lu, H.B., and Yang, Y.L. (2007). Ultra-low-k polyimide hybrid films via copolymerization of polyimide and polyoxometalates. *Journal of Materials Chemistry* 17: 1258–1261.
71 Leu, C.M., Yaote, C.A., and Wei, K.H. (2003). Polyimide-side-chain tethered polyhedral oligomeric silsesquioxane nanocomposites for low-dielectric film applications. *Chemistry of Materials* 15 (19): 3721–3727.
72 Zhang, W., Xu, J., Li, X. et al. (2014). Preparation, characterization, and properties of poly(aryl ether sulfone) systems with double-decker silsesquioxane in the main chains by reactive blending. *Journal of Polymer Science Part A: Polymer Chemistry* 52 (6): 780–788.
73 Geng, Z., Huo, M.X., Mu, J.X. et al. Ultra low dielectric constant soluble polyhedral oligomeric silsesquioxane (POSS)-poly(aryl ether ketone) nanocomposites with excellent thermal and mechanical properties. *Journal of Materials Chemistry C* 2 (2014): 1094–1103.

74 Weiwei, Y. and Xiaoqing, W. (2013). Research progress of radome materials. *Advanced Ceramics* 34 (1): 3–8.

75 Radford, D.W. (2018). Comprehensive composite materials II || 3.13 application of high temperature polymer matrix composites to engine intake valves. *Comprehensive Composite Materials II*: 312–349.

76 Walton, J.D. (1970). *Radome Engineering Handbook*. New York: Marcel Dekker, Inc.

77 Yanmin, W. (2014). Microwave absorbing materials based on polyaniline composites: a review. *International Journal of Materials Research* 105 (1): 3–12.

78 Macdiarmid, A.G., Chiang, J.C., and Richter, A.F. (1987). Polyaniline: a new concept in conducting polymers. *Synthetic Metals* 18 (1/2/3): 285–290.

79 Pomposo, J.A., Ochoteco, E., Pozo, C. et al. (2006). Conductivity enhancement in raw polypyrrole and polypyrrole nanoparticle dispersions. *Polymers for Advanced Technologies* 17 (1): 26–29.

80 Sun, Y.P., Xiao, F., Liu, X.G. et al. (2013). *RSC Advances* 3 (44): 22554–22559.

81 Chen, X., Chen, J., Meng, F. et al. (2016). Hierarchical composites of polypyrrole/graphene oxide synthesized by in situ, intercalation polymerization for high efficiency and broadband responses of electromagnetic absorption. *Composites Science and Technology* 127: 71–78.

82 Yang, R.B., Reddy, P.M., Chang, C.J. et al. (2016). Synthesis and characterization of Fe_3O_4/polypyrrole/carbon nanotube composites with tunable microwave absorption properties: role of carbon nanotube and polypyrrole content. *Chemical Engineering Journal* 285: 497–507.

4

Fire-Retardant Polymer Composites for Electrical Engineering

Zhi Li, En Tang, and Xue-Meng Cao

China-Spain Collaborative Research Center for Advanced Materials, School of Materials Science and Engineering, Chongqing Jiaotong University, Chongqing, China

4.1 Introduction

Due to mechanical robustness, superior flexibility, and insulation protectiveness, the polymeric materials are broadly utilized in electrical appliances and power transportation cables. However, the thermal accumulation and runaway from the miniatured equipment or circuit shortcut toward flammable polymers pose remarkable fire threats. In order to endow electrical appliances and power transportation cables with the up-to-standard fire-safe requirements, the fire-retardant treatment is stringently executed. The industry standards from Underwriters Laboratories, American Society of Testing Materials (ASTM), and International Electrotechnical Commission (IEC) organizations (Table 4.1) focus on different fire-developing stages and fire-risk scenarios i.e. ignition, fire spread, smoke production, gas toxicity, heat release, and fire structural stability. In the fire scenario, the fire spread, smoke production, and heat release closely linked to fatality and economic loss turn out to be the priority focus in the industry and research community.

The initiation of fire essentially meets the basic requirements defined by the fire triangle (e.g. heat, oxygen, and evolved gas). The fire-retardant modes accordingly are governed by how to disrupt the balance of fire triangle. The vapor-phase mode of action involves in the reduction of combustion efficiency via radical quenching or cooling effect. The condensed-phase mode is associated with the suppression or termination of interchange of heat, oxygen, and evolved gas between condensed and vapor phase. The fire-retardant treatment is supposed to adhere to the basic mode of action toward the optimization of vapor-phase and condensed-phase fire-retardant contribution.

Both additive fire retardants and intrinsic incorporation of fire-retardant elements are capable to impart electrical engineering polymers with fire retardancy. The commercialized fire retardants include halogenated fire retardants, organophosphorus fire retardants, and metal hydroxides (or relevant minerals). On account of Stockholm Convention, Restriction

Polymer Composites for Electrical Engineering, First Edition. Edited by Xingyi Huang and Toshikatsu Tanaka.

Table 4.1 The general fire tests for electrical engineering polymeric composites.

Fire tests	Standard	Definition
Limiting oxygen index (LOI)	ASTM D 2863-2013	Ignitability
UL-94 burning test	ASTM D 3801-2010	Fire propagation
Cone calorimetry	ISO 5660-1	Heat, smoke release
VW-1 burning test	UL1581	Fire propagation of cables
Glowing wire ignition temperature (GWIT)	IEC 60695-2-10	Ignitability of electrical engineering polymers
Comparative tracking index (CTI)	IEC-60112-2009	Sensitivity of surface tracking concerning insulation

of Hazardous Substances (RoHS), Waste Electrical and Electronic Equipment(WEEE), and Registration Evaluation and Authorization of Chemicals (REACH) regulatory directives, some halogenated fire retardants (e.g. polybrominated biphenyls, polybromo diphenyl ethers, and tris [2-chloroethyl] phosphate) are banned or listed as substances of very high concern. The halogen-free trend shifts the focus to the phosphorous-based molecular fire-fighter and metal hydroxides (or relevant minerals). In addition, the nitrogen-based triazine compounds, boron-based fire retardants, and metal salts are widely applied. The development of other high-performance alternatives (e.g. nanofillers and silicon-based compounds) or strategies (e.g. intumescence) for future technology preparation are upsurging.

The industrial applications of fire-retardant polymer composites for electrical engineering lie in cables, domestic electrical appliances, outdoor devices, voltage transformer, etc. The present obstruct is that the high-loading additive fire retardants deteriorate comprehensive properties such as insulation, processability, mechanical strength, thermal stability, etc. Currently, the balance of fire retardancy and other comprehensive property is constantly desirable.

4.2 Fire-Retardant Cables and Wires

4.2.1 Fundamental Overview

The cables in the cable trough suffer from fire risk due to the unexpected shortcut and thermal damage. The typical three-layer structural cables include the insulation layer, sheath layer, and metallic inner core. Under the exposure of external heat source, the flammable sheath layer turns out to be dominant fire-initiating point. If the external heat flux is over a transit value, the insulation layer becomes the following fire attack point [1]. Distinct from the sheath layer, the metallic core embedded in the insulation layers favors the massive conductivity of heat outside. Therefore, the decomposition and fire intensity of insulation layer is theoretically weaker than those of sheath layer.

Polyethylene (PE) [2, 3], ethylene-vinyl acetate (EVA) [4], linear low-density PE/polyolefin elastomer (e.g. LLDPE/EVA) [4], and thermoplastic polyurethane (TPU) [5] are ideal polymer matrices for insulation layer. In most cases, the crosslinking treatments (e.g. peroxide crosslinking, silane crosslinking [6], and electron radiation [7]) of insulation layer are

Table 4.2 Basic mechanical and flammability parameters of polymer matrices.

Polymer	Tensile strength (MPa)	Elongation at break (%)	LOI (%)
PE	~23	~450	~17
EVA	~21	~900	~17
TPU	~11	~500	~17
Flexible PVC	~12	~75	~24

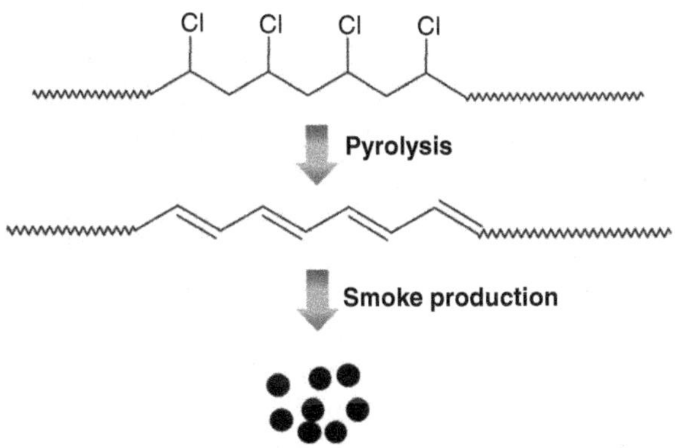

Scheme 4.1 Schematic illustration of smoke production of PVC.

required for improving mechanical property, antiaging property, and other comprehensive property. Parallelly, the robust but flexible polymers such as high-density PE (HDPE), TPU, EVA, as well as flexible polyvinyl chloride (fPVC) are used in sheath layer. All relevant polymers are subjected to the flammability issue (Table 4.2).

In terms of smoke, fPVC tends to generate dense smoke since the decomposition of PVC is initiated by HCl removal and ends up with formation of massive trans-polyene structures. Trans-polyene structures are transformed into the soot and smoke particles (Scheme 4.1). PE gives rise to a tiny amount of smoke. EVA, located between the charring and non-charring polymer, releases a certain concentration of smoke.

In terms of insulation and sheath layers, the fire-retardant evaluation involves not only in the generalized measurement of respective polymers but also in the assessment of entire cables. The generalized measurement of respective parts includes limiting oxygen index (LOI), UL-94 test, cone calorimeter test (CCT), and smoke density chamber. The integral evaluation of the entire cables consists of UL1581 test and CCT [8].

4.2.2 Understanding of Fire-Retardant Cables and Wires

4.2.2.1 Polyethylene Composites

The fire retardancy of PE is achieved via the addition of halogenated compounds, metal hydroxides, or intumescent fire retardants. The organophosphorus fire retardants are not suitable for PE due to the intensive noncompatibility. Even though the high efficiency of halogenated fire retardants is revealed, halogen-free trend is accepted.

Aluminum trihydrate ($Al(OH)_3$, ATH) and magnesium hydroxide ($Mg(OH)_2$, MDH) are two commercial fire retardants. ATH and MDH decomposes dominantly over the range of 220–330 °C and 300–500 °C with the heat adsorption of 1050 and 1300 kJ/kg, respectively. The fire-retardant mode is based on the cooling effect toward vapor-phase flaming zone. The remaining fire chars exert the barrier effect against heat, oxygen, and volatiles to some degrees. In a normal case, more than 60 wt% ATH is required to achieve UL-94 V-0 rating of PE. In comparison, 40–60 wt% MDH is essential to pass UL-94 V-0. The lower loading of MDH relative to ATH is attributed to the higher decomposition temperature to match polymer degradation. In addition, MDH presents a stronger smoke extinction than ATH due to the partial charring capability of MDH. Even though ATH and MDH favor the modulus enhancement, the excessive addition results in the remarkable deterioration of mechanical and insulation property.

Two generalized strategies, i.e. surface modification and synergy, are adopted to alleviate the negative effect on the mechanical property and comprehensive property. The surface modification promotes the cohesion between microscale fillers and polymer. As a result, the tensile strength and elongation at break are simultaneously improved. The stearic acid bearing polar carbonyl groups and long-chain alkyls as well as silane coupling agent containing polar groups (e.g. epoxy, amine, and isocyanate group) are used to treat ATH and MDH (Scheme 4.2). Additionally, the stearic acid modification of ATH and MDH reduces peak heat release rate (pHRR) in CCT. Evidently, the strengthened cohesion between fillers and polymer on one hand promotes the dispersion of high-loading fillers in polymer matrix for stronger dilution effect. On the other hand, the adhesion between fillers and polymer favors the formation of chars with higher continuousness and stronger fracture resistance.

Another effective strategy is to add fire-retardant synergistic agents. The higher fire-retardant efficiency permits a lower addition of ATH and MDH, which benefits the weakening of negative effect on the mechanical property and comprehensive property. The fire-retardant synergistic agents contain clays, metal compounds, phosphorous-based materials, boron-containing compounds, expandable graphite (EG) [9], carbon nanomaterials, etc. Two or more combination of synergistic agents constitutes the synergistic system. Phosphorous-containing materials (e.g. red phosphorous [RP]) exert a synergistic effect via the esterification charring reaction. In comparison to no rating and LOI of 21.6% of HDPE/40 wt%ATH, HDPE/35 wt%ATH/5 wt%RP passes UL-94 V-2 with LOI of 24.4% [10], clays normally function via the weak catalytic carbonization as well as participation of itself into charring process. As a special clay, two-dimensional nanoclay (e.g. montmorillonite (MMT) and layered double hydroxides (LDH)) possesses an extra fire-retardant mechanism. The two-dimensional nanoclay generates an intensive barrier effect and greatly suppress the penetration of heat, oxygen, and volatiles through fire chars [11]. The strengthening of condensed-phase barrier from nanoclay greatly optimizes the fire-retardant behavior of PE/ATH and PE/MDH. Even though the other carbon nanomaterials such as graphene, carbon nanotubes, carbon black, and C_{60} enable the enhancement of fire-retardant efficiency, the high electrical conductivity of carbon nanotubes destroys the insulation feature of PE matrix [12].

Impressively, the zinc borate ($ZnBO_3$) and borate present a highly efficient synergy toward ATH and MDH filled PE composites. $ZnBO_3$ possesses a lower melting point of 400 °C, which drives the melting and formation of continuous and glassy structure. Also,

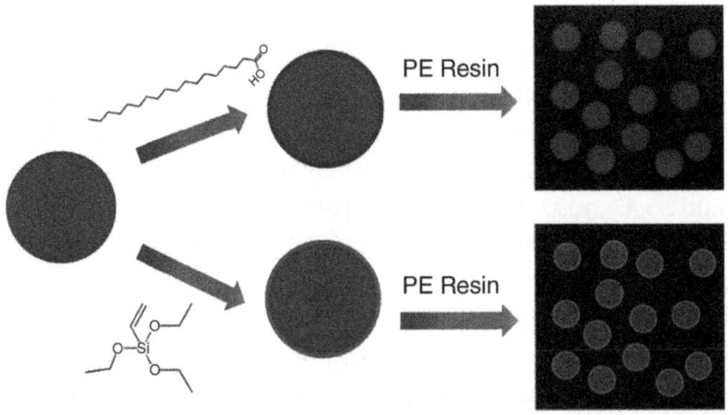

Scheme 4.2 Surface treatment of ATH and MDH by stearic acid and silane.

$ZnBO_3$ participates in the reaction with ATH and MDH to form glassy $Al_2(BO_3)_3$ and $MgBO_3$. The glassy substances are certified as the brilliant barrier structure for strengthening and optimizing the char structure [13].

The processing property of PE composites for cables is another consideration. The high-loading addition of MDH and ATH into PE significantly increases the melt viscosity. As a result, the melt elastic effect including extrudes expansion, shark skin, and melt fracture occurs. Additionally, the surface modification of ATH and MDH as well as the addition of synergistic agents increase the melt viscosity to certain degree.

In addition to metal hydroxides, the intumescent strategy on the basis of the conventional ammonia polyphosphate (APP) is capable to enhance the flame retardancy rigorously with more than 30 wt% for UL-94 V-0 rating [14]. The intumescent fire retardants contain three functional components, i.e. acid agent, carbon agent, and blowing agent, which interact physically and chemically to form protective cellular structures. The notable obstacle to block the commercialization is the high moisture-adsorption property of APP, which results in the surface fogging of PE composites and the failure of fire retardancy. Actually, the addition of classical intumescent fire retardants deteriorates the mechanical property and insulation of polymer matrix.

4.2.2.2 Ethylene-Vinyl Acetate (EVA) Copolymer

As a derivative of PE, EVA contains PE backbones with some structural units grafted by acetate groups. The incorporation of vinyl acetate results in the disruption of the molecular stereoregularity and geometry, which suppresses crystalline capability. Relative to EP, EVA resin presents a better flexibility with some sacrifice of tensile strength and modulus [15]. The polymer melt viscosity is remarkably lower than that of PE, which permits a higher addition of fire retardants.

EVA resin with low ratio of vinyl acetate behaves as PE, which generates few smokes and release enormous heat. Comparatively, EVA with higher ratio of vinyl acetate release the acetate acid during combustion at 300 °C, leaving the double-bond compounds or conjugated backbones. The acetate acid fuels the flaming vapor-phase zone, and the double-bond compounds are transformed into the soot and smoke. In parallel, with the higher content of vinyl acetate, the insulation property lowers due to the polarity of vinyl acetate.

ATH, MDH, and its derived minerals act as the prior alternatives in the fire retardancy of EVA resin. The addition of MDH and ATH increases LOI of EVA resin in almost linear manner [16]. MDH presents a slightly higher fire-retardant efficiency than ATH. However, the mechanical property of EVA is deteriorated seriously after the massive addition of ATH and MDH. A 50 wt% ATH or MDH decreases tensile strength by 60.0 and 65.3% as well as elongation at break by 47.5 and 55.4%, respectively [17]. In parallel, the volume electrical resistance and damp-heating aging behavior of EVA resin is significantly affected by MDH and ATH. Anyhow, the volume electrical resistance of EVA composites with ATH and MDH meets the requirements of 10^{14} ohm•cm in the case of appropriate formulation.

The synergistic fire-retardant systems represent a higher fire-retardant efficiency toward EVA, which allows the lower addition percentage and less impact on mechanical and insulation property. Nanofillers [18, 19], RP, siloxane [20], melamine [21], melamine borate [21], zinc hydroxystannate (ZHS) [22], and clays [23] synergistically work with ATH or MDH in fire retardancy of EVA. All involved synergistic agents favor the formation of structurally intact and mechanical robust carbonaceous layers. The nanofillers MMT [24] and carbon nanotube [25] assist to connect the microchars to an entity with higher strength and integrity. Resultantly, the fire propagation by UL-94 test as well as heat and smoke release in CCT are suppressed. Different from nanofillers, $ZnBO_3$ melts and flows to form the glassy network to stabilize the char structure in mechanics and improve the continuousness and integrity of carbonaceous layers. As a result, the heat release and fire propagation are hampered in the entire combustion [26]. In CCT of both cases, the second peak of heat release is delayed and suppressed in the release intensity.

Microencapsulated red phosphorous (MRP) [27] and siloxane participate in the charring reaction toward optimizing the char structure, which reduces the heat and smoke release. In parallel, the synergistic combination of EG and ATH or MDH implies that the intumescent process and vapor-phase cooling process interplay during the entire combustion [28]. The intumescent feature from EG is merged into the loose and porous char structure of ATH- or MDH-filled EVA. Hence, the heat release rate of EVA/MDH/EG and EVA/MDH/MRP composites is suppressed at the whole combustion stage remarkably.

In order to eliminate the kind of fire-retardant additives in processing operation and improve the fire-retardant efficiency, the interfacial manipulation on ATH or MDH or their respective clays works. ATH and MDH (15 μm) supply enormous surface area, which is a prominent platform for surface functionalization. $ZnBO_3$ is deposited on the exterior surface of brucite via the simple heterogenous sedimentation. EVA composites with brucite/$ZnBO_3$ manifests a lower pHRR and lower critical loading (45 wt%) to pass UL-94 V-0 compared with EVA/brucite composites. $ZnBO_3$ significantly improves the char structure and strength of EVA composites via the formation of glassy and continuous skeleton. Specifically, $ZnBO_3$ at the interface of EVA and brucite contributes to a stronger function, which results in a remarkable reduction of the second peak in CCT [29].

In parallel, brucite with alternatively deposited nickel ions and alginate via layer by layer route (B/A/Nia/A) reveals a fire-retardant enhancement relative to virgin brucite. EVA composites with 60 wt% B/A/Nia/A displays a reduced pHRR and peak smoke production rate (pSPR) by 41.5 and 66.7% separately in comparison to EVA/brucite. The graphene-like ordered char structure is formed via the catalysis by nickel compounds. The interface location of the catalysts between EVA and brucite expands the contact area of catalysts toward

EVA and alginate. In addition, the interface modification of brucite contributes to the enhancement of elongation at break preferentially [30].

Intumescent strategy exemplified by APP and EG seems as a prevailing solution for fire retardancy of EVA. The intumescent fire retardants generate the intumescent closed-pore or open-pore cellular char structure, which accounts for the improved fire retardancy (Scheme 4.3). EVA/APP composites give rise to arc-like chars with explicit cells. EVA/EG composites generate flat chars with notable open cells [31]. The in-depth reason is attributed to the different working modes of EG and APP. EG expands separately via the decomposition and release of inserted sulfuric acid molecules with the sheet-like structure expanded. APP works via the physical and chemical reaction with melt EVA matrix.

A 30 wt% conventional intumescent fire retardant decreases tensile strength and elongation at break by 35.4 and 22.0% separately [32]. The priority task is how to reduce the critical addition percentage. The diversified synergy between APP and other components is revealed in Table 4.3 on the basis of either the optimization of cellular char structure toward more compactness or the tailoring of the condensed-phase and vapor-phase fire behavior. Impressively, the incorporation of the charring-foaming compounds into EVA/APP formulation enables to reduce the heat release significantly at the whole stage.

Another barrier for the broader application is the high moisture adsorption of APP microparticles. Aiming to tackle with the high moisture adsorption and improve interfacial adhesion, different surface modifications by silica [39, 40], polyurethane [32], urea-formaldehyde [41], etc. are adopted. The surface microencapsulation of APP positively suppresses the moisture adsorption with a lower solubility and a less impact on flammability and mechanical performance [40]. Additionally, the interfacial microencapsulation effectively enhances the fire retardancy due to the internal synergistical effect between APP and microencapsulated polymers. The interfacial incorporation of heterogenous polymer layer increases the adhesion between EVA and APP, which accounts for the optimized mechanical property, particularly tensile strength. The volume resistance is decreased and then increased due to the subsequent incorporation of polar APP and weak-polar interface layers (Figure 4.1). In cables and wires, the employment of APP and microencapsulated APP do not exceed the electrical resistance limitation ($\geq 10^{14}$ ohm•cm).

As a physically intumescent fire retardant, the addition of EG particles increases fire retardancy (30 wt% EG to pass V-0 rating) but undesirably decreases the mechanical property. However, the isolated employment of EG in fire retardancy of EVA composites is not executed in consideration of low-quality chars. In normal cases, the modification of EG or synergistic combination between EG and other components (e.g. MDH, ATH and phosphorous-based compounds) contributes to the fire-retardant efficiency. Anyhow, the electrical conductivity of EG restricts the application in the insulation layers of cables.

4.2.2.3 Polyvinyl Chloride Composites

Flexible polyvinyl chloride (fPVC) with plasticizing agents (approximately 30–40 phr) is a critical alternative for cable sheath layer. In the molecular structure of PVC, the chloride atoms substitute hydrogen atoms of ethylene structural unit, which brings about a completely different combustion behavior from PE matrix. The high char yield (10%) and quenching effect of Cl radicals directly incite the high LOI value of 48% [42]. However, the massive addition of plasticizing agents transforms nonflammable PVC matrix to the

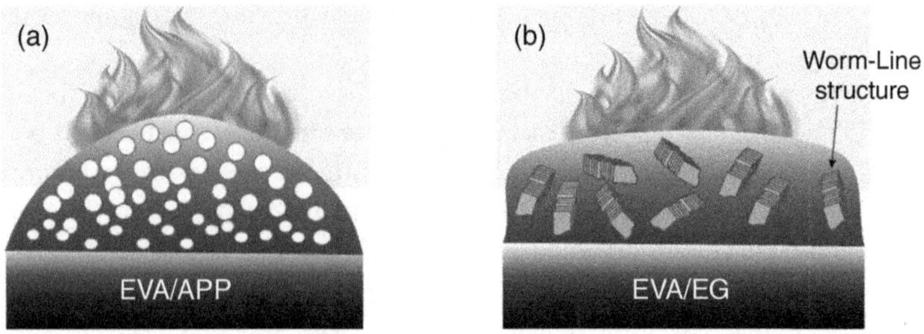

Scheme 4.3 Typical morphology of intumescent char structure from (a) EVA/APP and (b) EVA/EG.

Table 4.3 Synergistic effect of EVA/APP composites.

Resin	Synergistic agents	Formulations	Fire retardancy	Ref.
EVA (28% VA)	Charing-foaming agent (CFA), LDH	EVA/13.7 wt%APP/ 4.6 wt%CFA/1.8 wt%LDH	LOI = 29.7%, V-0 rating	[33]
EVA (18% VA)	Triazine charring-foaming agent (HTCFA)	EVA/22.4 wt%APP/ 5.6 wt%HTCFA	LOI = 32.6%, V-0 rating	[34]
EVA (18% VA)	Triazine charring-foaming agent (CNCA-DA)	EVA/13.3 wt%APP/ 6.7 wt% CNCA-DA	LOI = 27.0%, V-0 rating	[35]
EVA (16% VA)	P/N charring agent (PPPA)	EVA/16.5 wt%APP/ 5.5 wt% PPPA	LOI = 30.5%, V-0 rating	[36]
EVA (28% VA)	Cyclic phosphate (MOPO)	EVA/20 wt%APP/ 10 wt% MOPO	LOI = 28.4%, V-0 rating	[37]
EVA (18% VA)	LDH	EVA/12.5 wt%APP/ 37.5 wt% LDH	LOI = 22.9%, V-0 rating	[38]

flammable fPVC with LOI of 25% (depending on different plasticizing agent and its content). The general plasticizing agents in fPVC formulation are dioctyl phthalate (DOP), trioctyl trimellitate (TOTM), and didecyl phthalate (DIDP).

The fire risk of fPVC originates from heat release and smoke production. In principle, the heat relevant fire-retardant approach should be based on (i) how to strengthen the vapor-phase radical quenching effect of chloride atoms and (ii) how to optimize the condensed-phase barrier effect by charring reaction. Antimony oxide (ATO), phosphorous-based compounds, MDH, ATH, and metal compounds (e.g. tin- and zinc-based compounds) are intensively investigated. As a commercial fire retardant, ATO functions dominantly via enhancing the radical quenching effect (Scheme 4.4). Sb_2O_3 is transferred to $SbCl_3$ with the participation of HCl, which captures H and OH radicals to kill flame. Even though the combustion efficiency is reduced, the release of HCl and dense smoke is not decreased due to the radical quenching mode [43]. The intensive synergistic effect between ATO and ZHS exists since Zn^{2+} catalyzes dehydrochlorination and crosslinking and Sn^{4+} participates into the charring reaction [43]. Some solid solutions (e.g. ZnO/MgO, ZnO/CaO, and ZnO/CaO/MgO) promote the fire retardancy of fPVC/ATO [44].

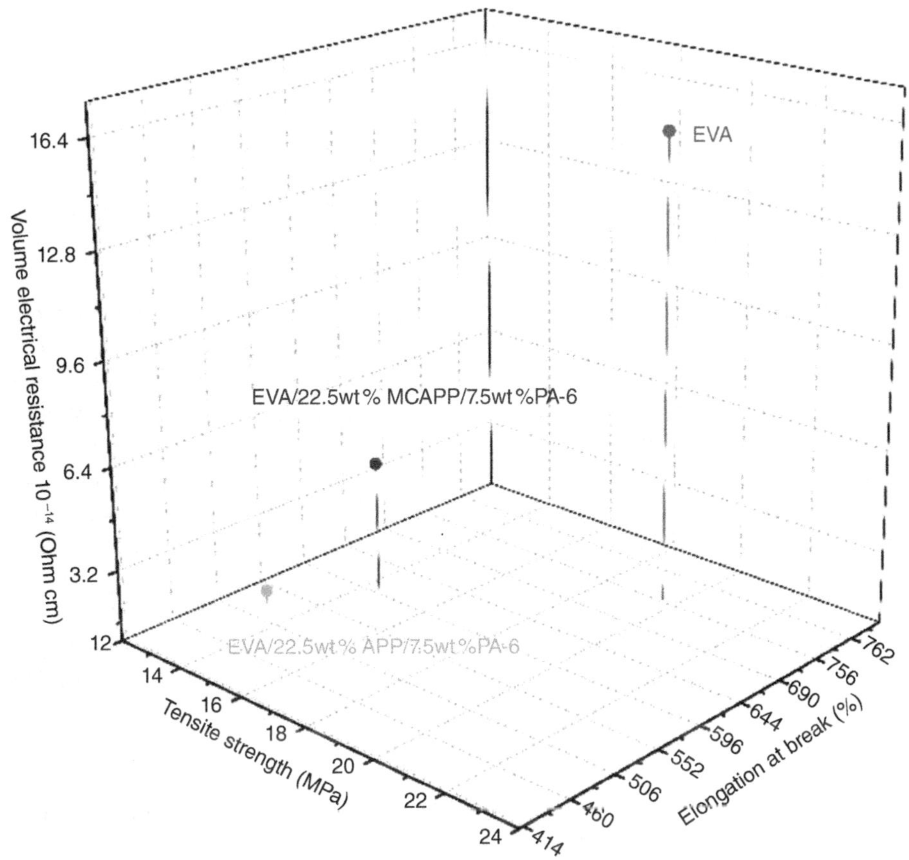

Figure 4.1 Impact of APP microencapsulation on mechanical property and volume electrical resistance. *Source:* Based on Wang et al. [32].

The incorporation of MDH or ATH enables to enhance LOI values as well as reduce pHRR and smoke production of fPVC in a nearly linear manner [45]. The use of hydroxystannate (ZHS) coating ATH or MDH (ZHS/ATH or ZHS/MDH) promote the fire retardancy significantly due to the increased char yield [45, 46]. The incorporation of Zn^{2+} and Sn^{4+} (in the form of SnO_3^{2-}) improves the char microstructure and macrostructure of fPVC/MDH or fPVC/ATH via catalyzing and participating the charring reaction. A 50 wt% ZHS/MH and ZHS/ATH impart fPVC with LOI of 41.7 and 45.8% compared with 30.3 and 33.5% of fPVC/MH and fPVC/ATH, respectively.

The other metal compounds exhibit the acceptable fire-retardant efficiency toward fPVC via the condensed-phase fire-retardant mode. The isolated use of SnO_2, ZHS, and zinc stannate (ZS) outperforms ATH and MDH in the fire retardancy, smoke suppression, and thermal stability of fPVC composites [47]. On basis of the prominent catalytic fire retardancy of tin-based compounds, the combination of tin compounds and other fire retardants (e.g. nano-fire retardant and iron compounds) gives rise to remarkable enhancement of fire retardancy of fPVC. The construction of tin-loaded mesoporous MCM-41 (TOS) [48] and

$$Sb_2O_3\ (s) + 2HCl(g) \xrightarrow{c.\ 250^{\circ}C} 2SbOCl\ (s) + H_2O(g)$$

$$5SbOCl\ (s) \xrightarrow{c.\ 250^{\circ}C} Sb_4O_5Cl_2\ (s) + SbCl_3\ (g)$$

$$4Sb_4O_5Cl\ (s) \xrightarrow{c.\ 450^{\circ}C} 5Sb_3O_4Cl\ (s) + SbCl_3\ (g)$$

$$3Sb_4O_5Cl_2\ (s) \xrightarrow{c.\ 500^{\circ}C} 4Sb_2O_3\ (s) + SbCl_3\ (g)$$

Scheme 4.4 Transformation reaction of solid Sb_2O_3 into gaseous $SbCl_3$.

nanoporous SnO_2@Fe_2O_3 0D/1D hybrid [49] improves fire retardancy and smoke suppression of fPVC without notable loss of thermal stability and mechanical property (Figure 4.2). Notably, the excessive addition of certain types of metal oxides (e.g. iron oxide) is not able to improve fire retardancy of fPVC due to the catalytic oxidation of newly formed polyaromatic chars by metal oxide.

The phosphorous-based fire retardants bring about the vapor-phase fire-retardant behavior via radical quenching as well as the condensed-phase fire-retardant behavior via the participation of charring reaction (Scheme 4.5). Actually, the integration of plasticizing molecules and phosphorous-containing molecules inspires researchers to design two-in-one structures. The merit lies in not only the reduction of processing complexity but also the compatibility enhancement of fire retardant in fPVC composites. However, the migration of plasticizing agent outside in fPVC composites keeps a tough problem [42, 50]. The intrinsic grafting of P/N-containing groups into PVC chain not only favors the self-plasticizing but also completely solves the migration problem, accompanied by the improved fire retardancy [51].

In addition to fire retardancy of fPVC, the fatality loss from the dense smoke of cables constitutes the major percentage. The metal compounds (e.g. ammonium octamolybdate, MoO_3, and Fe_2O_3) suppress the smoke production of PVC matrix via the catalytic crosslinking and charring reaction in the condensed phase. Some copper-based compounds (e.g. Cu(0) and CuOAc) catalyze the crosslinking reaction of allyl structure at a low temperature (200 °C) via the cross-coupling reaction. In parallel, organometallic compounds (e.g. ferrocene and benzoyl ferrocene) highly efficiently suppress the smoke production of PVC via the vapor-phase smoke-suppression mode. In the process, the catalytic thermal-oxidative combustion of soot in the vapor phase occurs [52]. The ferrocene possesses the low evaporation of c. 100 °C, which restricts the application. The use of the polymeric ferrocene provides a practical solution [53].

4.2.2.4 Other Polymers

Other polymers such as polypropylene (PP) and TPU are the proper items of insulation and sheath layers. Compared with PE resin, PP resin shows a notably higher tensile strength (~36.2 MPa) and modulus (~70.0%), which determines the application in the high-load unit. TPU illustrates a higher elongation (~835.9%) relative to PE resin, which imparts cables (insulation and sheath layer) with excellent bending resistance. The fire retardancy of TPU and PP is accomplished by halogenated fire retardant, MDH, ATH, and the intumescent strategy [54]. The nanofillers (e.g. LDH [55] and polyhedral oligomeric silsesquioxane [POSS]) [56]act as the fire-retardant nano-reinforcer in TPU and PP resin.

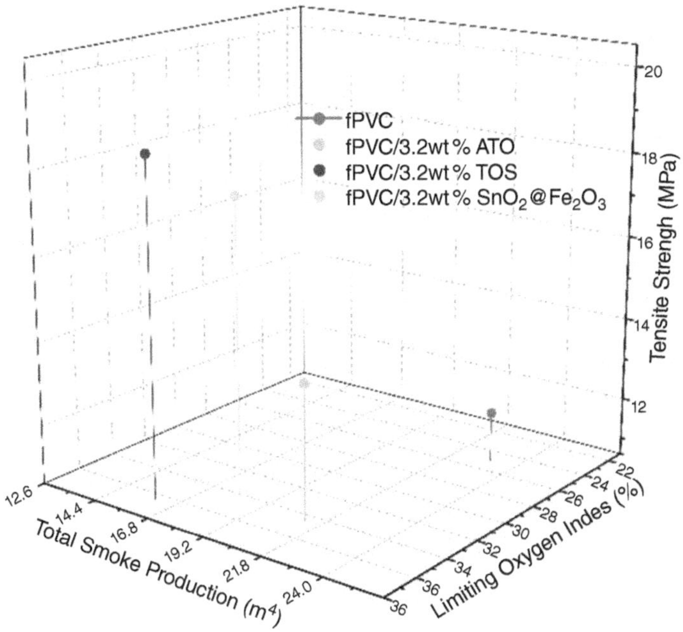

Figure 4.2 Effect of ATO, TOS, SnO_2@Fe_2O_3 on LOI, total smoke production (TSP), and tensile strength. *Sources*: Based on [48, 49].

Scheme 4.5 Schematic illustration of condensed-phase and gas-phase fire-retardant modes of fPVC composites.

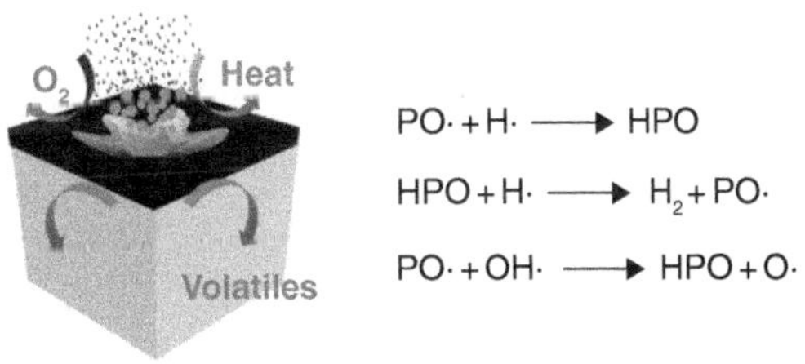

4.3 Fire-Retardant Polymer Composites for Electrical Equipment

4.3.1 Fundamental Overview

The domestic appliances (e.g. air conditioner, television set, and washing machine) and electrical devices (e.g. charging pile and charger) employ a diversity of polymers (e.g. high-impact polystyrene (HIPS), acrylonitrile-butadiene-styrene copolymer (ABS), polycarbonate/acrylonitrile-butadiene-styrene copolymer (PC/ABS), polycarbonate (PC), and polybutylene terephthalate (PBT)) as the main structural units. In consideration of the heat accumulation risk from circuit shortcut and electrical malfunctions, the fire-retardant treatment is essential. The fire-retardant characterizations are imposed on the basic aspect (e.g. LOI, UL-94, and CCT) and electricity relevant aspects (e.g. glowing wire ignition temperature (GWIT) and comparative tracking index (CTI)). HIPS and ABS show a similar

Table 4.4 Basic combustion profiles of polymeric composites of electrical equipment.

Polymers	LOI (%)	Smoke production
HIPS	18.1	Heavy
ABS	19.1	Heavy
PC/ABS	~23.6	Moderate
PC	25.9	Light
PBT	20.9	Light

combustion behavior with the massive release of dense smoke and nearly no char yields. PC/ABS present the nonflammability endowed by PC matrix and smoke-generating feature from ABS. PBT generates few smokes but sustainable flaming owing to the fast decomposition and release of volatiles (Table 4.4).

4.3.2 Understanding of Fire-Retardant Polymer Composites for Electrical Equipment

4.3.2.1 HIPS and ABS Composites

In order to overcome the brittleness of general polystyrene (PS), polybutadiene (PB) elastomer is incorporated into PS matrix via the grafting of PS chains on PB chains. Eventually, PB content is kept within 3–5% and form the dispersion phase with salami structure in the continuous PS matrix. In order to further improve the stiffness and chemical resistance of HIPS, the acrylonitrile is copolymerized with styrene on PB chain to form ABS. As a result, the PB phase is dispersed in the SAN phase (i.e. styrene and acrylonitrile copolymer). However, HIPS and ABS are not able to accommodate the application of electrical equipment due to their flammability.

Halogenated fire retardants for HIPS and ABS preferentially deteriorate the impact strength as well as induce the fogging phenomenon and ultraviolet (UV) sensitivity at a relative low loading (<25 wt%). Organophosphorus compounds including triphenyl phosphonate (TPP), oligomeric resorcinol bis(diphenyl phosphate) (RDP), and bisphenol A bis(diphenyl phosphate) (BDP) enjoy the utilization. However, the use of liquid organophosphorus compounds does not illustrate the high efficiency due to the absence of oxygen atom in the polymer matrix for the condensed-phase fire-retardant mode (25 wt% TPP into ABS, LOI increment = ~1%) [57]. The addition of liquid organophosphorus compounds results in the unexpected fogging phenomenon due to the poor compatibility. The low hydrolytic stability and the ease of evaporation of TPP, RDP, and BDP limits the massive application in HIPS and ABS matrix. MDH and ATH are utilized to impart the satisfied fire safety of HIPS and ABS. The critical content of metal hydroxides is so high (>30 wt%) that the mechanical property is deteriorated significantly [58]. Similar situation exists in terms of intumescent fire retardants. RP obtains certain applications in HIPS and ABS in spite of the color problem, the toxic gas evolution, as well as the reduced fire-retardant efficiency due to the oxygen-absence essence of HIPS and ABS. At present, the newly emerging metal phosphinate and metal hypophosphite present a promising prospect. Other fire-retardant approaches associated with nanotechnology and nanohybridization improve the fire-suppression performance in reducing heat and smoke release. In order to effectively alleviate the deterioration of toughness, the impact elastomer is usually demanded.

Aiming to decrease the effect of fire retardants on the comprehensive property, the key is to decrease the addition percentages and optimize types of fire retardants. Hence, the pursuit of the vigorous synergy between different fire-retardant species offers the feasible route. In RP-filled polymer composites, RP decomposes into white phosphorous at 400–450 °C, followed by the oxidation to oxygen-containing pyrophosphoric acid. The resultant pyrophosphoric acid reacts with the beneath polymers and gives rise to the chars, which shields against the penetration of heat, oxygen, and volatiles. Occasionally, the oxidation production of RP enters the vapor phase for flame poisoning. The higher fire-retardant efficiency is desirable if the condensed-phase char structures are optimized or the vapor-phase cooling effect/flame poisoning is incorporated. The isolated incorporation (>50 wt%) of RP or MRP is not able to effectively pass the UL-94 V-0 rating due to the oxygen-absence feature of HIPS. The intensive synergistic effect between MRP and MDH significantly improves the fire retardancy of HIPS with reduced heat release. The critical percentage for UL-94 V-0 is reduced from above 50.0 to 29.5 wt% with mass ratio of MRP and MDH as 5:2 [59].

Melamine cyanurate (MCA) synergistically promotes the vapor-phase dilution function and the condensed-phase charring effect of HIPS/RP concurrently. The addition of 30 wt% RP/MCA (mass ratio = 2:1) endows HIPS with UL-94 V-1 (1.6 mm). In parallel, RP synergistically participates in fire retardancy of ABS/aluminum isobutylphosphinate (APBu) with remarkably reduced heat and smoke release. The mere addition of APBu do not suppress smoke production due to the radical quenching effect identified by reduced efficient heat of combustion (EHC) [60]. The synergistic addition of RP optimizes the char structure via participation into the charring reaction with increased thermal stability and barrier effect even in the intumescence mode. Besides, the combination of low-thermal-stable intumescent formulation APP and high-thermal-stable RP favors the two-stage fire-retardant process (intumescence prior to contraction) of ABS resin toward a stronger barrier effect [61].

Metal (alkyl substituted) phosphinate (e.g. aluminum diethyl phosphinate [AlPi]) presents the potential in fire retardancy. The isolated incorporation of AlPi into ABS matrix do not induce UL-94 V-0 even at a high loading (~25 wt%). AlPi enables to synergistically function with APP with UL-94 V-0 attained at the loading of 25 wt% APP/AlPi (mass ratio = 1:1) [62]. The metal hypophosphite (e.g. aluminum hypophosphite [AP], magnesium hypophosphite (MP), and calcium hypophosphite (CP)) is the potential alternative in fire-retardant HIPS and ABS [63]. At the identical P loading of 10.5%, AP imparts ABS matrix with UL-94 V-0 compared with MP and CP due to the relatively lower thermal stability. During combustion, AP releases P_4, PH_3, and H_3PO_2 species into the vapor phase. The remaining $Al_4(PO_4)_3$ participates in the charring reaction in the condensed phase. In addition, MCA and silicon oligomer are certified as synergistic agents for ABS/AP composite [64].

Mono-component intumescent fire retardants present a better processing adaption relative to the conventional one. A 30 wt% poly(4,4-diaminodiphenyl methane spirocyclic pentaerythritol bisphosphonate) (PDSPB) reduces the flammability with LOI value of ABS from 19.1 to 28.6% and pHRR from 930 to 388 kW/m^2 [65]. The polymeric mono-component intumescent fire retardants enable to resist the moisture etching problem. P/N-containing polymeric mono-component intumescent fire retardant (PSPTR) imparts ABS with UL-94 V-2 and LOI of 25.2% at the loading of 30 wt%. The synergistic combination of PSPTR and phenol formaldehyde (PF) (mass ratio = 1:1) improves UL-94 rating to V-0 based on the joint charring reaction [66].

The isolated addition of TPP is not able to endow ABS with satisfied fire retardancy due to the absence of the condensed-phase charring behavior. The combination of TPP and charring agent (e.g. novolac phenol-phenyl isocyanate) effectively retains the phosphorous in the condensed phase for higher fire-retardant efficiency [57]. In order to suppress the high evaporation of TPP, the transformation of TPP into oligomer tetra-2,6-dimethyl phenyl resorcinol diphosphate (DMP-RDP) works [67]. The other oligomeric phosphate fire retardants such as RDP and BDP do not illustrate effective fire retardancy toward HIPS and ABS even though RDP and BDP present higher thermal stability and hydrolytic resistance than TPP.

4.3.2.2 PC/ABS Composites

PC/ABS blend combines the mechanical robustness, toughness, and heat resistance of PC phase as well as the processing ability of ABS phase. PC decomposes in anaerobic atmosphere in the manner of Fries arrangement prior to the crosslinking reaction (Scheme 4.6) [68]. The presence of trace of moisture catalyzes the chain scissor, the product phenol of which promotes the successive chain scissor. In parallel, the aerobic atmosphere contributes to the attack by oxygen on the methylene of isopropylidene moiety as well as the following rearrangement, chain scissor, and crosslinking. The final result is the gelation and the chain scissor of PC matrix with the generation of massive CO_2. The intensive charring feature and dilution effect of CO_2 endows PC matrix with excellent self-extinguishment. PC/ABS blend combines the charring and self-extinguishment of PC and flammability of ABS. The combustion property of PC/ABS depends on the mass ratio of PC to ABS, with higher PC percentage, the stronger will be the nonflammability.

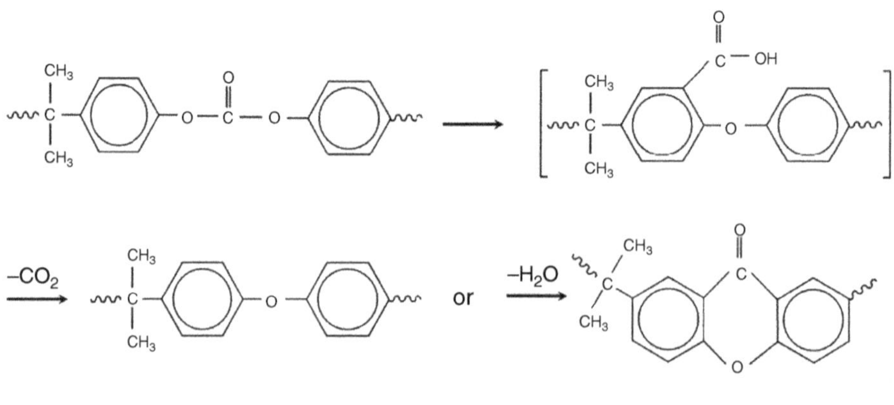

Scheme 4.6 Fires arrangements of PC chain.

Halogenated flame retardant (e.g. tetrabromobisphenol A) or halogenated phosphorous-containing fire retardants (e.g. tris[2,4-dibromophenyl] phosphate [PB-460]) based on phosphorous-halogen synergy show a highly efficient fire retardancy as well as excellent compatibility toward PC/ABS, which does not deteriorate the mechanical property and physical performance seriously. PC/$ABS_{3/1}$ (3:1 represents mass ratio of PC to ABS) with 6 wt% PB-460 presents UL-94 V-0 rating with heat distortion temperature (HDT) of 105 °C, which is 15 °C higher than that of PC/ABS/14 wt% TPP (UL-94 V-0 rating).

ATH and MDH impart PC/$ABS_{7/3}$ with UL-94 V-0 rating and LOI of 30% at the loading of 50 and 30 wt%, respectively. The tensile strength and notched impact strength are decreased by 69.1 and 93.6% for ATH as well as 63.6 and 90.9% for MDH. The similar case is illustrated in the APP-filled PC/ABS. More than 18 wt% APP endows PC/ABS matrix with UL-94 V-0 rating and LOI of 30%. The tensile strength and notched impact strength are reduced by 49.1 and 83.6%, respectively. The seriously deteriorated mechanical property, low thermal stability, and poor compatibility restricts the APP application. Apart from it, RP or MRP illustrate satisfied fire retardancy toward PC/ABS. The isolated addition of MRP increases fire retardancy of PC/$ABS_{3/1}$ with LOI of 27.7% and UL-94 of V-0 rating at the loading of 15 wt% [69]. The more incorporation of MRP decreases LOI value above 21 wt% attributed to the self-flammability of MRP. PC/$ABS_{3/1}$ composite with 18 wt% MRP retains the tensile strength and notched impact strength of 76.4 and 45.5%, which enables to be toughened by EVA and methyl methacrylate-butadiene-styrene copolymer (MBS), etc.

The commercialized and regulation-compliant phosphorous-based fire retardant imparts PC/ABS with positive fire retardancy as well as balanced mechanical and insulation property. The cost-effective phosphate is TPP. In PC/ABS with polytetrafluoroethylene (PTFE) anti-dripping agent, 12–18 wt% loading of TPP gives rise to UL-94 V-0 rating. The detailed TPP loading for UL-94 V-0 rating depends on ABS percentage in PC/ABS blend. The incorporation of TPP increases LOI in a linear mode with LOI of 27.2% at 14 wt% loading in terms of PC/ABS_{PTFE} (7/3). Nonflammable PTFE possesses a high melt viscosity, which fixes the melt polymer matrix into a networked form and retards the dripping occurrence at c. 0.5 wt% loading. To some degree, PTFE favors the fire retardancy in the condensed phase [70].

TPP is readily evaporated into the vapor zone and evolved into phosphorous-based radicals. The absence of condensed-phase reaction between PC and TPP permits the incorporation of some condensed-phase components to improve fire-retardant efficiency [70]. The synergistic combination of TPP and BDP enhances the fire-retardant efficiency, which involves in the crosslinking reaction between PC and BDP. Also, the highly efficient fire-retardant PC/ABS composite is attained via the synergistic combination of 6 wt% TPP and 2 wt% 9,10-dihydro-9-oxa-10-phosphaphenanthrene-10-oxide (DOPO) grafting MCM-41 mesoporous silica (DM) [71]. TPP presents a high evaporation and low melting point (50 °C), which induces the feeding hardship in extrusion processing.

The bridged aromatic phosphate such as RDP and BDP illustrates a higher thermal stability and lower volatilization feature. RDP is a phosphate oligomer with around two to five phosphorous atoms in the single molecules. The RDP loading for PC/ABS_{PTFE} to achieve UL-94 V-0 is 8–12 wt% [68], which displays a higher fire-retardant efficiency in UL-94 rating compared with TPP. The addition of RDP increases the LOI value in the linear manner until 15 wt%, above which the increase of LOI value is not notable. LOI of PC/ABS/RDP is located between PC/RDP and ABS/RDP with the proximity to ABS/TPP [72]. The fire-retardant mechanism of PC/ABS/RDP consists of the vapor-phase and condensed-phase modes. The substitution of 5 wt% RDP by TPP in PC/ABS_{PTFE} increases LOI and reduces the flammability via strengthening vapor-phase action [70]. The substitution of RDP by the charring polymer (e.g. novolac and epoxy resin), catalyst (e.g. Fe_2O_3 and MoO_3), and $ZnBO_3$ positively improves the fire retardancy via the condensed-phase mode [72]. However, similar to TPP, RDP possesses the demerit of low hydrolytic stability, which deteriorates the aging stability and crack resistance of PC/ABS. Some acid scavengers like epoxies and oxazolines enable to eliminate the acid.

Compared with RDP, BDP shows a higher thermal stability and hydrolytic stability yet possesses a higher viscosity. The viscous BDP causes the feeding and processing hardship of PC/ABS blend. Relative to RDP, BDP illustrates a lower phosphorous content and thus less effective in fire retardancy of PC/ABS_{PTFE}. The fire-retardant mechanism of BDP in PC/ABS_{PTFE} is the strong crosslinking reaction between BDP and PC, which promotes shielding effect of chars. Additionally, BDP enters the vapor phase and evolves to quench radicals. The isolated addition of 20 wt% BDP retains the tensile strength and notched impact strength by 84.1 and 81.8%, respectively. In addition to the above-mentioned synergy between BDP and TPP, other agents to strengthen the char structure could be the alternative. The synergistic fire retardancy between BDP and $ZnBO_3$ is revealed in LOI and heat release. The incorporation of $ZnBO_3$ favors the glassy borate network and generates the alpha-zinc phosphate and borophosphates in the condensed phase [73]. The similar case is illustrated in the synergy between BDP and AlOOH in LOI [74].

Resorcinol bis(di-2,6-xylyl phosphate) (RBXP) manufactured on a commercial scale is a solid diphosphate with a stronger hydrolytic stability due to the steric hindrance (Scheme 4.7). RBXP imparts PC/ABS with UL-94 V-0 at 12–16 wt% loading. Other phosphates including bisphenol S phosphate, biphenyl bis(diphenyl phosphate), phosphamidate, and cycloalkyl derivatives enable to improve fire retardancy. Most of them cannot enjoy the commercialization due to either the volatilization or deteriorating effect toward PC/ABS. P/N-containing compounds exhibit an impressive fire retardancy in PC/ABS. In parallel, hexaphenoxytricyclophosphazene (HPPZ) is found effective in PC/ABS with UL-94 V-0 rating at the loading of 12 wt%. In comparison to PC/ABS/TPP, PC/ABS/HPPZ illustrates a better fire retardancy with a lower negative effect on the mechanical property. The linear phosphazene (benzene disubstituted phosphazene, PDPP) is also capable to endow PC/ABS with UL-94 V-0 rating at the loading of 12%. The notched impact strength with incorporation of PDPP surprisingly increased due to the intrinsic elasticity of PDPP.

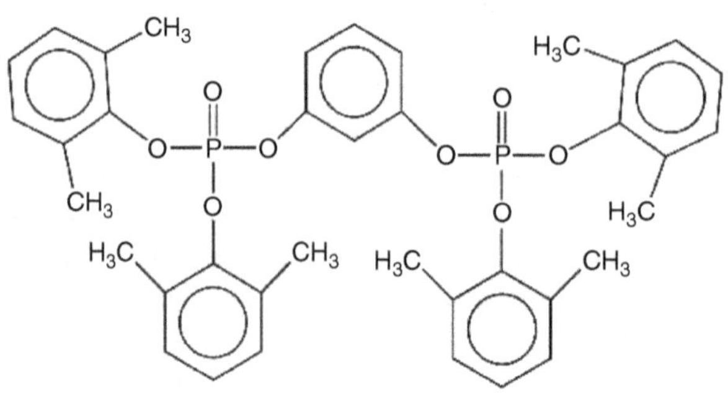

Scheme 4.7 RBXP structure.

4.3.2.3 PC Composites

PC manufactured from bisphenol A possesses a glass transition temperature (T_g) of 140–150 °C and HDT of 132–138 °C, which enables to work at the elevated temperature as the structural elements. During combustion, PC self-crosslinks with the release of CO_2, which infers PC with UL-94 V-2 rating. The phosphate fire retardants represent the non-halogen trend in spite of the loss of clarity and stress crack due to the poor hydrolytic resistance. PC composites with 6 wt% RDP or BDP pass UL-94 V-0 rating with the fire-retardant action in both condensed phase and vapor phase. The synergistic combination of TPP and RDP in the fire retardancy of PC resin is revealed based on the phosphorous-phosphorous synergy.

The alkali or alkali earth metal of certain aryl phosphates illustrate the highly efficient fire retardancy of PC. Potassium diphenylsulfone sulfonate (KSS) and potassium perfluorobutane sulfonate (KPFBS) provide UL-94 V-0 rating at a rather low loading of 0.05–0.1 wt%. Actually, the alkali or alkali earth metal aryl phosphates affect the haze of PC resin to some extent with KSS most selected species. The fire-retardant mode is associated with the condensed-phase barrier effect, which involves in the acceleration of Fries arrangement and crosslinking reaction by aromatic sulfonate. The combination of phosphonium and sulfonate into phosphonium sulfonate (Scheme 4.8) presents the excellent fire retardancy toward PC_{PTFE}. A 5 wt% phosphonium sulfonate imparts PC_{PTFE} with UL-94 V-0 rating and LOI of 33.7% [75].

Si-based compounds present a positive improvement of fire retardancy of PC. The addition of 2 wt% branched siloxane in combination with PTFE endows PC resin with UL-94 V-0 rating. The incorporation of $ZnBO_3$ into PC/polysiloxane illustrates the synergistic effect between $ZnBO_3$ and polysiloxane. $ZnBO_3$ functions via the fast formation of B-O-B and B-O-Si, which effectively restricts the dripping of PC matrix [75]. The grafting of sulfonate salt onto nanoscale polyhedral oligomeric silsesquioxane (S-POSS) (Scheme 4.9) endows PC resin with UL-94 V-0 rating at the thickness of 3.2 and 1.6 mm and LOI value of 33.3% at the loading of 0.25 wt% [76].

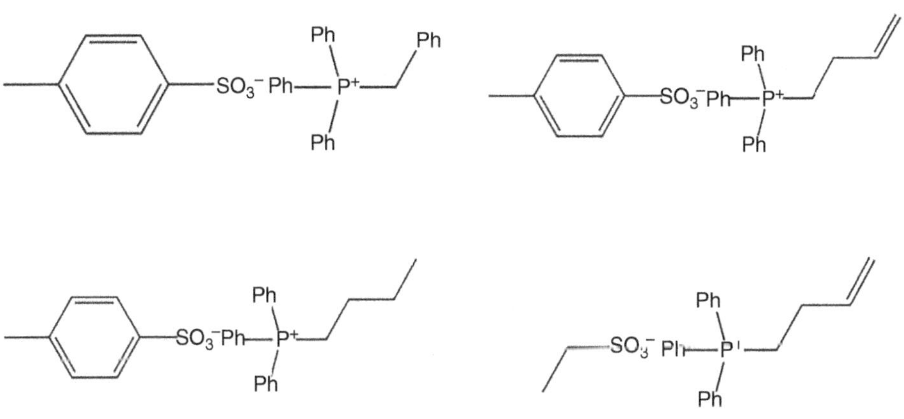

Scheme 4.8 Structure of phosphonium sulfonate. *Source:* Based on Hou et al. [75].

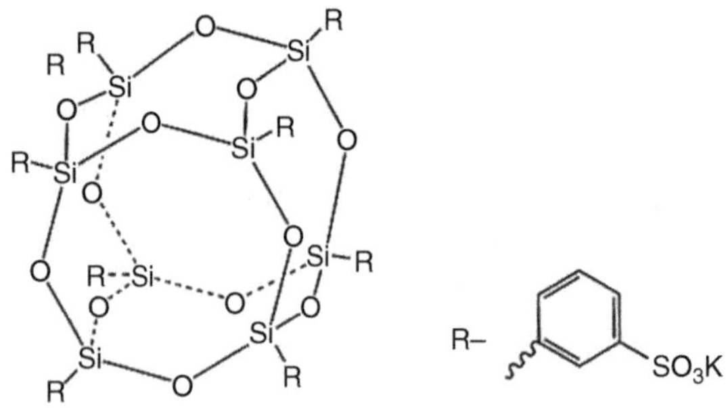

Scheme 4.9 Structure of S-POSS. *Source:* Based on Li and Yang [76].

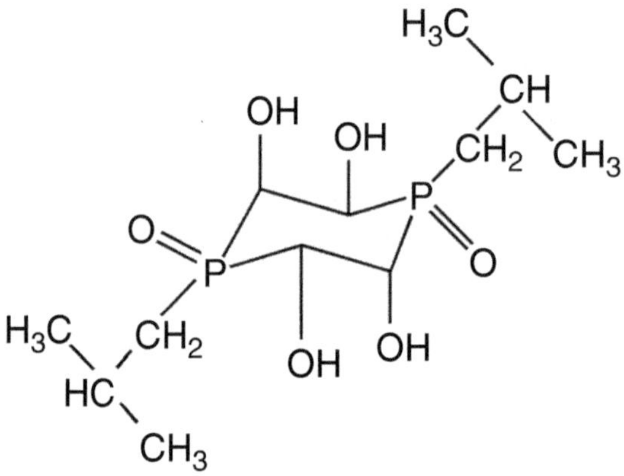

Scheme 4.10 Structure of Cyagard RF 1204.

4.3.2.4 PBT Composites

As a typical polyester, PBT possesses the semi-aromatic molecule structure and flexible segments, which determines a relatively higher impact strength and tensile strength (53–55 MPa). In combination with the excellent electrical insulation (400 V of CTI in 100 drops; 750 °C of GWIT at 3 mm thickness plague) [77], PBT attains a wider application (e.g. charging piles, charging guns, capacitor cases, and electrical connectors). The commercialized P-containing fire retardant 1,4-diisobutyl-2,3,5,6-tetrahydroxy-1,4-diphosphacyclohexane 1,4-dioxide (Cyagard RF 1204) (Scheme 4.10) imparts PBT matrix with UL-94 V-0/V-1 at 10–15 wt% loading [78].

Metal (alkyl substituted) phosphinates (e.g. aluminum dialkylphosphinate) present an excellent fire-retardant enhancement toward PBT. Aluminum diethylphosphinate (AlPi), zinc diethylphosphinate (ZnPi), etc. are used as commercialized products. AlPi imparts plain PBT with UL-94 V-0 at a loading of 20 wt% with LOI increment of 24.4% [79]. The dominant fire-retardant mode is associated with the vapor-phase radical quenching effect of diethylphosphonic acid and the secondary mode of the condense-phase barrier effect by the formation of the thermal-stable intermediate aluminum phosphinate terephthalate and aluminum phosphate.

In order to decrease the percentage of AlPi and alleviate the effect on the comprehensive property, the synergistic agents such as MCA, melamine polyphosphate (MPP) [80],

Fe_2O_3 [81], Al_2O_3 [82], sepiolite [83], silicon resin [84], and other charring components promote the fire-retardant efficiency via either vapor-phase dilution or condensed-phase barrier effect or both. The incorporation of nanometric Fe_2O_3 particles remarkably reduces the loading of UL-94 V-0 rating from 20 to 7 wt%. The synergistic combination of AlPi and epoxy-functional polysiloxane (EPM) reduces the critical loading of fire retardant from 20 to 12 wt% for UL-94 V-0 rating and increased LOI value of 17.4%. Notably, the incorporation of epoxy-functional polysiloxane elastomer retains the tensile strength and notched impact strength at 54.7 MPa and 49.1 J/m relative to virgin PBT (52.6 MPa and 59.6 J/m) [84]. However, the incorporation of AlPi keeps CTI values (400 V) but increases GWIT marginally, which is related to the pyrolysis behavior and char yield, respectively [85]. Also, the synergistic combination of AlPi with MCA or MPP enables to pass UL-94 V-0 rating at 20 wt% loading.

Phosphate, despite the great effort has been made, is not commercialized in the fire retardancy of PBT resin. The underlying reason is presumably that PBT tends to generate massive combustible gases without the formation of char [78]. The liquid aromatic phosphate (RDP, BDP, and bis[dixylenyl phosphates]) shows the "blooming" phenomenon in PBT. A 10 wt% BDP enables to impart PBT with self-extinguishment but no UL-94 vertical burning rating. Once the charring agent such as polyphenylene oxide (PPO) and PC is incorporated, UL-94 vertical burning rating is obtained. A 15 wt% BDP plus 15 wt% PC imparts plain PBT with UL-94 V-0 rating. A 15 wt% BDP (or RDP) coupled with 10 wt% novolac provides UL-94 V-0 rating. In addition, the aryl phosphate presents a vigorous synergy with the N-containing compounds such as MCA, MP, and MPP. The isolated addition of aryl phosphate and MCA does not give rise to excellent fire retardancy. The combination of 7.5 wt% RDP and 12.5 wt% MCA (or 20 wt% TPP and 10% MA) endows PBT with UL-94 V-0 rating [78].

4.4 Fire-Retardant Fiber Reinforced Polymer Composites

4.4.1 Fundamental Overview

The short glass fibers or mineral fillers are incorporated to enhance tensile property, flexural property, and HDT. In consideration of balanced comprehensive property of PBT and polyethylene terephthalate (PET), the glass fiber reinforced PBT and PET with better mechanical property (twice to forth increment relative to neat PBT and PET) and elevated working temperature are used as electrical connecter, motor casing, and switches. The incorporation of short glass fiber does not change the degradation route of polymer matrix. However, the micro-millimeter scale glass fiber interrupts the char structure and weakens the condensed-phase barrier effect [86]. The incorporation of glass fiber decreases melt flowing and restricts dripping, which deteriorates the flammability. Moreover, the "wick" effect emerges, in which the melt polymer flows along fiber and continuously supports the sustainable flaming. The underlying reason for the "wick" effect is that the lower density of glass fiber results in floating on the flaming frontier. Therefore, the fire-retardant treatment in glass reinforced PET and PBT is (i) to strengthen vapor-phase radical quenching or dilution effect, (ii) to strengthen condensed-phase shielding effect via optimized charring structure, and (iii) the removal of "wick" effect.

4.4.2 Understanding of Fire-Retardant Fiber Reinforced Polymer Composites

4.4.2.1 Reinforced PBT and PET Composites

AlPi imparts PBT/glass fiber (GF) (30 wt%) composite with satisfied fire retardancy (UL-94 V-0) at a loading of 14%. The excessive incorporation of AlPi reduces LOI due to the self-oxidation peculiarity of AlPi. PBT/GF with 14% AlPi retains the tensile strength and impact strength of 79.8 and 69.2%, respectively. The synergistic agents such as MCA, melamine, and MPP contribute to the fire retardancy of PBT/GF. The incorporation of MCA [87, 88], melamine, and MPP does not influence the decomposition of PBT matrix but participates in condensed-phase charring reaction and mainly dilutes the vapor phase at the early stage. In parallel, AP imparts PBT/GF (30 wt%) with UL-94 V-0 rating (1.6 mm) and LOI of 28.5% at the loading of 20 wt% [89]. Additionally, the synergistic combination of AP and MCA (2:1) improves LOI and shortens the sustainable flaming time in vertical burning test of PET/GF(30 wt%) [90].

Phosphonate fire retardants are widely used in glass fiber reinforced PET and PBT composites. UL-94 V-0 rating is achieved for PBT/GF (30 wt%) with the combination of 10 wt% cyclic phosphonate Antiblaze 1045 and 15 wt% MCA. The addition of 20 wt% RDP (or TPP) in combination with 15 wt% MCA generates UL-94 V-0 rating of PBT/GF. The addition of ZnS improves GWIT value to 960 °C with the formulation of 9 wt% BDP, 7 wt% MCA, and 6 wt% ZnS. The addition of CTI agent CaO enhances CTI value significantly in aryl phosphate filled PBT/GF composites. In addition, RP, in the masterbatch with PBT, increases the fire retardancy of PBT/GF (20 wt%) with UL-94 V-0 rating attained at the thickness of 3.2 and 1.6 mm at the loading of 14 wt%. The addition of RP in PBT/GF retains tensile strength and impact strength of 107.2 and 87.5%, respectively. The further incorporation of MDH, CaO, $ZnBO_3$, and ZnO increases CTI value marginally.

If the "wick" effect is alleviated, the fire retardancy of PBT/GF is enhanced to some extents. The feasible strategy is to coat the phosphorous/nitrogen-containing silane on the surface of glass fiber. Compared with the common silane, the P/N-containing silane modification on the glass fiber imparts PBT/GF (30 wt%) in combination with 15 wt% AlPi attains UL-94 V-0 and LOI of 33.0%. The interfacial combustion between PBT and GF is restricted by the interfacial charring reaction. The interface modification by P/N-containing silane attains a higher tensile strength and modulus compared with the common silane [91].

4.5 Conclusion and Outlook

The fire retardancy of electrical engineering polymeric composites is required without notable sacrifice of mechanical and insulation property. Even though halogenated fire retardants enjoy magnificent applications, halogen-free fire-retardant trend is essential by regulatory directives in fire-risk electrical engineering. In cables and wires, ATH, MDH, and derived minerals in combination with synergistic agents or surface catalyst deposition present notable advantages. Intumescent fire retardants with microencapsulation treatments possess a satisfied fire-suppression efficiency. In electrical appliances, AP and RP with synergistic agents impart HIPS and ABS with impressive fire retardancy. Silicon-based sulfonates coupled with nano reinforcer demonstrate a highly efficient fire retardancy with balanced comprehensive property. The bridged aromatic phosphates with hydrolytic resistance and matrix solubility are essential in PC/ABS. Aluminum phosphinate in combination with synergistic agents enhances fire retardancy of PBT and reinforced PBT with balanced CTI and GWIT. The surface fire-retardant functionalization of glass fiber provides an effective strategy for removing "wick" effect.

References

1 Gong, T., Xie, Q., and Huang, X. (2018). *Fire Safety Journal* 95: 113–121.
2 Xiao, M., Liang, D., and Shen, H. (2016). *Procedia Engineering* 135: 243–247.
3 Sener, A.A. and Demirhan, E. (2008). *Materials and Design* 29: 1376–1379.
4 Guo, Y., Xue, Y., Zuo, X. et al. (2017). *Polymer Degradation and Stability* 144: 155–166.
5 Liu, L., Xu, Y., He, Y. et al. (2019). *Polymer Degradation and Stability* 167: 146–156.
6 Wang, Z., Hu, Y., Gui, Z., and Zong, R. (2003). *Polymer Testing* 22: 533–538.
7 Guo-hui, W., Man-li, H., Fan-chao, C. et al. (2019). *Progress in Nuclear Energy* 112: 225–232.
8 Meinier, R., Sonnier, R., Zavaleta, P. et al. (2018). *Journal of Hazardous Materials* 342: 306–316.
9 Wang, Z., Qu, B., Fan, W., and Huang, P. (2001). *Journal of Applied Polymer Science* 81: 206–214.
10 Xu, Z., Yan, L., and Chen, L. (2016). *Procedia Engineering* 135: 631–636.
11 Lenża, J., Merkel, K., and Rydarowski, H. (2012). *Polymer Degradation and Stability* 97: 2581–2593.
12 Pan, Y., Han, L., Guo, Z., and Fang, Z. (2017). *Journal of Applied Polymer Science* 134 (9): 44551.
13 Kim, S. (2003). *Journal of Polymer Science Part B: Polymer Physics* 41: 936–944.
14 Khanal, S., Zhang, W., Ahmed, S. et al. (2018). *Composites Part A: Applied Science and Manufacturing* 112: 444–451.
15 Henderson, A.M. (1993). *IEEE Electrical Insulation Magazine* 9: 30–38.
16 Rothon, R.N. and Hornsby, P.R. (1996). *Polymer Degradation and Stability* 54: 383–385.
17 Camino, G., Maffezzoli, A., Braglia, M. et al. (2001). *Polymer Degradation and Stability* 74: 457–464.
18 Zhang, G., Peng, D., Zhang, M., and Qu, B. (2007). *Polymer Degradation and Stability* 92: 1715–1720.
19 Fu, M. and Qu, B. (2004). *Polymer Degradation and Stability* 85: 633–639.
20 Lei, Y., Miao, Y., Hong, Y. et al. (2013). *Polymer Degradation and Stability* 98: 868–874.
21 Hoffendahl, C., Duquesne, S., Fontaine, G. et al. (2015). *Polymer Degradation and Stability* 113: 168–179.
22 Cross, M.S., Cusack, P.A., and Hornsby, P.R. (2003). *Polymer Degradation and Stability* 79: 309–318.
23 Yen, Y.-Y., Wang, H.-T., and Guo, W.-J. (2012). *Polymer Degradation and Stability* 97: 863–869.
24 Zhang, Y., Hu, Y., Song, L. et al. (2010). *Polymers for Advanced Technologies* 19: 960–966.
25 Ye, L., Wu, Q., and Qu, B. (2009). *Polymer Degradation and Stability* 94: 751–756.
26 Carpentier, F., Bourbigot, S., Le Bras, M. et al. (2000). *Polymer Degradation and Stability* 69: 83–92.

27 Lv, J.-P. and Liu, W.-H. (2007). *Journal of Applied Polymer Science* 105: 333–340.

28 Li, Z. and Qu, B. (2003). *Polymer Degradation and Stability* 81: 401–408.

29 Wang, X., Pang, H., Chen, W. et al. (2014). *ACS Applied Materials & Interfaces* 6 (10): 7223–7235.

30 Yiliang, W., Xiaomei, Y., Hui, P. et al. (2016). *ACS Applied Materials & Interfaces* 8 (15): 9925–9935.

31 Moradkhani, G., Fasihi, M., Parpaite, T. et al. (2020). *Journal of Materials Research and Technology* 9: 7341–7353.

32 Wang, B., Tang, Q., Hong, N. et al. (2011). *ACS Applied Materials & Interfaces* 3: 3754–3761.

33 Wang, L., Xu, M., Shi, B., and Li, B. (2016). *Applied Sciences* 6: 255.

34 Xu, B., Ma, W., Wu, X. et al. (2018). *Materials Research Express* 5 (4): 045309.

35 Feng, C., Liang, M., Chen, W. et al. (2015). *Journal of Analytical and Applied Pyrolysis* 113: 266–273.

36 Dong, L.-P., Deng, C., Li, R.-M. et al. (2016). *RSC Advances* 6: 30436–30444.

37 Wang, D.Y., Cai, X.X., Qu, M.H. et al. (2008). *Polymer Degradation and Stability* 93: 2186–2192.

38 Li, L., Qian, Y., and Jiao, C. (2014). *Polymer Engineering & Science* 54: 766–776.

39 Wang, B., Wang, X., Gang, T. et al. (2012). *Composites Science and Technology* 72: 1042–1048.

40 Zhang, Y., Wang, B., Sheng, H. et al. (2016). *RSC Advances* https://doi.org/10.1039/C6RA15314C.

41 Wu, K. and Wang, Z. (2008). *Journal of Macromolecular Science, Part D: Reviews in Polymer Processing* 47: 247–254.

42 Jia, P., Zhang, M., Hu, L. et al. (2015). *RSC Advances* https://doi.org/10.1039/C5RA10509A.

43 Qu, H., Wu, W., Zheng, Y. et al. (2011). *Fire Safety Journal* 46: 462–467.

44 Tian, C., Hai, W., Liu, X. et al. (2010). *Journal of Applied Polymer Science* 89: 3137–3142.

45 Baggaley, R.G., Hornsby, P., Yahya, R. et al. (1997). *Fire and Materials*.

46 Qu, H., Wu, W., Xie, J., and Xu, J. (2009). *Fire and Materials*.

47 Wu, W., Qu, H., Zhi, L., and Yuan, H. (2010). *Journal of Vinyl & Additive Technology* 14: 10–15.

48 Trempont, C. and Wang, D.Y. (2016). *Chemical Engineering Journal* 295: 451–460.

49 Pan, Y.T., Wang, X., Li, Z., and Wang, D.Y. (2017). *Composites Part B Engineering* 110: 46–55.

50 Jia, P., Hu, L., Feng, G. et al. (2017). *RSC Advances* 7: 897–903.

51 Jia, P., Ma, Y., Zhang, M. et al. (2019). *Scientific Reports* 9 (1): 1–8.

52 Li, Z., Expósito, D.F., González, A.J., and Wang, D.Y. (2017). *Fire and Materials* 42 (3): 286–295.

53 Neuse, E.W., Woodhouse, J.R., Montaudo, G., and Puglisi, C. (1988). *Applied Organometallic Chemistry* 2.

54 Lubin, L., Yue, X., Sheng, L. et al. (2019). *Composites Part B Engineering* 176: 107218.

55 Wang, Q., Undrell, J.P., Gao, Y. et al. (2013). *Macromolecules* 46: 6145–6150.

56 Bourbigot, S., Turf, T., Bellayer, S., and Duquesne, S. (2009). *Polymer Degradation and Stability* 94: 1230–1237.

57 Ji, Y., Kim, J., and Bae, J.Y. (2010). *Journal of Applied Polymer Science* 102: 721–728.

58 Isitman, N.A. and Kaynak, R. (2010). *Polymer Degradation and Stability* 95: 1759–1768.

59 Liu, J., Yu, Z., Chang, H. et al. (2014). *Polymer Degradation and Stability* 103: 83–95.

60 Jian, R.K., Chen, L., Chen, S.Y. et al. (2014). *Polymer Degradation and Stability* 109: 184–193.

61 Yin, H.-Q., Yuan, D., and Cai, X.-F. (2013). *Polymer Degradation and Stability* 98: 288–296.

62 Haurie, L., Formosa, J., and Velasco, J.I. (2018). *Polymer Degradation and Stability* 155: 208–219.

63 Wu, N. and Li, X. (2014). *Polymer Degradation and Stability* 105: 265–276.

64 Jian, R.K., Chen, L., Zhao, B. et al. (2014). *Industrial and Engineering Chemistry Research* 53: 2299–2307.

65 Tong, L., Xu, Z., Fang, Z. et al. (2007). *Polymer Degradation and Stability* 92 (4): 720–726.

66 Hu, X., Guo, Y., Li, C. et al. (2012). *Polymer Degradation and Stability* 97: 1772–1778.

67 Yoon, K., Kim, J., Bae, J. et al. (2003). *Polymer Degradation and Stability* 81 (1): 173–179.

68 Levchik, S.V. and Weil, E.D. (2005). *Polymer International* 54: 981–998.

69 Jian, R.K., Li, C., Zhi, H., and Wang, Y.Z. (2011). *Journal of Applied Polymer Science* 123: 2867–2874.

70 Pawlowski, K.H. and Schartel, B. (2010). *Polymer International* 56: 1404–1414.

71 Wei, P., Tian, G., Yu, H., and Qian, Y. (2013). *Polymer Degradation and Stability* 98: 1022–1029.

72 Murashko, E.A., Levchik, G.F., Levchik, S.V. et al. (1998). *Journal of Fire Sciences* 16: 278–296.

73 Pawlowski, K.H., Schartel, B., Fichera, M.A., and Jäger, C. (2010). *Thermochimica Acta* 498: 92–99.

74 Pawlowski, K.H. and Schartel, B. (2008). *Polymer Degradation and Stability* 93: 657–667.

75 Hou, S., Zhang, Y.J., and Jiang, P. (2016). *Polymer Degradation and Stability* 130: 165–172.

76 Li, Z. and Yang, R. (2015). *Polymer Degradation and Stability* 116: 81–87.

77 Acquasanta, F., Berti, C., Colonna, M. et al. (2011). *Polymer Degradation and Stability* 96: 566–573.

78 Levchik, S.V. and Weil, E.D. (2010). *Polymer International* 54: 11–35.

79 Brehme, S., Schartel, B., Goebbels, J. et al. (2011). *Polymer Degradation and Stability* 96: 875–884.

80 Sullalti, S., Colonna, M., Berti, C. et al. (2012). *Polymer Degradation and Stability* 97: 566–572.

81 Gallo, E., Schartel, B., Braun, U. et al. (2010). *Polymers for Advanced Technologies* 22 (11): 2382–2391.

82 Gallo, E., Braun, U., Schartel, B. et al. (2009). *Polymer Degradation and Stability* 94: 1245–1253.

83 Bakirtzis, D., Ramani, A., Delichatsios, M.A., and Zhang, J. (2009). *Fire Safety Journal* 44: 1023–1029.

84 Han, X., Zhao, J., Liu, S., and Yuan, Y. (2014). *RSC Advances* 4: 16551–16560.

85 Acquasanta, F., Berti, C., Colonna, M. et al. (2011). *Polymer Degradation and Stability* 96: 2098–2103.

86 Köppl, T., Brehme, S., Wolff-Fabris, F. et al. (2012). *Journal of Applied Polymer Science* 124 (1): 9–18.

87 Ulrike, B. and Bernhard, S. (2008). *Macromolecular Materials and Engineering* 293 (3): 206–217.

88 Braun, U., Bahr, H., Sturm, H., and Schartel, B. (2010). *Polymers for Advanced Technologies* 19 (6): 680–692.

89 Chen, L., Luo, Y., Hu, Z. et al. (2012). *Polymer Degradation and Stability* 97: 158–165.

90 Yang, W., Song, L., Hu, Y. et al. (2011). *Composites Part B Engineering* 42: 1057–1065.

91 Yang, W., Zhang, Y.R., Yuen, C.Y. et al. (2017). *Chemical Engineering Journal* 321: 257–267.

5

Polymer Composites for Power Cable Insulation

Yoitsu Sekiguchi

Sumitomo Electric Industries, Ltd., Osaka, Japan

5.1 Introduction

In recent years, growing awareness of global environmental conservation has increased the demand for peak shift through inter-grid connections between nations or islands, and the introduction of renewable energy raises more necessity of peak shift and remote connections between power generation facilities and power demand areas. There is a lot of market interest in high-voltage direct current (HVDC) technology, which is advantageous for long-distance power transmission. HVDC power cables are playing an important role in this to connect HVDC generation facilities and the existing power network efficiently. Solid insulation power cables in particular are attracting attention worldwide, and various insulation materials have been proposed. Though crosslinked polyethylene (XLPE) cable is widely used for alternative current (AC) transmission, if attempts are made to adapt AC-XLPE cables for direct current (DC) transmission as is significant, electric field enhancement by space charge accumulation may occur, then breakdown would takes place especially during surge voltage in opposite polarity and polarity reversal of the cable system in case of line commutated converter (LCC) system. Therefore, it is desired to innovate and commercially realize insulating material for DC transmission to suppress the accumulation of space charge, and various studies have been conducted so far.

Insulation materials for power cables are classified in Table 5.1 [1]. Insulation materials for DC cables are being investigated and proposed in all areas from Ⓐ to Ⓓ (e.g. Ⓐ [2], Ⓑ [3], Ⓒ [4], and Ⓓ [5], respectively). Cable manufacturers are competing with each other to develop better materials in the field, and many researchers are taking interest in them as an academic research theme. Since this chapter deals with nanocomposite materials, we will mainly deal with materials described as Ⓑ and Ⓓ in Table 5.1; materials in Ⓐ and Ⓑ are also described for comparison with that in Ⓒ and Ⓓ as required.

Polymer Composites for Electrical Engineering, First Edition. Edited by Xingyi Huang and Toshikatsu Tanaka.

Table 5.1 Technical trend of insulation materials for power cable.

	Non-crosslinked	Crosslinked
Without fillers	Ⓒ Blends and copolymers of PP	Ⓐ XLPE
With fillers	Ⓓ PP nanocomposites	Ⓑ XLPE nanocomposites

Source: Sekiguchi [1]. © IEEJ.

An essential part of the story of solid insulated power cables is the development of polymerization technology. Various kinds of polymers were manufactured and improved corresponding to the rapid development of the chemical industry following World War II. At first, low-density polyethylene (LDPE) and butyl rubber, which have excellent electrical insulation properties, were used as insulators for solid insulated power cables. From a historical viewpoint, the polymer was merely coating the conductor in early practice, but the heat resistance of the cables improved drastically after crosslinking of polyethylene was invented and the process contributed to subsequent development of power cable technology. In addition, polypropylene (PP) is known to be a good insulating material with high heat resistance, but conventional PP is not suitable for cable insulation because of its large spherulite growth and lack of flexibility. In recent years, improvements in catalytic technology have led to the development of copolymers with fewer impurities such as catalyst residues and copolymerization with ethylene and other vinyl monomers that provides flexibility while maintaining a high melting point. A number of such PP-based materials have been introduced to the market. Non-crosslinked materials in area Ⓒ and Ⓓ of Table 5.1 are such PP-based materials.

Another key technology that is indispensable to nanocomposite materials is a blending technology. To provide properties that polymers alone lack, additives such as antioxidants, softeners, and ultraviolet (UV) absorbents are added to the insulating material. Fillers are another such additive and are often used to improve the mechanical strength of polymers. Nanofillers used in nanocomposites are fillers with nanosize particles but are characterized by the addition of very small amounts of filler compared to conventional composite materials with larger fillers. The idea to use a nanocomposite as a dielectric and insulating material was a milestone proposed by J. Lewis in the famous article "Nanometric Dielectrics" in 1994 [6]. Though the concept was not clear, an investigation to improve the DC properties of cable by the addition of a fine inorganic filler was considered experimentally in Japan around the same time [7]. The combination of these earlier experimental investigations and Lewis's concept has led to the discovery of the superior properties of nanocomposite insulation, especially as an insulator for DC cables, and a number of nanocomposite materials are now being investigated.

In *IEEE Transactions on Dielectrics and Electrical Insulation*, four times of special features related to the nanodielectrics [8–11] and twice of that related to insulation for HVDC cables [12, 13] are carried out and many articles in the field were gathered. The fact indicates that many researchers are highly concerned with this field.

5.2 Trend in Nanocomposite Materials for Cable Insulation

Many excellent books and review articles about nanocomposite materials for cable application have been released. Some of the books and review articles published in the past 10 years in which insulation and dielectric properties are discussed are summarized in Table 5.2.

5.2.1 Overview

According to these books and review articles, trend in nanocomposite study occurred in roughly three stages. In the first stage, the addition of various kinds of nanofillers to various kinds of polymers was widely studied, and the electrical properties were investigated up to early 2000. In the second stage, several attempts were made to theoretically explain the improvement in electrical properties due to the addition of nanofillers from the experimental data. And in the third stage, we are looking ahead to concrete targets, candidates for insulation materials that are available for practical use and theoretical studies on how the nanocomposites get the superior properties.

Following this trend, we first refer to the polymers and fillers used as raw materials. Polymers are divided into two groups, in which one is crosslinked polymers and the other is non-crosslinked polymers. Polyethylene is a typical polymer in the crosslinked polymer group, and PP in the non-crosslinked group. Fillers are divided into metal oxide and nitride, nanocarbons, and molecular-type cage complexes such as silsesquioxanes. After the description of raw materials, their nanocomposites will be described.

5.2.2 Polymer Materials as Matrix Resin

The varieties of polyolefins that are used as matrix resins of nanocomposites and their typical properties are summarized in Table 5.3, and their schematic structures are shown in Figure 5.1. LDPE is used as an insulator for electrical power cables as an XLPE. LDPE is divided, according to polymerization technique, into high-pressurized LDPE, which is synthesized by causing a radical reaction of ethylene monomer in a high-pressure tubular reactor or a vessel reactor, and linear low-density polyethylene (LLDPE) in which ethylene monomer and α-olefin are copolymerized using catalysts (Ziegler–Natta or metallocene catalysts). High-pressurized LDPE has many branches in the molecule, while the structure of LLDPE is linear and there are fewer and shorter branches than in high-pressurized LDPE. High-density polyethylene (HDPE) is a polyethylene polymerized with ethylene monomer alone using a similar catalyst, which is also characterized by a linear growth of the molecular chain.

In the category of PP, conventional homo-PP that is obtained by polymerization of propylene monomer only using Ziegler–Natta or metallocene catalyst is well known. Though homo-PP has high crystallinity and superior electrical properties, because of its poor flexibility, homo-PP was not applied to the cable insulation. In recent years, advances in catalytic technology and copolymerization techniques have led to the development of a number of new PP-based elastomers by copolymerization of propylene and ethylene monomers that retain the heat resistance of PP while adding the flexibility of polyethylene. These PP

Table 5.2 List of books and reviews about nanocomposites.

Title	Authors	Year	Reference
Nanocomposite science and technology	Pulickel M. Ajayan, Linda S. Schadler, and Paul V. Braun	2003	[14]
Dielectric polymer nanocomposites	J.K. Nelson	2010	[15]
A review on the importance of nanocomposite processing to enhance electrical insulation	Christopher Calebrese, Le Hui, Linda S. Schadler, and J. Keith Nelson	2011	[16]
Extruded cables for high-voltage direct-current transmission – advances in research and development	Giovanni Mazzanti and Massimo Marzinotto	2013	[17]
Nanoparticles – nanocomposites – nanomaterials An introduction for beginners	Dieter Vollath	2013	[18]
Advanced nanodielectrics – fundamental and application	IEEJ Investigation Committee on Advanced Polymer Nanocomposite Dielectrics	2014	[19]
Review of high voltage direct current cables	George Chen, Miao Hao, Zhiqiang Xu, Alun Vaughan, Junzheng Cao, and Haitian Wang	2015	[20]
A review on nanocomposite based electrical insulations	Ashish S. Paramane and K. Sathish Kumar	2016	[21]
Polymer nanocomposites – electrical and thermal properties	Edited by Xingyi Huang and Chunyi Zhi	2016	[22]
Polymeric insulation materials for HVDC cables: development, challenges and future perspective	Yao Zhou, Simin Peng, Jun Hu, and Jinliang He	2017	[23]
Tailoring of nanocomposite dielectrics	Edited by Toshikatsu Tanaka and Alun S. Vaughan	2017	[24]
Polymeric insulation for high-voltage DC extruded cables: challenges and development directions	Zhonglei Li and Boxue Du	2018	[25]
Past and future on nanodielectrics	Shaolong Zhong, Zhimin Dang, Wenying Zhou, and Huiwu Cai	2018	[26]
A review of thermoplastic elastomeric nanocomposites for high voltage insulation applications	Nurul Hidayah Ismail and Mariatti Mustapha	2018	[27]
Polypropylene nanocomposite for power equipment: a review	Lu Cheng, Xiaohong Chi, Chenyu Yan, Dongli Xie, Xiaowei Liu, Yinian Wen, Wenfeng Liu, and Shengtao Li	2018	[28]
Polyethylene nanocomposites for power cable insulations	Ilona Pleşa, Petru V. Notingher, Christina Stancu, Frank Wiesbrock, and Sandra Schlögl	2019	[29]
Polymer nanodielectrics – short history and future perspective	L.S. Schadler and J.K. Nelson	2020	[30]

Table 5.3 Typical properties of polyethylene and polypropylene.

	Polyethylene (PE)			Polypropylene (PP)			
				Homopolymer	Copolymer with ethylene		Thermoplastic polyolefin
	LDPE	HDPE	XLPE	Isotactic PP	Random PP	Block PP	TPO, POE
Density (kg/m^3)	910–925	941–965	~920	900–910			~880
Melting point (°C)	108–126	126–136	—	164–170	135–150	160–165	
Tensile strength (MPa)	0.4–1.6	2.1–3.9		3.0–3.9	<iPP	<iPP	<iPP
Elongation (%)	90–800	20–1300	>350	200–700			
Volume resistivity (Ω·cm)	$>10^{16}$	$>10^{16}$	$>10^{15}$	$>10^{16}$			
Relative permittivity	2.25–2.35	2.30–2.35	2.3	2.2–2.C6			

Source: Based on [32], Edited by Weber [33]. © John Wiley & Sons.

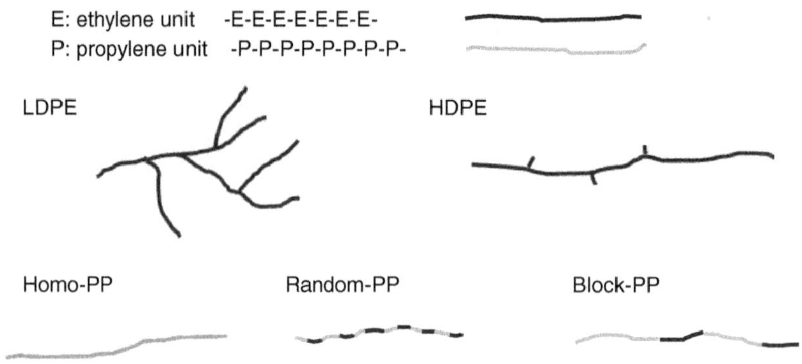

Figure 5.1 Schematic image of polyolefins.

Table 5.4 Properties of metal oxides and nitrides.

	Metal oxide					Metal nitride	
	SiO_2	TiO_2	ZnO	Al_2O_3	MgO	BN	AlN
Molar mass (g/mol)	60.1	79.9	81.4	102.0	40.3	24.8	41.0
Density (g/cm^3)[a]	2.65	4.26	5.61	3.98	3.58	3.48	3.26
Relative permittivity[b]	3.5–4.5	8	4.0	9.5	9.8	3.6–4.2	8.8
Thermal conductivity at 300 K[a] (W/mK)	6.2	7.4 (rutile)	30	46	59	760	320
Bandgap (eV)[a]	8.4	3.5 (rutile)	3.35	9.9	7.8	7.5	6.23
M-X distance (Å)			1.98			1.57	

[a] *Source:* Ref. [33].
[b] *Source:* Ref. [34].

elastomers are categorized, according to the formation of the molecular sequence in the copolymer, into random copolymer (rPP, ethylene monomer and a propylene monomer copolymerize at random) and block copolymer (bPP, ethylene sequence and propylene sequence each form a block in the copolymer). These PP-based polymers and related technologies are described in detail, e.g. in Ref. [31]. Their features and typical properties are summarized in Table 5.3.

Also, these polymers are often used in blends of two or more types.

5.2.3 Fillers

Fillers that are examined as nanocomposite insulation materials for cable use are shown in Table 5.4. They are classified roughly into metal oxide, metal nitride, and nanocarbon materials. To get an overview of the materials, the characteristics of each in the bulk state are summarized in Table 5.4. As the characteristics given here are those of crystals, they may differ from those of nano-sized particles.

Considering the materials, metal oxides and nitrides are so-called semiconducting and insulating materials. The synthesis of these nanoparticles is performed by sol–gel and gas-phase methods (Physical and chemical vapor deposition [PVD and CVD] methods), although the details are left to technical publications (e.g. [18]). Here, let us consider the sense of size of a nanoparticle. Assuming a cubic model for a nanoparticle with a 50 nm edge and a bond distance between the metal and oxygen of 2 Å (0.2 nm), there are 100 repeating units of metal and oxygen unit in an edge.

Properties shown in Table 5.4 are those of materials in the solid and bulk state. Processing the materials into nanoparticles, the properties would be affected and characteristics peculiar to nanoparticles, such as larger surface area, zeta potential [35], diameter of primary particle, and aggregate, would be actualized. These properties are not intrinsic to the materials and might be changed according to the process of synthesis, surface treatment, and distribution process, and they should be evaluated individually considering the purpose of usage of nanoparticles.

In recent years, advances in the science of nanocarbons have revealed the existence of various allotropes composed of carbon atoms. In addition to conventional amorphous carbon (e.g. carbon black), nanocarbons include fullerenes, carbon nanotubes, and graphene. In the case of fullerene (C_{60}), the diameter of the molecule is around 1 nm. The structure of these nanocarbons is based on a graphite structure that consists of sp^2 hybrid orbital structures of carbon. Graphene is a monolayer of graphite, carbon nanotubes are tubes made of rolled graphene, and fullerene is a sphere of rounded graphene. The π-orbital nature of graphene is preserved in any case, but the electronic characteristics of the nanocarbons might be affected by the structure. Graphite structure is associated with conductivity due to the free movement of electrons in π orbitals, but in the case of nanotubes, conductivity varies depending on which direction the graphene is wound, and three types of structures (armchair, zigzag, and chiral) exhibit metallic to semiconducting properties [36]. In the case of molecular fullerene (C_{60}), due to the very high symmetry of its icosahedral structure, which results in highly degenerate π-molecule orbitals, its electronic structure is unique. Even within the same family of fullerenes, the properties of C_{60} differ from those of the others [37]. Thus, as there are many kinds of electronic structures and characteristics in nanocarbons, the properties of nanocomposite materials in which nanocarbons are added might be affected by the selection of nanocarbons. The properties of nanocarbons are summarized in Table 5.5.

It is well known that macromolecules with metal centers in the active site of enzymes exist in living organisms, which conducts an oxidation–reduction reaction in the enzymatic reaction. In such macromolecules, the center of the active site often form a metal cluster or a cage compound [38].

As for metal oxides, reducing the diameter of the material might lead to such cluster or cage models. Such a similarity can be understood easily when silica and silsesquioxane are compared. Both of them have a frame including a Si-O bond, but the repetition of the bond is several hundred or more in silica and one or two in silsesquioxane. In cluster or cage molecules, as chemical modification by various functional groups is possible, there is a potential to control the compatibility to the matrix polymer [39]. They might be considered as an ultimate nanocomposite material. A polyhedral oligomeric silsesquioxane (POSS) and Fe-S cluster as samples of cage compounds are shown in Figure 5.2.

Table 5.5 Properties of nanocarbons.

	Amorphous carbon	Nanocarbon			
		Fullerene[a]			
	Acetylene black	C_{60}	C_{70}	Nanotube[b]	Graphene
Molecular weight	—	720	840	—	—
Diameter (nm)	50	<1	<1	0.4–40 ϕ 1 μm–10 mm L	
Ionization potential (eV)	—	7.58	—	—	—
Electron affinity (eV)	—	2.65	—	—	—
Thermal conductivity (W/mK)	150			3000	5000
Relative permittivity		4.4	4.4		
Bandgap (eV)		1.85	1.0	0–2 (metal semiconductor)	0 (metal like)

a:[37] and b:[36].

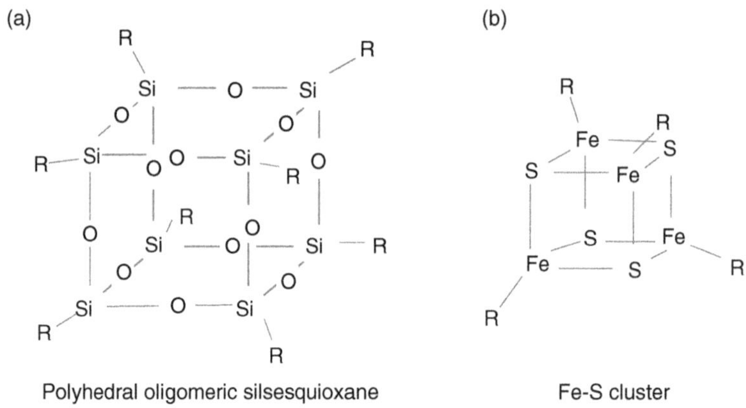

Figure 5.2 Some examples of cage complexes.

5.2.4 Nanocomposites

For the sake of improvement or as an alternative to XLPE, various kinds of nanocomposite materials are being considered and developed, especially as an insulation material for DC cables. Here, the raw matrix resins are categorized into two groups: polyethylene (LDPE, XLPE) and PP, for each of which the dielectric and insulating properties were mainly evaluated in a paper published in *IEEE Transactions on Dielectrics and Electrical Insulation*. As the unified understanding of these materials is difficult because of the difference in methods and conditions of treatment, evaluation examples have been given for consideration only.

5.2.4.1 XLPE Nanocomposites

XLPE and LDPE nanocomposites that have been used for the purpose of improvement or replacement of XLPE for cable insulation are summarized in Table 5.6. Adopted materials (base resin and filler); the type, size, and concentration of the filler's surface treatment agent; the electrical properties being evaluated; and the year of publication are summarized.

5.2.4.2 PP Nanocomposites

Nanocomposites based on PP and PP elastomers are also being explored. The reason why PP-based materials are selected can be attributed to their high melting point, which is approximately 160 °C. In the case of LDPE, the melting point of which is approximately 110 °C, crosslinking between molecular chains of polymer is necessary to add heat resistance, but in the case of PP materials, they have potential to be adopted as an insulator for power cables without crosslinking. To use the insulation materials without crosslinking might lead to improvement in the recyclability of insulation materials and to environmentally friendly material design. The examples under consideration are shown in Table 5.7. The summary is based on Table 5.6.

5.2.4.3 Nanocomposite with Cluster/Cage Molecule

Nanocomposites in which "silsesquioxane" is used are introduced in this section. Table 5.8 shows an example of polyolefin nanocomposites with POSS in the silsesquioxanes. POSS is a kind of cage complex that has a chemical formula $Si_8O_{12}R_8$, a central cage of Si_8O_{12} with eight functional groups (–R) from every Si atom, which might control compatibility and reactivity of the POSS within the matrix resin. The kinds of functional groups –R are described in Table 5.8. For POSS composite materials with resins other than polyolefins, there are review papers, for example [73].

5.2.4.4 Copolymer and Polymer Blend

In nanocomposites, additives such as inorganic fillers are actively added to polymers. Considering the higher order structure of polymers, polyethylene for example can be viewed as a kind of nanocomposite with amorphous and crystalline regions and microcrystals floating in the amorphous region. Lewis considered the conduction phenomena of polyethylene based on these ideas [74]. In the same manner, blended materials in which polymers with different crystallinity and copolymers can be regarded as "nanocomposite material" in the broadest sense. Such blend materials and copolymers are described in this section.

Table 5.9 summarizes examples of polyolefin blends and copolymers. The base polymer, the polymers to be mixed with, their concentrations, and mixing methods are described.

Polymer blending is a technique of mixing polymers that have different characteristics in order to compensate for properties that are lacking on their own. It is one of the conventional techniques to improve the characteristics of polymers. For insulating materials used under high voltage, it should be noted that the dielectric and insulating properties may change if phase separation or denaturation of the phase interface occurs. Recent research appears to be oriented toward PP-based materials to build non-crosslinked insulating systems. To add a flexibility and to keep the balance of mechanical properties and electrical properties, various kinds of PP blends with elastomers are examined. Copolymers are

Table 5.6 Reference list of LDPE and XLPE nanocomposites.

Base resin	Filler	Coupling agent	Size (nm)	Contents	Evaluated properties	Year	Reference
LDPE (XLPE)	SiO_2	KH570, TC9	30	2 wt%	DC conductivity, permittivity, space charge distribution (PEA), thermally stimulated depolarization current (TSDC)	2014	[40]
PE (XLPE)	SiO_2	TC, VI	30	2 wt%	DC-BDS, space charge distribution (PEA [−220∼−270 kV/mm])	2014	[41]
LDPE	TiO_2	KH550	20 ± 5	1, 2, 5, 10 phr	Space charge distribution (PEA [10, 20, 30, 50 kV/mm])	2015	[42]
LDPE	ZnO		30–50	0.1, 0.5, 1, 5, 7 wt%	Charging current, space charge distribution (PEA [10, 30, 50 kV/mm]), trap level distribution (isothermal discharge current)	2015	[43]
LDPE	MgO	KH550	50	0.1, 0.5, 1, 2, 5 wt%	Space charge distribution (PEA), trap level distribution, thermally stimulated current (TSC), DC-BDS	2015	[44]
XLPE	Fullerene (C_{60}), PCBM			0.1 wt%	DC-BDS		[45]
LDPE	MgO	C8, C18	10	0.1, 1, 3, 6, 9 wt%	DC conductivity	2016	[46]
LDPE	Al_2O_3	C1, C8, C18	50	1, 3, 5, 10 wt%	DC conductivity	2017	[47]
LDPE (80%), HDPE (20%)	SiO_2 (Calcined at 600 and 900 °C)		20	1, 4, 8 wt%	Dielectric spectroscopy, AC- and DC-BDS	2019	[48]
LDPE, LDPE with 10% HDPE, LDPE with 10% PP	Graphene			0.002 and 0.02 wt%	Electrical treeing, DC conductivity, space charge distribution, and trap energy distribution (PEA)	2019	[49]

LDPE	SiO_2	(Hydrophilic)	16	0.5, 1, 2, 3, 5 wt%	Space charge distribution and trap energy distribution (PEA, conduction current)	2019	[50]
PE(XLPE)	Graphene oxide			0.001, 0.01, 0.1 wt%	DC conductivity, DC-BDS, space charge distribution (PEA), trap level distribution	2020	[51]
XLPE	Surface functionalized graphene oxide	Functionalized by FAC		0.01 wt%	DC conductivity, space charge distribution (PEA), DC-BDS	2020	[52, 53]

TC and TC9: nonpolar titanate coupling agent; VI: vinyl-triethoxysilane; BDS: breakdown strength; KH570: γ-methacryroxypropyl-trimethoxy-silane; KH550: 3-amino-propyl-triethoxy-silane; C1: methyl-trimethoxysilane; C8: *n*-octyltriethoxysilane; C18: *n*-octadecyltrimethoxysilane; PAC: polycyclic aromatic compounds (4,4′-dihydroxy diphenyl ketone was used here); PCBM: 6,6-phenyl-C_{61}-butylic acid methyl ester; PEA, pulsed electro acoustic method; TSDC, thermally stimulated depolarization current; DC-BDS, breakdown strength under DC voltage.

Table 5.7 Reference list of polypropylene nanocomposites.

Base resin	Filler	Coupling agent	Size (nm)	Contents	Evaluated properties	Year	Reference
iPP	MgO		30	0.5, 1, 3, 6 wt%	AC and DC BDS, space charge distribution (PEA), permittivity, and tanδ	2016	[54]
PP/POE	ZnO		40	0.5 and 2 wt%	Space charge distribution (PEA), isothermal discharge current	2016	[55]
iPP	Fullerene		(1)	0.1–1 wt%	Space charge distribution (PEA), DC-BDS, volume resistivity, trap level distribution	2017	[56]
iPP	MgO, TiO_2, ZnO, Al_2O_3	KH570	50	1, 3, and 5 phr	Permittivity, tanδ, volume resistivity, space charge distribution (PEA), DC-BDS	2017	[57]
iPP/POE	SiO_2	Polydimethylsiloxane (PDMS), OMS, and MCLS	30	0.3, 0.6, 1.2, and 2.4 wt%	Space charge distribution (PEA), volume resistivity, DC-BDS	2017	[58]
iPP	MgO	APTMS		0.5, 1, 3, and 6 wt%	DC-BDS, space charge distribution (PEA), permittivity, and tanδ	2017	[59]
PP/styrene-ethylene-butylene-ethylene copolymer (SEBS) (35 phr)	MgO	KH550	50	0, 0.5, 1, and 1.5 phr	Space charge distribution (PEA), DC-BDS, volume resistivity, permittivity, and tanδ	2017	[60]
iPP/POE	SiO_2	KH570	30	0.5, 1, 3, and 5	DC-BDS, space charge distribution (PEA)	2018	[61]
iPP (85%) Ultra-low density polyethylene (ULDPE) (15%)	Graphene			0.005, 0.01, 0.05 wt%	Space charge distribution (PEA), trap level distribution, isothermal discharge current (IDC), DC-BDS	2018	[62]

iPP	Al_2O_3	KH570	50	1, 7, 15, and 20 wt%	Permittivity, conductivity, TSDC, DC-BDS	2019	[63]
bPP/SEBS	ZnO particle and nanowire		50 (particle), 80ϕ; 600–800 L (nanowire)	0, 0.5, 1.0, 1.5 wt%	Dielectric spectroscopy, space charge distribution (PEA), DC-BDS, electrical conductivity	2019	[64]
PP (85 wt%) ULDPE (15 wt%)	Graphene			0.01, 0.05 wt%	Space charge distribution (PEA), DC-BDS, DC conductivity	2019	[65]
iPP	MgO	MPS	50	1, 3, and 5 phr	TSDC, conductivity, space charge distribution (PEA), DC-BDS	2019	[66]

POE: polyolefin elastomer; KH550: 3-amino-propyl-triethoxy-silane; APTMS: aminopropyl-trimethoxy-silane; PDMS, polydimethylsiloxane; OMS: octyl-trimethoxy-silane; MCLS: dimethyl-dichloro-silane; KH570: γ-methacryroxypropyl-trimethoxy-silane; MPS: γ-methacryroxypropyl-trimethoxy-silane.

Table 5.8 Reference list of polyolefin-polyhedral oligomeric silsesquioxane nanocomposites.

Base resin	Filler	Function group R	Contents	Evaluated properties	Year	Reference
PP	POSS	R=methyl, isooctyl	1, 2, 3, 5, 10 wt%	AC and LI BDS, dielectric spectroscopy	2008	[67]
LDPE (by Ziegler–Natta catalysts)	POSS	R=vinyl	0.5, 1, 2, 4, 6 wt%	Dielectric spectroscopy, AC-BDS	2009	[68]
LDPE (XLPE)	POSS	R=vinyl	0.2, 0.5, 2.0 wt%	Rheological properties	2014	[69]
LDPE	POSS, MgO	R=isobutyl	1 wt% (POSS) 2, 3 wt% (MgO)	Dielectric spectroscopy, space charge distribution, thermal step method (TSM),	2017	[70]
LDPE	POSS	R=methyl, isopropyl, isooctyl	1, 5 wt%	Permittivity	2017	[71]
LDPE	POSS	R=methyl, isopropyl, isooctyl	1, 5 wt%	Permittivity, AC-BDS	2017	[72]

Table 5.9 Reference list of polyolefin blend composites and copolymers.

Base polymer/content wt%	Mixing method	Evaluated properties	Year	Reference
iPP, rPP (2.2, 3.4, 4.9, and 10% of ethylene content)		BDS	2009	[75]
Ethylene-vinyl acetate, polyethylene/20% HDPE in LDPE or ethylene-vinyl acetate copolymer (EVA)	Solution blending	AC-BDS	2010	[76]
iPP, sPP, propylene copolymer with ethylene (2, 12, 40%) or butene (12%)/blend1: 20% iPP in sPP, blend2: 50% iPP in sPP, blend3: 20% iPP in PE40, blend4: 50% iPP in PE40	Solution blending	AC-BDS	2011	[77]
HDPE (20%), LDPE (80%)		AC-BDS (film), DC-BDS (minicable)	2013	[78]
PP (Ziegler–Natta), metallocene-catalyzed propylene-ethylene copolymer (9, 12, 15 mol% of ethylene)/various mixing ratio within 0–100 wt% of copolymers	Mixing in a xylene solution, melt mixing and extrusion	BDS by film and minicable	2015	[79]
PP, POE (ethylene-octene copolymer with 25% of octane)/0–30 wt% of POE		DC BDS, volume resistivity, space charge distribution (PEA)	2015	[80]

Table 5.9 (Continued)

Base polymer/content wt%	Mixing method	Evaluated properties	Year	Reference
MAH grafted iPP	Reactive processing in a mixer at 200 °C	Space charge distribution (PEA), trap level distribution (TSDC), conduction current, BDS	2016	[81]
iPP, bPP, rPP		Space charge behavior (PEA), dielectric spectroscopy, DC-BDS	2017	[82]
iPP, POE, PBE		Isothermal surface potential decay, DC-BDS	2018	[83]
HDPE (20%), LDPE (80%)	Melt mixing	Electrical conductivity, DC-BDS, space charge distribution (PEA)	2018	[84]
XLPE with aromatic voltage stabilizers (0.5 wt%)	Bunbury mixer at 110 °C	Charging current, DC conductivity, DC-BDS, space charge distribution (PEA)	2019	[85]
bPP (9% of ethylene), rPP (2.8% of ethylene)		DC-BDS, space charge distribution (PEA), DC volume resistivity	2019	[86]
LLDPE (70%), HDPE (30%)	Melt mixing by twin screw extruder	DC conductivity, BDS	2019	[87]
Homogeneous PP, ethylene based copolymer (EBC), propylene based copolymer (PBC)/10, 20, 30, and 40 wt% of POE or PBC with homo-PP	Melt mixing at 175 °C	Leakage current, resistivity, space charge distribution (PEA)	2020	[88]
iPP, POE (ethylene-octene copolymer), PBE/0, 10, 30 wt% of elastomers with iPP	Melt mixing	Space charge distribution (PEA), trap level distribution (isothermal surface potential decay)	2020	[89]
bPP, SEBS/0, 5, 10, 15, 20, and 25 wt% of SEBS	Melt mixing	Space charge distribution (PEA), DC-BDS	2020	[90]

iPP: isotactic polypropylene; rPP: random copolymer of propylene and ethylene; bPP: block copolymer of propylene and ethylene; MAH: maleic acid anhydride; BDS: breakdown strength; PBE: propylene-based elastomer; EVA, ethylene-vinyl acetate copolymer; EBC, ethylene based copolymer; PBC, propylene based copolymer.

based on the same considerations, the difference between copolymers and blend materials being the existence of a PP segment that has high heat resistance and an ethylene segment that has high flexibility in the same molecular chain, which has the advantage of stability against long-term changes mentioned above. In addition, graft polymerization is also a method to improve the characteristics of polymers. Pioneering studies on graft polymerization for high voltage insulation materials were performed in the 1990s, in which styrene monomer or carbonyl groups were grafted to polyethylene. These studies did not result in a practical application [91, 92]. In recent years, "high-performance thermoplastic elastomer (HPTE)" is developed for a cable without curing. HPTE has been shown to be a kind of PP-based compound [93, 94]. It is a pioneering development that led to the practical use in which PP is adopted as an insulator for power cables. Since then, various studies in which PP have been used have been performed, and non-crosslinked insulation materials appear to be a major development target.

5.3 Factors Influencing Properties

Many nanocomposite materials have been studied, but inconsistency was often seen between material systems and the observed characteristics at the beginning of the development, especially in the measurement results of insulation breakdown strength. One factor that determines the characteristics of nanocomposites is the dispersion state of nanoparticles in the matrix resin. It is important to consider the dispersion of the nanoparticles from the material and processing process.

Calebrese et al. [16] focused attention on the importance of control in the process of manufacturing nanocomposites. In their work, they pointed out dispersibility, water adsorption of the nanoparticle, and their suppression as parameters to be noted.

Roy et al. [95, 96] examined the influence of surface treatment agents on nano silica on dielectric properties. XLPE nanocomposites with nano silica particles treated with triethoxyvinylsilane (TES), *n*-(2-aminoethyl) 3-aminopropyltrimethoxysilane (AEAPS), and hexamethyldisilazane were examined. Compared to XLPE without fillers and micrometer-sized silica, XLPE nanocomposite with surface treated nano silica revealed improved insulation breakdown strength and reduction of leakage current. In the composite with micro-sized silica, defects might be evoked on or around the particles, and thus insulation breakdown strength might be reduced. Surface treatment of the nanofiller could create a chemical bond between the particles and the polymer, weakening the defects or introducing deep trapping sites in the polymer due to the functional groups of the surface treatment agent.

Zhang et al. [40, 41] treated nano silica with 3-(methacryloxy) propyltrimethoxysilane, TSE, and titanate coupling agents. XLPE nanocomposites with treated and untreated nano silica were examined. Nanocomposite with untreated nano silica revealed higher electric conductivity, more space charge accumulation, and lower DC breakdown strength than that with treated nano silica, which is attributed to poor dispersion caused by incompatibility with LDPE.

There are also many studies on the effect of the type of nanoparticles. But, it is not easy to compare the data taken by different research laboratories because of inconsistency in the evaluation items and conditions. Some works in which different materials are examined by the same research laboratory are extracted here.

Zhou et al. [57] fabricated and evaluated nanocomposites using MgO, TiO_2, ZnO, and Al_2O_3 with an average particle size of 50 nm as nanoparticles, surface treated with γ-methacryloxypropyltrimethoxysilane and mixed with iPP at concentrations of 1, 3, and 5 phr. Good dispersion of nanoparticles in the matrix polymer was observed in all of the nanocomposite materials. DC electric conductivity, space charge behavior, and DC breakdown strength were compared with iPP without fillers, and improvement in these electric properties were observed in all nanocomposite materials. Some differences are shown within nanocomposites; MgO and TiO_2 nanocomposites showed much better characteristics in their performance.

5.4 Issues in Nanocomposite Insulation Materials Research

As mentioned in the previous section, many nanocomposite materials are being considered for use as cable insulators. However, the materials in practical use are very few, and materials that have been evaluated for their tolerance in actual usage sufficiently are also very few. There is some skepticism about nanocomposite insulating materials with respect to their high temperature properties and lifetime [97]. Challenges relating to the research of nanocomposite insulating materials are considered here.

In general, the features of high voltage and ultra-high voltage power cable are (i) a thick insulator and (ii) long distance. The current flows through the central conductor, which generates heat due to its transmission capacity and becomes hot. In contrast, the outside is in air or water and the ambient temperature is generally low, resulting in a temperature difference between the inside and outside. This temperature gradient might lead to a concentration of the electric field to the outer side of insulator because of the temperature dependence of resistivity of the polymer insulation material. The first issue is to control the temperature dependence and electric field dependence of resistivity.

Being a long cable reminds us of the importance of mechanical stress to the insulator under installation and operation. Stress of the cable bending, stress due to the difference of the coefficient of thermal expansion, and interface between cable insulation and accessory insulation are under consideration. Although the resistance to this combination of mechanical stresses and electrical properties characteristic of cables will eventually need to be tested with cables, it is important to refer to the literature that describes cables in detail [17, 98] to understand at least what properties are needed. Thus, to keep the electrical properties affected by mechanical stress under control is the second issue.

Focusing on DC power cables, there does not seem to be enough long-term evaluation, compared with initial features such as space charge behavior and conductive phenomena that show brilliant and lively data. As a power cable is a product that will be used continuously for a long period of time, more than 30 years, it is not possible to evaluate it for the same period; the third issue is to evaluate the properties during long-term loading in combination with the mechanical behavior mentioned above. Evaluation by cable sample is also indispensable for the issue ultimately.

Another issue that we should remember is the condition monitoring technique. Condition changing and ageing under DC voltage are not clear, and there is no established test method for them. Basic research related to condition changing and ageing is important and necessary all the more, and it should be recognized that it is one of the major issues in the development of insulating materials for DC cables. At the same time, for the evaluation of initial dielectric properties, researchers are currently setting conditions such as electric field, temperature, and time at their own discretion, making it difficult to compare the data obtained in this way. It is often seen that the scientific significance and the practical significance of cable technology do not necessarily match, and evaluations in condition that are far from usage of the cable were carried out. Of course, extensive and in-depth evaluations are necessary to understand the capabilities of the materials and the nature of the phenomena taking place, but it is time to bring such phenomenological considerations to a conclusion.

5.5 Understanding Dielectric and Insulation Phenomena

To summarize the issues in previous section, and to get a deeper understanding of the dielectric and insulation properties of nanocomposite materials, the dielectric and insulating phenomena will be discussed briefly.

5.5.1 Electromagnetic Understanding

In general, insulation material is understood as a material that does not conduct electricity that electrically separates the conductors that conduct electricity from the external environment that has a zero potential. There are so many insulation materials depending on usage. Taking power cables as an example, there are also significant differences between AC and DC cables. While it goes without saying that high dielectric strength is desirable for high voltage insulation materials, AC materials must be stable with respect to periodically varying voltage polarity, have a small dielectric tanδ, and have fewer defects that cause partial discharge. Meanwhile, materials for DC cable must have the ability to suppress the transient phenomenon of space charge accumulation and its transfer when DC voltage is applied.

Space charges are charged particles that accumulate in and migrate through a sample when a voltage is applied to it, and their morphology can be classified into several categories.

The first is the charge associated with the inflow and outflow of electrons from an electrode, such as the injection of electrons and holes, which is a concept close to semiconductor technology. However, as polymer materials that are used as main insulation material and have a "soft structure with complex movement of molecular chains" are quite different from semiconductor materials that have a "periodic and hard structure," the same theoretical explanation as semiconductor materials cannot be adopted for polymer materials.

The second is a case where the presence of foreign substances such as impurities and additives in the polymer material cannot be ignored, and these substances become carriers of charge. A typical example is the by-product of a chemical crosslinking agent.

The third one is an interface charge that is known as the "Maxwell–Wagner effect." In the interface of two different materials that have different electric conductivity and permittivity, space charges will accumulate. This is especially important for understanding the interfacial phenomena between the cable insulator and the rubber connection components used in the cable connections and the rubber stress cone at the cable end.

Equation (5.1) is derived from Gauss's law, the condition of current continuity and Ohm's law, which shows a general solution in the case that material is inhomogeneous (grad(ε/κ)≠0), where ρ is a charge density in the material of permittivity ε and electric conductivity κ and i is the current density flowing through the sample [99].

$$\rho(t) = -\frac{\varepsilon}{\kappa}\frac{d\rho(t)}{dt} + i \cdot grad\frac{\varepsilon}{\kappa} \tag{5.1}$$

In the case that the material is homogeneous (grad(ε/κ)=0), differential Eq. (5.2) is obtained.

$$\frac{\kappa}{\varepsilon}\rho + \frac{\partial}{\partial t}\rho = 0 \tag{5.2}$$

Equation (5.3) is obtained as a solution of Eq. (5.2). Equation (5.3) expresses that charges accumulated in the insulation material reduce exponentially according to relaxation time τ (=ε/κ).

$$\rho = \rho_0 \exp\left(-\frac{\kappa}{\varepsilon}t\right) \tag{5.3}$$

The first term on the right side of Eq. (5.1) shows a behavior of accumulating charges in the sample, while the second term on the right side shows the behavior of the conduction current after the charge movement is stabilized (i.e. $d\rho/dt$=0) (see Eq. [5.4]).

$$\rho(t) = i \cdot grad\frac{\varepsilon}{\kappa} \tag{5.4}$$

When considering the properties of a material as an insulating material, these equation helps to get a complete picture of the phenomenon.

5.5.2 Understanding Space Charge Behavior by *Q*(*t*) Method

Dielectric and insulating properties of cable insulation materials include dielectric constant, dielectric tanδ, electrical conductivity, space charge characteristics, and dielectric breakdown strength. Conventional characterizations of these properties are discussed elsewhere (e.g. [17]), but here we will mainly discuss space charge accumulation behavior.

Currently, the pulsed electroacoustic method (PEA method) is widely adopted as a method of measuring space charge behavior [100, 101]. PEA is an excellent method to detect positional information of charges in the thickness direction of a sample. But the subjects of measurement of the PEA method are limited to relatively slow charges, and the fact it has difficulty to detect fast charges needs to be considered. As there is much research using the PEA method, it is easy to be misled if data is only interpreted in this way.

To make up for the disadvantage of the PEA method, weak current measurements are often done by picoammeter in conjunction with it. As resistivity increases higher, current becomes smaller, and measurement of current is a battle with noise. Recently, $Q(t)$ method has been developed, and it is increasingly used as a complementary method to the PEA method. The $Q(t)$ method is a method to measure a charge amount as an integration of current by an integration capacitor connected with a sample in series, as it were, kind of ammeter by integral calculus. Usually, only conduction current is measured in ammetry, while total of currents is measured in a procedure from application of voltage to grounding in the $Q(t)$ method to grasp all phenomena related to the current exhaustively. Another advantage of the $Q(t)$ method is that noise can be reduced by integration [102].

Normally, when a square wave voltage is applied to a sample, $I_{disp}(t)$ that is a charging current corresponding to the initial rise of voltage application, $I_{abs}(t)$ that is an absorption current due to the accumulation and migration of space charge under constant voltage, and conduction current $I_{cond}(t)$ are flowing in the sample. Since the time integral of the current is the amount of charge, this relationship can be expressed as Eq. (5.5).

$$Q(t) = \int I(t)dt = \int \left[I_{\text{disp}}(t) + I_{\text{abs}}(t) + I_{\text{cond}}(t) \right] dt \tag{5.5}$$

Among them, $I_{disp}(t)$ is a charging current that flows corresponding to the initial rise of applied voltage, a product of capacitance of the sample C_s and applied voltage V. When the value is expressed as $Q(0)$, the integration of $I_{disp}(t)$ is expressed as $\int I_{disp}(t)=C_sV=Q(0)$, after all Eq. (5.5) is modified to Eq. (5.6), which suggests to us that measuring $Q(t)$ corresponds to measuring the integration of $I(t)$ and integration constant $Q(0)$ simultaneously. The $Q(t)$ method has the advantage of obtaining more information compared with ammetry.

$$Q(t) = Q(0) + \int \left[I_{\text{abs}}(t) + I_{\text{cond}}(t) \right] dt \tag{5.6}$$

Usually, these characteristics are measured by different methods according to each characteristic, but the $Q(t)$ method is unique in that it provides a comprehensive understanding of these characteristics. A short explanation of the $Q(t)$ method will be given here.

The $Q(t)$ method (direct current integrated charge method) is, as mentioned previously, a method to measure a charge amount, which is integrated in an integration capacitor connected to a sample in series, and enables the understanding of physical phenomena that are related to the current flowing as expressed in Eq. (5.6) by an integration of currents within the duration from application of voltage to grounding of the integration capacitor [102].

A schematic image of a measurement with a current integration meter ($Q[t]$ meter) is shown in Figure 5.3 (a). Here, the $Q(t)$ meter is connected to a sample in series on the ground side. When a square wave voltage V as shown in Figure 5.3 (b) is applied to the sample, to form an electrode charge $Q(0)$ ($=C_sV$, C_s is a capacitance of the sample), a charging current $I_{disp}(t)$ is flowing instantaneously with a rise of voltage. A rise of $Q(t)$ in Figure 5.3 (c) and a steep peak of $I(t)$ in Figure 5.3 (d) could be observed.

When both sides of Eq. (5.6) are divided by $Q(0)$, Eq. (5.7) is obtained. The meaning of Eq. (5.7) is that charge ratio of $Q(t)/Q(0)$ is "1" when both $I_{abs}(t)$ that is accompanied with space charge accumulation and migration and $I_{cond}(t)$ that is accompanied with conduction

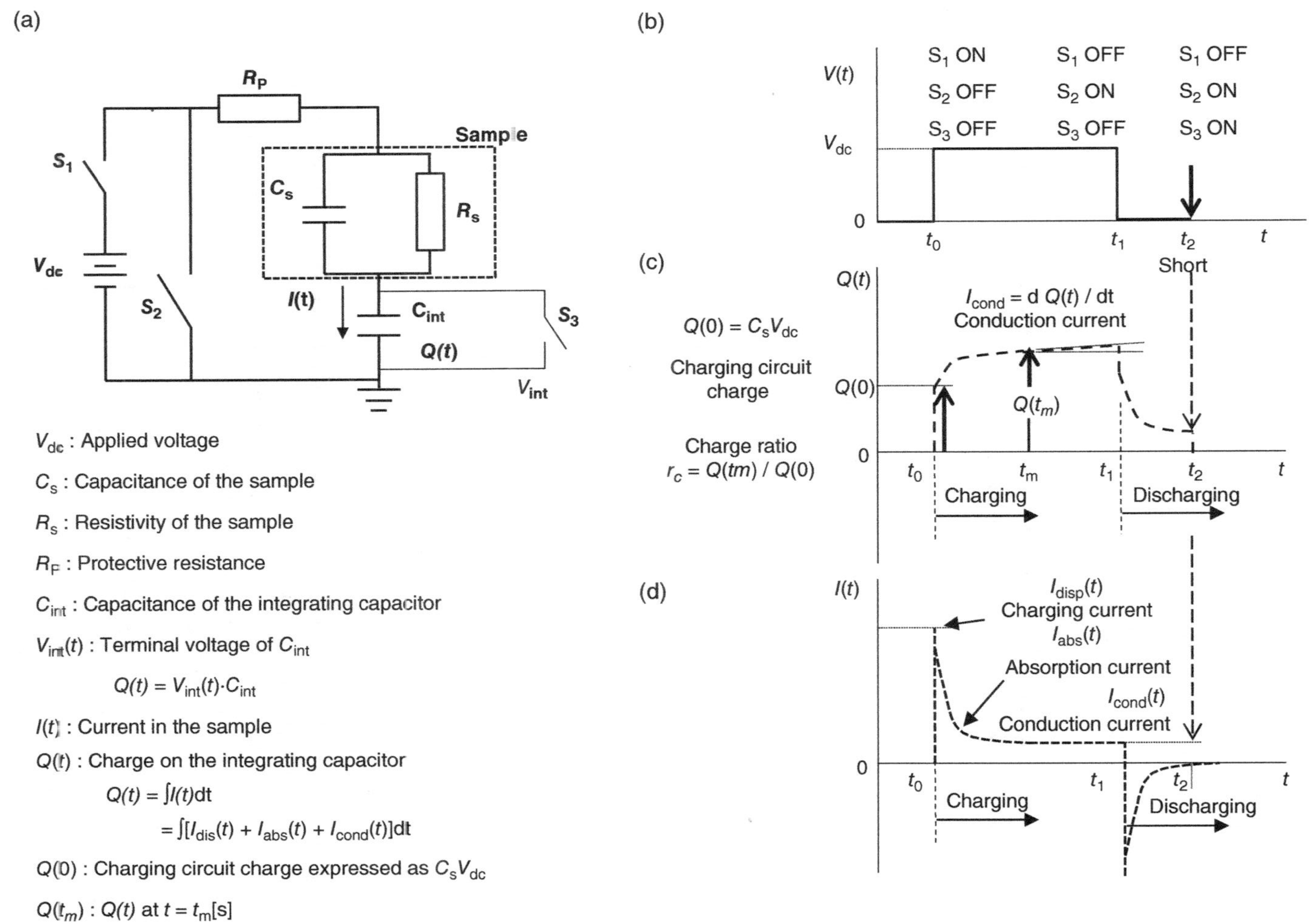

Figure 5.3 Measurement circuit and conceptual image of Q(t) and I(t). (a) Measurement circuit, (b) Applied voltage, (c) Charge Q(t) as integrated current and (d) Current I(t) *Source:* Sekiguchi et al. [102]. © Sumitomo Electric Industries, Ltd.

are not flowing, and deviates from "1" with the progression of space charge accumulation and conduction.

$$\frac{Q(t)}{Q(0)} = 1 + \frac{1}{Q(0)} \int \left[I_{\text{abs}}(t) + I_{\text{cond}}(t) \right] dt \tag{5.7}$$

Using Ohm's law and expression that current is differential of charge ($I[t] = dQ/dt$), Eq. (5.8) is obtained.

$$\frac{dQ}{dt} = \frac{\kappa}{\varepsilon} Q(0) \tag{5.8}$$

Though the original meaning of dQ/dt is a gradient of $Q(t)$ in a minute section, assuming the linearity of $Q(t)$, dQ/dt could be modified as Eq. (5.9).

$$\frac{dQ}{dt} = \frac{\Delta Q}{\Delta t} = \frac{Q(t_m) - Q(0)}{t_m} \tag{5.9}$$

Where t_m is a certain measuring time, and $Q(t_m)$ is a $Q(t)$ value when $t = t_m$.

From Eqs. (5.8) and (5.9), Eq. (5.10) is obtained.

$$\frac{Q(t_m)}{Q(0)} = 1 + \frac{\kappa}{\varepsilon} t_m \tag{5.10}$$

A time when ρ is $1/e$ in Eq. (5.3) is introduced as τ that is dielectric relaxation time, $\tau = \varepsilon/\kappa$, after all the charge ratio r_c is expressed by a simple relation of t_m and τ as shown in Eq. (5.11) [103].

$$r_c = \frac{Q(t_m)}{Q(0)} = 1 + \frac{t_m}{\tau} \tag{5.11}$$

Comparing Eq. (5.11) with Eq. (5.7), it is clear that t_m/τ, the second term on the right side of Eq. (5.11), is the ratio of changes of charges by space charge accumulation and conduction and $Q(0)$ as shown in Eq. (5.7). Though a more thorough explanation might be necessary about the relation of t_m and τ, they are omitted here. From Eq. (5.11), we can see that, in general, the following is true:

a) $r_c = 1$When t_m is small enough for τ, the right side of Eq. (5.10) is "1." Neither charge injection nor transfer has occurred.
b) $1 < r_c < 2$When t_m is getting close to τ, r_c increases gradually. The case in which space charge accumulation and migration are observed.
c) $r_c = 2$When t_m is equal to τ, the right side of Eq. (5.11) is "2." Space charges are just filled within the sample.
d) $2 < r_c$When t_m is larger than τ, charges cannot stay in the sample; t_m exceeds τ and is dominated by the conduction current flowing through the sample.

These relations will be considered taking AC-XLPE as a sample [104].

A compound of AC-XLPE is mixed by a roll mixer to homogenize sufficiently, and pressed by a hot-press machine at 120 °C as preliminary forming, then pressed at 180 °C to obtain a final form of a sample with a 0.2 mm thickness. After drying in a vacuum thermostatic

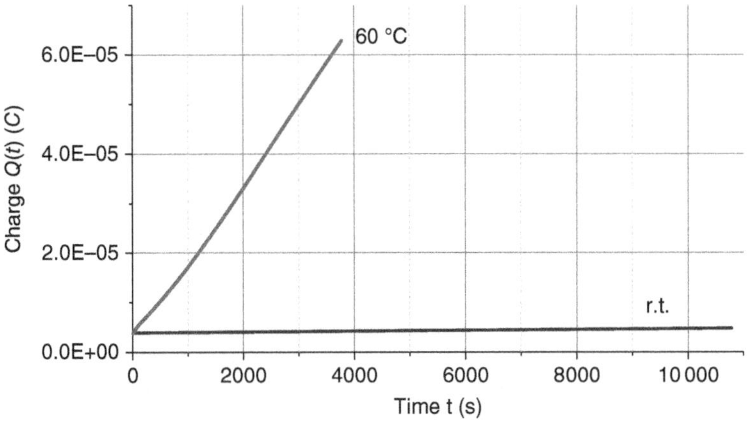

Figure 5.4 $Q(t)$ profile of AC-XLPE at room temperature and 60 °C under 60 kV/mm.

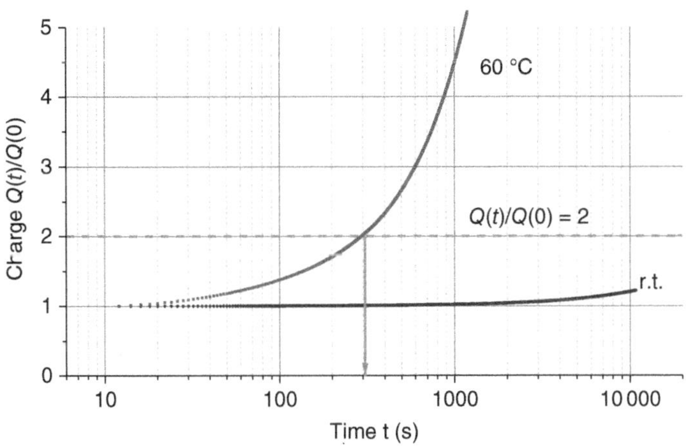

Figure 5.5 $Q(t_m)/Q(0)$ profile of AC-XLPE at room temperature and 60 °C under 60 kV/mm.

chamber at 60 °C for 24 h to remove the by-products of crosslinking agent, gold electrode is sputtered on the sample.

$Q(t)$ curves measured at room temperature and 60 °C under 60 kV/mm are shown in Figure 5.4. The gradient of $Q(t)$ is small at room temperature, while that at 60 °C is increasing rapidly. Figure 5.5 shows the relation of the charge ratio ($r_c = Q(t)/Q(0)$) with time, the x-axis is expressed in logarithmic scale. In Figure 5.5, deviation of r_c from "1" is not so large even after three-hour application of voltage at room temperature, while r_c rises rapidly at 60 °C, and reaches "2" only after 300 s from the start. Thus, under 60 kV/mm, the r_c of AC-XLPE at room temperature is in the range of "1–2," which means space charge accumulation and migration are dominant in the duration, and that at 60 °C it reaches "2" after 300 s, which means space charges are filled in the sample at 300 s and after that conduction is dominant.

References

1 Sekiguchi, Y. (2019). History and recent trend of technologies on insulation materials of transmission cable. *IEEJ Trans. Fundam. Mater.* 139 (9): 400–405. (in Japanese).

2 Farkas, A., Olsson, C.O., Dominguez, G.Z. et al. (2011). Development of high performance polymeric materials for HVDC cables, Jicable'11, A. 2.5.

3 Murata, Y., Sakamaki, M., Kazutoshi, A.B.E. et al. (2013). Development of high voltage DC-XLPE cable system. *SEI Technical Review* 76: 55–62.

4 Belli, S., Perego, G., Caimi, L. et al. (2011). P-Laser: 10000 km of installed cables based on a worldwide new technology, Jicable'11, E. 7.2.21.

5 Montanari, G.C., Seri, P., Karttunen, M. et al. (2017). Nanocomposite polypropylene for DC cables and capacitors: a new European project. *Int. Symp. Electr. Insul. Mater.* I4.

6 Lewis, T.J. (1994). Nanometric dielectrics. *IEEE Trans. Dielectr. Electr. Insul.* 1 (5): 812–825.

7 Maekawa, Y., Yamaguchi, A., Hara, M., and Sekii, Y. (1992). Research and development of XLPE insulated DC cable. *IEEJ Trans. Power Energy* 112 (10): 905–913. (in Japanese).

8 Dissado, L.A. and Fothergill, J.C. (2004). Dielectrics and nanotechnology. *IEEE Trans. Dielectr. Electr. Insul.* 11 (5): 737–738.

9 Keith Nelson, J. (2008). Nanodielectrics. *IEEE Trans. Dielectr. Electr. Insul.* 15 (1): 1.

10 Keith Nelson, J. and Schadler, L.S. (2014). Nanodielectrics – III. *IEEE Trans. Dielectr. Electr. Insul.* 21 (2): 411.

11 Huang, X. and Andritsch, T. (2020). Nanodielectrics. *IEEE Trans. Dielectr. Electr. Insul.* 27 (2): 335.

12 Cavallini, A., Montanari, G.C., and Morshuis, P. (2015). HVDC insulation and diagnostics. *IEEE Trans. Dielectr. Electr. Insul.* 22 (1): 4–5.

13 He, J. and Chen, G. (2017). Insulation materials for HVDC polymeric cables. *IEEE Trans. Dielectr. Electr. Insul.* 24 (3): 1307.

14 Ajayan, P.M., Schadler, L.S., and Braun, P.V. (2003). *Nanocomposite Science and Technology*. Wiley.

15 Nelson, J.K. (ed.) (2010). *Dielectric Polymer Nanocomposites*. Springer.

16 Calebrese, C., Le, H., Schadler, L.S., and Keith Nelson, J. (2011). A review on the importance of nanocomposite processing to enhance electrical insulation. *IEEE Trans. Dielectr. Electr. Insul.* 18 (4): 938–945.

17 Mazzanti, G. and Marzinotto, M. (2013). *Extruded Cables for High-Voltage Direct-Current Transmission – Advances in Research and Development*. Wiley.

18 Vollath, D. (2013). *Nanoparticles – Nanocomposites – Nanomaterials An introduction for Beginners*. Wiley.

19 Tanaka, T. and Imai, T. (eds.) (2017). *Advanced Nanodielectrics – Fundamental and Application*. Jenny Stanford Publishing.

20 Chen, G., Hao, M., Xu, Z. et al. (2015). Review of high voltage direct current cables. *CSEE J. Power Energy Syst.* 1 (2): 9–21.

21 Paramane, A.S. and Kumar, K.S. (2016). A review on nanocomposite based electrical insulations. *Trans. Electr. Electron. Mater.* 17 (5): 239–251.

22 Huang, X. and Zhi, C. (eds.) (2016). *Polymer Nanocomposites – Electrical and Thermal Properties*. Wiley.

23 Zhou, Y., Peng, S., Hu, J., and He, J. (2017). Polymeric insulation materials for HVDC cables: development, challenges and future perspective. *IEEE Trans. Dielectr. Electr. Insul.* 24 (3): 1308–1318.
24 Tanaka, T. and Vaughan, A.S. (eds.) (2017). *Tailoring of Nanocomposite Dielectrics*. Jenny Stanford Publishing.
25 Li, Z. and Du, B. (2018). Polymeric insulation for high-voltage DC extruded cables: challenges and development directions. *IEEE Electr. Insul. Mag.* 34 (6): 30–43.
26 Zhong, S.-L., Dang, Z.-M., Zhou, W.-Y., and Cai, H.-W. (2018). Past and future on nanodielectrics. *IET Nanodielectr.* 1: 41–47.
27 Cheng, L., Chi, X., Yan, C. et al. (2018). Polypropylene nanocomposite for power equipment: a review. *IET Nanodielectr.* 1 (2): 92–103.
28 Ismail, N.H. and Mustapha, M. (2018). A review of thermoplastic elastomeric nanocomposites for high voltage insulation applications. *Polym. Eng. Sci.*: E36–E63.
29 Pleşa, I., Noţingher, P.V., Stancu, C. et al. (2019). Polyethylene nanocomposites for power cable insulations. *Polymers* 11 (1): 1–61.
30 Schadler, L.S. and Nelson, J.K. (2020). Polymer nanodielectrics – short history and future perspective. *J. Appl. Phys.* 128: 120902.
31 Pasquini, N. (ed.) (2005). *Polypropylene Handbook*, 2e. Carl Hanser Verlage.
32 IEEJ (2013). *Electrical Engineering Handbook*, 7e. The Institute of Electrical Engineering of Japan (IEEJ) (in Japanese).
33 Weber, M.J. (ed.) (2003). *Handbook of Optical Materials*. CRC Press.
34 IEEJ DEI (2019). Technical report No. 1455: advanced tailor-made composite insulation materials – towards innovative composite materials to meet demand in electric power and electronics sectors (in Japanese).
35 Butt, H.-J., Graf, K., and Kappl, M. (2013). *Physics and Chemistry of Interfaces: Third, Revised and Enlarged Edition*. Wiley.
36 NEDO report (2016). *Carbon Nanotube*. Nikkan Publication in Japanese.
37 Shinohara, H. and Saito, Y. (1997). *Chemistry and Physics of Fullerenes*. Nagoya University Press (in Japanese).
38 Atkins, P., Overton, T., Rourke, J. et al. (2006). *Shriver & Atkins Inorganic Chemistry*, 4e. Oxford University Press.
39 Auner, N. and Weis, J. (eds.) (2000). *Organosilicon Chemistry IV*. Wiley.
40 Zhang, L., Zhou, Y., Huang, M. et al. (2014). Effect of nanoparticle surface modification on charge transport characteristics in XLPE/SiO2 nanocomposites. *IEEE Dielectr. Electr. Insul.* 21 (2): 424–433.
41 Zhang, L., Zhou, Y., Cui, X. et al. (2014). Effect of nanoparticle surface modification on breakdown and space charge behavior of XLPE/SiO2 nanocomposites. *IEEE Dielectr. Electr. Insul.* 21 (4): 1554–1564.
42 Li, S., Zhao, N., Nie, Y. et al. (2015). Space charge characteristics of LDPE nanocomposite/LDPE insulation system. *IEEE Trans. Dielectr. Electr. Insul.* 22 (1): 92–100.
43 Tian, F., Yao, J., Li, P. et al. (2015). Stepwise electric field induced charging current and its correlation with space charge formation in LDPE/ZnO nanocomposite. *IEEE Trans. Dielectr. Electr. Insul.* 23 (2): 1232–1239.
44 Peng, S., He, J., Hu, J. et al. (2015). Influence of functionalized MgO nanoparticles on electrical properties of polyethylene nanocomposites. *IEEE Trans. Dielectr. Electr. Insul.* 22 (3): 1512–1519.

45 Jarvid, M., Johansson, A., Kroon, R. et al. (2015). A new application area for fullerenes: stabilizers for power cable insulation. *Adv. Mater.* 27: 897–902.

46 Pallon, L.K.H., Hoang, A.T., Pourrahimi, A.M. et al. (2016). The impact of MgO nanoparticle interface in ultra-insulating polyethylene nanocomposites for high voltage DC cables. *J. Mater. Chem. A* 4: 8590–8601.

47 Liu, D., Hoang, A.T., Pourrahimi, A.M. et al. (2017). Influence of nanoparticle surface coating on electrical conductivity of LDPE/Al2O3 nanocomposites for HVDC cable insulation. *IEEE Trans. Dielectr. Electr. Insul.* 24 (3): 1396–1404.

48 Rahim, N.H., Lau, K.Y., Muhamad, N.A. et al. (2019). Effects of filler calcination on structure and dielectric properties of polyethylene/silica nanocomposites. *IEEE Trans. Dielctr. Electr. Insul.* 26 (1): 284–291.

49 Chen, X., Jiang, C., Hou, Y. et al. (2019). Polyethylene blends with/without graphene for potential recyclable HVDC insulation. *IEEE Trans. Dielectr. Electr. Insul.* 26 (3): 851–858.

50 Zhou, G., Wu, J., Zhang, Y. et al. (2019). Investigation of charge transport in LDPE/SiO2 nanocomposite based on the simultaneous observation of charge and current behaviour. *IEEE Trans. Dielectr. Electr. Insul.* 26 (6): 1981–1988.

51 Boxue, D., Han, C., Li, J., and Li, Z. (2020). Temperature-dependent DC conductivity and space charge distribution of XLPE/GO nanocomposites for HVDC cable insulation. *IEEE Trans. Dielectr. Electr. Insul.* 27 (2): 418–426.

52 Han, C., Du, B.X., Li, J. et al. (2020). Investigation of charge transport and breakdown properties in XLPE/GO nanocomposites part 1: role of functionalized GO quantum wells. *IEEE Trans. Dielectr. Electr. Insul.* 27 (4): 1204–1212.

53 Han, C., Du, B.X., Li, J. et al. (2020). Investigation of charge transport and breakdown properties in XLPE/GO nanocomposites part 2: effect of polarity reversal. *IEEE Trans. Dielectr. Electr. Insul.* 27 (4): 1213–1221.

54 Li, Z., Cao, W., Sheng, G. et al. (2016). Experimental study on space charge and electrical strength of MgO nano-particle/polypropylene composite. *IEEE Trans. Dielectr. Electr. Insul.* 23 (3): 1812–1819.

55 Boxue, D., Hang, X., Li, J., and Li, Z. (2016). Space charge behaviors of PP/POE/ZnO nanocomposites for HVDC cables. *IEEE Trans. Dielectr. Electr. Insul.* 23 (5): 3165–3174.

56 Dang, B., Jun, H., Zhou, Y., and He, J. (2017). Remarkably improved electrical insulating performances of lightweight polypropylene nanocomposites with fullerene. *J. Phys. D. Appl. Phys.* 50: 455303.

57 Zhou, Y., Hu, J., Dang, B., and He, J. (2017). Effect of different nanoparticles on tuning electrical properties of polypropylene nanocomposites. *IEEE Trans. Dielectr. Electr. Insul.* 24 (3): 1380–1389.

58 Diao, J., Huang, X., Jia, Q. et al. (2017). Thermoplastic isotactic polypropylene/ethylene-octene polyolefin copolymer nanocomposite for recyclable HVDC cable insulation. *IEEE Trans. Dielectr. Electr. Insul.* 24 (3): 1416–1429.

59 Cao, W., Li, Z., Sheng, G., and Jiang, X. (2017). Insulating property of polypropylene nanocomposites filled with nano-MgO of different concentration. *IEEE Trans. Dielectr. Electr. Insul.* 24 (3): 1430–1437.

60 Zha, J.-W., Wang, Y., Li, W.-K. et al. (2017). Electrical properties of polypropylene/styrene-ethylenebutylene-styrene block copolymer/MgO nanocomposites. *IEEE Trans. Dielectr. Electr. Insul.* 24 (3): 1457–1464.

61 Chi, X., Cheng, L., Liu, W. et al. (2018). Characterization of polypropylene modified by blending elastomer and nano-silica. *Materials* 11: 1321.

62 Du, B.X., Hou, Z.H., Li, J., and Li, Z.L. (2018). Effect of graphene nanoplatelets on space charge and breakdown strength of PP/ULDPE blends for HVDC cable insulation. *IEEE Trans. Dielectr. Electr. Insul.* 25 (6): 2405–2412.

63 Xie, D., Min, D., Huang, Y. et al. (2019). Classified effects on nanofillers on DC breakdown and partial discharge resistance of polypropylene/alumina nanocomposites. *IEEE Trans. Dielectr. Electr. Insul.* 26 (3): 698–705.

64 Zha, J., Qin, Q., and Dang, Z. (2019). Effect of multi-dimensional zinc oxide on electrical properties of polypropylene nanocomposites for HVDC cables. *IEEE Trans. Dielectr. Electr. Insul.* 26 (3): 868–875.

65 Du, B.X., Hou, Z.H., Li, Z.L., and Li, J. (2019). Temperature dependent space charge and breakdown strength of PP/ULDPE/Graphene nanocomposites for HVDC extruded cable insulation. *IEEE Trans. Dielectr. Electr. Insul.* 26 (3): 876–884.

66 Zhao, Y., Yuan, C., Li, C. et al. (2019). Temperature dependent electrical properties of thermoplastic polypropylene nanocomposites for HVDC cable insulation. *IEEE Trans. Dielectr. Electr. Insul.* 26 (5): 1596–1604.

67 Takala, M., Karttunen, M., Salovaara, P. et al. (2008). Dielectric properties of nanostructured polypropylene polyhedral oligomeric silsesquioxane compounds. *IEEE Trans. Dielectr. Electr. Insul.* 15 (1): 40–51.

68 Huang, X., Xie, L., Jiang, P. et al. (2009). Morphology studies and ac electrical property of low density polyethylene/octavinyl polyhedral oligomeric silsesquioxane composite dielectrics. *Eur. Polym. J.* 45: 2172–2183.

69 Wu, J., Wu, Z., Yang, H., and Zheng, Q. (2014). Crosslinking of low density polyethylene with octavinyl polyhedral oligomeric silsesquioxane as the crosslinker. *RSC Adv.* 4: 44030–44038.

70 Castellon, J., Yahyaoui, H., Guille, O. et al. (2017). Dielectric properties of POSS/LDPE and MgO/LDPE nanocomposites compounded by different techniques. *IEEE CEIDP*: 457–460.

71 Guo, M., Fréchette, M., David, É. et al. (2017). Polyethylene polyhedral oligomeric silsesquioxanes composites: electrical insulation for high voltage power cable. *IEEE Trans. Dielectr. Electr. Insul.* 24 (2): 798–807.

72 Guo, M., David, É., Fréchette, M., and Demarquette, N.R. (2017). Polyethylene/polyhedral oligomeric silsesquioxane composites: dielectric, thermal and rheological properties. *Polymer* 115: 60–69.

73 Zhou, H., Ye, Q., and Xu, J. (2017). Polyhedral oligomeric silsesquioxane-based hybrid materials and their applications. *Mater. Chem. Front.* 2017 (1): 212–230.

74 Lewis, T.J. (2014). Charge transport in polyethylene nano dielectrics. *IEEE Trans. Dielectr. Electr. Insul.* 21 (2): 497–502.

75 Hosier, I.L., Cozzarini, L., Vaughan, A.S., and Swingler, S.G. (2009). Propylene based systems for high voltage cable insulation application. *J. Phys. Conf. Ser.* 183: 012015.

76 Hosier, I.L., Vaughan, A.S., and Swingler, S.G. (2010). An investigation of the potential of ethylene vinyl acetate/polyethylene blends for use in recyclable high voltage cable insulation systems. *J. Mater. Sci.* 45: 2747–2759.

77 Hosier, I.L., Vaughan, A.S., and Swingler, S.G. (2011). An investigation of the potential of polypropylene and its blends for use in recyclable high voltage cable insulation systems. *J. Mater. Sci.* 46: 4058–4070.

78 Green, C.D., Vaughan, A.S., Stevens, G.C. et al. (2013). Recyclable power cable comprising a blend of slow-crystallized polyethylenes. *IEEE Trans. Dielectr. Electr. Insul.* 20 (1): 1–9.

79 Green, C.D., Vaughan, A.S., Stevens, G.C. et al. (2015). Thermoplastic cable insulation comprising a blend of isotactic polypropylene and a propylene-ethylene copolymer. *IEEE Trans. Dielectr. Electr. Insul.* 22 (2): 639–648.

80 Zhou, Y., He, J., Hu, J. et al. (2015). Evaluation of polypropylene/polyolefin elastomer blends for potential recyclable HVDC cable insulation applications. *IEEE Trans. Dielectr. Electr. Insul.* 22 (2): 673–681.

81 Zhou, Y., Hu, J., Dang, B., and He, J. (2016). Mechanism of highly improved electrical properties in polypropylene by chemical modification of grafting maleic ahhydride. *J. Phys. D: Appl. Phys.* 49: 415301.

82 Huang, X., Fan, Y., Zhang, J., and Jiang, P. (2017). Polypropylene based thermoplastic polymers for potential recyclable HVDC cable insulation application. *IEEE Trans. Dielectr. Electr. Insul.* 24 (3): 1446–1456.

83 Gao, Y., Li, J., Yuan, Y. et al. (2018). Trap distribution and dielectric breakdown of isotactic polypropylene/propylene based elastomer with improved flexibility for DC cable insulation. *IEEE Access* 6: 58645–58661.

84 Hosier, I.L., Vaughan, A.S., Pye, A., and Stevens, G.C. (2018). Polymer blend systems for HVDC cable applications. IEEE 2nd International Conference on Dielectrics (ICD).

85 Du, B.X., Han, C., Li, J., and Li, Z. (2019). Effect of voltage stabilizers on the space charge behavior of XLPE for HVDC cable application. *IEEE Trans. Dielectr. Electr. Insul.* 26 (1): 34–42.

86 Meng, P., Zhou, Y., Yuan, C. et al. (2019). Comparisons of different polypropylene copolymers as potential recyclable HVDC cable insulation materials. *IEEE Trans. Dielectr. Electr. Insul.* 26 (3): 674–680.

87 Zhang, K., Zhong, L., Gao, J. et al. (2019). Temperature dependence of crystalline structure and DC performance in LLDPE/HDPE blending material. *IEEE Trans. Dielectr. Electr. Insul.* 26 (3): 754–759.

88 Wu, J., Dang, B., and Hu, J. (2020). Comparison of effects of ethylene-based and propylene-based copolymer on tailoring the properties of polypropylene. *IEEE Access* 8: 123507–123513.

89 Gao, Y., Li, J., Chen, G. et al. (2020). Compatibility dependent space charge accumulation behavior of polypropylene/elastomer blend for HVDC cable insulation. *IEEE Trans. Dielectr. Electr. Insul.* 27 (3): 947–955.

90 Liang, Y., Weng, L., Zhang, W., and Li, C. (2020). Block polypropylene/styrene-butylene-styrene tri-block copolymer blends for recyclable HVDC cable insulation. *Mater. Res. Express* 7: 085301.

91 Tanaka, Y. and Ohki, Y. (1992). Morphology of ethylene-styrene copolymer and its effect on dielectric strength. *IEEE Trans. Electr. Insul.* 27 (3): 432–439.

92 Ikeda, M., Yamamoto, Y., and Kawabata, H. (2002). Development of new type polyethylene with high resistivity for power cable insulator. *IEEJ Trans. Fundam. Mater.* 122 (11): 987–995. (in Japanese).

93 Belli, S., Perego, G., Bareggi, A. et al. (2010). P-Laser: breakthrough in power cable systems. *IEEE Int. Symp. Electr. Insul.*

94 Albertini, M., Bareggi, A., Caimi, L. et al. (2015). Development and high temperature qualification og innovative 320 kV DC cable with superiorly stable insulation system. 9th International Conference on Insulated Power Cable, A7.3.

95 Roy, M., Nelson, J.K., MacCrone, R.K. et al. (2005). Polymer nanocomposite dielectrics – the role of the interface. *IEEE Trans. Dielectr. Electr. Insul.* 12 (4): 629–643.

96 Roy, M., Nelson, J.K., MacCrone, R.K., and Schadler, L.S. (2007). Candidate mechanisms controlling the electrical characteristics of silica/XLPE nanocomposites. *J. Mater. Sci.* 42: 3789–3799.

97 Montanari, G.C., Seri, P., Naderiallaf, H. et al. (2019). Adding nanofillers in polymeric insulating materials: so far so good? The case of polypropylene for DC cables. *Proceedings of 2019 Electrical Insulation Conference (EIC)*, 477–480.

98 Worzyk, T. (2012). *Submarine Power Cables: Design, Installation, Repair, Environmental Aspects*. Springer.

99 Kojima, K. and Matsuura, K. (1970). Electrical stress in oil-filled and solid type high voltage direct current cables under thermal and electrical transient conditions. *Direct Current* 1 (2): 55–72.

100 IEC Technical specification, IEC/TC 62758 (2012). Calibration of space charge measuring equipment based on the pulsed electro-acoustic (PEA) measurement principle.

101 IEEE Std. 1732-2017 (2017). IEEE recommended practice for space charge measurements on high-voltage direct-current extruded cables for rated voltage up to 550 kV, IEEE.

102 Sekiguchi, Y., Takada, T., Miyake, H., and Tanaka, Y. (2018). Evaluation of insulation materials under high voltage by direct current integrated charge method. *SEI Tech. Rev.* 87: 59–63.

103 Takada, T., Sekiguchi, Y., and Tanaka, Y. (2021). Classification if Q(t) profiles to induced electrode charge, accumulated charge and leakage charge current. *IEEJ Trans. Fundamentl. Mater.* 141(4): 245–251 (in Japanese).

104 Sekiguchi, Y., Hosomizu, K., Yamazaki, T. et al. (2020). Space charge accumulation and conduction phenomena of XLPE considered by relaxation time. Annual Conference of Fundamentals and Mater. IEEJ, 2-D-a1-5. (in Japanese).

6

Semi-conductive Polymer Composites for Power Cables

Zhonglei Li, Boxue Du, Yutong Zhao, and Tao Han

Key Laboratory of Smart Grid of Education Ministry, School of Electrical and Information Engineering, Tianjin University, Tianjin, China

6.1 Introduction

Semi-conductive polymer composites, together with metal materials and polymer insulating materials, are important materials for extruded cables. The shielding layer, made by the extruded semi-conductive polymer composites, plays significant roles of high-voltage (HV) and grounding electrodes and acts as the interfaces for both conductor and metallic shields to provide a uniform electrode [1].

6.1.1 Function of Semi-conductive Composites

For the power cables with extruded insulation, the semi-conductive composites provide a smooth, continuous, conductive, and isopotential interface between the electrodes and insulation, which eliminates the interface defects, including metal fibers, protrusions, and cavities (air gap). Without the semi-conductive layer, the contact impedance consumes energy and causes high-temperature heat spots, which will result in nonuniform potential gradient [1]. The excessive electric gradients can cause partial discharges within these defects that break the polymer molecules into low molecular weight fragments and oxidation productions, and eventually result in electrical trees [2, 3]. On the other hand, the semi-conductive polymer composites act as "potential buffer layers." The injection and extraction of electrons and holes occur at the interfaces between electrodes and insulation. The semi-conductive material can mitigate the velocity of charge carriers and absorb the shock caused by electrons passing through insulation to metal electrodes [1].

Polymer Composites for Electrical Engineering, First Edition. Edited by Xingyi Huang and Toshikatsu Tanaka.

6.1.2 Development of Semi-conductive Composites

At the beginning of the semi-conductive shield application, the cotton and nylon taps impregnated with carbon black or graphite (GT) are used in early cables insulated with butyl and oil-based rubbers. With the development of cable extrusion insulation, extruded shielding layers have been developed to replace the tape shields. The extruded semi-conductive polymer composites mainly consist of polymer matrix (including thermoplastic and thermosetting polymer matrix), conductive or semi-conductive fillers (such as carbon black commonly used), antioxidants, and other additives.

In the late 1980s, triple extrusion technology was applied to ensure that insulation, inner semi-conductive layers, and outer semi-conductive layers were continuously extruded and then crosslinked at the same time in the vulcanization tube. Compared with the extrusion layer by layer, the triple extrusion process can greatly reduce the contamination level at the interface between the insulation and semi-conductive shields, thus enhancing the long-term electrical performance of power cables.

In the 1990s, the major breakthrough was the development of super-smooth semi-conductive shields [4]. It is known that the surface protrusion of semi-conductive shields will lead to significant distortions of electric field at the insulation-shield interface, according to Mason's equation [5],

$$\delta = \frac{2d}{r \ln\left(1 + \frac{4d}{r}\right)} \tag{6.1}$$

where δ is the localized electrical stress enhancement factor, d is a protrusion point to plane distance, and r is a radius of the protrusion. The concentrated electric stress will induce the electrical degradation of cable insulation and even undesired treeing. Typical sources of the protrusions are grit from carbon black, poor dispersion of carbon black, and contamination from the environments during the storage and transportation processes.

Furnace carbon black and acetylene carbon black are conductive materials commonly used for semi-conductive polymer composites of power cables. Furnace black is produced by partial oxidation of hydrocarbon oil or gas, which has been used in conventional semi-conductive shields. Acetylene black is produced by continuous thermal decomposition of acetylene gas, which has much higher purity and less contamination than furnace black [6, 7], resulting in a smoother surface of super-smooth semi-conductive shields than that of conventional shields [8, 9]. Figure 6.1 shows the surface smoothness of conventional semi-conductive shields with furnace carbon black and the super-smooth shields with acetylene black, which were measured by the laser profilometer surface excellence evaluation device (SEED) [2]. For the super-smooth shields, there are no protrusions with a height of 40 microns and even above, and the number of 30-micron protrusions is 3–4 orders of magnitude smaller than that of the traditional shielding material. It is indicated that the surface protrusion height of the super-smooth material is significantly reduced by using acetylene black.

The super-smooth, extra clean semi-conductive shields could uniform the interface electric field and extend cable life, which has been further improved and used up to now [7]. Over the years, chemical impurities, sulfur, and ash content levels of carbon black have

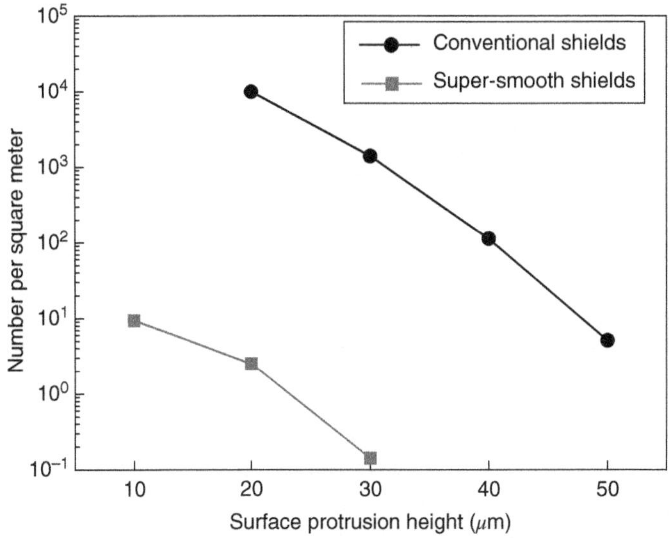

Figure 6.1 Surface smoothness of conventional semi-conductive shields with furnace carbon black and the super-smooth shields with acetylene black, which were measured by the laser profilometer SEED. *Source:* Based on [2]. © John Wiley & Sons.

been further improved for both furnace black and acetylene black. Nevertheless, acetylene black is still preferred to be used in semi-conductive composites for both high-voltage alternating-current (HVAC) and high-voltage direct current (HVDC) cables. Figure 6.2 presents the surface morphology of the semi-conductive shields (Borealis LE0592S) for HVAC cables with a rated voltage of 220 kV. The surface morphology, with an area of $1.22 \times 0.95\,\text{mm}^2$, is profiled by the three-dimensional (3D) surface profilometer based on scanning white light interferometry (Bruker Contour GT-K 3D). It is observed that the maximum of protrusion height (peak-to-peak value) is less than 5 μm, and the surface smooth has been markedly improved compared with the 1990s super-smooth shields.

6.2 Conductive Mechanism of Semi-conductive Polymer Composites

The volume resistivity of semi-conductive polymer composites is one of the key parameters for cable shields. According to the IEC-62067 standard for the HV power cable, the volume resistivity of semi-conductive layer, both before and after aging, shall not exceed 1000 Ω·m for the conductor shield (inner shield) and 500 Ω·m for insulation shield (outer shield), at a temperature within ±2 K of the maximum conductor temperature [10]. For the medium- and low-voltage cable, the volume resistivity of semi-conductive composite materials without aging shall not exceed 100 Ω·cm at 20 °C and 5000 Ω·cm at 90 °C [11]. Carbon black is known as intrinsic conductors with a conductivity of 10^{-1}~10^{2} S/cm. When the carbon black is dispersed into the polymer matrix, the conductive mechanism can be concluded as the following.

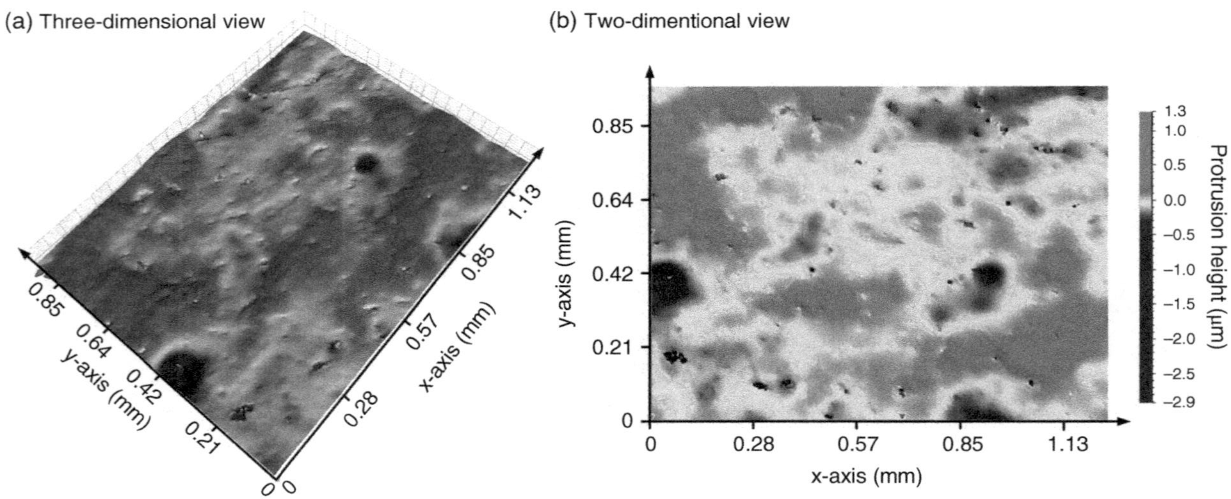

Figure 6.2 Surface morphology of the semi-conductive shields (Borealis LE0592S) for 220 kV HVAC cables tested by the 3D surface profilometer based on scanning white light interferometry (Bruker Contour GT-K 3D).

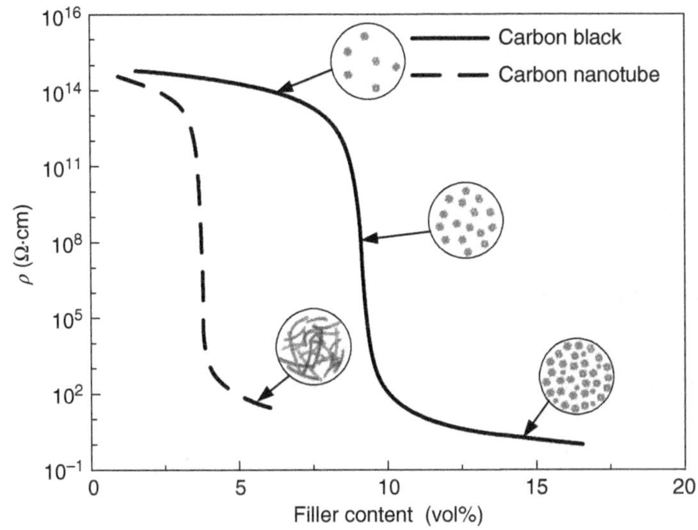

Figure 6.3 Schematic diagram of the volume resistivity varying with the conducting filler content and structure.

6.2.1 Percolation Theory

As the carbon black or other conducting fillers are added into the insulating polymer, the conducting fillers are randomly dispersed into the matrix and form distributed conductive regions, in which the charge carriers (including electrons and holes) could migrate locally under the coulomb force. However, for the composites with low filler fractions, the mean distance between conducting fillers is large [12], the conduction of composites are dominated by the hopping conduction process in the polymer matrix [13, 14].

When the filler content reaches a certain threshold, the volume resistivity (ρ) decreases by order of magnitude with increasing filler content, as shown in Figure 6.3. In this case, the fillers are in contact with each other, and charge carriers could transport along the percolation paths.

The corresponding critical value of filler content is known as the percolation threshold [15], which is presented as Φ_c. It is directly related to the formation of conductive networks throughout the composite, depending on the properties of matrix, the shape of fillers [16], and dispersion processes [17]. Generally, for the random dispersion of zero-dimensional particles, the Φ_c in volume fraction is about 16%. However, for the conductive fibers, nanotubes, and nanoplatelets with high aspect ratios, the threshold will be much less than 10%, due to the advantage in forming the conductive network structure.

6.2.2 Tunneling Conduction Theory

When the de Broglie wavelength of microparticles is close to the barrier height, which is the quantum barrier, carrier charges will tunnel through the barrier with the wave-like behavior. As the localized electric field is high enough, for instance, higher than 100 MV/m, electron flow occurs when the carbon black aggregates are in contact or separated by a distance of 100 Å or less than the percolation distance [18]. In this case, electrons tunnel through the resistive polymer from aggregate to aggregate [19].

Actually, the conduction theories mentioned above, including hopping, percolation, and tunneling conduction theory, all exist in the conductive processes. Among them, which conductive mechanism dominates depends on the shape and content of the filler and the characteristics of the polymer matrix.

6.2.3 Mechanism of Positive Temperature Coefficient

For the commonly used insulation materials, their volume resistivity characteristics show negative temperature coefficient (NTC) phenomena, indicating an increasing DC conductivity with the temperature rising. However, semi-conductive composites have positive temperature coefficient (PTC) properties under the operating temperature from 20 to 90 °C, especially near the melting point (T_m) [20]. As the ambient temperature is above the melting point, the volume resistivity turns to the NTC region. The temperature with the largest volume resistivity value is called the critical temperature ($T_{critical}$), which is close to T_m of the polymer matrix. Figure 6.4 presents a schematic diagram of the relationship between volume resistivity and temperature.

The rise in temperature causes the volume expansion of the polymer resin, which increases the distance between the neighboring aggregates of conductive fillers [21, 22], especially when the temperature is close to the melting temperature. The increased distance between filler aggregates within the polymer matrix breaks the conductive network and leads to a weakness of percolation and tunneling conductive processes of charge carriers [23]. This process is considered as the main cause of the PTC effect. It is well known that the PTC phenomenon markedly depends on the specific type of conductive fillers and

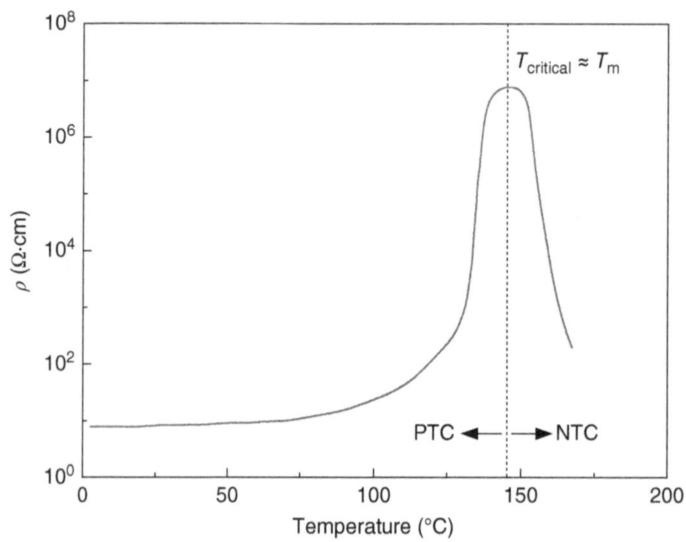

Figure 6.4 Schematic diagram of the volume resistivity varying with temperature.

characteristics of the polymers including chemical structure, crystallization, processing conditions, and thermal history to which they were exposed [24–26].

On the other hand, according to the NTC phenomenon [27], as the melting of resin crystals occurs, the conductive fillers, such as carbon black, with fluidity are able to rearrange themselves, forming new percolation and tunneling conductive networks. This phenomenon can be explained by the electrophoresis effect, which accelerates the charge carrier transport, and therefore explaining how the volume resistivity value decreases [28, 29].

6.3 Effect of Polymer Matrix on Semi-conductivity

Nowadays, the extruded insulation used for HV cables can be divided into thermoset insulation (crosslinked polyethylene, XLPE, commonly used) and thermoplastic insulation, for instance, polypropylene (PP) [30]. Correspondingly, the semi-conductive polymer composites for power cables also have thermoset and thermoplastic polymer resins, in order to be suitable for the triple extrusion of insulation and inner and outer semi-conductive shielding layers.

The most common thermoset polymers are ethylene-vinyl acetate (EVA) copolymer, ethylene-ethyl acrylate (EEA) copolymer, and ethylene-butyl acrylate (EBA) copolymer [31, 32]. The EEA copolymers bond well to XLPE insulation and are generally employed for conductor shields and bonded insulation shields. EVA copolymers are most often used as the base for strippable insulation shields and some conductor shields [23]. Compared with EVA and EEA [33], EBA copolymers have better toughness especially in the case of high filler content and a lower melt index. The copolymers, as well as the monomer content in copolymers, have significant influences on the electrical, mechanical, and other properties of semi-conductive composites.

The thermoplastic polymers for semi-conductive composites include PP, high-density polyethylene (HDPE), and other blended composites with the above copolymers. The melting temperature of the thermoplastic resin or their blends is directly related to the critical point between PTC and NTC effect as discussed in the following.

6.3.1 Thermoset Polymer Matrix

In reference [34], EVA, EEA, and EBA were used as the basic materials. The carbon black fillers with contents of 20, 30, and 40 wt% were dispersed into the polymer matrix and hot-pressed to obtain the flat specimens. Figure 6.5a shows the volume resistivity varying with the content of carbon black at 25 °C of semi-conductive shield materials. The volume resistivity decreased in the range of 15~218 Ω·cm as the carbon black content increased. Figure 6.5b shows the measurements of volume resistivity at 90 °C of semi-conductive shield materials according to the content of carbon black. The volume resistivity decreased in the range of 135~561 Ω·cm as the carbon black content increased, which presents the PTC phenomenon involving with volume expansion of the polymer resin. Just like at 25 °C, the volume resistivity decreased as the carbon black content increased, but the volume resistivity increased more dramatically than at 25 °C. Among the semi-conductive composites, the carbon black/EEA samples have lower volume resistivity with a much lower PTC

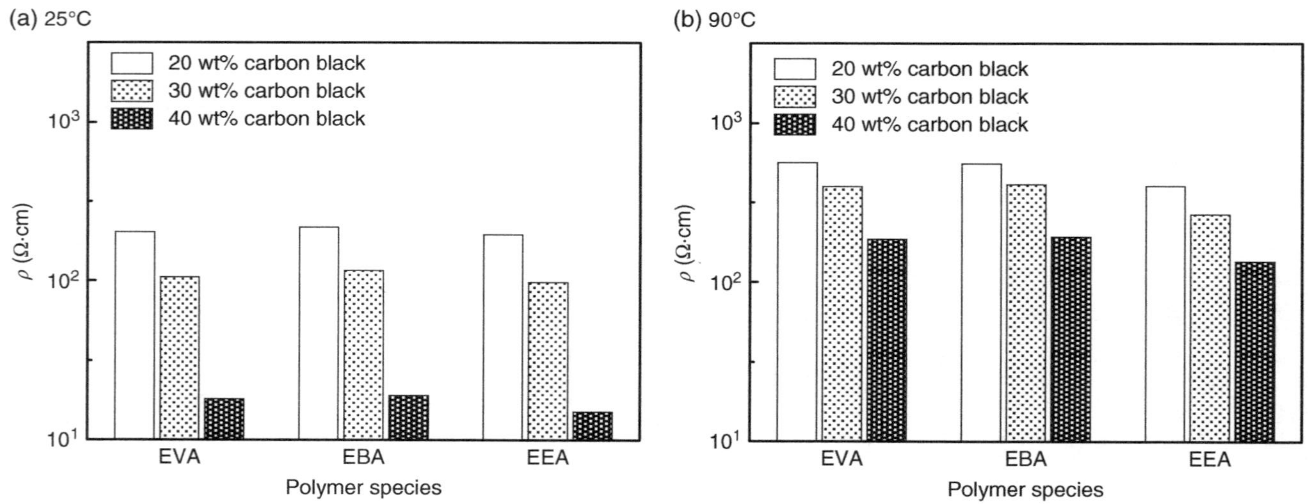

Figure 6.5 Effect of polymer species on volume resistivity of semi-conductive shields filled with 20, 30, and 40 wt% carbon black. *Source:* Based on [34]. © John Wiley & Sons.

(herein defined as the ratio of volume resistivity maximum to the room temperature resistivity), which are attributed to the better dispersion of carbon black in EEA base resin.

The copolymer is made from two or more monomers via continuous bulk polymerization or an emulsion process. Taking EVA as an example, the vinyl acetate (VA) content in EVA copolymer ranges from 1 to 50 wt%, depending on the desired mechanical and physical properties. The inclusion of higher VA content in EVA decreases the average molecular weight of EVA copolymer, which consequently changes the copolymer properties [35]. For instance, the stiffness modulus, surface hardness, crystallinity, melting point, and softening point of the EVA polymer are reduced due to the higher VA content [34, 36]. A significant effect of VA content (from 0 to 70 wt%) on the electrical and rheological properties of thermally reduced graphite oxide (TRGO)/EVA composites are reported by [37]. Figure 6.6 presents the volume resistivity of EVA-based semi-conductive composites with different VA content. The curves show that, for the composites with VA weight percent of 9.3, 18, and 28%, the volume resistivity dramatically drops as the volume fraction of TRGO exceeds 0.5%. With the increase of the TRGO content, the resistance decreases exponentially and then gradually enters the saturation region until the filler content reaches 2 vol%. The absolute values reached for the highest degrees of filling are less than 1000 Ω·cm. However, the composites with neat LDPE base resin and higher-VA monomer show lower end conductivities and a relatively flat slope in the drop region. Similar behaviors in semi-conductive composites with carbon black, carbon nanotubes, and GT as fillers are reported, proving that the VA contents play an important role in the formation of conductive paths in the polymer matrix [38]. The main reason is attributed to the influence of the polarity of the polymer matrix on the dispersion and percolating behaviors, which will be more prominent for the high specific surfaces of conductive fillers. This phenomenon discussed above is of great importance to choose the polymer matrix of semi-conductive composites for power cables.

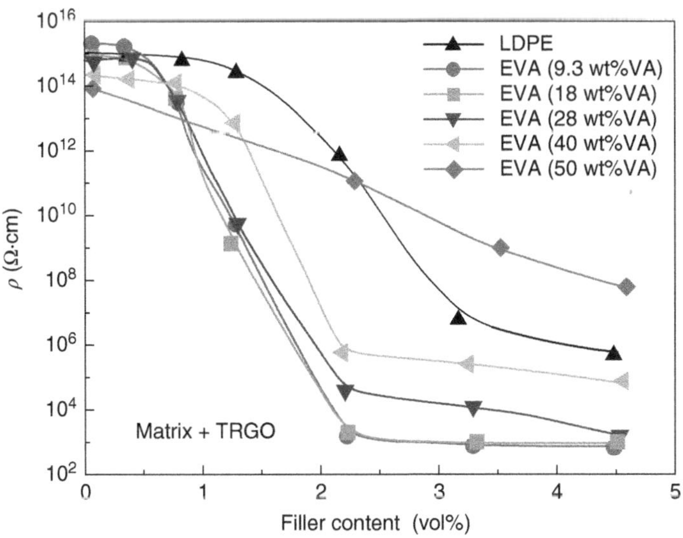

Figure 6.6 Volume resistivity of EVA-based semi-conductive composites with different VA content. *Source:* Based on [37]. © John Wiley & Sons.

6.3.2 Thermoplastic Polymer Matrix

It has been mentioned that the semi-conductive composites with a thermoplastic polymer, for instance PP or HDPE, have significant PTC and NTC effects as the temperature exceeds their melting temperature [39–41]. For the HV power cable, the long-term operating temperature is 90 °C (70 °C for some HVDC cables), but the conductor temperature will rise to 135 °C under an emergency load. In this case, the cable shields, especially the conductor shield, are subject to an annealing process of heating and slow cooling [42]. Generally, the annealing might improve crystallinity and enlarge the crystal size of the semicrystalline matrix, which might be critical for the enhancement of both stability and long-term service life [43]. Nevertheless, the annealing at temperatures above the melting temperature could cause the redispersion of filler particles in the polymer matrix, further influencing the PTC and NTC behaviors of the composites.

Song et al. studied the effect of annealing on volume resistivity of HDPE/carbon black composites [44]. The annealing processes labeled with treatment A, B, C, and D were set as follows. Treatment A was done at temperatures located in the PTC transition region yet below the melting temperature (T_m) of HDPE, after heating the sample from room temperature. Treatment B was done at temperatures located in the NTC region (above T_m), after heating the sample from room temperature. Treatment C was done at temperatures above crystallization temperature (T_c), after cooling the sample from 149 °C. Treatment D was done at three different temperatures during a heating–cooling cycle. The sample was annealed at 122 °C after heating from room temperature, at 149 °C after heating from 122 °C, and at 122 °C after cooling from 149 °C.

Figure 6.7 presents the relationship between the volume resistivity and temperature of the annealing-treated semi-conductive composites, as well as the virgin samples [44]. It is

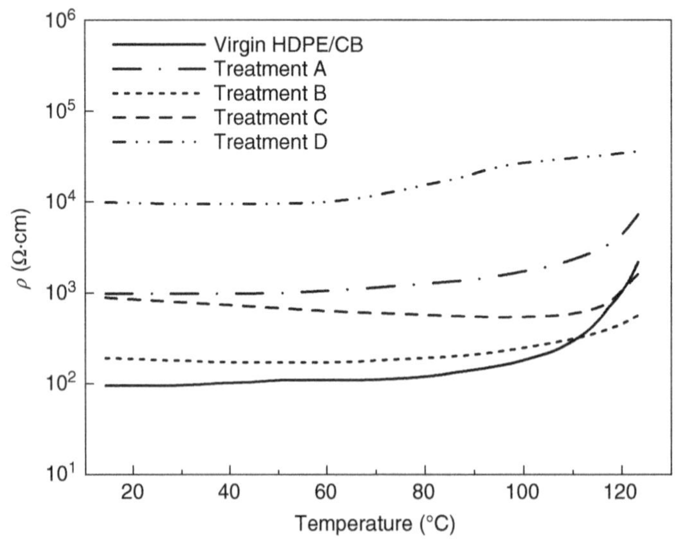

Figure 6.7 Effect of annealing-treated on volume resistivity of CB/HDPE semi-conductive composites. *Source:* Based on [44]. © John Wiley & Sons.

obvious for the samples experienced treatments A, B, and C, the resistivity exhibits an obvious PTC effect during the heating process. Besides, the resistivity at room temperature increases by about 1 order of magnitude. However, as the samples were subjected to treatment D, the room temperature resistivity enhances by 2~3 orders of magnitude followed by a much slighter PTC effect.

For the virgin sample and those treated by the annealing process of A, B, and C, the strong PTC phenomena are attributed to the expansion of HDPE matrix with the temperature rising, especially near the T_m about 125 °C. When the carbon black (CB)/HDPE composite is annealed at temperatures in the PTC region, partial melting of HDPE crystals leads to the short-distance migration of the adjacent particles. CB aggregates tend to form clusters in the amorphous phase to produce larger secondary structures, which destroys the virgin percolation paths and influences the conduction significantly [44, 45]. As the annealing temperature reaches the NTC region, the higher temperature leads to a reduced viscosity to redistribute the CB aggregates in the polymer matrix as recrystallization proceeds, thus forming the percolation networks. Accordingly, the treatment C has two competing processes including the destruction and reformation of percolation clusters, resulting in a volume resistivity between the PTC curves A and B.

The annealing treatment D causes a distinct increase in room temperature resistance. Differential scanning calorimetry (DSC) measurements were employed to evaluate the melting/crystallization temperature as well as the crystallinity. The results showed that the samples treated by A, B, and C processes have similar crystallinities with the virgin composites, varying from 58 to 62%. However, the annealing treatment D results in a significant decrease in crystallinity by 28% in comparison with the virgin composite. The distinct increase in volume resistivity behavior is mainly related to a high degree of agglomeration of CB aggregates and to the volume dilution effect due to the increase of the volume fraction of the amorphous phase in HDPE.

6.3.3 Blended Polymer Matrix

Mixing the conductive fillers in the polymer matrix in a molten state will result in a higher percolation threshold. Therefore, if the conductive particles are not uniformly separated in the polymer matrix, an extremely high concentration of conductive fillers is required to achieve long-range connectivity [46–48]. Using polymer blends as the matrix resin is a feasible method to reduce the concentration of conductive fillers in the composites. Due to the penetration effect, whether double or multiple, the critical value of the resistivity transition of the composite material may tend to a lower value. This concept was first introduced into carbon black blends [49–51] and then proved to be applicable to other carbon-based nanofillers, including carbon nanotubes, graphene, and so on. The distribution of nanoparticles in the blended matrix depends on the thermodynamics of the blend system, which is related to the particle size, shape, and chemical components [52, 53]. Different polymers will cause different interfacial migration phenomena of nanoparticles [52]. The greater binding energy between nanoparticles and polymers means that they have a stronger tendency to attract, while the particles tend to remain in this phase. If this is not the case, the nanoparticles tend to enter other phases or remain in the interface zone.

Many studies have focused on the electrical properties of semi-conductive composites with the blended polymer matrix. Xu et al. proved that the composites in which CB particles are located at the interface of two polymer components have higher electric conductivity when the content of the two polymers is near to 1:1 [54]. Cheah et al. found that the percolation threshold was dramatically reduced by the isolation of the CB particles at the continuous interface of a binary polystyrene and poly(styrene-co-acrylonitrile) matrix [55]. Kurusu et al. chose linear low-density polyethylene (LLDPE) as the second resin due to a melting point close to that of EVA, and graphene nanoplatelets (GN) were premixed with LLDPE and EVA [56]. The results indicated that the former premixed with GN presented a much lower percolation threshold (2.2 vol%) than the EVA/GN composites (10 vol%). It is proved by scanning electron microscopy (SEM) and transmission electron microscopy (TEM) that a selective location of GN within the ultra low density polyethylene (ULDPE) and the ULDPE/EVA continuous interfaces leads to a formation of conductive networks.

The relationship between the resistivity of the HDPE/EVA/CB composites with a CB content of 12 wt% and the EVA content is shown in Figure 6.8, which shows a valley-shaped curve [39]. When the EVA content is less than 10 wt%, the carbon black particles tend to migrate to the EVA phase, making the conductive network density higher. However, due to the low EVA content, a continuous EVA phase cannot be formed, and even if the density of the conductive network is higher than the percolation threshold, the complete conductivity cannot be formed. With the increase of EVA content, the HDPE/EVA co-continuous phase gradually increases, which makes a large amount of conductive network formed in the blend system. This is the reason for the sharp decrease in the resistivity of the composite material, which is called the insulator–conductor transition, the first transition of the valley-shaped curve. With the further increase of EVA content, the resistivity of the composite material does not change significantly. The migration of carbon black particles from

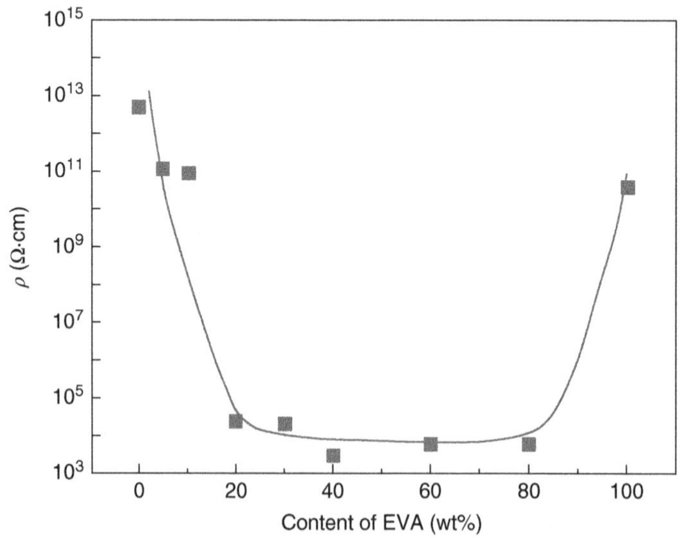

Figure 6.8 Relationships between room temperature resistivity and EVA content of HDPE/EVA/CB composites with a CB content of 12 wt%. *Source:* Based on [39]. © John Wiley & Sons.

the HDPE phase to the EVA phase compensates for the dilution of the carbon black concentration caused by the expansion of the amorphous phase. In addition, the interaction between the EVA molecules and the polar functional groups on the surface of the carbon black particles inhibits the aggregation of carbon black. When the EVA content reaches 80 wt%, there are too few carbon black particles to form a conductive network in the blend system, so the resistivity of the composite material rises sharply, which is called the conductor–insulation transition, the second transition of the valley-shaped curve.

6.4 Effect of Conductive Fillers on Semi-conductivity

Compared with the polymer matrix, the conductive fillers in the semi-conductive composites have more significant influences on the volume resistivity, depending on the filler content, shape, size, and distribution condition. Carbon black is commonly used fillers in the semi-conductive layer of power cable, in addition to GT, single-wall/multiwall carbon nanotubes (SWNTs/MWNTs), GN, etc. The effect of conductive filler properties will be discussed in this section.

6.4.1 Carbon Black

Carbon black is an intrinsic semiconductor with a resistivity of 10^{-2}~10^{1} Ω·cm [57]. The carbon black is added to the polymer matrix, forming a continuous path in it. If the carbon black aggregate spacing is less than 100 Å [5], electron flow will be induced. When the agglomerate content increases and the spacing becomes smaller, the tunneling effect will occur, resulting in a significant decrease in the resistivity of composites. Therefore, the semi-conductive characteristics markedly depend on the filler aggregates.

Figure 6.9 presents the effect of carbon black properties on electrical and physical properties of semi-conductive composites [5]. With the decrease of the filler size, the volume resistivity of composites decreases, following with a higher viscosity and a poor processing performance. On the contrary, increasing the particle size of carbon black will enlarge the volume resistivity, whereas lower viscosity is reached to obtain a better processing property. In addition, the larger filler aggregate structure with larger filler size will lead to an improved dispersibility, thus forming a localized conductive path for charge carriers. Nevertheless, too large filler size will also cause a decrease in surface smoothness. Therefore, the properties of carbon black aggregates are crucial to balance the electrical, mechanical, processing, and other macro properties of semi-conductive polymer composites for power cables.

The purity of carbon black is a key factor affecting the performance of the semi-conductive shielding layer, in terms of conductive properties and surface smoothness. Furnace black is produced by partial oxidation of hydrocarbon oil or gas, while acetylene black comes from continuous thermal decomposition of acetylene gas. Due to the nature of feedstocks, acetylene black is known to have much higher purity and less contamination than furnace black. Belhadfa et al. carried out proton-induced X-ray emission (PIXE) and neutron activation analysis (NAA) measurements to obtain the quantitative concentration and distribution of mineral impurities in three different semi-conductive composites used

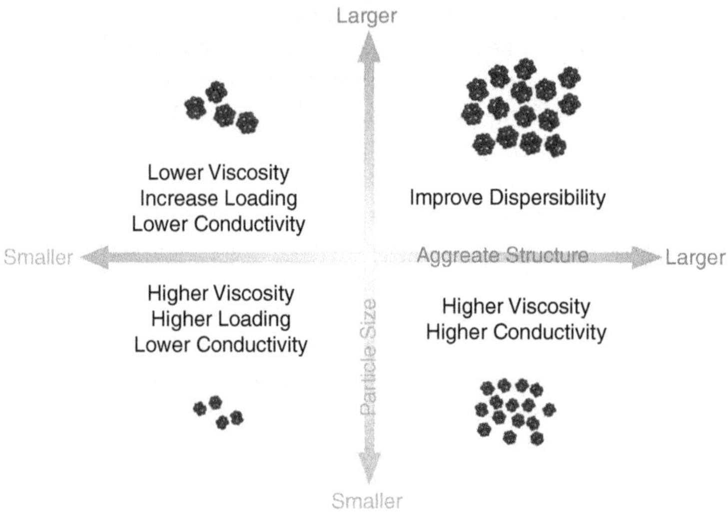

Figure 6.9 Effect of carbon black properties on electrical and physical properties of semi-conductive composites. *Source:* Based on [5]. © John Wiley & Sons.

as HV cable shields [9]. Both sets of data showed that the conventional semi-conductive composites based on furnace black contain a large number of mineral impurities (4000~8000 ppm), which could potentially migrate into the polyethylene insulation. The use of acetylene black significantly reduces the mineral impurity concentrations in cable shields by 1–2 orders of magnitude, resulting in an enhancement in surface smoothness as seen in Figure 6.1.

Merzouki et al. researched the electrical conductivity of PP-based semi-conductive composites filled with traditional furnace black and acetylene black [58]. The results showed that the percolation threshold is located at a filler content of ~2.6 vol%, while the threshold of furnace black is around 5.3 vol%. As the filler content reaches 10 vol%, the composites with acetylene black are 1 order of magnitude higher than that of the latter. Nilsson et al. from Borealis studied the influence of semi-conductive EBA composites with different carbon black on space charge injection and accumulation in HVDC cable insulation [6]. It is also concluded that the superior space charge properties are obtained by the addition of acetylene black, instead of furnace carbon black, which significantly suppress the electric field distortion induced by space charge.

6.4.2 Carbonaceous Fillers with One- and Two-Dimensions

High conductive nanofillers including carbon nanotubes (CNTs) and GN are considered as alternative additives for semi-conductive composites [59]. More importantly, CNTs and GN with one- and two-dimensional structures have a much higher specific surface area than the spherical-shaped carbon black. Owing to these advantages, their composites show much lower percolation thresholds, as compared to other relevant sorts such as carbon black.

George et al. carried out a comparative study of EVA nanocomposites based on expanded GT, multiwalled carbon nanotubes, and carbon nanofibers to investigate the effect of carbon-based nanofillers on the electrical properties of the corresponding composites [60]. The results showed that carbon nanofibers provided maximum conductivity as well as the lowest percolation threshold (8.2 vol%) compared to expanded GT and multiwalled carbon nanotubes filled composites. The improvement in both electrical conductivity and dielectric constant was attributed to the high filler aspect ratio and the formation of conducting network.

Ratzsch et al. investigated the electrical properties and percolation characteristics of EVA-copolymer composites filled with TRGO, GT, and MWNTs [37]. The ratio of volume resistivity of semi-conductive composites (ρ) to the matrix resistivity (ρ_m) is employed to describe the effect of different conductive fillers. Figures 6.10(a)–(c) present the log (ρ/ρ_m) varying with the filler content from 0 to 5 vol%, and Figure 6.10(d) illustrates the percolation thresholds of the semi-conductive composites. The curves show that, for the composites filled with TRGO, the volume resistance dramatically drops as the filler content exceeds 1.0~2.0 vol%, which is also dependent on the polymer matrix. With the increase of the TRGO content, the resistance decreases exponentially and then gradually enters the saturation region until the filler content exceeds 3 vol%. However, the LDPE/EVA blends with GT and MWNTs have a higher percolation threshold and a flatter slope of volume resistivity to the filler content. Especially for the GT-filled composites, the resistivity has not entered the saturation region until the filler content reaches 5%. As shown in Figure 6.10(d), for the LDPE/EVA blends with an EVA content of 18%, the percolation thresholds decrease from ~4.5 to ~1.1 vol% as the GT fillers are replaced by TRGO. Similar behaviors in LDPE with the conductive fillers are presented, which are attributed to the formation of more conductive network paths in the TRGO filled composites. It is concluded that the TRGO is appropriate as the conductive fillers to obtain superior electrical performance.

6.4.3 Secondary Filler for Carbon Black Filled Composites

As discussed in Section 6.4.1, carbon black fillers with a content up to 20~40 wt% are dispersed into polymer resins to obtain excellent semi-conductive characteristics, which may also result in poor mechanical and rheological properties. Hence, the incorporation of one- and two-dimensional nanomaterials (for instance, MWNTs) into the CB-filled composites as secondary fillers can reduce the required contents of CB, but also improve the electrical properties, due to their high specific surface area and aspect ratio [61].

For comparison, the volume resistivity at room temperature and PTC intensity of MWNTs/CB/polymer composites in references [61–64] are given in Figure 6.11. Generally, the PTC intensity is defined as the ratio of peak resistivity (ρ_P) to the room temperature resistivity (ρ_{RT}). The arrows are employed to emphasize the effect of MWNTs as secondary fillers. It is presented for the different polymer resin, the MWNTs, with a concentration of only 0.5–5%, can dramatically decrease the PTC intensity, following with the drop of volume resistivity at room temperature. The addition of MWNTs can form more conductive networks, bridging and connecting the carbon black aggregates dispersed in the polymer matrix. More importantly, its one-dimensional structure can maintain a conductive network when the matrix expands as the temperature rises.

In addition to the electrical properties, one- and two-dimensional nanomaterials also have a significant effect on improving the mechanical properties of composite materials.

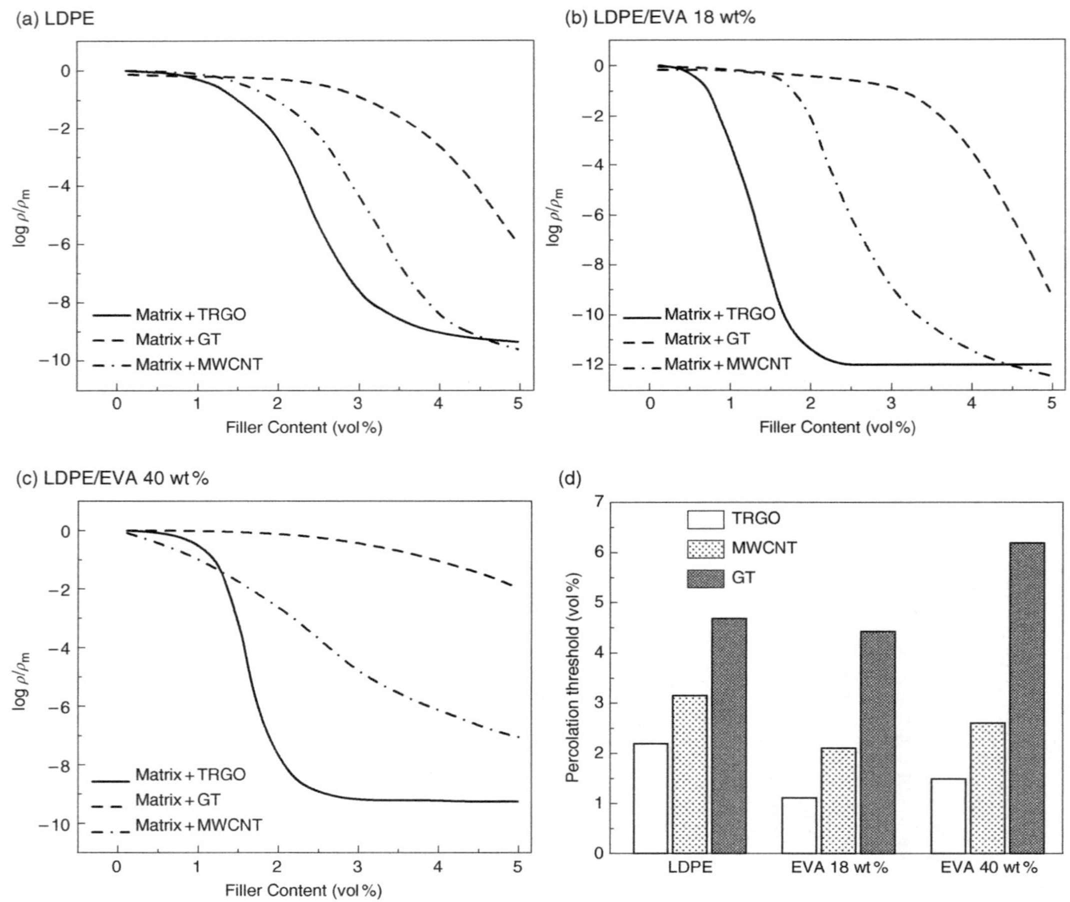

Figure 6.10 Effect of carbon black properties on electrical and physical properties of semi-conductive composites. *Source:* Based on [37]. © John Wiley & Sons.

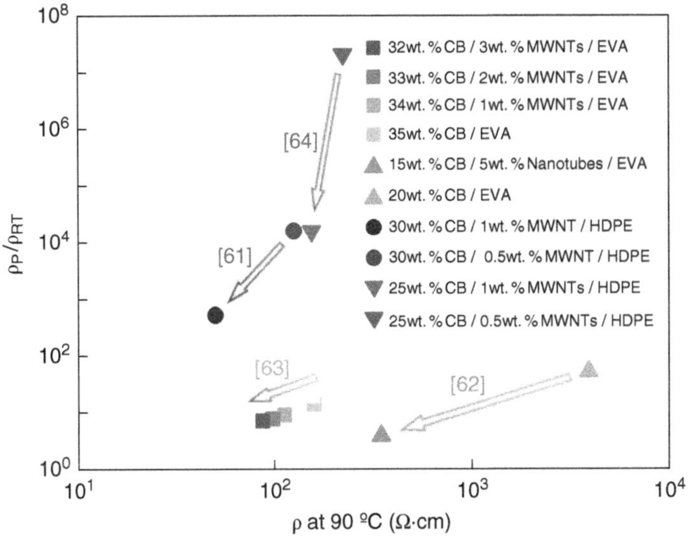

Figure 6.11 Volume resistivity at room temperature and PTC of MWNTs/CB/polymer composites with semi-conductive characteristics. *Source:* Based on [61–64]. © John Wiley & Sons.

However, the content of the secondary filler should not be too much. The introduction of MWNTs increases the viscosity and shear force during melt processing, which damages the carbon black structure. The increase in the content of MWNTs will also reduce its dispersibility, which reduces the effect of MWNTs on the improvement of mechanical properties.

6.5 Effect of Semi-conductive Composites on Space Charge Injection

Injection and accumulation of space charge in HV and extra-high voltage (EHV) DC XLPE power cables are one of the prime problems leading to the degradation and aging of electrical insulation [65–67]. The semi-conductive shield layer plays a significant role in affecting the space charge injection. This section herein introduces a graphene coating layer spread over the surfaces of the cable insulation. Graphene is of unique physical properties. The graphene layer shows semimetal characteristics with a zero bandgap, which can be slightly increased up to 0.25 eV in multilayer graphene [68, 69]. Therefore, the potential barrier height where graphene coating is in contact with insulation is large and this may affect space charge injection. The XLPE samples coating with the semi-conductive layer and graphene is made as shown in Figure 6.12 [65]. Pulsed electroacoustic (PEA) technique has been employed to measure the space charge dynamic behaviors under the applied field of 40 kV/mm at room temperature.

Figures 6.13(a)–(c) present the space charge dynamic behaviors in specimens A, B, and C at 40 kV/mm [65]. For the sample A, homocharge injected from anode increases and neutralizes the negative charge. The negative charges accumulated in the insulation bulk may originate from the impurities in materials and migrate from cathode to anode at the

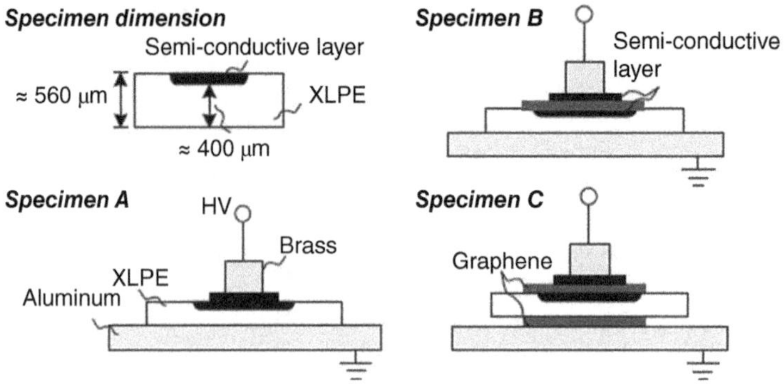

Figure 6.12 Specimen dimension and electrode configuration. Specimen A simulates a real interface between the semi-conductive layer and insulation. Specimen B has a graphene coating outside the semi-conductive layer. Specimen C employs graphene coating as the semi-conductive layers for both anode and cathode. *Source:* [68]. © IEEE.

initial polarization stage. At the same time, negative charges are injected into XLPE from cathode accumulating as homocharge.

The space charge of specimen B with a graphene coating outside the semi-conductive layer at a polarization field of 40 kV/mm is shown in Figure 6.13(b). Compared to specimen A, homocharge migrates from cathode to anode and accumulates in XLPE bulk, which leads to positive charges originating from anode injection to decrease.

Figure 6.13(c) shows the space charge profiles of specimens coated with graphene on both sides. During the polarization process, heterocharge at the cathode accumulates with time, and no clear evidence of injection can be seen. However, homocharge accumulation in the vicinity of anode is well observable during the depolarization process. During depolarization, heterocharge in the vicinity of cathode is dissipated quite fast while the homocharge accumulated near the anode remains for much longer times. It can be concluded that near the cathode, no injected charge can be seen, and heterocharge migrating from anode accumulates, in which case the electric field near anode decreases. During depolarization, injected homocharge can only be seen in the vicinity of anode. Therefore, graphene coating on the surface of XLPE can suppress the homocharge injection and block heterocharge.

It is possible to explain the effect of graphene coating on the XLPE side contacting the cathode in accordance with Figure 6.14 [68]. Figure 6.14(a) is the band structure of specimen before contact, where the bandgap E_g of XLPE is about 8~9 eV and the work function of aluminum W_m, semi-conductive layer W_s, and graphene is about 4.04, 5.6, and 4.48 eV, respectively [70, 71]. In a freestanding graphene layer, the Fermi energy of conduction and valence bands meets at the conical point, namely, the Dirac point.

Generally, when the electric field is applied as shown in Figure 6.14(b), the potential barrier of hole injection from anode lowers, which results in the positive charges being injected and hopping easily between shallow traps above valance band in XLPE, so the positive charges can be injected more easily than negative charges, and form homocharge accumulation in the vicinity of anode. On the contrary, at cathode, with the rise of the electric field, the Dirac point of graphene shifts upward, which provides a similar effect to the introduction of deep traps. Moreover, energy bands of XLPE also are distorted. In this case, only

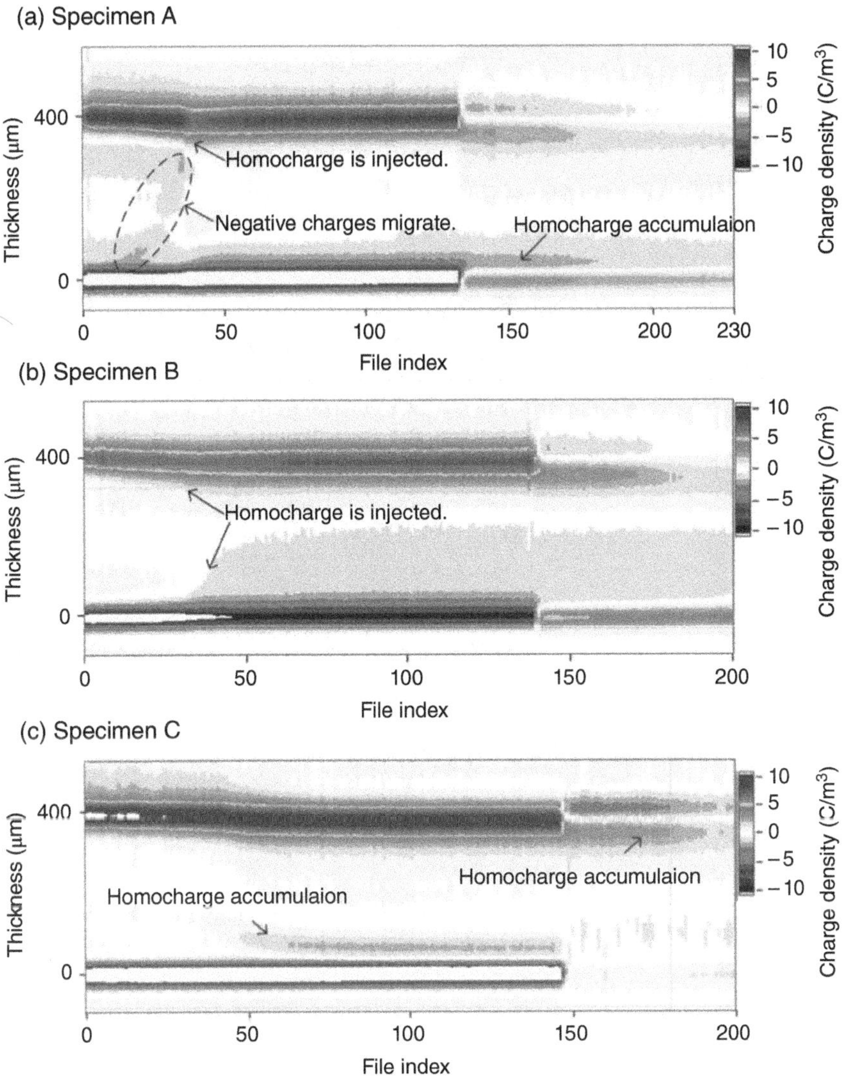

Figure 6.13 Space charge dynamic behaviors in specimens A, B, and C at 40 kV/mm. *Source:* [68].

when negative charges from cathode jump over both the Dirac point and potential barrier of XLPE, charges can be injected into XLPE, so negative charge injection is limited at the metal electrode. At 40 kV/mm, the conduction band is tilted, and the probability of hole passing through the barrier also increases, which leads to the positive charges further accumulating on the front of cathode. In addition, graphene coating on XLPE side suppresses the transport or tunneling effect of negative charges by the potential barrier, which explains why a small amount of negative charges accumulates in specimen B. From the above research, it is proved that the semiconducting layer has a significant effect on the injection and accumulation of space charges, but the current investigations are not yet detailed.

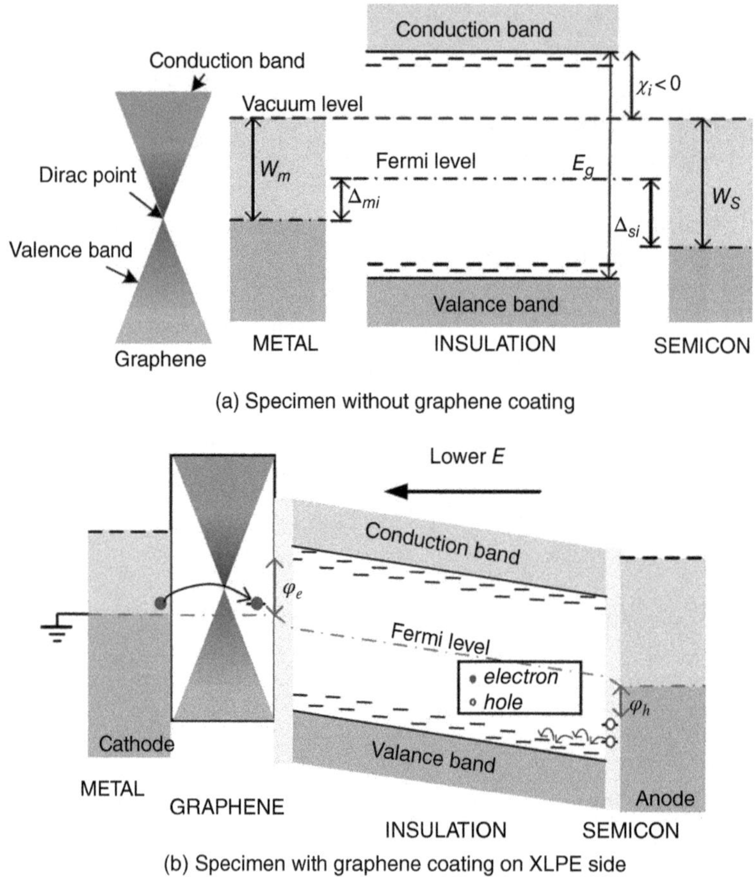

Figure 6.14 Schematic diagrams of the band structures of metal electrode, graphene coating, traditional semi-conductive layer, and insulation. (a) Specimen without graphene coating and (b) specimen with graphene coating. *Source:* [65]. © IEEE.

6.6 Conclusions

In this chapter, the function and development history of semi-conductive polymer composites for power cables are introduced, and the conductive mechanisms are discussed to give an insight into the PTC and NTC effect. Several recent research about the modification strategies of semi-conductivity characteristics are reviewed in this chapter. The effects of the polymer matrix on the semi-conductivity, as well as PTC, mainly include three factors: (i) the compatibility of the polymer matrix and the filler, (ii) the crystallization of polymer resin during the formation of conductive network, and (iii) the percolation network modification of the co-continuous phase in the blended polymer matrix. From the perspective of conductive fillers, the secondary additives including carbonaceous fillers with one-dimensional and two-dimensional structures are proved to be a significantly effective method to modify the electrical properties of carbon black filled composites, which can greatly reduce the percolation threshold and inhibit the PTC effect. For the semi-conductive composite shield used in HVDC cables, it can be noted that the space charge injection and extraction characteristics at the shield-insulation interface are markedly affected by the band structures of the metal and semi-conductive materials. Nevertheless, many problems are not solved yet.

Going into the future, semi-conductive composite insulation materials based on thermoplastic polymers should be further studied, with the development of environmentally friendly insulation materials such as PP. It is vital to balance the electrical, mechanical, processing, and other macro properties of semi-conductive polymer composites for HV power cables. Additionally, the charge emission mechanism at the interface between the metal or semiconducting electrode and the insulation needs to be further studied, which is of great significance for the improvement of the semi-conductive shields of HVDC cables.

References

1 Chang, G.W., Huang, H.M., and Lai, G.G. (1874–1878). A study on controlling electrical stress of underground cable by semi-conductive shielding. *IEEE/PES Transmission and Distribution Conference and Exhibition* 3: 2002. https://doi.org/10.1109/TDC.2002.1177743.

2 Burns, N.M., Eichhorn, R.M., and Reid, C.G. (1992). Stress controlling semiconductive shields in medium voltage power distribution cables. *IEEE Electrical Insulation Magazine* 8: 8–24. https://doi.org/10.1109/57.156940.

3 Han, S.J. and Kjellqvist, J. (2015). Aspects of semiconductive shields under accelerated wet electrical aging conditions of power cables. *IEEE Electrical Insulation Conference (EIC)* https://doi.org/10.1109/ICACACT.2014.7223565.

4 Li, W.K., Yan, H.D., Zhou, Y., et al. (2017). Supersmooth semiconductive shielding materials use for XLPE HVDC cables. 1st International Conference on Electrical Materials and Power Equipment (ICEMPE), 447–451. https://doi.org/10.1109/ICEMPE.2017.7982125.

5 Han, S.J., Mendelsohn, A., and Ramachandran, R. (2006). Overview of semiconductive shield technology in power distribution cables. IEEE/PES Transmission and Distribution Conference and Exhibition, 641–646. https://doi.org/10.1109/TDC.2006.1668572.

6 Nilsson, U.H. and Boström, J. (2010). Influence of the semiconductive material on space charge build-up in extruded HVDC cables. IEEE International Symposium on Electrical Insulation, 1–4. https://doi.org/10.1109/ELINSL.2010.5549811.

7 Li, W.K., Zhang, C., Zha, J.W., and Dang, Z. (2018). Dispersion of carbon blacks and their influence on the properties of semiconductive materials use for high-voltage power cables. IEEE International Conference on High Voltage Engineering and Application (ICHVE), 1–4. https://doi.org/10.1109/ICHVE.2018.8641842.

8 Brigandi, P.J., Person, T.J., Caronia, P.J., and Groot-Enzerink, E. (2014). Impact of semiconductive shield quality on accelerated aging cable performance. IEEE PES T&D Conference and Exposition, 1–5. https://doi.org/10.1109/TDC.2014.6863562.

9 Belhadfa, A., Houdayer, A.J., Hinrichsen, P.F. et al. (1989). Impurities in semiconductive compounds used as HV cable shields. *IEEE Transactions on Electrical Insulation* 24: 709–712. https://doi.org/10.1109/14.34206.

10 IEC 62067:2011. Power cables with extruded insulation and their accessories for rated voltages above 150 kV (Um = 170 kV) up to 500 kV (Um = 550 kV) – Test methods and requirements. International Electrotechnical Commission.

11 JB/T10738-2007. Semi-conductive shielding compound for power cables of rated voltages up to and including 35 kV.

12 Niu, F.Y., Zhang, Y.W., An, Z.L. et al. (2011). Space charge injection in LDPE by semi-conductive electrode with different carbon black filling rates. Proceedings of 2011 International Symposium on Electrical Insulating Materials, 12–15, 2011. https://doi.org/10.1109/ISEIM.2011.6826264.

13 Xiao, C., Zhang, Y.W., An, Z.L. et al. (2011). Interface electric field of carbon black loaded electrode and its significant influence on charge injection into polyethylene. *Journal of Applied Polymer Science* 123: 3017–3022. https://doi.org/10.1002/app.34950.

14 Wei, Y.H., Han, W., Li, G.C. et al. (2020). Research progress of semiconductive shielding layer of HVDC cable. *High Voltage* 5: 1–6. https://doi.org/10.1049/hve.2019.0069.

15 Dudić, D., Škipina, B., Dojčilović, J. et al. (2011). Effects of charge trapping on the electrical conductivity of low-density Polyethylene-Carbon black composites. *Journal of Applied Polymer Science* 121: 138–143. https://doi.org/10.1002/app.33421.

16 Yang, J.S., Shin, D.H., Yang, H. et al. (2007). A study on the properties of semiconductive shield materials for power cables in accordance with content of multi-walled carbon nanotubes. IEEE International Conference on Solid Dielectrics, 322–325. https://doi.org/10.1109/ICSD.2007.4290817.

17 Han, P., Zha, J.W., Wang, S., and Dang, Z.M. (2017). Theoretical analysis and application of polymer-matrix field grading materials in HVDC cable terminals. *High Voltage* 2: 39–46. https://doi.org/10.1049/hve.2016.0067.

18 Wang, X., Nelson, J.K., Schadler, L.S., and Hillborg, H. (2010). Mechanisms leading to nonlinear electrical response of a nano p-SiC/silicone rubber composite. *IEEE Transactions on Dielectrics and Electrical Insulation* 17: 1687–1696. https://doi.org/10.1109/TDEI.2010.5658218.

19 Stepashkina, A.S., Tsobkallo, E.S., and Alyoshin, A.N. (2014). Electrical conductivity modeling and research of polypropylene composites filled with carbon black. *Journal of Physics Conference* 572: 012032. https://doi.org/10.1088/1742-6596/572/1/012032.

20 Yu, G., Zhang, M.Q., Zeng, H.M. et al. (1999). Effect of filler treatment on temperature dependence of resistivity of carbon-black-filled polymer blends. *Journal of Applied Polymer Science* 73: 489–494. https://doi.org/10.1002/(SICI)1097-4628.

21 Hou, Y.H., Zhang, M.Q., Rong, M.Z. et al. (2010). Improvement of conductive network quality in carbon black-filled polymer blends. *Journal of Applied Polymer Science* 84: 2768–2775. https://doi.org/10.1002/app.10574.

22 Hou, Y.H., Zhang, M.Q., and Rong, M.Z. (2003). Performance stabilization of conductive polymer composites. *Journal of Applied Polymer Science* 89: 2438–2445. https://doi.org/10.1002/app.12465.

23 Henderson, A.M. (1993). Ethylene-vinyl acetate (EVA) copolymers: a general review. *IEEE Electrical Insulation Magazine* 9: 30–38. https://doi.org/10.1109/57.249923.

24 Zhang, F.W. (2013). Investigating on mLLDPE/LDPE/CB composite with PTC. 2nd International Conference on Measurement, Information and Control (ICMIC). https://doi.org/10.1109/MIC.2013.6758205.

25 Zhang, J.F., Zheng, Q., Yang, Y.Q., and Yi, X.S. (2010). High-density polyethylene-carbon black conductive composites. I. Effect of CB surface modification on its resistivity–temperature behavior. *Journal of Applied Polymer Science* 83: 3112–3116. https://doi.org/10.1002/app.10049.

26 Jia, W.T. and Chen, X.F. (2010). PTC effect of polymer blends filled with carbon black. *Journal of Applied Polymer Science* 54: 1219–1221. https://doi.org/10.1002/app.1994.070540904.

27 Xie, H.F., Deng, P.Y., Dong, L.S., and Sun, J.Z. (2010). LDPE/carbon black conductive composites: influence of radiation crosslinking on PTC and NTC properties. *Journal of Applied Polymer Science* 85: 2742–2749. https://doi.org/10.1002/app.10720.

28 Boggs, S.A. (2001). 500 Ω·m-low enough resistivity for a cable ground shield semicon. *IEEE Electrical Insulation Magazine* 17: 26–32. https://doi.org/10.1109/57.948981.

29 Boggs, S. and Kuang, J. (1998). High field effects in solid dielectrics. *IEEE Electrical Insulation Magazine* 14: 5–12. https://doi.org/10.1109/57.730802.

30 Zhao, W.J., Siew, W.H., and Given, M.J. (2013). The electrical performance of thermoplastic polymers when used as insulation in cables. 48th International Universities' Power Engineering Conference (UPEC). https://doi.org/10.1109/UPEC.2013.6714868.

31 Marlard, T., Bezille, J., Janah, H., and Becker, J. (1993). Influence of migration on the electrical properties of HV extruded cables. 3rd International Conference on Power Cables and Accessories 10 kV–500 kV.

32 Kim, E.J., Park, D.H., Kim, H.S., and Lee, G.J. (1998). An investigation of the influence of semiconductive electrode materials in the ac breakdown and the charge accumulation in XLPE. *1998 Annual Report Conference on Electrical Insulation and Dielectric Phenomena* 2: 546–549. https://doi.org/10.1109/CEIDP.1998.732956.

33 An, B., Li, C.M., Ma, F.L. et al. (2015). Research on Effect of Modified Semi Conductive Material on the Space Charge Behaviour in XLPE. *4th International Conference on Mechatronics, Materials, Chemistry and Computer Engineering (ICMMCCE)* 39: 2281–2287. https://doi.org/10.2991/icmmcce-15.2015.438.

34 Lee, K.Y., Nam, J.C., Park, D.H., and Park, D.H. (2005). Electrical and thermal properties of semiconductive shield for 154 kV power cable. *International Symposium on Electrical Insulating Materials (ISEIM)* 3: 616–619. https://doi.org/10.1109/ISEIM.2005.193444.

35 Azizi, S., David, E., Fréchette, M.F. et al. (2018). Electrical and thermal conductivity of ethylene vinyl acetate composite with graphene and carbon black filler. *Polymer Testing* 72: 24–31. https://doi.org/10.1016/j.polymertesting.2018.09.031.

36 Wang, Q., Meng, Q.G., Wang, T.L., and Guo, W.H. (2017). High-performance antistatic ethylene-vinyl acetate copolymer/high-density polyethylene composites with graphene nanoplatelets coated by polyaniline. *Journal of Applied Polymer Science* 134: 45303. https://doi.org/10.1002/app.45303.

37 Ratzsch, K.F., Cecen, V., Tölle, F. et al. (2014). Rheology, electrical properties, and percolation of TRGO-filled EVA-copolymers. *Macromolecular Materials and Engineering* 299: 1134–1144. https://doi.org/10.1002/mame.201300440.

38 Sumita, M., Asai, S., Miyadera, N. et al. (1986). Electrical conductivity of carbon black filled ethylene-vinyl acetate copolymer as a function of vinyl acetate content. *Colloid and Polymer Science* 264: 212–217. https://doi.org/10.1007/BF01414955.

39 Yang, Q.Q. and Liang, J.Z. (2010). Electrical properties and morphology of carbon black-filled HDPE/EVA composites. *Journal of Applied Polymer Science* 117: 1998–2002. https://doi.org/10.1002/app.32117.

40 Park, J.S., Kang, P.H., Nho, Y.C., and Suh, D.H. (2003). Effects of thermal ageing treatment and antioxidants on the positive temperature coefficient characteristics of carbon black/polyethylene conductive composites. *Journal of Applied Polymer Science* 89: 2316–2322. https://doi.org/10.1002/app.11953.

41 Tang, H., Liu, Z.Y., Piao, J.H. et al. (2010). Electrical behavior of carbon black-filled polymer composites: effect of interaction between filler and matrix. *Journal of Applied Polymer Science* 51: 1159–1164. https://doi.org/10.1002/app.1994.070510701.

42 Hou, Y.H., Zhang, M.Q., Mai, K.C. et al. (2010). Heat treatment-induced multiple melting behavior of carbon black-filled polymer blends in relation to the conductive performance stabilization. *Journal of Applied Polymer Science* 80: 1267–1273. https://doi.org/10.1002/app.1212.

43 Luo, Y.L., Wang, G.C., Zhang, B.Y., and Zhang, Z.P. (1998). The influence of crystalline and aggregate structure on PTC characteristic of conductive polyethylene/carbon black composite. *European Polymer Journal* 34: 1221–1227. https://doi.org/10.1016/S0014-3057(98)00099-8.

44 Song, Y.H. and Zheng, Q. (2007). Influence of annealing on conduction of high-density polyethylene/carbon black composite. *Journal of Applied Polymer Science* 105: 710–717. https://doi.org/10.1002/app.26076.

45 Miyasaka, K., Watanabe, K., Jojima, E. et al. (1982). Electrical conductivity of carbon-polymer composites as a function of carbon content. *Journal of Materials Science* 17: 1610–1616. https://doi.org/10.1007/BF00540785.

46 Feng, J.Y. and Chan, C.M. (2000). Positive and negative temperature coefficient effects of an alternating copolymer of tetrafluoroethylene–ethylene containing carbon black-filled HDPE particles. *Polymer* 41: 7279–7282. https://doi.org/10.1016/S0032-3861(00)00095-1.

47 Shemesh, R., Siegmann, A., Tchoudakov, R., and Narkis, M. (2006). Electrical behavior of high impact polystyrene/liquid crystalline polymer blends containing low content of carbon black. *Journal of Applied Polymer Science* 102: 1688–1696. https://doi.org/10.1002/app.24249.

48 Sumita, M., Sakata, K., Asai, S. et al. (1991). Dispersion of fillers and the electrical conductivity of polymer blends filled with carbon black. *Polymer Bulletin* 25: 265–271. https://doi.org/10.1007/BF00310802.

49 Moreno, V.M. and Gorur, R.S. (2001). Effect of long-term corona on non-ceramic outdoor insulator housing materials. *IEEE Transactions on Dielectrics and Electrical Insulation* 8: 117–128. https://doi.org/10.1109/94.910434.

50 Hirano, Y., Inohara, T., Toyoda, M. et al. (2001). Accelerated weatherability of shed materials for composite insulators. *IEEE Transactions on Dielectrics and Electrical Insulation* 8: 97–103. https://doi.org/10.1109/94.910431.

51 Zhang, H. and Hackam, R. (1999). Influence of fog parameters on the aging of HTV silicone rubber. *IEEE Transactions on Dielectrics and Electrical Insulation* 6: 835–844. https://doi.org/10.1109/94.822025.

52 Wang, X. and Yoshimura, N. (1999). Fractal phenomena. Dependence of hydrophobicity on surface appearance and structural features of SIR insulators. *IEEE Transactions on Dielectrics and Electrical Insulation* 6: 781–791. https://doi.org/10.1109/94.822015.

53 Chang, R.J. and Mazeika, L. (2000). Analysis of electrical activity associated with inclined-plane tracking and erosion of insulating materials. *IEEE Transactions on Dielectrics and Electrical Insulation* 7: 394–400. https://doi.org/10.1109/94.848923.

54 Xu, Z.B., Zhao, C., Gu, A.J. et al. (2007). Effect of morphology on the electric conductivity of binary polymer blends filled with carbon black. *Journal of Applied Polymer Science* 106: 2008–2017. https://doi.org/10.1002/app.26827.

55 Cheah, K., Forsyth, M., and Simon, G.P. (2000). Processing and morphological development of carbon black filled conducting blends using a binary host of poly(styrene co-acrylonitrile) and poly(styrene). *Journal of Polymer Science: Part B: Polymer Physics* 38: 3106–3119. https://doi.org/10.1002/1099-0488.

56 Kurusu, R.S., Helal, E., Moghimian, N. et al. (2018). The role of selectively located commercial graphene nanoplatelets in the electrical properties, morphology, and stability of EVA/LLDPE blends. *Macromolecular Materials and Engineering* 303: 1800187. https://doi.org/10.1002/mame.201800187.

57 Donnet, J.B. and Voet, A. (1977). Carbon black, physics, chemistry, and elastomer reinforcement. *Journal of Polymer Science, Polymer Letters Edition* 15: 631–632. https://doi.org/10.1002/pol.1977.130151013.

58 Merzouki, A. and Haddaoui, N. (2012). Electrical conductivity modeling of polypropylene composites filled with carbon black and acetylene black. *ISRN Polymer Science* 2012: 1–7. https://doi.org/10.5402/2012/493065.

59 Yang, H., Yang, J.S., Lee, K.Y. et al. (2018). Thermal properties of CNT reinforced semiconductive shield materials used in power cables. International Symposium on Electrical Insulating Materials (ISEIM). https://doi.org/10.1109/ISEIM.2008.4664477.

60 George, J.J., Bhadra, S., and Bhowmick, A.K. (2010). Influence of carbon-based nanofillers on the electrical and dielectric properties of ethylene vinyl acetate nanocomposites. *Polymer Composites* 31: 218–225. https://doi.org/10.1002/pc.20788.

61 Kalappa, P., Lee, J.H., Rashmi, B.J. et al. (2008). Effect of polyaniline functionalized carbon nanotubes addition on the positive temperature coefficient behavior of carbon black/high-density polyethylene nanocomposites. *IEEE Transactions on Nanotechnology* 7: 223–228. https://doi.org/10.1109/TNANO.2007.914994.

62 Han, S.J. and Wasserman, S.H. (2010). Agglomeration and percolation network behavior of semiconductive polymer composites with carbon nanotubes. IEEE International Symposium on Electrical Insulation. https://doi.org/10.1109/ELINSL.2010.5549784.

63 Fang, Y., Zhao, J., Zha, J.W. et al. (2012). Improved stability of volume resistivity in carbon black/ethylene-vinyl acetate copolymer composites by employing multi-walled carbon nanotubes as second filler. *Polymer* 53: 4871–4878. https://doi.org/10.1016/j.polymer.2012.08.035.

64 Lee, J.H., Kim, S.K., and Kim, N.H. (2006). Effects of the addition of multi-walled carbon nanotubes on the positive temperature coefficient characteristics of carbon-black-filled high-density polyethylene nanocomposites. *Scripta Materialia* 55: 1119–1122. https://doi.org/10.1016/j.scriptamat.2006.08.051.

65 Lei, Z.P., Fabiani, D., Li, C.Y. et al. (2019). Effect of graphene coating on space charge characteristic of XLPE and semiconductive layer at different temperatures. *IEEE Access* 7: 124540–124547. https://doi.org/10.1109/ACCESS.2019.2938539.

66 Mazzanti, G., Chen, G., Fothergill, J.C. et al. (2015). A protocol for space charge measurements in full-size HVDC extruded cables. *IEEE Transactions on Dielectrics and Electrical Insulation* 22: 21–34. https://doi.org/10.1109/TDEI.2014.004557.

67 Fabiani, D., Montanari, G.C., Laurent, C. et al. (2008). HVDC cable design and space charge accumulation. Part 3: effect of temperature gradient. *IEEE Electrical Insulation Magazine* 24: 5–14. https://doi.org/10.1109/MEI.2008.4473049.

68 Lei, Z.P., Fabiani, D., Palmieri, F. et al. (2020). Space charge characteristics of XLPE and semiconductive layer coated with graphene. *IEEE Transactions on Dielectrics and Electrical Insulation* 27: 128–131. https://doi.org/10.1109/TDEI.2019.008347.

69 Lin, S.S., Lu, Y.H., Xu, J. et al. (2017). High performance graphene/semiconductor van der Waals heterostructure optoelectronic devices. *Nano Energy* 40: 122–148. https://doi.org/10.1016/j.nanoen.2017.07.036.

70 Giovannetti, G., Khomyakov, P.A., Brocks, G. et al. (2018). Doping graphene with metal contacts. *Physical Review Letters* 101: 026803. https://doi.org/10.1103/PhysRevLett.101.026803.

71 Takada, T. (2019). Studies on space charge accumulation properties in dielectric materials. International Conference on Electrical Materials and Power Equipment (ICEMPE). https://doi.org/10.1109/ICEMPE.2019.872725

7

Polymer Composites for Electric Stress Control

Muneaki Kurimoto

Nagoya University, Nagoya, Japan

7.1 Introduction

The electric field stress inside power apparatus must be reduced to fabricate compact apparatus with high reliability. Figure 7.1 shows the distribution of the electric field stress around a solid spacer of gas-insulated apparatus [1]. The electric field stress in and around the solid insulator causes partial discharge and eventually leads to dielectric breakdown. The shapes of solid insulators and electrodes are optimized to reduce the electric field stress. In particular, the surface of an insulator is coated with an electrically conductive material to reduce the electric field stress on the surface. However, a new method is required to further reduce the electric field stress.

As a novel concept, functionally graded solid insulators were proposed in the International Symposium on High Voltage Engineering 2001 [2]. The structure of these insulators exhibits a graded permittivity distribution. Unlike conventional homogeneous materials, these insulators reduce the electric field stress while maintaining the simple shapes of solid spacers. The rest of this chapter is organized as follows: The numerical and experimental verification of the electric field stress control by functionally graded solid insulators is described in Section 7.2. The development of functionally graded solid insulators for gas-insulated switchgear (GIS) spacer applications is presented in Section 7.3. The application of functionally graded solid insulators to various types of equipment is described in Section 7.4.

7.2 Functionally Graded Solid Insulators and Their Effect on Reducing Electric Field Stress

Functionally graded solid insulators, which are a type of functionally graded materials (FGMs), show a spatial distribution of permittivity or conductivity. The application of functionally graded solid insulators with a permittivity distribution (ε-FGMs) to a GIS

Polymer Composites for Electrical Engineering, First Edition. Edited by Xingyi Huang and Toshikatsu Tanaka.

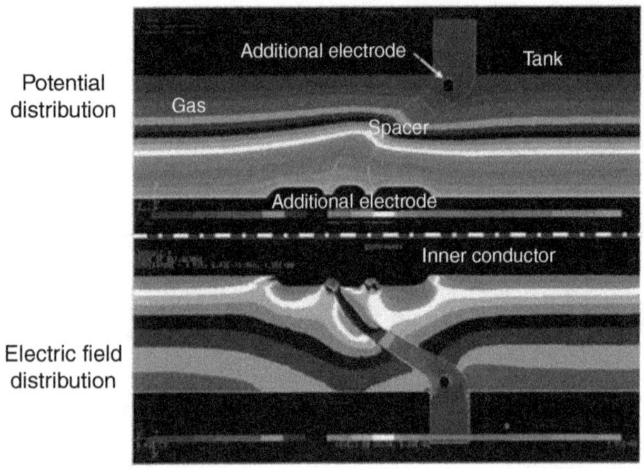

Figure 7.1 Distribution of electric field stress around GIS spacer. *Source:* Kurimoto [1].

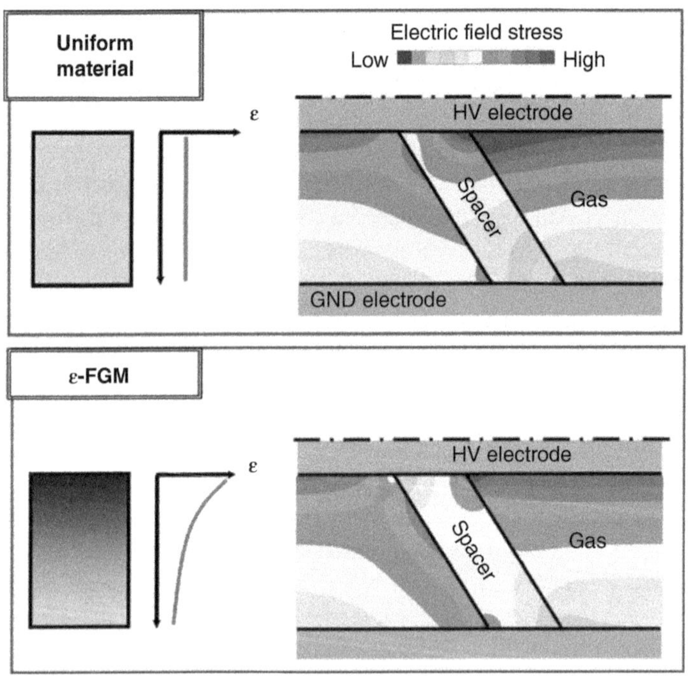

Figure 7.2 Concept of reduction in electric field stress by applying ε-FGM to GIS spacer (HV: high voltage, GND: ground).

spacer reduces the electric field stress around the spacer (Figure 7.2). The reduction effect has been verified through numerical simulation [3, 4] and breakdown tests [3–5].

In ε-FGMs, a solid material with high permittivity is used in the parts with large stress and a solid material with low permittivity is used in the other parts. Such a permittivity distribution in a solid insulator controls the electrical potential distribution to reduce the

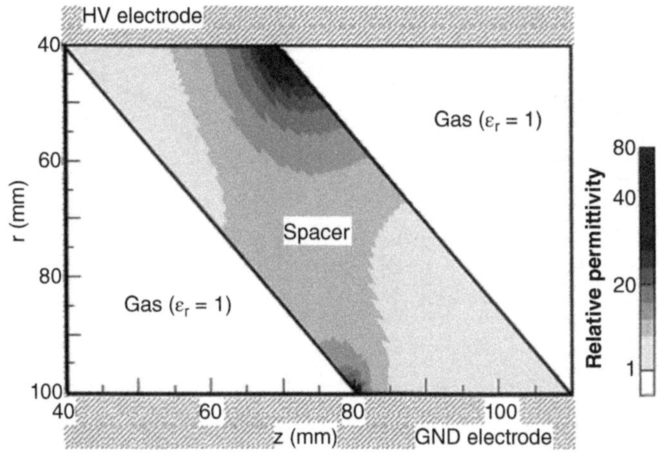

Figure 7.3 Optimized permittivity distribution in ε-FGM spacer (HV: high voltage, ε_r: relative permittivity, 4 < relative permittivity < 80).

maximum electric field stress (Figure 7.3). The stress reduction effect can be obtained in a continuously graded permittivity distribution and a layered permittivity distribution [3]. Additionally, this effect can be obtained using a conductivity distribution. A computer-aided optimization technique can be used to obtain a novel permittivity distribution that provides the strongest electric field reduction effect in GIS spacers [6].

A model spacer sample with a continuously graded permittivity distribution has been used to experimentally verify the stress reduction (Figure 7.4) [3, 4]. The sample is composed of an epoxy resin filled with TiO_2 particles (Figure 7.5a). FGM spacer is compared with uniform spacer whose relative permittivity is constant at 4. The continuously graded permittivity distribution is provided by controlling the density distribution of the particles through centrifugation [3, 4]. Various permittivity distributions can be obtained by controlling the permittivity, size, and initial arrangement of the particles and centrifugation conditions (Figure 7.5b and 7.5c) [7].

The variation in the surface breakdown voltage (BDV) with the electric field stress at a triple junction has been evaluated using a parallel plate electrode sandwiched between the model spacer samples, as shown in Figure 7.6. The BDV of the ε-FGM spacer is higher than that of the uniform spacer [3]. The increase in the BDV is caused by the reduction in the electric field stress due to the continuously graded permittivity distribution, and it can be estimated based on the volume time theory [8]. The ε-FGM spacer may have a long lifetime because of the low electric field stress at the interface between the spacer and electrode (Figure 7.7) [5].

7.3 Practical Application of ε-FGMs to GIS Spacer

The GIS spacer material is an epoxy resin with ceramic fillers (e.g. SiO_2, Al_2O_3). The permittivity distribution can be controlled based on the density distribution of high-permittivity and low-permittivity fillers. However, the coefficient of thermal expansion (CTE) is

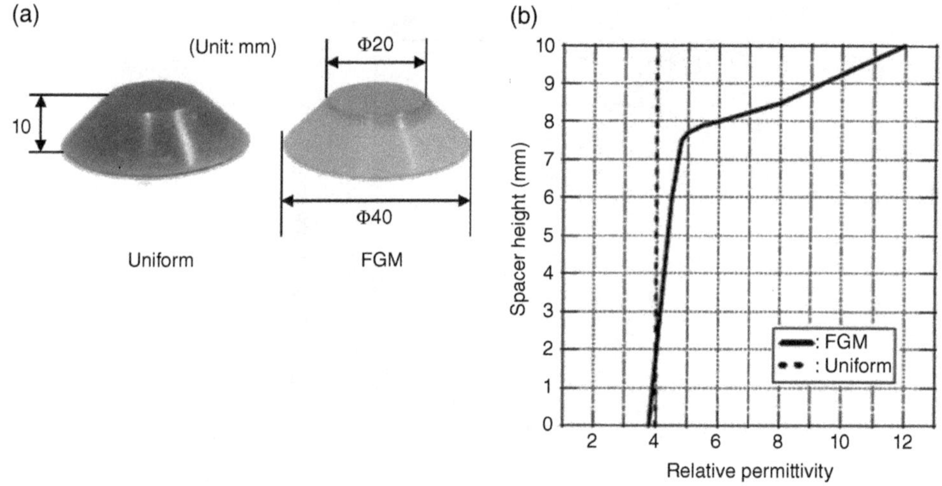

Figure 7.4 Uniform and ε-FGM spacer samples for breakdown test. (a) Spacer sample. *Source:* Kato et al. [3]. (b) Permittivity distribution.

distributed from high to low depending on the density distribution of fillers, which causes practical problems such as delamination between an insulator and a metal conductor (Figure 7.8). A previous study proposed the use of an epoxy resin with a spatial distribution of the filling ratio of a high-permittivity filler (e.g. TiO_2) and a low-permittivity filler (SiO_2) while maintaining a constant total filler volume to obtain an ε-FGM whose CTE was as low as that of a metal conductor [9] (Figure 7.9).

A large-sized ε-FGM spacer may be manufactured using a conventional casting method. A flexible mixture casting (FMC) method has been proposed [8] and demonstrated [10] (Figure 7.10). In this method, mixtures with various mixing ratios of high-permittivity and low-permittivity resins are sequentially injected into a metal mold. The FMC method can produce small-sized ε-FGM spacer samples with a continuously graded permittivity distribution. The ε-FGMs can be applied to GIS spacers if we can solve problems such as the deterioration of surface insulation performance due to the adhesion of foreign particles and the nondestructive measurement of permittivity distribution (Figure 7.11).

7.4 Application to Power Apparatus

The reduction in the electric field stress is a common issue for various types of electric power apparatus such as gas-insulated apparatus, power cable joints, and power electronics modules. The electric field distortion in a complex insulation system can be improved by applying an ε-FGM to a solid insulator. However, it is difficult to optimize the permittivity distribution. The permittivity distribution in a solid insulator can be optimized by optimizing the topology of the insulator (e.g. the use of stress cones in a cable joint system [11]) (Figures 7.12 and 7.13).

The method of producing a functionally graded solid insulator depends on its application. High-permittivity magnetron sputtering (Figure 7.14) is effective for fabricating surface FGM insulators for gas-insulated power lines [12]. Electrophoresis is effective for

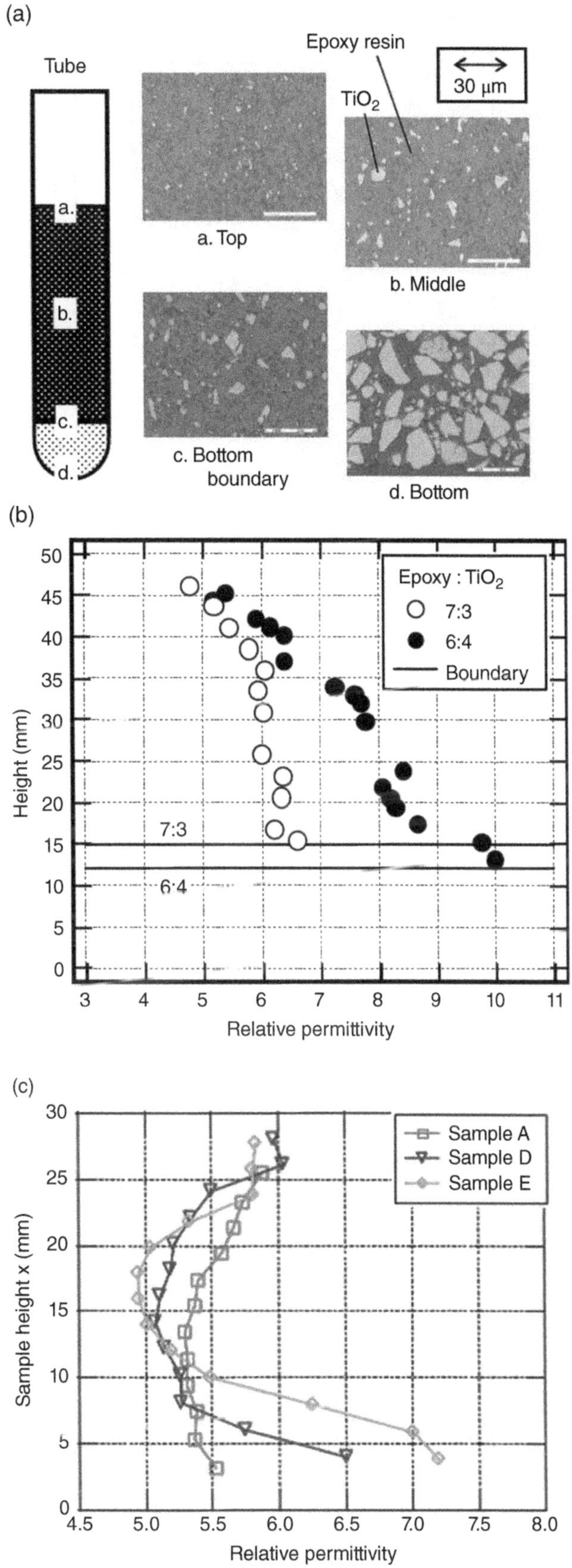

Figure 7.5 Continuously graded permittivity distribution obtained via centrifugation. (a) Density distribution of TiO_2 particles in epoxy resin under centrifugal force. *Source:* Kato et al. [3]. (b) Linear-shaped permittivity distribution. (c) U-shaped permittivity distribution.

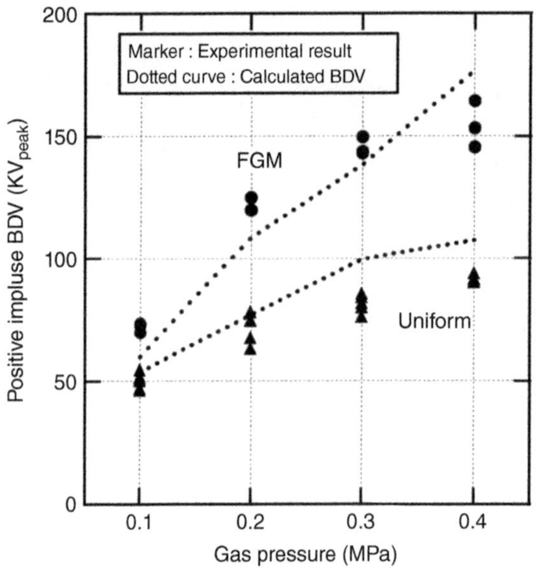

Figure 7.6 BDV of uniform and ε-FGM spacer samples.

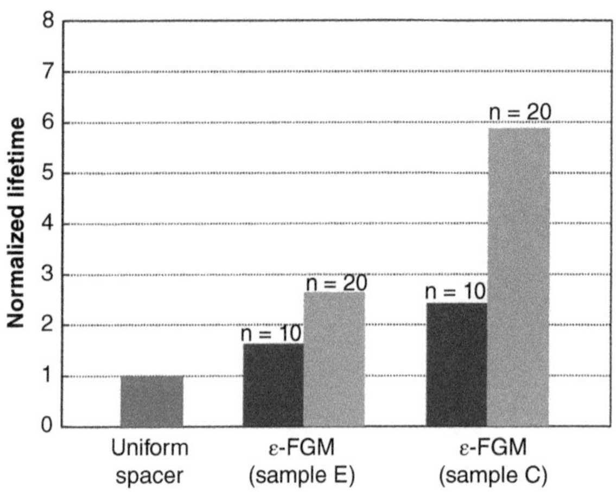

Figure 7.7 Estimated lifetime of uniform and ε-FGM spacer samples (n is a slope of the lifetime curve like the voltage-time [V-t] or electric field-time [E-t]).

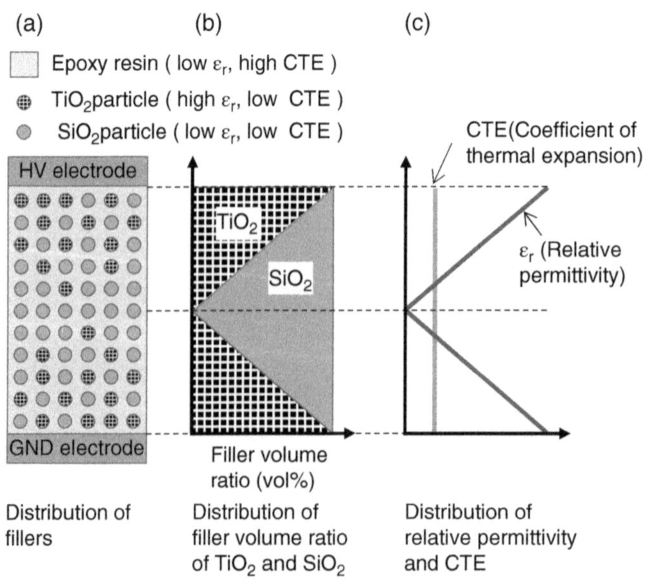

Figure 7.8 Conceptual diagram of ε-FGM whose CTE is as low as that of a metal conductor (HV: high voltage, GND: ground, ε_r: relative permittivity).

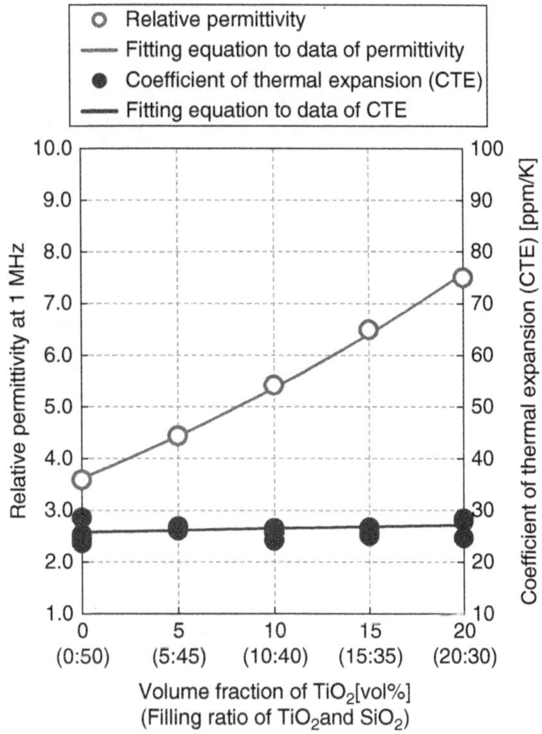

Figure 7.9 Relative permittivity and the CTE for different filler volume fractions of TiO_2/SiO_2 epoxy composites.

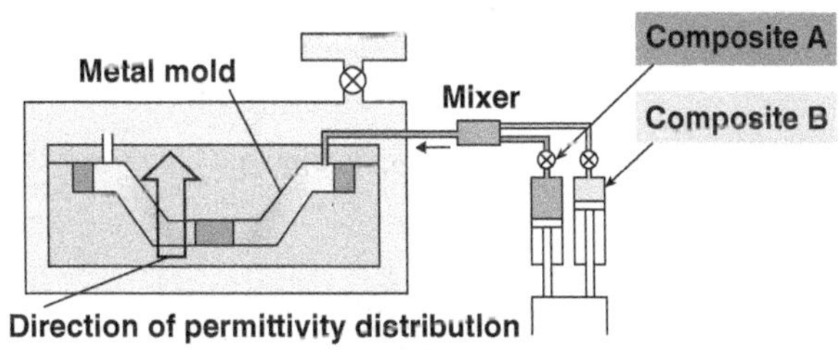

Figure 7.10 Concept of FMC method for ε-FGM.

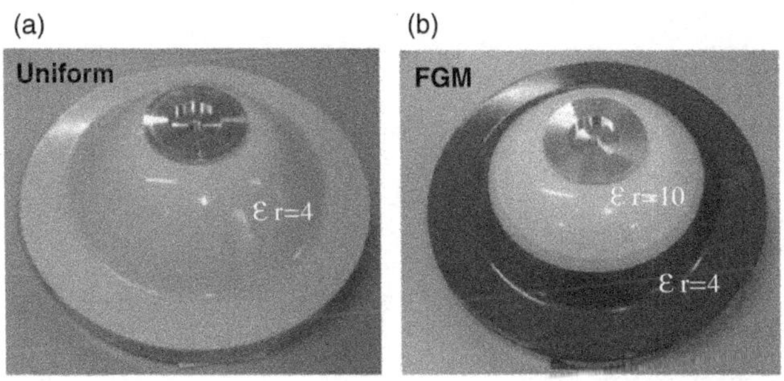

Figure 7.11 Cone-type GIS spacer samples (ε_r: relative permittivity). (a) Uniform spacer; (b) ε-FGM spacer.

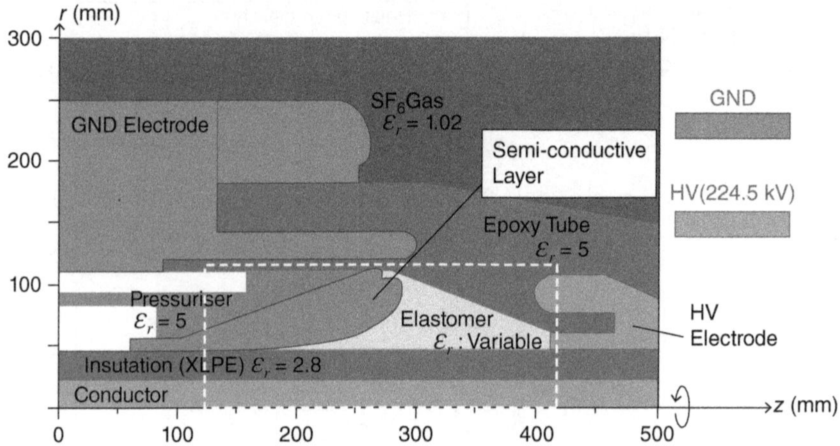

Figure 7.12 Insulation system in power cable termination (HV: high voltage, GND: ground, XLPE: crosslinked polyethylene, ε_r: relative permittivity).

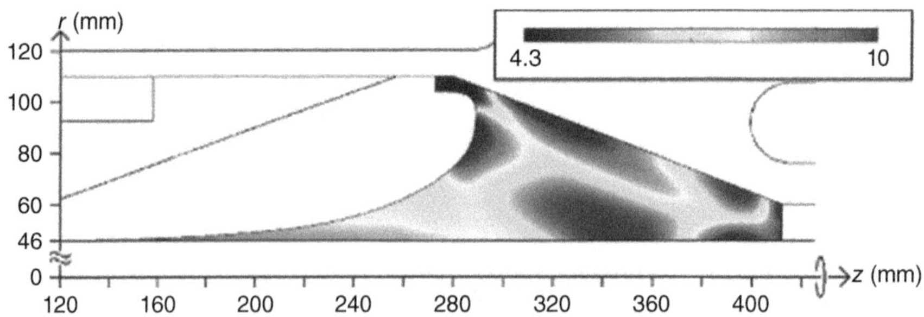

Figure 7.13 Graded permittivity distribution obtained by topology optimization in a stress cone.

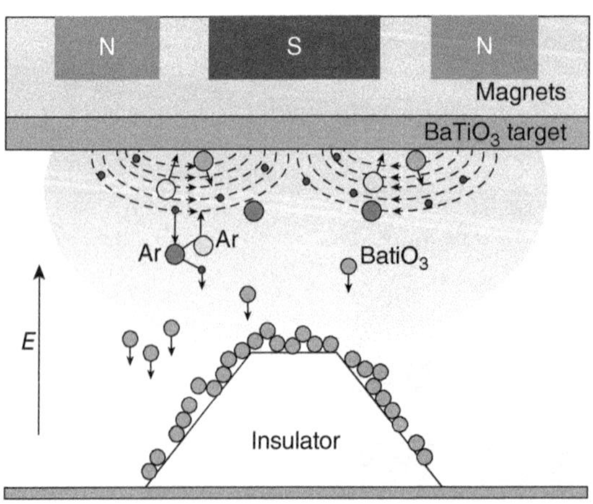

Figure 7.14 Principle of $BaTiO_3$ magnetron sputtering.

creating the FGMs that are used between the contact electrodes on a substrate in a power module (Figure 7.15) [13]. Three-dimensional printing can form FGMs with various permittivity distribution (Figure 7.16) [14, 15].

A functionally graded solid insulator is a technology for reducing the electric field stress, that is, a material technology close to the insulation design technology for power apparatus. The researchers and engineers in the material and design fields can contribute to the development of functionally graded solid insulators to fabricate innovative power apparatus.

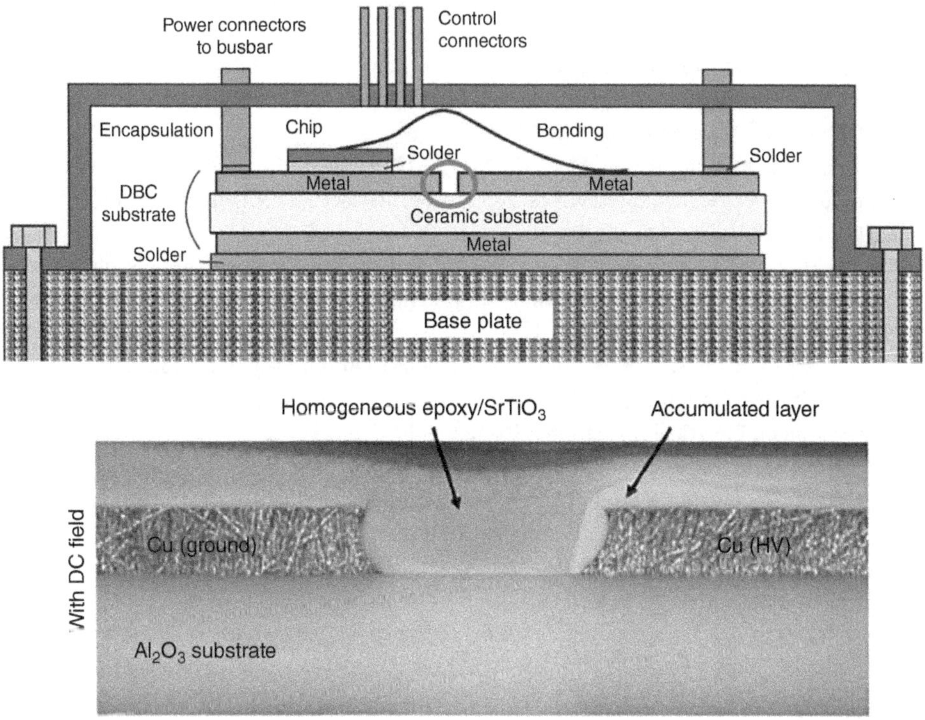

Figure 7.15 Cross-section of a high-voltage (HV) power module and the application of ε-FGM to metallized alumina direct-bonded copper (DBC) substrates. *Source:* Diaham et al. [13].

Figure 7.16 Conical insulating spacer with height of 10 mm using an alumina composite with micrometric alumina fillers with an average diameter of 15 μm and a two-layered ε-FGM insulator. *Source:* Kurimoto et al. [14, 15].

References

1 Kurimoto, M. (2010). With functionally graded materials and nanocomposites research on improving the electrical insulation performance of electric power equipment. Ph.D thesis. https://ci.nii.ac.jp/naid/500000509577, https://nagoya.repo.nii.ac.jp/?action=repository_uri&item_id=11554&file_id=17&file_no=1.

2 Kato, K., Kimura, K., Sakuma, S., and Okubo, H. (2001). Functionally gradient material (FGM) for GIS spacer insulation. *proceedings of International Symposium on High Voltage Engineering* 2: 401–404.

3 Kato, K., Kurimoto, M., Shumiya, H. et al. (2006). Application of functionally graded material for solid insulator in gaseous insulation system. *IEEE Transactions on Dielectrics and Electrical Insulation* 13 (2): 362–372.

4 Kurimoto, M., Kato, K., Hanai, M. et al. (2010). Application of functionally graded material for reducing electric field on electrode and spacer interface. *IEEE Transactions on Dielectrics and Electrical Insulation* 17 (1): 256–263.

5 Okubo, H., Kato, K., Hayakawa, N. et al. (2009). Functionally graded materials and their applications to high electric field power equipment. CIGRE SC D1 – Colloquium in Hungary Budapest, D1–101.

6 Okubo, H., Shumiya, H., Ito, M., and Kato, K. (2006). Optimization techniques on permittivity distribution in permittivity graded solid insulators. IEEE International Symposium on Electrical Insulation, 519–522.

7 Hayakawa, N., Ishiguro, J., Kojima, H. et al. (2016). Fabrication and simulation of permittivity graded materials for electric field grading of gas insulated power apparasut. *IEEE Transactions on Dielectrics and Electrical Insulation* 23 (1): 547–554.

8 Hayakawa, N., Miyaji, Y., Kojima, H., and Kato, K. (2018). Simulation on discharge inception voltage improvement of GIS spacer with permittivity graded materials (ε-FGM) using flexible mixture casting method. *IEEE Transactions on Dielectrics and Electrical Insulation* 25 (4): 1318–1323.

9 Kurimoto, M., Ozaki, H., Sawada, T. et al. (2018). Filling ratio control of TiO2 and SiO2 in epoxy composites for permittivity-graded insulator with low coefficient of thermal expansion. *IEEE Transactions on Dielectrics and Electrical Insulation* 25 (3): 1112–1120.

10 Imai, T., Hoshina, Y., Kojima, H., and Hayakawa, N. (2019). Resin injection behavior in process of manufacturing insulation spacer with permittivity gradient using computational fluid dynamics. *IEEE Transactions on Dielectrics and Electrical Insulation* 26 (5): 1669–1677.

11 Kurimoto, M. (2020). Topology optimization and 3D printing: toward a functionally graded solid insulator. *2020 International Symposium on Electrical Insulating Materials*, 107–110.

12 Du, B.X., Wang, Z.H., Li, J. et al. (2020). Epoxy insulator with surface graded-permittivity by magnetron sputtering for gas-insulated line. *IEEE Transactions on Dielectrics and Electrical Insulation* 27 (1): 197–205.

13 Diaham, S., Valdez Nava, Z., Lévêque, L. et al. (2018). An original in-situ way to build field grading materials (FGM) with permittivity gradient using electrophoresis. *2018 IEEE 2nd International Conference on Dielectrics*, Budapest, Hungary, (1–5 July 2018).

14 Kurimoto, M., Ozaki, H., Yamashita, Y. et al. (2016). Dielectric properties and 3d printing of uv-cured acrylic composite with alumina microfiller. *IEEE Transactions on Dielectrics and Electrical Insulation* 23 (5): 2985–2992.

15 Kurimoto, M., Sawada, T., Kato, T., and Suzuoki, Y. (2018). 3D Printing of 2 layered permittivity-graded material using UV-cured-resin/alumina composite. *2018 12th International Conference on the Properties and Applications of Dielectric Materials (ICPADM)*, Xi'an, China (20–23 May 2018), pp. 1010–1013.

8

Composite Materials Used in Outdoor Insulation

Wang Xilin, Jia Zhidong, and Wang Liming

Engineering Laboratory of Power Equipment Reliability in Complicated Coastal Environments, Tsinghua Shenzhen International Graduate School, Tsinghua University, Shenzhen, Guangdong, China

8.1 Introduction

External insulation refers to the insulation of air gaps and surface insulation of insulators exposed to the atmosphere. Insulators for overhead transmission lines and substations constitute the main electrical equipment using external insulation. Overhead line transmission is a prime type of high-voltage power transmission. In China, the maximum voltage level of ultra-high-voltage overhead transmission lines has now reached 1100 kV. As important parts of overhead transmission lines, insulators can enable electrical insulation and maintain mechanical connections between wires and towers. High-voltage electrical appliances in substations also use many insulation devices such as bushings and bar posts, which can also be called insulators. With rapid technological development, more and more insulators are composed of composite materials, and devices using silicone rubber (SIR) as the main body, such as composite insulators, coatings, and composite bushings, have been widely used. The D1 working group of the International Council on Large Electric Systems (CIGRE) research committee concluded that outdoor polymer materials should meet performance requirements in 12 aspects [1], namely, arc resistance, breakdown field strength, glass transition temperature, hydrophobicity, chemical and physical aging resistance, chemical corrosion resistance, corona and ozone resistance, flame retardancy, weather and ultraviolet (UV) radiation resistance, resistance to tracking and erosion, tear strength, and volume resistivity.

At present, the most widely used composite materials for outdoor high-voltage insulation mainly include ethylene propylene diene monomer (EPDM), SIR, cycloaliphatic epoxy resin (CEP), and hydrophobic CEP (HCEP). The performances of these composites are substantially better than those of conventional porcelain and glass insulation materials. A performance comparison of several types of insulation materials [2] are illustrated in Table 8.1.

Polytetrafluoroethylene (PTFE) insulators exhibit excellent aging resistance and long-term stability, low surface energy, and strong surface hydrophobicity at the initial stage of

Polymer Composites for Electrical Engineering, First Edition. Edited by Xingyi Huang and Toshikatsu Tanaka.

Table 8.1 Performance comparison of different insulation material types.

Type of materials	Porcelain	Ethylene propylene rubber	SIR	HCEP
Design flexibility	Poor	Moderate	Moderate	Very good
Weight	Poor	Good	Good	Good
Manufacturing process	Very poor	Poor	Poor	Very good
Ease of operation	Poor	Good	Good	Good
Hydrophobicity	Very poor	Good	Excellent	Very good
Hydrophobic mobility	Very poor	Poor	Excellent	Very good
Thermal impact resistance	Poor	Very good	Very good	Very good
Damage resistance	Poor	Good	Good	Good
Bird damage resistance	Good	Poor	Poor	Good
Flashover resistance	Good	Good	Good	Good
Tracking resistance	Excellent	Very good	Very good	Excellent
Maintenance	Poor	Good	Very good	Very good
Insulator cost	Good	Poor	Very poor	High
Overall cost	Moderate	Moderate	Moderate	High

use. However, their prices are high, and their relevant processing is challenging. In addition, the surfaces of PTFE insulators will be adversely affected by pollution after a period of operation, which is difficult to eliminate [3].

EPDM insulators are cost-effective and perform well in aging resistance, but their long-term anti-pollution flashover performance is not ideal. In natural pollution conditions, EPDM insulators, after a certain period of operation, experience severe bushing and shed deterioration; their surface leakage current is even greater than that of the porcelain insulators, and the pollution flashover voltage decreases significantly.

SIR materials have the same excellent insulation performance as porcelain and glass media and possess unique hydrophobicity and hydrophobic mobility as well. The so-called hydrophobic mobility means that after surface contamination, SIR will cause the polluted layer deposited on its surface to be hydrophobic [4]. Therefore, after SIR and its surface polluted layer get damp, no water film will come into being on the surface, but water exists in the form of isolated, dispersed water droplets. This damp state helps inhibit the development of discharge and current leakage along the surface, which significantly improves the anti-pollution flashover capability of external insulation equipment [5, 6].

This chapter introduces the current SIR composite materials used in high-voltage power transmission and transformation equipment, the assessment of their aging characteristics and states, and the development and application of new composite materials.

8.2 Overview of SIR Materials

In the field of external insulation of power systems, commonly used SIR materials can be divided into three categories: high-temperature vulcanized silicone rubber (HTV SIR),

Table 8.2 Comparison of the main specifications of three types of SIR.

Main specifications	RTV	HTV	ALSR
Raw rubber form	Semi-solid	Liquid	Liquid
Component	/	One- or two-component	One- or two-component
End group	$-CH_3$ and $-CH{=}CH_2$	$-CH{=}CH_2$	-Si-H and $-CH{=}CH_2$
Active crosslinking group	$-CH{=}CH_2$	$-CH{=}CH_2$	-Si-H and $-CH{=}CH_2$
Viscosity (255)/mPaco	Approx. 2×10^7	200–10000	5×10^3–1×10^4
Molar mass/($\times 10^4$ g/mol)	40–80	1–10	1–10
Degree of polymerization/($\times 10^3$)	5–10	0.1–1.0	0.1–1.0
Crosslinking mechanism	Organic peroxide vulcanization	Condensation- or addition-type vulcanization	Addition reaction

room temperature vulcanized silicone rubber (RTV SIR), and liquid silicone rubber (LSR). HTV SIR is mainly used for composite insulators umbrella sheathing, as well as the external insulation shed of lightning arresters, circuit breakers, and transformers. RTV SIR is mainly used for manufacturing anti-pollution flashover coatings. Such coatings can be applied on the surfaces of porcelain and glass insulators, thereby improving the anti-pollution flashover performance of external insulation. The production and manufacturing processes of HTV SIR and RTV SIR have been relatively mature. Meanwhile, manufacturers of these two materials currently occupy larger market shares in China. LSR is mainly used for hollow composite insulators of substation equipment. LSR products are mainly manufactured outside China, such as Dow Corning's Xiameter RBL series products. In the preparation of all the above materials, polydimethylsiloxane (PDMS) is used as the raw rubber and supplemented by reinforcing fillers, flame-retardant fillers, and other additives until it is vulcanized under certain conditions. However, differences can be observed in their preparation processes. In the preparation of HTV SIR, PDMS with a higher molecular weight (about 300 000–700 000) and a large amount of vinyl serves as the raw rubber; under high-temperature conditions, peroxide is used as a vulcanizing agent to crosslink with PDMS until it is mold. In the preparation of RTV SIR, hydroxyl-terminated PDMS with a smaller molecular weight (generally only about 10 000) serves as the raw rubber, and it cures and molds under room temperature conditions (Table 8.2).

8.2.1 RTV Coatings

RTV coatings are currently known as excellent solid coatings for stain resistance in the second sector of industry. The Pacific Gas and Electric Company began to conduct research on the field operations of RTV coatings in 1973 and published a public report in 1982. Its research results revealed that under fairly severe pollution conditions, the RTV coatings remained hydrophobic for six years, showing excellent anti-pollution flashover performance. At the 354 kV switching station of Milltone nuclear power plant in the United States in 1987, all the insulators and porcelain bushings were applied with RTV coatings, leading to good operating results. The High Voltage Laboratory of Tsinghua University was the first in China to start the research on RTV coatings in 1984 and started a small-scale trial run in 1986 at the Tianjin Electric Power Bureau. Until now, the longest service life of the trial RTV coatings has reached 30 years, during which no maintenance measures were taken, in addition to which, they survived the major pollution flashovers in 1990, 1996, and 2001 [7].

RTV SIR uses hydroxyl-terminated poly-organic siloxane with a low molecular weight (10 000–80 000 or so) as the basic rubber. By reacting with the crosslinking agent and catalyst, the basic rubber can be crosslinked at room temperature into a three-dimensional network structure. RTV SIR can be single-component or two-component depending on the process used. In terms of single-component RTV SIR, the basic rubber, filler, and crosslinking agent or catalyst are mixed evenly in a certain condition and then airtight packaged; the mixture starts a crosslinking reaction when it encounters the moisture in the atmosphere. In terms of two-component RTV SIR, the basic rubber and the crosslinking agent or catalyst are packaged separately; when required for use, they are mixed in a certain ratio, after which crosslinking occurs, which has no relation with environmental moisture. Due to the small intermolecular forces of poly-organic siloxane, the mechanical performance of the raw rubber after vulcanization is poor. Filler reinforcement must be applied to improve the mechanical properties of the raw rubber. Fillers are mainly synthetic silica (commonly known as the white carbon black), with a specific surface area above 200 m^2/g. Additionally, they can increase the strength of the vulcanized rubber by 20–30 times. In polluted areas and bad meteorological conditions, coatings may produce dry band discharges. The arc ablation resistance of the coatings will undoubtedly affect their use, and it can be improved by adding inorganic flame-retardant materials (generally alumina trihydrate [ATH]) to the coatings.

The physical form of RTV coatings is usually a flowable viscous liquid due to the low molecular weight of the raw rubber. The RTV coatings are applied, brushed, or sprayed onto the surface of conventional porcelain and glass external insulators. After a certain period of curing (vulcanization) time, the coating cures from a liquid state to a film that adheres to the surface of the external insulator. At this time, the RTV coating after curing will turn the original surface of porcelain and glass media from hydrophilic into hydrophobic, which is important to improve and enhance the anti-pollution flashover capability of traditional porcelain and glass external insulators. The hydrophobicity of the coated surface is not constant. It will be temporarily weakened or even disappear with the changes in environmental conditions of the surface. Due to the characteristics of SIR materials, the surface hydrophobicity disappearing temporarily can be restored. After RTV coatings are soaked for a certain period of time, the surface hydrophobicity will disappear temporarily.

When the surface of a coating is covered from pollution, the hydrophobicity of the coating can transfer to the surface of the polluted layer, which means that the polluted layer is also characterized by hydrophobicity. This phenomenon is called hydrophobic mobility. Hydrophobic mobility is unique to RTV coatings, and some hydrophobic materials, such as polyethylene, polyvinyl chloride, PTFE, epoxy resin, nylon, and polyester resin, do not exhibit this characteristic. It is believed that when the pollution accretes on the RTV coating surface, the chain segments of some SIR molecules and uncrosslinked molecular substances migrate into the pollution from inside the RTV coating. This migration changes the structure and characteristics of the pollution particles, thus making the pollution particles hydrophobic. Hydrophobic mobility is the result of the diffusion of bulk phase molecules toward outside the boundary. When the SIR molecules diffuse out of the boundary from the interface, they receive little resistance from the bulk phase molecules; hence, they can easily diffuse out of the body. When there is a pollution layer on the surface, they can easily enter the pollution layer. In this case, they are hijacked by the pollution layer and stay in the pollution layer. At the same time, the strength of diffusion follows the basic theory of adsorption and coalescence. That is, the diffusion of SIR molecules toward outside the body is related to the nature of the surface pollution layer, and this process follows the principle of minimum energy. When adsorption coalescence reaches a state of equilibrium, the bulk phase molecules will no longer diffuse. Therefore, if the pollution layer on the surface is not removed, the bulk phase molecules will not diffuse.

In the case of the same salt deposit density, the power transmission equipment with RTV long-lasting coatings should be applied. Its resistance to fouling can be doubled compared to uncoated insulators. Besides, its pollution flashover voltage also witnesses a substantial increase and so do its initial discharge voltage, the fog flashover voltage, and wet flashover voltages in clean state. RTV long-lasting coatings can adapt to a wide variety of natural pollution (e.g. alkali and salt pollution in coastal areas and heavy saline wastelands) and industrial pollution (e.g. acidic and alkaline dust pollution from chemical plants, cement factories, etc.).

At present, the State Grid Corporation of China requires every porcelain insulator to be coated when used on high-voltage transmission lines above 110 kV. The State Grid also requires the application of RTV coatings to porcelain insulators used on 1100 kV ultra-high-voltage lines at delivery. This is a factory compounding process, which no longer requires onsite application of coatings to the lines.

8.2.2 Composite Insulators

The idea of using composite insulators emerged internationally in the 1850s, but the first composite insulators were made of epoxy resin. SIR composite insulators were first used in North America around 1965. Since then, the number of various types of SIR insulators in the power system has grown exponentially. In present North America, composite insulators account for as much as 60–70% of the total, and grow steadily.

In order to prevent the problem of pollution flashover of porcelain and glass insulators, the development of composite insulators in China began in the late 1970s and early 1980s, under the auspices of Tsinghua University and other research institutions [8]. With a series of SIR composite insulators successively passing the appraisal and being put into trial operation, some transmission lines began to use SIR composite insulators in the mid- to late-1980s. Under the pressure of frequent flashover accidents, the power system expanded the use of composite insulators in polluted areas. Subsequently, the standards related to the use of composite insulators have been established and improved. Composite insulators have various advantages, such as lightweight, easy transportation, strong weather resistance, less maintenance, and excellent anti-pollution flashover performance. Especially with the improvement of SIR materials, the improvement of core rod fabrication process, and the advancement of processing and sealing process, composite insulators have been widely accepted by the power system. By the end of 2019, nine million composite insulators had been used on high-voltage transmission lines (110 kV and higher) in China, and the overall number of composite insulators has accounted for more than one-third of the total number of insulators.

The main components of composite insulator SIR include silicone resin, ATH, and silica, besides which, a small amount of other organic additives (e.g. vulcanizing agent, coupling agent, and various types of silicone oil) and inorganic additive components (e.g. zinc oxide and inorganic pigments) are added. These component materials and formulations have an important impact on the quality and service life of SIR sheds and even composite insulators. In Japan, S. Kumagai et al. [9] studied the corrosion of HTV SIR and the inhibitory effect of ATH on mechanical performance. They combined heat loss, differential thermal analysis, and mass spectrometry to investigate the high-temperature aging of HTV SIR with and without ATH fillers under the high-temperature conditions of air and argon atmosphere. Their experiments showed that the large amount of ATH fillers (>40% wt) reduced the small molecules present in the SIR that cause dry-band arcing and carbonization. This was because the water evaporation from ATH and the methyl group at the thermal decomposition temperature led to a change in the chemical structure.

8.2.3 Liquid Silicone Rubber (LSR)

LSR means that the raw rubber prior to vulcanization is in the liquid state. The LSR after vulcanization is not liquid, but solid. At present, all LSRs used in power systems are of addition type. LSR takes vinyl-containing polysiloxane as the basic polymer and hydrogen-containing polysiloxane as a crosslinking agent. Under the action of the platinum-based catalyst, they are crosslinked into an elastomer through the addition reaction of silicon hydrogenation.

In China, the sheds and bushings of the large-size composite insulators for extra- and ultra-high-voltage transmission lines (e.g. transformer bushings, hollow composite pillar insulators for substations, and current transformer bushings) are mostly made of addition-cure liquid silicone rubber (ALSR). However, after about 15 years of operation, the ALSR sheds of some composite insulators and bushings show obvious aging phenomena such as chalking and cracking, and may even encounter local arcing or flashover faults. In contrast, the ALSR composite insulators with the same operating life in Europe and the United States did not exhibit similar aging phenomena. The electric trace resistance and erosion resistance of SIR can be improved from the following aspects:

1) Flame-retardant fillers with a certain mass fraction can be introduced to improve the thermal decomposition characteristics and anti-tracking and anti-erosion performances of SIR [10–14]. This method is currently the most commonly used method. Flame-retardant fillers can be inorganic fillers (e.g. ATH [10, 15–20], calcium carbonate [21], and cyanuric acid melamine [22]) or organic fillers such as (G-ureido propyl) allyl ethoxysilane [23, 24].

2) Nanofillers with a certain mass fraction can be added to improve the thermal stability of SIR itself, thereby improving the anti-tracking and anti-erosion performances of SIR [25–27]. Most of the nanofillers are inorganic fillers, such as nano-silica and nano-alumina.
3) Highly thermally conductive inorganic fillers can be introduced to improve the thermal conductivity of SIR materials. In this manner, when SIR materials are subjected to electric arc burns, such fillers can more quickly transmit the arc heat out and reduce the erosion of the materials, thereby improving the anti-tracking and anti-erosion performances of SIR [28, 29]. Commonly used inorganic fillers with high thermal conductivity are silicon nitride [30, 31] and boron nitride [32, 33].
4) Surface modification can be applied on the filler to improve the bond strength between the filler and the matrix, thus improving the macroscopic properties such as breakdown field strength, mechanical properties, and arc resistance of SIR [34–37]. Modification of the filler surface can make the filler more evenly dispersed in the matrix and strengthen the interaction between the filler and the matrix as well.

8.2.4 Aging Mechanism and Condition Assessment of SIR Materials

SIR materials are a type of organic polymer materials. They will inevitably age as a result of withstanding the influences of various factors such as long-term electricity, heat, machinery, and acid rain during outdoor usage. SIR aging at the early stage manifests itself more in fading and hardening. There will even be skirt cracking and hydrophobicity decline, leading to increased pollution flashover probability of the composite insulators. In the worst case, the aging of SIR bushings will make the composite insulator core rod directly exposed to the outside environment, leading to core rod damage. This will cause internal insulation failure and even string dropping. In the mid-1990s, the CIGRE conducted statistics on composite insulator failures, of which composite insulator failures due to aging SIR accounted for up to 64%.

At present, researchers from various countries have carried out extensive research on the operating characteristics, aging process, and influencing factors of SIR materials. Research shows that SIR materials are mainly subject to various environmental stresses such as temperature, UV, oxidation, electric field, wind, moisture, and pollution during operation. These stresses may have an impact on the performance of SIR materials.

SIR materials have good thermal stability. The simple thermal action will generally not cause the aging of SIR materials. Kumagai and Lutz et al. conducted differential scanning calorimetry (DSC) and thermogravimetric analysis (TG) on new SIR specimens [15, 38]. The analysis results reveal that the temperature for thermal stability is above 200 °C. When the temperature reaches 300 °C, the weight of SIR specimens will gradually decrease, an obvious exothermic peak can be observed, and the internal ATH filler starts dehydration reaction. However, it is worth noting that high temperature tends to accelerate the internal crosslinking and degradation process of SIR and promote SIR aging. Therefore, temperature is the key influencing factor in the aging process.

Short-wave UV from sunlight is sufficient to break the molecular chains within the materials, but UV radiation can also cause deterioration of the materials due to photooxygenation and filler effects. Oxidation is an important influencing factor in the aging process of

SIR; Chen et al. [39] conducted an artificial accelerated thermal oxygen aging test on brand new SIR specimens and restored the aging characteristics of SIR such as discoloration and hardening. The analytical results revealed that after SIR was aged by thermal oxygen, the crosslinking degree increased and the organic groups were oxidized, which led to an increase in stiffness of the samples and the obvious discoloration.

Shed materials, especially high-voltage end shed materials, operate in a high electric field environment for a long time, which is why the effect of the electric field on their aging cannot be ignored. In a humid environment, discrete water droplets on the SIR surface may experience corona discharge; after continuous discharge, the surface hydrophobicity of the materials will exhibit a significant decrease [40]. Sima et al. [41] performed an infrared spectroscopic analysis on SIR materials after corona discharge of water droplets; the results showed that the corona discharge would enable the production of hydrophilic compounds such as silanol on the SIR surface and alter the main chain structure, thereby reducing the hydrophobicity of the materials.

Besides, wind has a significant influence on the mechanical properties of SIR materials in some extreme regions. Zhu et al. [42] analyzed the SIR cracking of composite insulators in areas with strong winds. The results showed that the SIR materials would oscillate under strong wind conditions, and this periodic fatigue vibration would lead to the degradation of the mechanical properties of the materials; the shed materials would crack if the vibration continued.

The aging of SIR materials tends to manifest itself in a variety of ways as follows:

1) Changes in appearance, including roughness changes, hardening, cracking, chalking, discoloration, and loss of gloss.
2) Changes in physical properties, including changes in the hydrophobicity, water and air permeability, swelling capacity, high-temperature resistance, and low-temperature performance of SIR materials.
3) Changes in mechanical properties, including changes in tensile strength, tear strength, elongation at break, and anti-fatigue properties.
4) Changes in electrical properties, including changes in volume resistance, surface resistance, dielectric constant, dielectric loss angle tangent, and electrical breakdown strength.

Regarding these types of changes, researchers have used a variety of experimental and analytical methods to investigate the operating state of SIR insulation materials to assess the actual performance of composite insulators using a number of practical, repeatable, and accurate measurements. Common measurement methods are based on stiffness, hydrophobicity, leakage current [43, 44], heat loss analysis, Fourier transform infrared (FTIR) spectroscopy [39], and thermoelectric stimulated current [45].

8.3 New External Insulation Materials

The operating environment of transmission lines is changeable and requires high reliability. The commonly used external insulation composites cannot meet the requirements under special environmental operating conditions. To cope with this problem, experts in the field of high voltage have been working hard to develop new external insulation composites, such as anti-icing materials to inhibit icing of transmission lines and alicyclic epoxy composites with better mechanical properties and trampling resistance.

8.3.1 Anti-icing Semiconductor Materials

Semiconductor SIR materials are prepared by adding conductive fillers to SIR, which can not only improve the mechanical properties and thermal stability of the composite materials but also reduce the resistivity of the SIR materials [46]. The concept of using semiconductor SIR coatings for insulator anti-icing was first proposed by Liao et al. [47]. After nearly 15 years of research and development, the first batch of coated insulators has been available for nearly four years and is currently the only successful onsite application of the insulator coating anti-icing method.

Semiconductor SIR-coated insulators combine the hydrophobicity and electrothermal properties of the coating materials to provide an anti-icing function. Under the low-temperature icing conditions, the good hydrophobicity of SIR materials can weaken the adhesion between ice and the coating, greatly reduce the density of ice, and promote the shedding of the ice coated on the insulator. Semiconductor SIR coatings have semiconductivity, which can increase the leakage current on the insulator surface. The Joule heat produced by the leakage current can improve the surface temperature of the insulator, thereby slowing down the growth of insulator icing. The test results of Liao et al. showed that when the surface leakage current of coated insulators is about 1 mA, the growth of icing can be effectively slowed down by more than 50%, compared to uncoated insulators. Xu et al. found that [48] an improvement of the semiconductor coating conductivity will increase the leakage current and temperature of the insulator surface and strengthen the anti-icing effect of the coated insulators. When the leakage current is greater than 1 mA, the insulator surface demonstrates a slower ice growth. When the leakage current reaches 5 mA or more, no ice is roughly formed on the insulator surface.

To solve the problem of long-term heat loss in coated insulators, Xu et al. proposed a semiconductor coating structure with a self-turn-off effect. This structure reduces the continuous heat loss of the coating while effectively slowing down and stopping the growth of ice on the insulator surface [49, 50]. As shown in Figure 8.1, the semiconductor coating is applied only on the lower surface of the insulator, while the upper surface is an uncoated area. In good weather, the coated insulator does not generate large leakage currents due to the presence of uncoated areas. Under icing conditions, the formation of a water film of supercooled water droplets can connect the coating and the metal fittings to increase the insulator leakage current and surface temperature, thereby slowing down the freezing of supercooled water droplets on the insulator surface [14]. As shown in Figure 8.2, coated insulators are very good at preventing the growth of ice in both clean and polluted conditions; compared to uncoated insulators, the rain flashover voltage of coated insulators is not reduced, and the pollution flashover voltage is significantly increased [14].

In the fall of 2010, semiconductor SIR-coated insulators were first piloted in icing-prone areas such as Qingyuan and Shaoguan in northern Guangdong, and Zhaotong in Yunnan, China. The voltage level of the test lines ranged from 10 to 220 kV. Their arrangement patterns included overhanging strings, V-shaped strings, and tension-resistant strings.

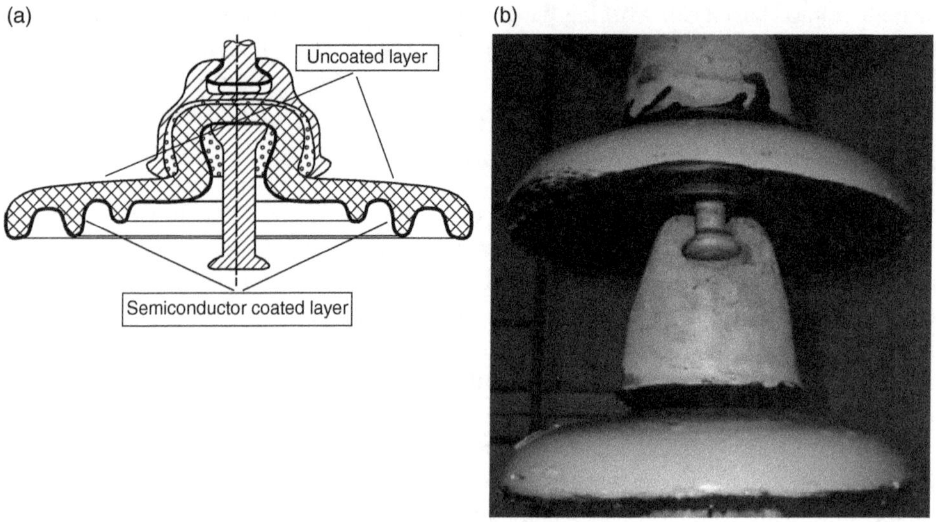

Figure 8.1 Configurations of semiconductor coatings on insulator surfaces. (a) Schematic diagram. (b) Picture.

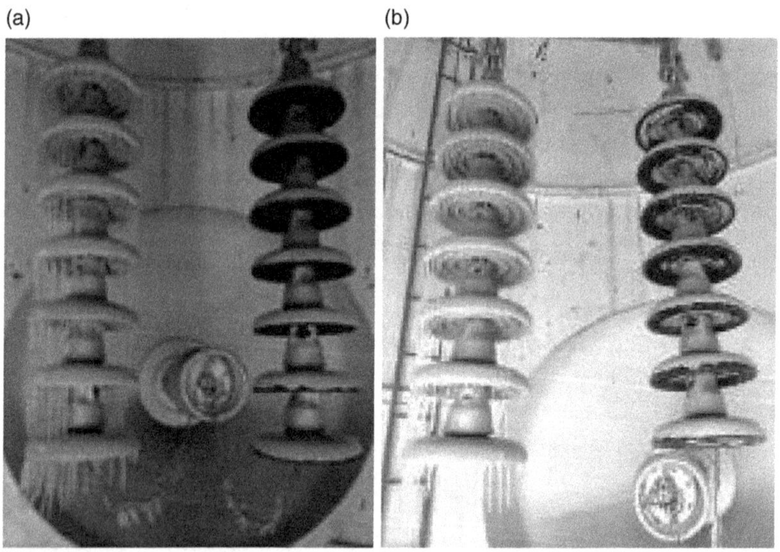

Figure 8.2 Anti-icing effect of semiconductor SIR-coated insulators at the laboratory. (a) Clean surfaces. (b) Polluted surfaces. *Source:* El-Hag et al. [14].

The insulator strings included ordinary insulator strings and interspersed insulator strings [51]. In the field test, the semiconductor-coated insulators achieved a certain anti-icing effect [14]. As shown in Figure 8.3 below, the coatings have been running safely for nearly 10 years, during which no ice flashover accident has occurred in this segment of the Guangdong power grid.

Figure 8.3 Onsite anti-icing effect of semiconductor SIR-coated insulators in the winter of 2010; 110 kV Damei line (linear string).

8.3.2 Hydrophobic CEP

The insulators with epoxy resin as a constituent material have been successfully used for decades, including bisphenol epoxy resin insulators, insulation bushings, and dry-type transformers. They are mainly featured by lightweight, low processing costs, impact resistance, and easy processing into a more complex insulation structure. On the other hand, bisphenol epoxy resin performed badly in the resistance to tracking, erosion, and aging; hence, it can hardly be used outdoors. For this reason, researchers developed HCEP, which is actually a hydrophobic product made by adding a special rheological additive to the commonly used CEP material.

Generally, the particle size of special additives added to the insulation material is approximately less than one millimeter to a few millimeters. Due to their special rheological properties, these additives can play the role of an emulsifying agent for insulation materials. After HCEP molding and curing process, these additives penetrate into the interior of the materials. When pollutants adhere to the material surface, these additives will change the state of the pollution layer, making the material hydrophobicity mobile. Therefore, insulators made of HCEP compounds have good hydrophobicity, hydrophobic mobility, and hydrophobic reducibility. They can be used in extra- and even ultra-high-voltage power systems.

A new type of insulator produced by a Chinese manufacturer is illustrated in Figure 8.4. This type of insulator uses hydrophobic cycloaliphatic compounds. Such insulators are

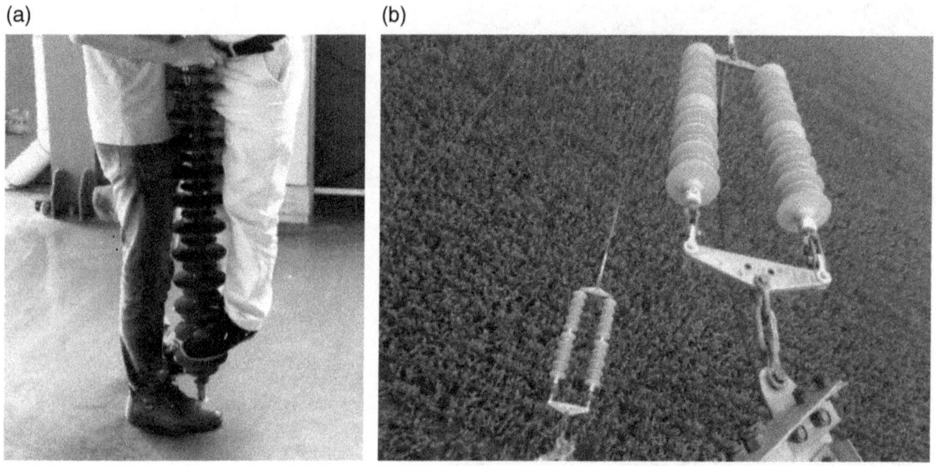

Figure 8.4 New type composite insulators. (a) High strength. (b) Operation.

generally made of HCEP. This can increase the insulation margin to a large extent, which prevents the occurrence of flashover under any circumstances. This also confirms the mechanical reliability of HCEP, indicating that HCEP can completely replace some common support materials. It can be seen that HCEP has outstanding mechanical properties and very high strength, and that its performance is insusceptible to artificial trampling. At present, this type of composite insulator has been in operation in Northwest China (Figure 8.4).

CEP is featured by good hydrophobicity. It has good hydrophobic mobility for diatomaceous earth but weak hydrophobic mobility to kaolin earth. It also possesses better resistance to tracking and erosion than common SIR materials. In addition, the stiffness of CEP makes the insulators resistant to trampling, bird pecking, and wind and sand erosion. Unlike common composite insulators, CEP insulators are used in transmission lines in short strings, so they can be easily transported. Besides, different numbers of strings can be hung on lines of different voltage levels according to the actual situation. The use of advanced production and development technology ensures excellent electrical and mechanical properties of CEP insulators. Especially, such insulators overcome the weakness of the short string structure and exhibit excellent internal insulation strength in the steep wave test. CEP also has excellent anti-pollution flashover performance and higher pollution flashover voltage gradient than ordinary SIR insulators. A layer of RTV coating onto the insulator surface has little effect on the changes in the pollution flashover voltage. References are also given for the selection of the number of strings under different pollution conditions and voltage levels.

8.4 Summary

The development and application history of composite insulation materials for external insulation shows that the composite materials used in field operation often require decades of research and development, continuous adjustment and optimization of formulations, and production processes. Their operating characteristics and aging state are also of great concern to power grid companies. Although new composite insulation materials continue to emerge, they fail to fundamentally change the absolute dominance of SIR composite insulation materials. In terms of cost, operational performance, and anti-pollution flashover requirements, SIR is still the dominant material for external insulation. However, if the specific operating environment requires it, the application of new composite materials can overcome the deficiencies of the existing SIR materials. Therefore, future research should pay more attention to the application of composite insulation materials in extreme, special environments and the development of new materials.

References

1 Gubanski, S. (2015). Outdoor polymeric insulators: role of corona in performance of silicone rubber housings. *Conference on Electrical Insulation and Dielectric Phenomena*, Ann Arbor, MI, USA (18–21 October 2015), pp. 1–9.

2 Xiu-Min, Z. and Xiao-Dong, Z. (2006). Recent progress of high voltage outdoor insulator materials. *Insulating Materials* 1: 46–51.

3 Tao, X. (2007). Development history and direction of composite insulators on important manufacture technique. *High Voltage Engineering* 1 (33): 106–110.

4 Hackam, R. (1999). Outdoor HV composite polymeric insulators. *IEEE Transactions on Dielectrics and Electrical Insulation* 6 (5): 557–585. https://doi.org/10.1109/TDEI.1999.9286745.

5 Karady, G.G., Shah, M., and Brown, R.L. (1995). Flashover mechanism of silicone rubber insulators used for outdoor insulation-II. *IEEE Transactions on Power Delivery* 10: 1972–1978.

6 Karady, G.G., Shah, M., and Brown, R.L. (1995). Flashover mechanism of silicone rubber insulators used for outdoor insulation-I. *Power Delivery IEEE Transactions on Power Delivery* 10: 1965–1971.

7 Gao, H., Jia, Z., Guan, Z. et al. (2007). Investigation on field-aged RTV-coated insulators used in heavily contaminated areas. *IEEE Transactions on Power Delivery* 22: 1117–1124.

8 Xidong, L. and Shaowu, W. (1999). Development of composite insulators in China. *IEEE Transactions on Dielectrics and Electrical Insulation* 6: 586–594.

9 Yoshimura, N. and Kumagai, S. (1999). Electrical and environmental aging of silicone rubber used in outdoor insulation. *IEEE Transactions on Dielectrics and Electrical Insulation* 6: 632–650.

10 Ghunem, R.A., Jayaram, S.H., and Cherney, E.A. (2015). Suppression of silicone rubber erosion by alumina trihydrate and silica fillers from dry-band arcing under DC. *IEEE Transactions on Dielectrics and Electrical Insulation* 22: 14–20.

11 Ansorge, S., Schmuck, F., and Papailiou, K.O. (2012). Improved silicone rubbers for the use as housing material in composite insulators. *IEEE Transactions on Dielectrics and Electrical Insulation* 19: 209–217.

12 Xue, Y., Li, X.-f., Zhang, D.-h. et al. (2018). Comparison of ATH and SiO2 fillers filled silicone rubber composites for HTV insulators. *Composites Science and Technology* 155: 137–143.

13 Ansorge, S., Schmuck, F., and Papailiou, K.O. (2015). Impact of different fillers and filler treatments on the erosion suppression mechanism of silicone rubber for use as outdoor insulation material. *IEEE Transactions on Dielectrics and Electrical Insulation* 22: 979–988.

14 El-Hag, A.H., Simon, L.C., Jayaram, S.H., and Cherney, E.A. (2006). Erosion resistance of nano-filled silicone rubber. *IEEE Transactions on Dielectrics and Electrical Insulation* 13: 122–128.

15 Kumagai, S. and Yoshimura, N. (2001). Tracking and erosion of HTV silicone rubber and suppression mechanism of ATH. *IEEE Transactions on Dielectrics and Electrical Insulation* 8: 203–211.

16 Kumagai, S. and Yoshimura, N. (2004). Polydimethylsiloxane and alumina trihydrate system subjected to dry-band discharges or high temperature part I: chemical structure. *IEEE Transactions on Dielectrics and Electrical Insulation* 11: 691–700.

17 Kumagai, S. and Yoshimura, N. (2004). Polydimethylsiloxane and alumina trihydrate system subjected to dry-band discharges or high temperature part II: electrical insulation. *IEEE Transactions on Dielectrics and Electrical Insulation* 11: 701–707.

18 Meyer, L.H., Cherney, E.A., and Jayaram, S.H. (2004). The role of inorganic fillers in silicone rubber for outdoor insulation alumina tri-hydrate or silica. *IEEE Electrical Insulation Magazine* 20: 13–21.

19 Ramirez, I., Cherney, E.A., and Jarayam, S. (2012). Comparison of the erosion resistance of silicone rubber and EPDM composites filled with micro silica and ATH. *IEEE Transactions on Dielectrics and Electrical Insulation* 19: 218–224.

20 Ramirez, I., Jarayam, S., and Cherney, E.A. (2010). Performance of silicone rubber nanocomposites in salt-fog, inclined plane, and laser ablation tests. *IEEE Transactions on Dielectrics and Electrical Insulation* 17: 206–213.

21 Koshino, Y., Umeda, I., and Ishiwari, M. (1998). Deterioration of silicone rubber for polymer insulators by corona discharge and effect of fillers. *1998 Annual Report Conference on Electrical Insulation and Dielectric Phenomena (Cat. No.98CH36257)*, vol. 71, Atlanta, GA, USA (25–28 October 1998), pp. 72–79.

22 Schmidt, L.E., Kornmann, X., Krivda, A., and Hillborg, H. (2010). Tracking and erosion resistance of high temperature vulcanizing ATH-free silicone rubber. *IEEE Transactions on Dielectrics and Electrical Insulation* 17: 533–540.

23 Fang, W., Lai, X., Li, H. et al. (2014). Effect of urea-containing anti-tracking additive on the tracking and erosion resistance of addition-cure liquid silicone rubber. *Polymer Testing* 37: 19–27.

24 Fang, W., Zeng, X., Lai, X. et al. (2015). Thermal degradation mechanism of addition-cure liquid silicone rubber with urea-containing silane. *Thermochimica Acta* 605: 28–36.

25 Tsekmes, I.A., Morshuis, P.H.F., Smit, J.J., and Kochetov, R. (2015). Enhancing the thermal and electrical performance of epoxy microcomposites with the addition of nanofillers. *IEEE Electrical Insulation Magazine* 31: 32–42.

26 Yong, L., Zhonglei, L., and Boxue, D. (2014). Effects of nano-SiO 2 particles on surface tracking characteristics of silicone rubber composites. *Applied Physics Letters* 105: 102905.

27 Loganathan, N., Muniraj, C., and Chandrasekar, S. (2014). Tracking and erosion resistance performance investigation on nano-sized SiO_2 filled silicone rubber for outdoor insulation applications. *IEEE Transactions on Dielectrics and Electrical Insulation* 21: 2172–2180.

28 Cherney, E.A. (2005). Silicone rubber dielectrics modified by inorganic fillers for outdoor high voltage insulation applications. *IEEE Transactions on Dielectrics and Electrical Insulation* 12: 1108–1115.

29 Meyer, L., Jayaram, S., and Cherney, E.A. (2004). Thermal conductivity of filled silicone rubber and its relationship to erosion resistance in the inclined plane test. *IEEE Transactions on Dielectrics and Electrical Insulation* 11: 620–630.

30 Zha, J.W., Dang, Z.M., Li, W.K. et al. (2014). Effect of micro-Si_3N_4-nano-Al_2O_3 co-filled particles on thermal conductivity, dielectric and mechanical properties of silicone rubber composites. *IEEE Transactions on Dielectrics and Electrical Insulation* 21: 1989–1996.

31 Zha, J.-W., Zhu, Y.-H., Li, W.-K. et al. (2012). Low dielectric permittivity and high thermal conductivity silicone rubber composites with micro-nano-sized particles. *Applied Physics Letters* 101: 062905.

32 Donnay, M., Tzavalas, S., and Logakis, E. (2015). Boron nitride filled epoxy with improved thermal conductivity and dielectric breakdown strength. *Composites Science and Technology* 110: 152–158.

33 Fang, L., Wu, C., Qian, R. et al. (2014). Nano–micro structure of functionalized boron nitride and aluminum oxide for epoxy composites with enhanced thermal conductivity and breakdown strength. *RSC Advances* 4: 21010–21017.

34 Ramirez, I., Jayaram, S., Cherney, E.A. et al. (2009). Erosion resistance and mechanical properties of silicone nanocomposite insulation. *IEEE Transactions on Dielectrics and Electrical Insulation* 16: 52–59.

35 Stöckelhuber, K.W., Svistkov, A.S., Pelevin, A.G., and Heinrich, G. (2011). Impact of filler surface modification on large scale mechanics of styrene butadiene/silica rubber composites. *Macromolecules* 44: 4366–4381.

36 Dongling, M., Treese, A.H., Richard, W.S. et al. (2005). Influence of nanoparticle surface modification on the electrical behaviour of polyethylene nanocomposites. *Nanotechnology* 16: 724.

37 Zhang, L., Zhou, Y., Cui, X. et al. (2014). Effect of nanoparticle surface modification on breakdown and space charge behavior of XLPE/SiO_2 nanocomposites. *IEEE Transactions on Dielectrics and Electrical Insulation* 21: 1554–1564.

38 Lutz, B., Cheng, L., Guan, Z. et al. (2012). Analysis of a fractured 500 kV composite insulator – identification of aging mechanisms and their causes. *IEEE Transactions on Dielectrics and Electrical Insulation* 19: 1723–1731.

39 Chen, C., Jia, Z., Wang, X. et al. (2015). Micro characterization and degradation mechanism of liquid silicone rubber used for external insulation. *IEEE Transactions on Dielectrics and Electrical Insulation* 22: 313–321.

40 Lopes, I.J.S., Jayaram, S.H., and Cherney, E.A. (2001). A study of partial discharges from water droplets on a silicone rubber insulating surface. *IEEE Transactions on Dielectrics and Electrical Insulation* 8: 262–268.

41 Sima, W.X., Liu, Z.Y., Jiang, X.L., and Yang, Q. (2005). Effects on surface performance of partial discharges due to discrete water droplets on the silicone rubber. *Proceedings of the CSEE* 6 (25): 113–118.

42 Zhu, Z. et al. (2015). Fatigue fracture of composite insulator sheds utilized in strong wind areas. *IEEE Transactions on Dielectrics and Electrical Insulation* 22: (3): 1636–1643. https://doi.org/10.1109/TDEI.2015.7116360.

43 Vosloo, W.L., Holtzhausen, J.P., and Roediger, A.H.A. (1996). Leakage current performance of naturally aged non-ceramic insulators under a severe marine environment. *IEEE Africon* 1: 489–495. https://doi.org/10.1109/AFRCON.1996.563162.

44 El-Hag, A.H., Jayaram, S.H., and Cherney, E.A. (2003). Fundamental and low frequency harmonic components of leakage current as a diagnostic tool to study aging of RTV and HTV silicone rubber in salt-fog. *IEEE Transactions on Dielectrics and Electrical Insulation* 10: 128–136.

45 Liang, Y., Ding, L., Li, C.R., and Yang, K. (2006). The TSC characteristics of HTV composite insulators. Electrical Insulation, 1988. *Conference Record of the 1988 IEEE International Symposium on*, Toronto, ON, Canada (11–14 June 2006), pp. 190–192.

46 Momen, G. and Farzaneh, M. (2011). Survey of micro/nano filler use to improve silicone rubber for outdoor insulators. *Reviews on Advanced Materials Science* 27: 1–13.

47 Liao, W., Jia, Z., Guan, Z.C. et al. (2007). Reducing ice accumulation on insulators by applying semiconducting RTV silicone coating. *IEEE Transactions on Dielectrics and Electrical Insulation* 14: 1446–1454.

48 Xu, Z., Jia, Z., Li, Z. et al. (2011). Anti-icing performance of RTV coatings on porcelain insulators by controlling the leakage current. *IEEE Transactions on Dielectrics and Electrical Insulation* 18: 760–766.

49 Wei, X., Jia, Z., Sun, Z. et al. (2012). Study of anti-icing performance of insulator strings bottom-coated with semiconductive silicone rubber coating. *IEEE Transactions on Dielectrics and Electrical Insulation* 19: 2063–2072.

50 Qin, Y., Wang, Z., Xu, Z. et al. (2009). Reducing ice coating on insulators by combining RTV with different volume resistivity. *Conference on Electrical Insulation and Dielectric Phenomena, IEEE Conference*, Virginia Beach, VA, USA (18–21 October 2009), pp. 446–449.

51 Wei, X., Xu, Z., Zhao, Y. et al. (2012). Anti-icing mechanism and operation performance research of insulator string by using semiconductive coating based on switch-off effect. *Gaodianya Jishu/High Voltage Engineering* 38: 2549–2558.

9

Polymer Composites for Embedded Capacitors

Shuhui Yu, Suibin Luo, Riming Wang, and Rong Sun

Shenzhen Institute of Advanced Electronic Materials, Shenzhen Institute of Advanced Technology, Chinese Academy of Sciences, Shenzhen, China

9.1 Introduction

9.1.1 Development of Embedded Technology

The opening of the information age relies on its physical foundation of electronic and electrical systems. The electronic circuit, consisting of individual electronic components, represents a typical functional unit of these systems. The constituent electronic components include resistors, capacitors, inductors, transistors, diodes, etc. From a historic point of view, the application of electronics dates back to the first radio transmission in 1895 by Marconi. Since the 1950s, the electronic industry has undergone explosive growth, characterized by the continuous miniaturization, functionalization, and integration of electronic circuit components. The huge leap in the development of electronics was enabled by lots of inventions, among which diodes, triodes, bipolar transistors, and integrated circuits are the most remarkable milestones in the evolution of electronics. Nowadays, electronic circuits are pervasive in our daily lives, from home appliances, such as cell phones, TVs and refrigerators, etc., to the controlling system in factories.

A traditional electronic circuit usually contains active and passive components, which are mounted on the surface of the printed wire board and electrically interconnected with each other. To define a component as an active or passive component, the power-related behavior of the components in the electronic circuit is considered. Passive components, like resistors, capacitors, and inductors, absorb and dissipate electrical power provided by active devices or external power supply. They perform various important functions, such as bias, attenuating, decoupling, filtering, tuning, and termination in an electronic circuit. Active components rely on external energy supply and can inject power into a circuit. Typical actives include transistors, triode vacuum tubes, tunnel diodes, and so on.

Polymer Composites for Electrical Engineering, First Edition. Edited by Xingyi Huang and Toshikatsu Tanaka.

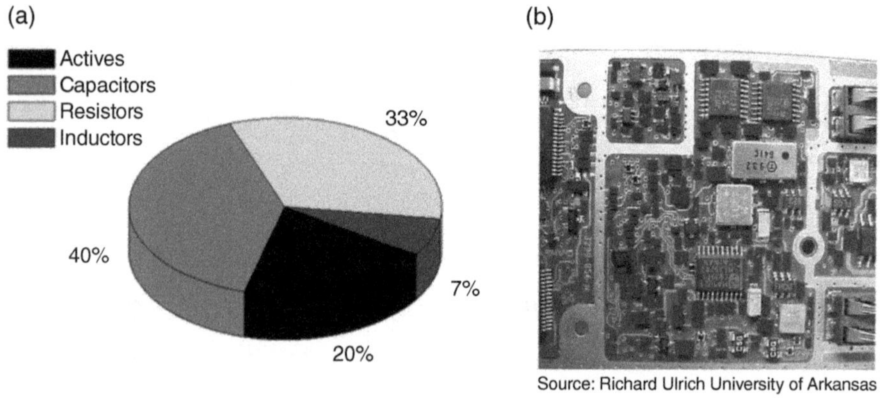

Figure 9.1 Capacitors in the integrated circuit. (a) The number proportion of actives, capacitors, resistors, and inductors in a typical electronic circuit. (b) A photo image of a typical electronic circuit.

The total number of actives account for about 20% of all the components in a typical electronic circuit as illustrated in Figure 9.1a. Apart from the actives, about 80% of the components in number are passives which cover more than 40% of the surface area on a printed circuit board (PCB). As shown in Figure 9.1b, the surface of the board is covered by a large number of passives. The passive is connected onto the board by soldering and separated in a long distance between each other. The passive consumes lots of soldering actions and introduces large parasitic inductance. With the demand for miniaturization, low cost, and high performance, the technology of embedding passives inside the PCB has emerged as an alternative to surface-mounted art. [1–4]

Capacitors are the most frequently used passive components; thus, the development of embedded capacitors (Figure 9.2) can significantly increase the integration of electronic circuits. [5] The embedded capacitance has the following advantages:

- Significantly enhances the electric performance of the circuit while reduces impedance, power bus noise, and electromagnetic interference (EMI).
- Allows for removal of large numbers of discrete capacitors.
- Increases usable board area, allowing for board size reduction.
- Performs better decoupling function than discrete capacitors in independent testing.

The embedded capacitor materials possess a sandwich-like structure with a dielectric layer in the middle and covered by a copper foil on each side. The performance of the embedded capacitor material mainly depends on the dielectric layer. The properties of this dielectric layer that are of our concern will be discussed in detail in the following section. The copper foil has two types, electrodeposited copper foil and rolled copper foil. The electrodeposited copper foil has one rough surface while the other is flat. Generally, the rough surface is used for laminating with the dielectrics and the flat surface for the formation of the electric pattern. The two surfaces of the rolled copper foil are the same with high flatness. Both two types of copper foils can be used to fabricate embedded capacitor materials. The key is to choose copper foil with suitable roughness. High roughness results in high

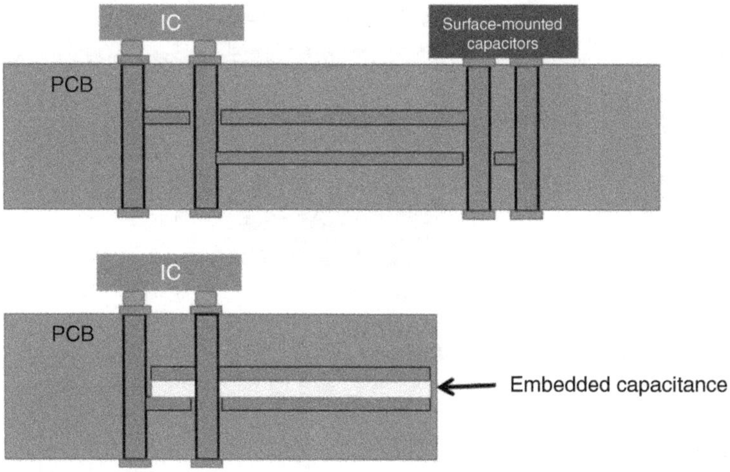

Figure 9.2 Schematic illustration of surface-mounted capacitors and embedded capacitors.

peel strength between the dielectric layer and copper foil which, in turn, enhances reliability. But it has the risk of an easy electric breakdown when the ratio of copper roughness to dielectric layer thickness is too high. Recently, high-end electrodeposited copper foil with an Rz profile of <1 um has been developed. The rolled copper foil has an ultralow Rz profile, and the embedded capacitor materials fabricated with rolled copper foils can achieve higher electric breakdown strength than that uses electrodeposited copper foils. But the peel strength of the embedded capacitor materials fabricated with rolled copper foil is usually lower than 0.6 N/mm, while those fabricated with electrodeposited one has a much higher peel strength.

Dielectric materials and the capacitor fabrication process will be described in details in the following part. Figure 9.3 displays the integration process of the embedded capacitor with PCB. Based on the self-strength of the embedded capacitor materials, a sequential process (Figure 9.3a) or both side process (Figure 9.3b) can be employed. On the one hand, for the embedded capacitor material with low self-strength, typically the dielectric material with high filler loading and ultrathin thickness, the application of the sequential process is appropriate. On the other hand, for the embedded capacitor material with high self-strength, typically the material with low filler loading and high thickness, or having a strengthening layer, both side process can be used.

The sequential process in Figure 9.3a can be briefly described as follows. The embedded capacitor materials are heat-treated to remove the moisture and washed to remove the surface oil dirt. Then the electric patterns are formed on one Cu foil of the embedded capacitor material by laminating photosensitive dry film, exposing, developing, and etching in sequence. After that, a prepreg and a copper foil are placed on the side of the formed electric pattern. At high temperature and high pressure, the prepreg fills the pattern and attaches it with the etched embedded capacitor material and copper foil. Then, the other copper foil is etched to form the electric pattern. After laminating with another prepreg and copper foil, the laminated board is formed and can be used as a core board to form the outside patterns.

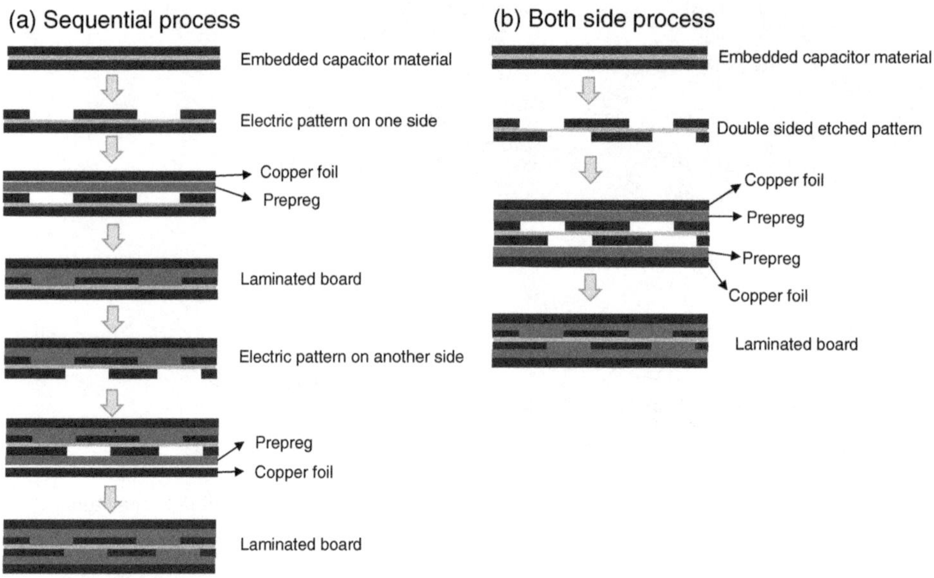

Figure 9.3 Fabricating process of printed circuit board with embedded capacitor materials. (a) sequential process, (b) both side process.

Compared with the sequential process, the both side process in Figure 9.3b is more convenient. The electric pattern can be formed simultaneously on both copper foils of the embedded capacitor material. After laminating with the prepreg and copper foils, the laminated board is formed for fabricating outside patterns.

The performance of both sequential process and both side process highly depends on the dielectric properties of the embedded capacitor materials; so, the development of high-performance-embedded capacitors urges the solution to the material and process issues. Therefore, it is necessary to develop dielectric materials that satisfy the requirements of embedded capacitor fabrication.

9.1.2 Dielectric Materials for Commercial Embedded Capacitors

The material which meets the requirements for embedded capacitor application should possess high dielectric permittivity, low dielectric loss, high thermal stability, high breakdown strength, high peel strength, simple processing technique, and so on. Besides, it is of primary importance to stand the PCB processing conditions. Up to now, the PCB can be divided into three types, ceramic-based PCB, metal-based PCB, and organic PCB. Organic-based PCB is the most widely used PCB in high-end equipment in telecommunication, computing, military, and medical industries, as well as portable electronics. Organic-based PCB becomes the mainstream owing to its easy processing at a relatively low temperature. In this chapter, we focus on the polymer-based dielectrics that are compatible with the organic-based PCB.

Among polymer-based dielectrics, polymer composites filled with ceramic fillers are widely investigated and used in electronics. Pure polymer dielectrics, like epoxy and PI which satisfy the organic PCB process, are not directly used due to their low dielectric permittivity. Other polymers with high dielectric permittivity, such as the ferroelectric polymer, polyvinylidene fluoride (PVDF), P(VDF-TrFE), and P(VDF-TrFE-HFP), are not suitable as embedded capacitor materials due to their weak thermal stability at high temperature. [6–10] The polymer composites can be divided into different types based on the polymer matrix. Epoxy resin is the most popular and important polymer matrix for embedded capacitor materials. Most of the organic-based PCBs are composites with epoxy matrix, such as FR-4 and FR-5. A typical organic-based PCB mainly contains epoxy resin, copper foil, glass fiber cloth, and ceramic fillers.

The first-generation polymeric dielectric was invented by Sanmina-SCI, i.e. BC2000 (dielectric layer of 25 μm) and BC1000 (dielectric layer of 50 μm), composed of low-k and low-thickness woven glass-reinforced FR-4 epoxy. The capacitance is intrinsically low, but because of the diminished parasitic inductance and shortened transportation length in comparison with the surface-mounted ones, the signal quality is effectively enhanced, and this material has been used for a long time. BC2000 and BC1000 are the most frequently used embedded capacitors in the current market. Further reduction of thickness and improvement of dielectric permittivity is hindered by the existence of woven glass. Thus, the second-generation polymeric dielectric emerged as resin film without woven glass and filled with high-k filler instead. Polymer/ceramic composites combine the high-k of ceramics and the excellent processability of polymer, and a high k value can be obtained through a low-temperature film fabrication technique. Polymer/ceramic composites attract a lot of attention from both academia and industry, represented by several commercially available products by OAK-MITSUI (epoxy-based BC24, BC16, BC12, BC8), DUPONT (PI-based HK4), and 3M (epoxy-based C-Ply) as listed in Table 9.1.

Most of the current-embedded capacitor materials except C-Ply comply with the both side process technology. But their capacitance density is much lower than that of the C-Ply due to the limited thickness and permittivity. Generally, materials applicable to the both side process should have a strengthening layer of the glass-fiber cloth, e.g. BC2000, FaradFlex, or enough mechanical strength, e.g. Interra HK. The C-Ply is a composite constructed by epoxy and $BaTiO_3$ particles without a strengthening layer. The permittivity of C-Ply is higher than that of the others owing to the removal of low-permittivity glass fiber cloth. Besides, the thickness of the C-Ply can be ultralow without glass fiber cloth. The disadvantages of the ultrathin embedded capacitor material include the low breakdown electric field and the complicated PCB processing. The Hi-K developed by the authors' group in SIAT is constructed with modified epoxy/$BaTiO_3$ with high permittivity and has enough mechanical strength, enabling the both side PCB processing even without a strengthening layer.

The development of high-performance electronic circuit requires increasingly higher capacitance. In the long term, the research on embedded capacitor materials with high performance is of vital importance.

Table 9.1 Commercially embedded capacitors and their dielectric properties.

Company	Product	Composition	Capacitance nF/in^2	Thickness µm	Df@1 kHz	Dk@1 kHz	PCB processing
Oak-Mitsui	FaradFlex	Epoxy/Y5V ceramic	0.9–4.5	8, 12, 16, 24	0.015–0.019	4.6–10	Both side
Sanmina	BC2000	Epoxy/glass fiber cloth	0.5	50	0.015	4.2	Both side
3M	C-Ply	Epoxy/$BaTiO_3$	6.4~40	3, 6, 12, 14	0.1	16, 22	Sequential
Dupont	Interra HK	Polyimide	0.8, 1.7	12, 25	0.005	3.5	Both side
SIAT[a]	Hi-K	Modified Epoxy/$BaTiO_3$	4.7~7.5	12, 14, 16, 19	0.01	16	Both side

[a] Shenzhen Institute of Advanced Technology.

9.2 Researches on the Polymer-Based Dielectric Nanocomposites

The fabrication of embedded capacitor materials starts from the preparation of varnish which contains inorganic filler particles and polymer matrix. The following parts review the efforts that have been made by our group and other researchers in obtaining polymer-based nanocomposites with the desired dielectric and mechanical performance.

9.2.1 Filler Particles

The filler is the main constituent of the epoxy varnish with a weight percentage of up to 50 wt%. To ensure a good filler dispersion in the varnish, surface modification is necessary. In the chemical industry, dispersants produced by BYK, BASF, and coupling agents produced by Shinetsu are widely used. Generally, the dispersant can directly attach to the filler. [11] After the mixing process, by means of ball milling, sand grinding, three-roll grinding, ultrasonic, or others, the fillers can be well dispersed in the mixture without agglomeration. This is a simple and effective method to prepare varnish. By using coupling agent, the surface of the fillers is treated before being added into the mixture. Compared with the application of dispersant, the use of coupling agents is more complicated. Some functional groups, such as amino-group, carboxyl-group, epoxy-group, and sulfhydryl group, etc., can react with the epoxy molecule and significantly enhance thermal stability and reliability of the composites. Hence, many publications focus on filler surface modification by using the coupling agent. [12–18]

The key characteristic of the chemical that is suitable for filler surface modification is that it has at least one bonding group to react with the filler surface and at least one dispersion group to react with the epoxy molecule, as shown in Figure 9.4. The dispersion group should be organic for better dispersion in the epoxy suspension. A solvent is another consideration when choosing a dispersant because the affinity of the dispersant with the filler and epoxy suspension will be influenced by the polarity of the solvent.

The filler surface modification with surfactants (dispersants and coupling agents) can improve the dispersion of fillers in polymer composites. The most important role of well-dispersed fillers in industry product is that they can ensure reproducibility among different batches. Besides, better filler dispersion in composites leads to better electric and mechanic properties. For example, the epoxy-based composites can obtain high electric breakdown strength, high elongation, low water absorption, low dielectric loss, etc., when the filler particles are well dispersed.

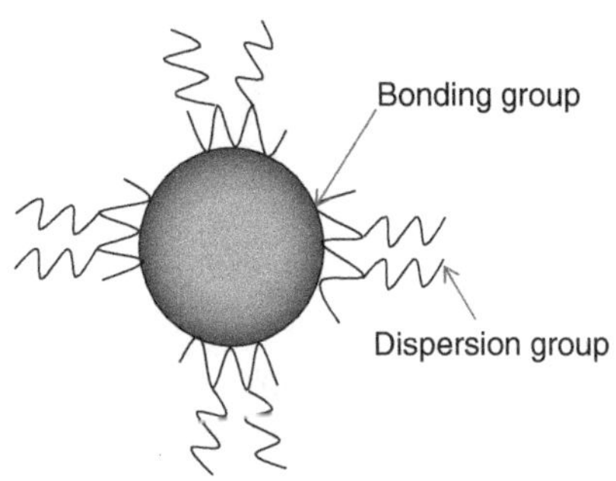

Figure 9.4 The schematic illustration of the surface-modified filler.

Embedded capacitor materials with high dielectric permittivity are highly desired in the industry to provide high capacitance density. To achieve a high dielectric permittivity, increasing the filler content is a simple and effective method. [19–23] However, high filler content

(>80 wt%) will significantly reduce the electric breakdown strength, peel strength, and increase water adsorption. Hence, in commercially embedded capacitor materials, the filler content is usually below 80 wt%, and the permittivity is limited to 20~30 even if the content is up to 80 wt%. Researches on filler and epoxy modification have been carried out in recent years to further enhance the dielectric permittivity.

Based on the percolation theory, high dielectric permittivity can be achieved by using conductive fillers with a filler loading near the percolation threshold. [24–29] However, two major disadvantages, high dielectric loss and narrow filler content window, limit the real application of this type of composites in embedded capacitor materials. To realize the possible application of percolative composites in embedded capacitor materials, our group designed a hybrid structure of $BaTiO_3$-Ag particles in which Ag nanoparticles with the size of 5~20 nm randomly deposited on the surface of 100 nm $BaTiO_3$ particles as shown in Figure 9.5.[30] The deposition of Ag nanoparticles leads to the enhancement of the polarization of the ferroelectric $BaTiO_3$ nanoparticles. Hence, the composites could achieve an improved permittivity compared with $BaTiO_3$-filled composites with the same filler loading in volume. Importantly, the randomly deposited Ag nanoparticles do not form a network even at a high filler loading. Moreover, the composite maintains a very low conductivity when the filler loading is up to 56.8 vol%. The low conductivity means low dielectric loss. With the filler contents of 33 and 43.4 vol%, the dielectric permittivities are 54 and 94.3, respectively, and the dielectric loss is around 0.06, endowing this hybrid structure a promising potential in embedded capacitor materials.

The $BaTiO_3$-Ag hybrid particles were also used to prepare a prototype of embedded capacitor material. [31] Figure 9.6a shows a photo of an etched $BaTiO_3$-Ag/epoxy-embedded capacitor material. As shown in Figure 9.6b, the dielectric permittivity of the $BaTiO_3$-Ag/epoxy composites is much higher than that of the $BaTiO_3$/epoxy composites when the content of the fillers is 80 wt%. The permittivity and dielectric loss are 48.2 and 0.035, respectively for $BaTiO_3$-Ag/epoxy, and 16 and 0.008 for BT-epoxy at 100 Hz. These results evidence that the conductive nanoparticles deposited on the surface of ferroelectric particles can achieve high dielectric permittivity and low loss. The disadvantage of using the $BaTiO_3$-Ag

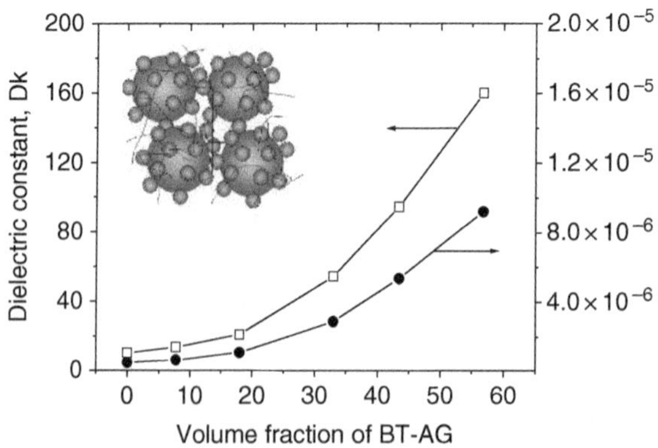

Figure 9.5 The hybrid structure of $BaTiO_3$-Ag particles and dielectric permittivity and conductivity of its PVDF composites. *Source:* Luo et al. [30]. © American Chemical Society.

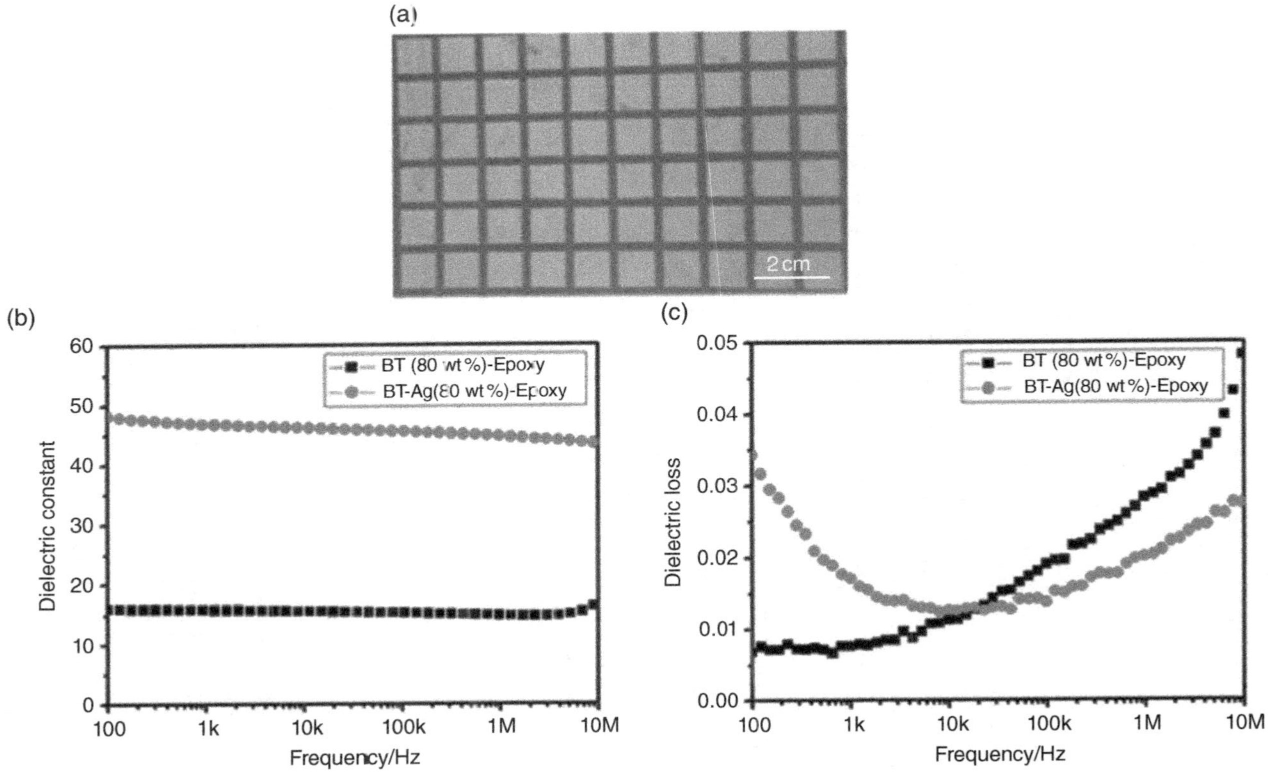

Figure 9.6 Picture and dielectric performance of the embedded capacitor prototype. (a) The picture of etched $BaTiO_3$-Ag/epoxy embedded capacitor material for dielectric properties testing. Dependence of (b) the dielectric permittivity and (c) dielectric loss of $BaTiO_3$-Ag/epoxy composites and BT/epoxy composites with the filler loading of 80 wt% on the frequency at room temperature. *Source:* Based on [31]. © John Wiley & Sons.

hybrid particles as fillers is that a large amount of $AgNO_3$ is consumed. As described in the experimental part, the weight ratio of the $AgNO_3$ and $BaTiO_3$ is 1. Hence, for practical applications, more work is needed to decrease the filler cost.

In this regard, Cu emerges as a low-cost alternative to Ag. Hybrid particles with Cu nanoparticles deposited on the surface of the $BaTiO_3$ nanoparticles were synthesized. [32] Due to the proneness to oxidation of the copper nanoparticles, a protected layer of polyethylene glycol (PEG) was coated on the surface of the hybrids to isolate oxygen. As shown in Figure 9.7, the size of the copper nanoparticles is 15~25 nm in diameter and the thickness of PEG was 2~4.2 nm. The XRD result in Figure 9.7e reveals that no copper oxide was formed after being stored for 60 days at room temperature in the ambient atmosphere. High dielectric permittivity (ε_r~150.6) was achieved on the BT-Cu-filled PVDF composites, which is four times the BT particles-filled PVDF composites. More interestingly, the BT-Cu/PVDF composites achieved high dielectric permittivity and low dielectric loss at high frequency. As shown in Table 9.2, at 1 GHz, the pure PVDF shows a dielectric permittivity of 2.62, which is much lower than the value (ε_r~10.1) obtained at 1 kHz. For the composite filled with 80 wt% BT-Cu hybrids, a high dielectric permittivity of 57.1 at 1 GHz was achieved, which could be explained by a synergistic effect between Cu, BT, and PVDF matrix, probably relating to an effective interface polarization maintaining to a wide frequency range up to GHz level.

So far, many other attempts have been proposed to enhance the dielectric permittivity of polymer-based composites with surface-modified fillers. But most of these fillers, including our work, stand out only in one or two aspects, and fail to demonstrate a balanced overall performance, thus they are far from practical application. Hence, the commercial embedded capacitors are still using the pure ferroelectric filler like $BaTiO_3$ particles. More constructive researches are needed to overcome the limitations neglected by publications.

9.2.2 Epoxy Matrix

Epoxy, as the polymer matrix, is another main part in embedded capacitor materials. Epoxy matrix is the cured composition between epoxy resin, curing agent, and curing accelerator. Many types of epoxy resin and curing agent are commercialized. The epoxy composition is usually designed based on the required performance, such as glass transition temperature, coefficient of thermal expansion, dielectric permittivity, dielectric loss, electric breakdown strength, peel strength, and water adsorption, etc. All the properties need to be considered at the same time, otherwise, the product may not be suitable for practical application. The effects of epoxy resin and curing agent on the performance and the relevant mechanism have been widely reported.[33–35] It will not be repeated in this chapter.

Some suggestions on epoxy composition design are given below.

1) More than one epoxy resins and curing agents may be needed to optimize the performance as required. The understanding of the relation between various curing mechanisms is very important.
2) The content of hydroxyl group of the cured epoxy influences the dielectric permittivity, dielectric loss, and peel strength. A high content of hydroxyl groups produces high permittivity, high dielectric loss, especially at high frequency, and high peel strength. Control over the hydroxyl group content of the cured epoxy composites is important.

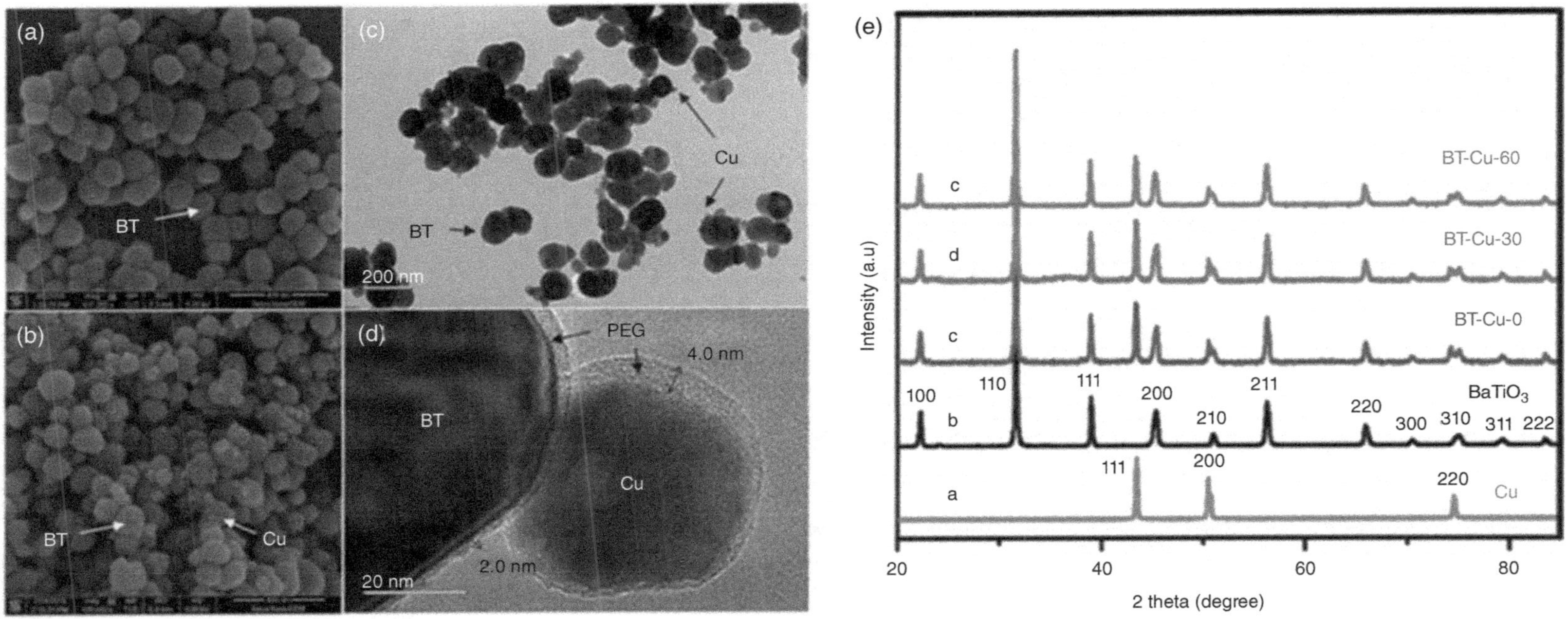

Figure 9.7 Physical characterization of BT particles. SEM images of (a) pure BT, (b) Cu-deposited BT hybrid particles, (c) TEM images of as-synthesized BT-Cu hybrid nanoparticles, (d) BT-Cu hybrid particle coated with PEG-4000. The PEG coating layer on surface of BT-Cu hybrid particles is indicated with arrows and (e) XRD patterns of Cu, $BaTiO_3$, and BT-Cu kept at room temperature for 0, 30 and 60 days.
Source: Li et al. [**32**]. © Royal Society of Chemistry.

Table 9.2 Contents of each component in the BT-Cu/PVDF composites and dielectric properties at 1 kHz, 100 kHz, and 1 GHz.[32]

Samples	BT-Cu (wt%)	BT-Cu (vol%)	BT (vol%)	Cu[a] (vol%)	ε_r/δ (1 kHz)	ε_r/δ (100 kHz)	ε_r/δ (1 GHz)
PVDF	0	0	0	0	10.1/0.03	9.61/0.05	2.62/0.083
BT-Cu/PVDF	5%	1.81%	1.55%	0.26%	12.8/0.024	11.9/0.065	—
	10%	4.1%	3.5%	0.6%	18.3/0.023	16.7/0.064	—
	20%	7.3%	6.3%	1.0%	24.2/0.025	22.3/0.061	9.13/0.091
	40%	17.9%	15.3%	2.6%	47.70.038	36.6/0.063	19.8/0.090
	60%	32.3%	27.6%	4.7%	66.6/0.067	60.5/0.074	25.7/0.084
	70%	42.3%	36.1%	6.2%	88.1/0.091	66.1/0.079	35.8/0.08
	80%	53.7%	45.8%	7.5%	150.6/0.16	97/0.084	57.1/0.064

[a] The calculated mass fraction of Cu in the BT-Cu hybrid fillers was 21.6 wt%.

Source: Li et al. [32]. © Royal Society of Chemistry.

3) High initial curing temperature (>100 °C) and mid curing peak temperature (150~200 °C) are favorable for many electronic packaging materials. Relatively high initial curing temperature provides a wide process time window and a mild storage condition. To obtain a high initial curing temperature, a latent curing agent and curing accelerator should be used together.
4) The coefficient of thermal expansion is related to the glass transition temperature. A high glass transition temperature usually provides a low coefficient of thermal expansion in the same temperature range. High crosslinking density and high content of rigid groups like benzene ring will result in high glass transition temperature.
5) Other polymers are needed to adjust the viscosity of the varnish to satisfy the coating requirements.
6) Other additives like an antifoaming agent and leveling agent are usually added in the varnish to control the defect on coating films.
7) The boiling point, polarity, and residual quantity (in dried films) of the solvent should be adjusted by mixing several types of solvents.

9.2.2.1 Modification to Improve Dielectric Properties

For embedded capacitor materials, the dielectric permittivity is the key parameter. Based on theoretical models, such as the mixing model, Logarithmic model, Jayasundere-Smith model, and Maxwell-Garnett model, the dielectric permittivity of polymer composites is significantly influenced by the dielectric permittivity of the epoxy matrix. As shown in Figure 9.8a, a high dielectric permittivity is easily achieved by polymer-based composites using polymer matrix with high dielectric permittivity. For example, compared with the epoxy matrix, a much higher dielectric permittivity is achieved for the PVDF composites owing to the high dielectric permittivity of the PVDF matrix. [23, 38] According to the published works as referenced in Figure 9.8b, the dielectric permittivity enhancement ratio ($\varepsilon_{eff}/\varepsilon_m$) is very limited and far below the predicted value based on the Logarithmic mixing (L-M) model and parallel model.[37] Hence, improving the dielectric permittivity of epoxy is critical for fabricating embedded capacitor materials with high capacitance density.

Generally, the dielectric permittivity of epoxy is usually 3~4 at 1 kHz. The dielectric loss is in a relatively wide range between 0.015 and 0.025 at 1 kHz. Focusing on the improvement of the dielectric permittivity of the epoxy matrix, our group has conducted some work about graft modification. [36] PEG and chrome acetylacetonate (CrAcac) were employed as graft modification reagents. Methyl tetrahydrophthalic anhydride (MeTHPA) was used as a curing agent for epoxy resin, which also acted as a precursor for the graft modification reagent. A small amount of polyol reacted with the curing agent of anhydride and triggered the curing chain reaction, which brought more oxygen atoms and produced more ester groups at the same time, as shown in Figure 9.9a. Cr Acac could accelerate the anhydride-curing process and form a new conjugating system with a strong polar center as shown in Figure 9.9b, which increases the molecular polarizability of the epoxy resin. Both of these effects lead to the improved permittivity of the derived epoxy resin.

The unmodified epoxy matrix showed a dielectric permittivity of 3.98 at 1 kHz. After modification with PEG, CrAcac, and both, as shown in Figure 9.10a, the dielectric permittivity of the epoxy matrix obtained a high value of 4.37, 4.74, and 5.16, respectively. Importantly, the dielectric loss of the modified epoxy is not increased in comparison to that

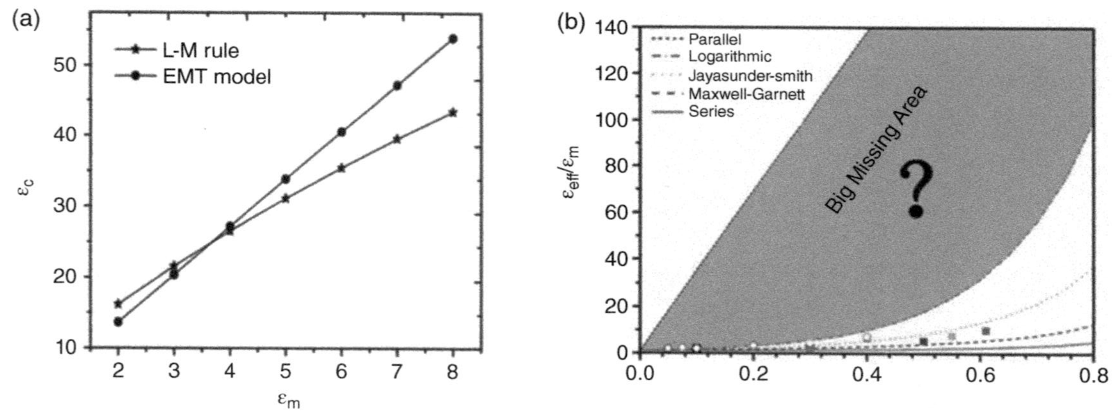

Figure 9.8 Predicted dielectric permittivity by theoretical models. (a) Simulated results of effective permittivity of composites versus permittivity of polymer matrix with L-M rule and EMT model, assuming that the permittivity and loading amount of filler particles are 3000 and 0.66 (volume fraction), respectively *Source:* Lai et al. [36]. © Elsevier. (b) Plots of $\varepsilon_{eff}/\varepsilon_m$ vs. f calculated using various models for polymeric composites. The permittivities of BT (ε_{eff} = 1235) and epoxy resin (ε_m = 3.6) are adopted as the permittivity of the filler and matrix for calculation, respectively. *Source:* Luo et al. [37]. © Royal Society of Chemistry.

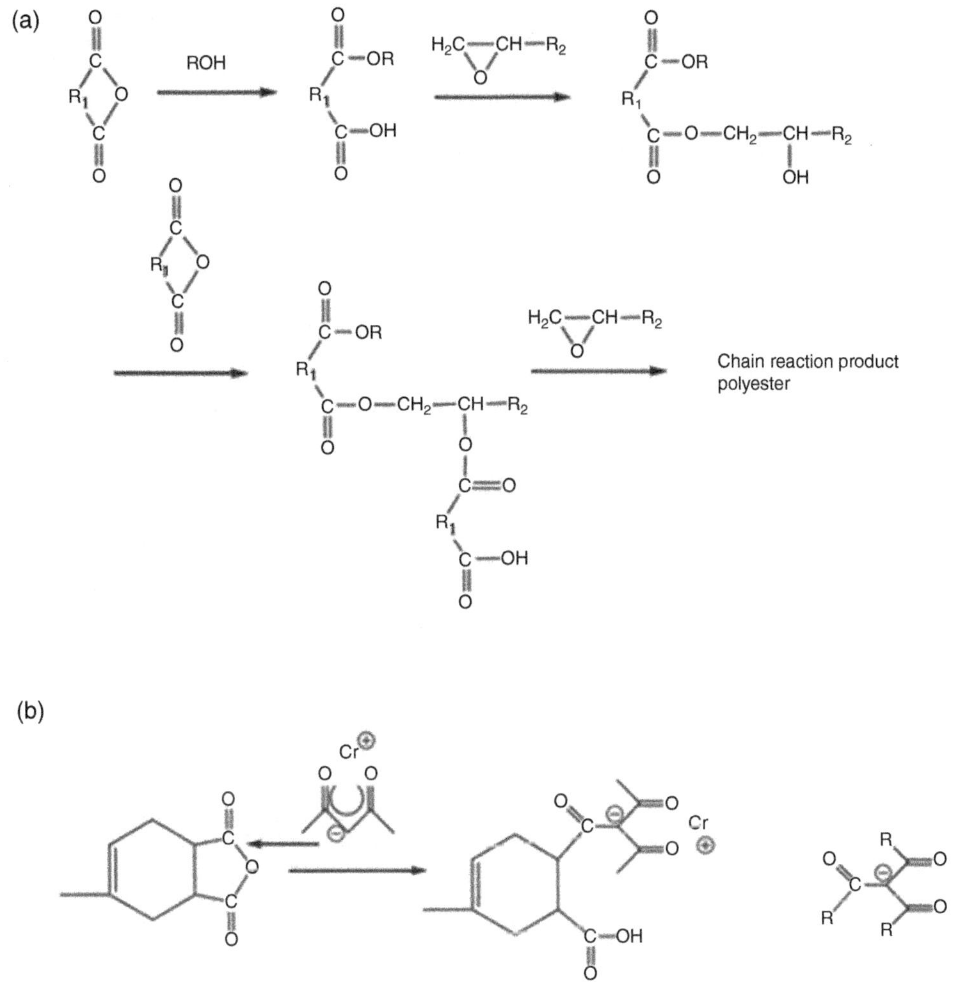

Figure 9.9 Reaction mechanisms involved in epoxy modification. (a) Reaction mechanism of epoxy cured by anhydride with polyol as initiator. (b) Reaction mechanism of MeTHPA and CrAcac. The acetylacetonate group reacted with anhydride and formed a seven-center-eight-electron conjugate system. *Source:* Lai et al. [36]. © Elsevier.

of the unmodified epoxy and provides a low value of 0.01 at 1 kHz as shown in Figure 9.10b. With the $BaTiO_3$ (BT) particles added into the epoxy matrix, the improvement of dielectric permittivity becomes more obvious, as displayed in Figure 9.10c, especially at high filler loadings. The thermal stability of the dielectric permittivity and loss of the modified epoxy/BT composites is the same as the unmodified composites.

The PEG and CrAcac graft-modified epoxy/BT composites have obtained significantly improved dielectric permittivity, but more work is still needed to maintain the required peel strength.

9.2.2.2 Modification to Improve Mechanical Properties

Another key aspect of epoxy-based composites as embedded capacitor materials is the mechanical behavior. Embedded capacitor materials with high toughness are required to

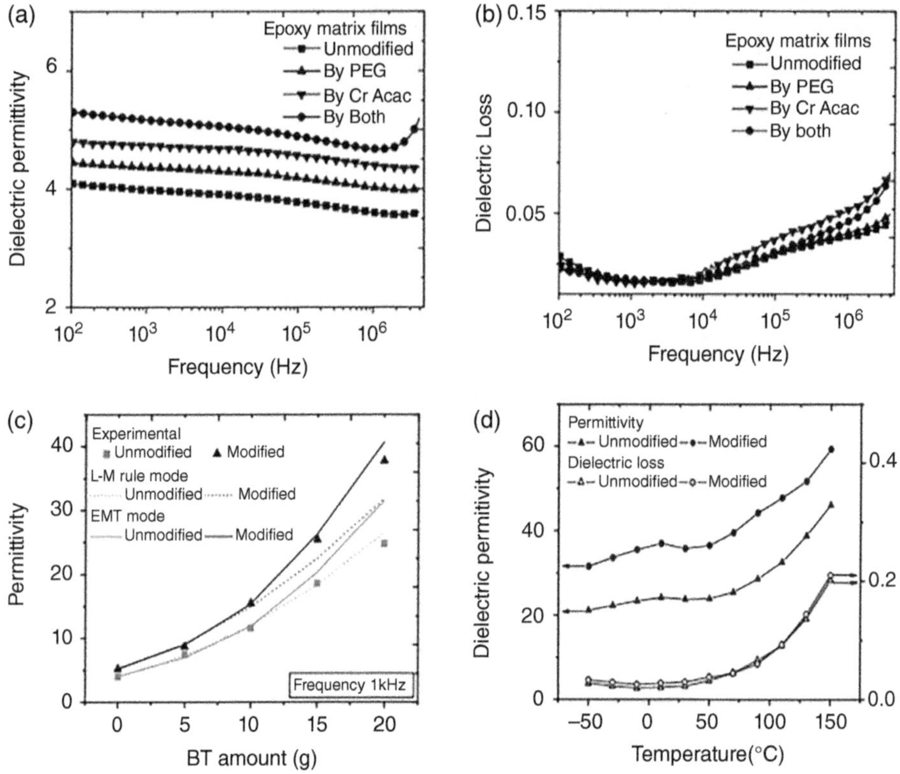

Figure 9.10 Dielectric performance of matrix-modified composites. Frequency dependence of (a) permittivity, and (b) dielectric loss of unmodified epoxy matrix film and films modified respectively by PEG, CrAcac, and both PEG and CrAcac. (c) Experimental and predicted permittivity of the unmodified and modified matrix respectively mixed with different amount of BT particles, at 1 kHz. (d) Temperature dependence of permittivity and dielectric loss of the graft modified and unmodified BT/ER composites. *Source:* Lai et al. [36]. © Elsevier.

withstand the PCB process and satisfy device reliability requirements. However, epoxy is a thermosetting resin with inherent brittleness. The toughness of the epoxy matrix is dependent on the chemical structure, content of toughener, and microstructure of the interpenetrating polymer networks. Most of the current work on improving the toughness of epoxy focuses on three aspects: (i) reducing the cross-link density and molecular weight of the epoxy system, (ii) chemical modification of the epoxy by incorporating flexible backbone, (iii) adding a second phase like rubber, hyper-branched polymers, inorganic nanoparticles with high aspect ratio, etc. Although methods (i) and (ii) can significantly improve the toughness of the epoxy system, the glass transition temperature and thermal stability will rapidly decrease. Hence, these two methods are not suitable for the design of PCB materials.

Using rubber as an additive to improve the toughness of epoxy-based composites has emerged for a long time. By grafting the rubber with end-groups, such as amino group, carboxyl group, epoxy group, and hydroxyl group, the added rubber phase can react with the epoxy and form a stable system. Table 9.3 shows five examples of the modified rubber-formulated epoxy composition with enhanced roughness. However, the thermal stability of

Table 9.3 Summary on rubber/epoxy composites.

Type of Epoxy	Modifier Rubber	Curing Agent	Rubber Addition (phr)	[a]K_{IC} (MPa/m$^{1/2}$)	Refs.
Bisphenol A epoxy	Amine-terminated butadiene acrylonitrile	Polyaminoimidazoline	12.5	1.49	[39]
Bisphenol A epoxy	Carboxyl-terminated acrylonitrile butadiene	Cycloaliphatic amine	15	2	[40]
Bisphenol A epoxy	Amine-terminated butadiene acrylonitrile	Cycloaliphatic polyamine	15	2.9[(a)]	[41]
Bisphenol A epoxy	Epoxy-terminated butadiene acrylonitrile	Trimethylene glycol di-p-aminobenzoate	15	1.65	[42]
Bisphenol A epoxy	Hydroxyl-terminated polybutadiene	Cyclic anhydride	15	1.5	[43]

[a] K_{IC} is the critical stress intensity factor. $K_{IC} = \frac{3PSa^{1/2}Y}{2tw^2}$, where P is the critical load for crack propagation (N), S is the length of the span (mm), a is the pre-crack length (mm), t is the specimen thickness (mm), w is the specimen width (mm), Y is the geometrical factor given by:

$$Y = 1.93 - 3.07\left(\frac{a}{w}\right) + 14.53\left(\frac{a}{w}\right)^2 - 25.11\left(\frac{a}{w}\right)^3 + 25.8C\left(\frac{a}{w}\right)^4.$$

the rubber-modified epoxy system shows a decrease. When the content of carboxyl terminated acrylonitrile butadiene (CTBN) is 15%, the degradation temperature of T_{10} decreases from 340 °C of the pure epoxy to 329 °C. [44]

As shown in Figure 9.11a, the elongation at break of the CTBN/epoxy composites increases with the content of the CTBN. When the content of CTBN is 5 wt%, the elongation is 0.9%. When the content of CTBN is further increased to 20 wt%, the elongation is improved to 2.4%, which constructs an improvement of 2.7 times. However, the storage modulus decreases rapidly with the increase of temperature as the content of CTBN reaches higher than 5 wt% as shown in Figure 9.11b. Besides, the glass transition temperature (T_g) of the CTBN/epoxy composites also reduces with the increase of the CTBN content as shown in Figure 9.11c. When the content of CTBN is 5 wt%, the T_g is 160 °C. As the content of CTBN is increased to 20 wt%, the T_g decreases to 148 °C. Due to the decreased T_g of the CTBN/epoxy composites, the coefficient of thermal expansion increases. In conclusion, the addition of CTBN improves the toughness of the epoxy composites but deteriorates their thermal stability.

Adding nanoparticles into epoxy is also an efficient method to improve the toughness of the composites. Figure 9.12 summarized researches on the application of the carbon nanotubes, graphene, nanoclay, and nanosilica particles as additives to improve the toughness of epoxy-based composites. The biggest increase in toughness can be achieved with the addition of graphene and nanosilica, both at an additional level of around 4 wt%. Other nanomaterials like nanoclays and carbon nanotubes seem to be less efficient as tougheners for epoxy resins at similar additional levels. Embedded capacitor materials are insulating dielectrics. Hence, the use of graphene is inappropriate.

The use of small portion silicate-based fillers to improve the toughness of epoxy-based composites has become a great research interest. For example, Zheng et al. used the bio-based silane-coupling agent surface-modified nano SiO_2 as fillers and diglycidyl ether of bisphenol A (DGEBA) and isophorone diamine (IPDA) as curing agents to prepare the epoxy-based composites with enhanced toughness. [46] Table 9.4 lists the flexural, impact, and fracture properties of the SiO_2/DGEBA/IPDA composites. The flexural and fracture properties of epoxy resin were gradually enhanced with the increasing nano-SiO_2 content, but the modified nano-SiO_2 composites have a more significant impact on these properties. When the modified nano-SiO_2 was filled up to 4%, the flexural strength was increased by about 22% over the neat epoxy resin. Impact property was also enhanced by using modified nano-SiO_2. When adding 4 wt% nano-SiO_2, the K_{IC} of modified samples is 6.3% higher than the unmodified ones, which was mainly attributed to the increased adhesion between nano-SiO_2 and epoxy matrix. The interaction between nano-SiO_2 and epoxy will cause more cracks near the particles and turn the cracks into scaly dimples, which can improve the toughening property. Hence, the modified nano-SiO_2 particles can significantly improve the toughness of the epoxy resin.

9.3 Fabrication Process of Embedded Capacitors

As mentioned above, the embedded capacitor materials can be divided into two types: glass fiber cloth strengthened and non-strengthened. For different types of embedded capacitor

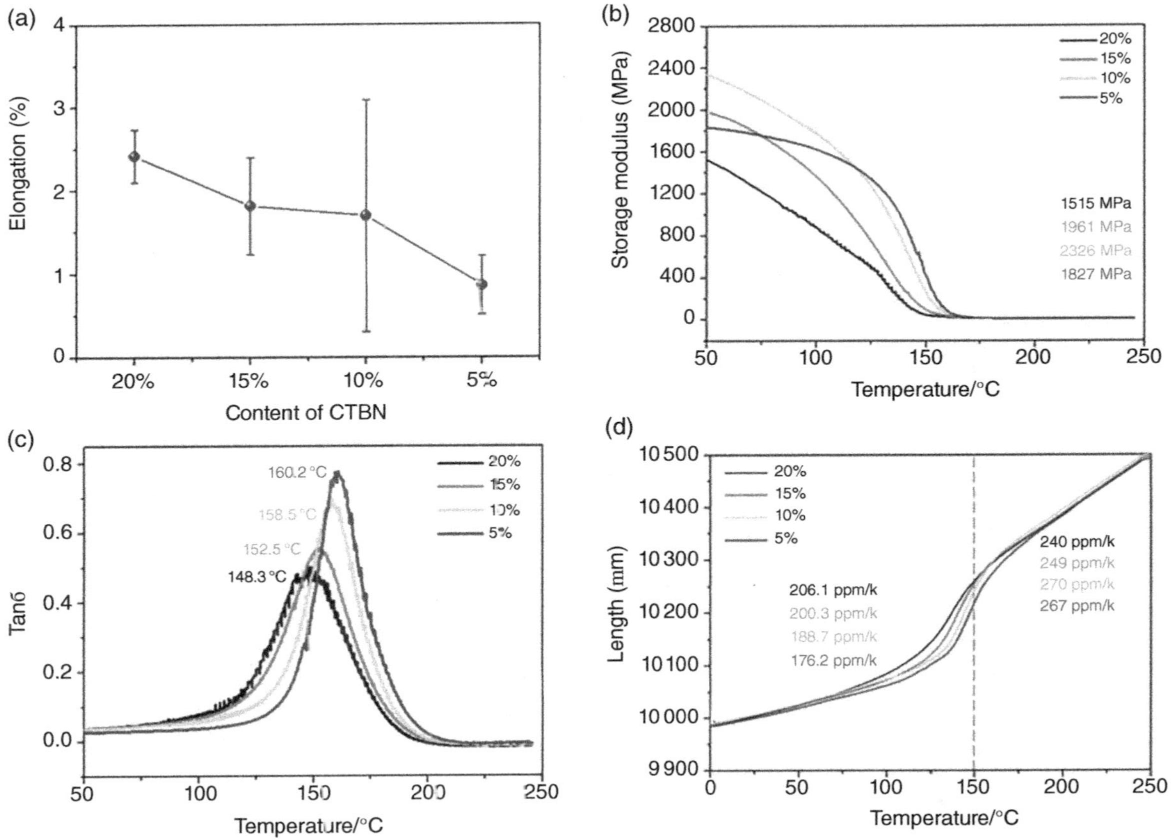

Figure 9.11 Mechanical property of the epoxy modified with different content of CTBN. (a) Elongation, (b) storage modulus, (c) Tanδ, and (d) coefficient of thermal expansion.

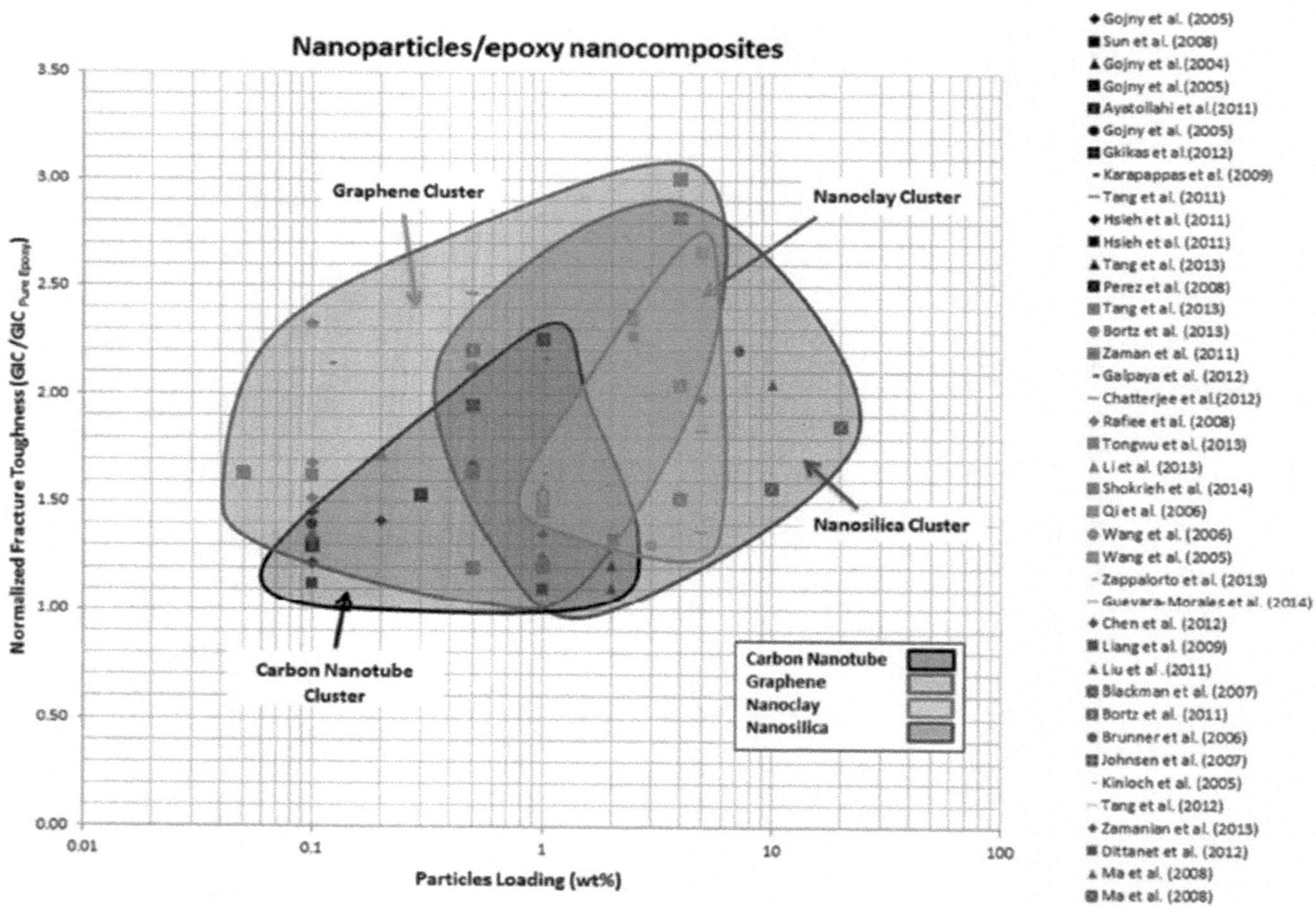

Figure 9.12 Map of fracture toughness of nanoparticles/epoxy nanocomposites with respect to particle loading. *Source:* Domun et al. [45]. © Royal Society of Chemistry.

Table 9.4 Mechanical properties of DGEBA/IPDA with different content of nano-SiO_2.[46]

Formulation	Flexural Strength/MPa	Flexural Modulus/MPa	Impact Strength kJ/m^2	K_{IC} MPa/m$^{1/2}$
DGEBA/IPDA	84.6 ± 5.2	2330 ± 100	13.9 ± 1.1	1.043 ± 0.045
1% unmodified	88.5 ± 5.4	2440 ± 100	14.1 ± 0.9	1.069 ± 0.066
2% unmodified	93.2 ± 4.8	2570 ± 100	14.8 ± 1.1	1.205 ± 0.082
4% unmodified	96.6 ± 3.2	2660 ± 100	15.5 ± 1.2	1.251 ± 0.068
1% modified	90.5 ± 2.9	2490 ± 100	14.8 ± 0.8	1.104 ± 0.086
2% modified	94.2 ± 3.5	2590 ± 100	16.3 ± 0.7	1.218 ± 0.042
4% modified	103.6 ± 4.0	2720 ± 100	18.2 ± 1.2	1.330 ± 0.061

Source: Zheng et al. [46]. © John Wiley & Sons.

materials, the fabrication process is different. To fabricate embedded capacitor materials, the first stage is to prepare the varnish containing all the chemicals. The difference lies in the second stage and the third stage. For the glass fiber cloth-strengthened embedded-capacitor materials, the film is formed by dipping the glass fiber cloth in the varnish, then being heated to remove the solvent. The dried prepreg is cut into pieces, laminated with copper foils, and cured in a hot-pressing machine in vacuum condition as displayed in Figure 9.13a.

The typical fabrication process of the embedded capacitor materials without strengthening layer is displayed in Figure 9.13b. After the preparation of varnish, the film is formed on the copper foil by coating. The coating methods include closed type coaters, die coaters, gap coating methods, reverse coating methods, gravure coating methods, and so on. To ensure the insulation of the embedded capacitor materials, the coated film should bear no defects, e.g. holes, pores, and pits. The formation of defects depends on the formulation and drying process. Adding the defoamer agent and preparing a homogeneously dispersed slurry can reduce the content of defects formed on the surface of the coated dielectric film.[47] Adjusting the drying process based on the boiling point of the solvents used in the formulation can also help avoid the formation of defects. Generally, the initial drying temperature should be about half of the boiling point of solvents. The fast volatilization is easy to introduce defects to the film. Besides, the surface of the film dries more quickly than that at the bottom. More solvent will reside within the film when the initial temperature is at a high value and thus reducing the reliability of the embedded capacitor materials.

Coating a dielectric layer with low thickness is beneficial for obtaining a high capacitance density of the embedded capacitor materials. As shown in Figure 9.14, the capacitance density increases rapidly with thickness when the thickness is lower than 10 um. But dielectric layer with low thickness has the risk of easy electric breakdown. Therefore, the defect control becomes more crucial as the thickness of the dielectric layer is reduced to a low value.

The lamination process involves the use of heated rollers with pressure. After lamination, the laminated film is wound up and cured in an oven. The final product is fabricated by slitting the rolled embedded capacitor materials into pieces.

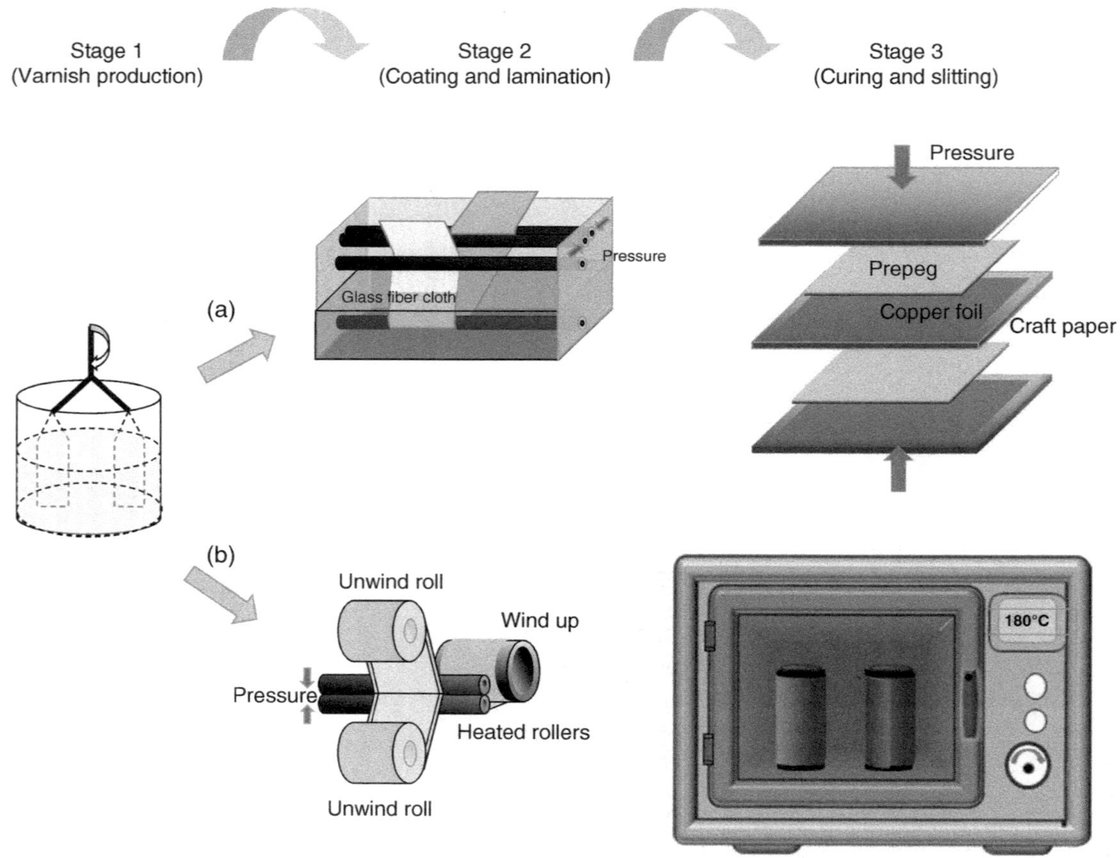

Figure 9.13 Typical fabrication process of embedded capacitor materials.

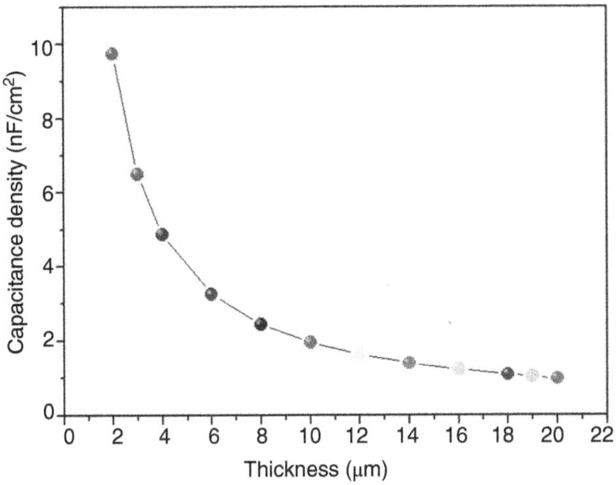

Figure 9.14 Relationship between the thickness of the dielectric layer and the capacitance density of the embedded capacitors.

9.4 Reliability Test of Embedded Capacitor Materials

Reliability evaluation of embedded capacitors in PCB contains a series of accelerated tests, [48, 49] including long-term thermal aging, thermal shocks, voltage stress, and modelling time to failure based on experimental data. Commercially available materials in the form of dielectrics with copper foil on both sides are applied for the manufacturing of test capacitors embedded inside PCB. Usually, many different ways are used to test the reliability of the capacitor in PCB, for example, the change of capacitance and dissipation factora with temperature, increasing from room temperature to 100 °C, long-term aging at 100 °C, thermal shocks between −55 and 125 °C, and so on. The test of breakdown voltage variation with the temperature is also used.

Polymer-based composites are sensitive to moisture, which could change the capacitance, dissipation factor, breakdown voltage, and leakage current of the capacitors. The stability of the material properties limit capacitor applications especially in the medium- and high-power system and under high humidity conditions. In general, the high dielectric permittivity is accompanied by unstable capacitance at elevated temperatures and voltage. High-k dielectric materials which can be realized for embedded passive applications are required to have high dielectric permittivity, low dielectric loss, high thermal stability, simple processability, and good dielectric properties over a broad frequency range.

After the capacitors are embedded in PCB substrates, many reliability tests will be carried out. Usually, for thermal properties, the tests in the PCB factory include thermal stress, thermal shock, and so on. There are many methods for the thermal stress test, such as hot oil test, reflow test, thermal cycling test, and thermal shock test. The global electronic industry association standard of IPC-TM-650 is a typical reference test method. For the packaging industry, the method of temperature cycling test of JESD22-A104E is used to examine the thermal reliability of the packaging materials.

In addition to the thermal stress test, there are many other kinds of reliability tests for the properties of the polymer-based composite materials embedded inside PCB. For example, for the 85 °C/85RH aging test, the embedded capacitor components were placed in an environmental chamber at 85 °C and 85% relative humidity (RH) for 1000 hours. The dielectric properties of embedded capacitor components after 85/85 aging are measured at room temperature. Air-to-air thermal cycling (AATC) is used to test the reliability of embedded capacitor components. The temperature is from −55 to 125 °C with a dwell time of 10 minutes at each extreme and a ramp time of 10 minutes. The embedded capacitors are subjected to 1000 thermal cycles for the thermal cycling test, and the dielectric properties of the embedded capacitor components were measured at room temperature. There are also bonding tests to examine the reliability of the adhesion between the polymer-based composites and copper foils and test of 155 °C baking for four hours, and so on.

9.5 Conclusions and Perspectives

Embedded capacitance provides quite some advantages in IC packaging field. However, its wide application is still restricted by low capacitance density. So far, the commercially embedded capacitors with polymeric composites as dielectrics can only afford small capacitance density, and the need for highly embedded capacitance drives the search for high-k dielectrics. Challenges still exist today on rationally designing optimum composites to realize a target high-k. The relationship between the structure of epoxy, fillers, composite, and the dielectric properties is far from being fully understood. From the perspective of engineering, reducing the thickness of the epoxy composites dielectric layer is another way to achieve high capacitance density. Challenges exist in controlling the defects in the dielectric layer. More innovative research work for practical application is expected in the future.

References

1 Park, K.-W., Cho, S.-D., Lee, J.-Y., and Hyun, J.-G. (2002). Low tolerance epoxy-$BaTiO_3$ composite embedded capacitor films(ECFs). 4th International Symposium on Electronic Materials and Packaging, 341–347.

2 Zhang, Z.-F., Bai, X.-F., Zha, J.-W. et al. (2014). Preparation and dielectric properties of $BaTiO_3$/epoxy nanocomposites for embedded capacitor application. *Composites Science and Technology* 97: 100–105.

3 Ogitani, S., Bidstrup-Allen, S.A., and Kohl, P. (1998). Development of high-k photo-definable epoxy composites for embedded capacitors. IEEE/CMPT Electronics Manufacturing Technology Symposium, 199–205.

4 Hanemann, T., Gesswein, H., and Schumacher, B. (2011). Development of new polymer–$BaTiO_3$-composites with improved permittivity for embedded capacitors. *Microsystem Technologies* 17: 195–201.

5 Alam, M.A., Azarian, M.H., and Pecht, M.G. (2012). Embedded capacitors in printed wiring board: a technological review. *Journal of Electronic Materials* 41: 2286–2303.

6 Fu, J., Hou, Y., Zheng, M. et al. (2015). Improving dielectric properties of PVDF composites by employing surface modified strong polarized $BaTiO_3$ particles derived by molten salt method. *ACS Applied Materials & Interfaces* 7: 24480–24491.

7 Niu, Y., Bai, Y., Yu, K. et al. (2015). Effect of the modifier structure on the performance of barium titanate/poly(vinylidene fluoride) nanocomposites for energy storage applications. *ACS Applied Materials & Interfaces* 7: 24168–24176.

8 Tang, H., Zhou, Z., and Sodano, H.A. (2014). Relationship between $BaTiO_3$ nanowire aspect ratio and the dielectric permittivity of nanocomposites. *ACS Applied Materials & Interfaces* 6: 5450–5455.

9 Xie, Y., Yu, Y., Feng, Y. et al. (2017). Fabrication of stretchable nanocomposites with high energy density and low loss from cross-linked PVDF filled with poly(dopamine) encapsulated $BaTiO_3$. *ACS Applied Materials & Interfaces* 9: 2995–3005.

10 Zhou, T., Zha, J.W., Cui, R.Y. et al. (2011). Improving dielectric properties of $BaTiO_3$/ferroelectric polymer composites by employing surface hydroxylated $BaTiO_3$ nanoparticles. *ACS Applied Materials & Interfaces* 3: 2184–2188.

11 Kim, P., Jones, S.C., Hotchkiss, P.J. et al. (2007). Phosphonic acid-modified barium titanate polymer nanocomposites with high permittivity and dielectric strength. *Advanced Materials* 19: 1001–1005.

12 Dang, Z.-M., Wang, H.-Y., and Xu, H.-P. (2006). Influence of silane coupling agent on morphology and dielectric property in $BaTiO_3$/polyvinylidene fluoride composites. *Applied Physics Letters* 89: 112902.

13 Iijima, M., Sato, N., Wuled Lenggoro, I., and Kamiya, H. (2009). Surface modification of $BaTiO_3$ particles by silane coupling agents in different solvents and their effect on dielectric properties of $BaTiO_3$/epoxy composites. *Colloids and Surfaces A: Physicochemical and Engineering Aspects* 352: 88–93.

14 Ramajo, L., Castro, M.S., and Reboredo, M.M. (2007). Effect of silane as coupling agent on the dielectric properties of $BaTiO_3$-epoxy composites. *Composites Part A: Applied Science and Manufacturing* 38: 1852–1859.

15 Dalle Vacche, S., Oliveira, F., Leterrier, Y. et al. (2014). Effect of silane coupling agent on the morphology, structure, and properties of poly(vinylidene fluoride–trifluoroethylene)/$BaTiO_3$ composites. *Journal of Materials Science* 49: 4552–4564.

16 Yuan, Y., Yin, Y., Yu, D. et al. (2016). Effects of compound coupling agents on the properties of PTFE/SiO2 microwave composites. *Journal of Materials Science: Materials in Electronics* 28: 3356–3363.

17 Li, G., He, Y., Zhu, P. et al. (2018). Tailored surface chemistry of SiO2 particles with improved rheological, thermal-mechanical and adhesive properties of epoxy based composites for underfill applications. *Polymer* 156: 111–120.

18 Li, G., Zhao, T., Zhu, P. et al. (2019). Structure-property relationships between microscopic filler surface chemistry and macroscopic rheological, thermo-mechanical, and adhesive performance of SiO2 filled nanocomposite underfills. *Composites Part A: Applied Science and Manufacturing* 118: 223–234.

19 Cho, S., Lee, S., Hyun, J., and Paik, K. (2005). Comparison of theoretical predictions and experimental values of the dielectric constant of epoxy-$BaTiO_3$ composite embedded capacitor films. *Journal of Materials Science: Materials in Electronics* 16: 77–84.

20 Jang, K.-W. and Paik, K.-W. (2008). Screen printable epoxy/$BaTiO_3$ embedded capacitor pastes with high dielectric constant for organic substrate applications. *Journal of Applied Polymer Science* 110: 798–807.

21 Rao, Y., Ogitani, S., Kohl, P., and Wong, C.P. (2002). Novel polymer-ceramic nanocomposite based on high dielectric constant epoxy formula for embedded capacitor application. *Journal of Applied Polymer Science* 83: 1084–1090.

22 Lee, S. and Paik, K. (2010). Reduced temperature coefficient of capacitance (TCC) of embedded capacitor films (ECFs) for organic substrates using $SrTiO_3$ and multifunctional epoxy. *Journal of Electronic Materials* 39: 1358–1363.

23 Xu, J., Bhattacharya, S., Pramanik, P., and Wong, C.P. (2006). High dielectric constant polymer-ceramic (epoxy varnish-barium titanate) nanocomposites at moderate filler loadings for embedded capacitors. *Journal of Electronic Materials* 35: 2009–2015.

24 Jiang, M.-J., Dang, Z.-M., and Xu, H.-P. (2007). Giant dielectric constant and resistance-pressure sensitivity in carbon nanotubes/rubber nanocomposites with low percolation threshold. *Applied Physics Letters* 90: 042914.

25 Deepa, K.S., Kumari Nisha, S., Parameswaran, P. et al. (2009). Effect of conductivity of filler on the percolation threshold of composites. *Applied Physics Letters* 94: 142902.

26 He, F., Lau, S., Chan, H.L., and Fan, J. (2009). High dielectric permittivity and low percolation threshold in nanocomposites based on poly(vinylidene fluoride) and exfoliated graphite nanoplates. *Advanced Materials* 21: 710–715.

27 Nan, C.W., Shen, Y., and Ma, J. (2010). Physical properties of composites near percolation. *Annual Review of Materials Research* 40: 131–151.

28 Wang, D., Zhang, X., Zha, J.-W. et al. (2013). Dielectric properties of reduced graphene oxide/polypropylene composites with ultralow percolation threshold. *Polymer* 54: 1916–1922.

29 Wang, D., Zhou, T., Zha, J.-W. et al. (2013). Functionalized graphene–$BaTiO_3$/ferroelectric polymer nanodielectric composites with high permittivity, low dielectric loss, and low percolation threshold. *Journal of Materials Chemistry A* 1: 6162.

30 Luo, S., Yu, S., Sun, R., and Wong, C.P. (2014). Nano Ag-deposited $BaTiO_3$ hybrid particles as fillers for polymeric dielectric composites: toward high dielectric constant and suppressed loss. *ACS Applied Materials & Interfaces* 6: 176–182.

31 Luo, S., Yang, W., Yu, S. et al. (2016). Development and application of polymer-based nanocomposites dielectrics. 17th International Conference on Electronic Packaging Technology, 411–416.

32 Li, C., Yu, S., Luo, S. et al. (2016). Enhancement of dielectric performance upto GHz of the composites with polymer encapsulated hybrid $BaTiO_3$–Cu as fillers: multiple interfacial polarizations playing a key role. *RSC Advances* 6: 36450–36458.

33 Pethrick, R.A. (2014). Design and ageing of adhesives for structural adhesive bonding – a review. *Proceedings of the Institution of Mechanical Engineers Part L: Journal of Materials: Design and Applications* 229: 349–379.

34 Cakic, S.M., Ristic, I.S., Jaso, V.M. et al. (2012). Investigation of the curing kinetics of alkyd-melamine-epoxy resin system. *Progress in Organic Coatings* 73: 415–424.

35 Srivastava, A.K. and Mohan, P. (2006). Synthesis, reactions and properties of modified epoxy resins. *Journal of Macromolecular Science, Part C: Polymer Reviews* 37: 687–716.

36 Lai, M., Yu, S., Sun, R. et al. (2013). Effects and mechanism of graft modification on the dielectric performance of polymer–matrix composites. *Composites Science and Technology* 89: 127–133.

37 Luo, S., Shen, Y., Yu, S. et al. (2017). Construction of a 3D-$BaTiO_3$ network leading to significantly enhanced dielectric permittivity and energy storage density of polymer composites. *Energy & Environmental Science* 10: 137–144.

38 Xu, J. and Wong, C.P. (2006). Effect of the polymer matrices on the dielectric behavior of a percolative high-K polymer composite for embedded capacitor applications. *Journal of Electronic Materials* 35: 1087–1094.

39 Chikhi, N., Fellahi, S., and Bakar, M. (2001). Modification of epoxy resin using reactive liquid (ATBN) rubber. *European Polymer Journal* 38: 251–264.

40 Hosseini, M., Esfandeh, M., Razavi-Nouri, M., and Rezadoust, A.M. (2018). Effect of hybridization of carboxyl-terminated acrylonitrile butadiene liquid rubber and alumina nanoparticles on the fracture toughness of epoxy nanocomposites. *Polymer Composites* 40: 2700–2711.

41 Bakar, M. and Szymanska, J. (2014). Property enhancement of epoxy resin using a combination of amine-terminated butadiene-acrylonitrile copolymer and nanoclay. *Journal of Thermoplastic Composite Materials* 27: 1239–1255.

42 Hwang, J.F., Manson, J.A., Hertzberg, R.W. et al. (1989). Structure-property relationships in rubber-toughened epoxies. *Polymer Engineering and Science* 29: 1466–1476.

43 Raju, T., Ding, Y., He, Y. et al. (2008). Miscibility, morphology, thermal, and mechanical properties of a DGEBA based epoxy resin toughened with a liquid rubber. *Polymer* 49: 278–294.

44 Puglia, P.C.P.D., Kenny, P.V.P.J.M., and Sabu, T. (2017). The role of clay modifier on cure characteristics and properties of epoxy/clay/carboxyl-terminated poly(butadiene-co-acrylonitrile) (CTBN) hyrid. *Materials Technology* 32: 171–177.

45 Sprenger, S. (2020). Nanosilica-toughened epoxy resins. *Polymers* 12: 1777.

46 Zheng, J., Zhang, X., Cao, J. et al. (2020). Behavior of epoxy resin filled with nano-SiO_2 treated with a Eugenol epoxy silane. *Journal of Applied Polymer Science* 138: e50138.

47 Luo, S., Gao, C., Cao, L. et al. (2020). Defect control in epoxy dry film with improved electric performances. 21st International Conference on Electronic Packaging Technology (ICEPT).

48 Muhammad, N., Fang, Z.G., Shah, S.Y., and Haider, D. (2020). Reliability and remaining life assessment of an electronic fuze using accelerated life testing. *Micromachines* 11: 272.

49 Pascal, Y., Labrousse, D., Petit, M. et al. (2018). Experimental investigation of the reliability of printed circuit board (PCB)-embedded power dies with pressed contact made of metal foam. *Microelectronics Reliability* 88–90: 707–714.

10

Polymer Composites for Generators and Motors

Hirotaka Muto, Takahiro Umemoto, and Takahiro Mabuchi

Advanced Technology Research and Development Center, Mitsubishi Electric Corporation, Tokyo, Japan
Coedited by Xingyi Huang and Toshikatsu Tanaka

10.1 Introduction

A large-sized generator is a key power apparatus for the stability of the electric power supply. The demand for a new generator or for upgrading and refurbishment of the existing units continues to increase due to the growth of electric power consumption worldwide. The renewals or refurbishments of aged machines are desired to avoid the risk of a sudden failure and to keep a stable supply of electric power. Efficiency improvement and power loss reduction are strongly required for the new machines from the viewpoint of energy saving and global warming prevention. On the other hand, the deregulated energy market leads to cost pressure and high operation reliability on the power plant assets of the utilities. This forces manufacturers to develop new generator designs with higher efficiency, higher reliability that allows lower lifetime operating costs. In large-sized rotating machines, i.e. generators and motors, the performance of the insulation greatly influences efficiency, reliability, and miniaturization. Hence, the improvement of electrical insulation materials is indispensable to realize low loss and miniaturization of rotating machines.

Figure 10.1 shows a cut view of a large-sized generator (a), the appearance of a stator winding coil (b), and a cross-sectional view of one stator coil bar (c). The basic configuration of a generator consists of a rotor, a stator core, and a stator winding coil. Dozens of stator coil bars are arranged concentrically on the stator core, each of which is sandwiched between the stator iron cores as shown in Figure 10.1 (c). An insulating material is formed around the coil conductor bars. The insulation part between the coil conductor and the iron core is called "main ground wall insulation" or simply "main insulation." It is composed of a multi-layer structure of mica and epoxy resin to be explained in detail later. Under an operation of a generator, a high voltage of about 20 kV is applied to the main ground wall insulation.

In order to realize loss reduction, it is effective to make the main ground wall insulation thinner and to increase its thermal conductivity. By thinning the insulation material, the cross section of the coil conductor can be expanded and the joule loss (IR2) is reduced. In

Polymer Composites for Electrical Engineering, First Edition. Edited by Xingyi Huang and Toshikatsu Tanaka.

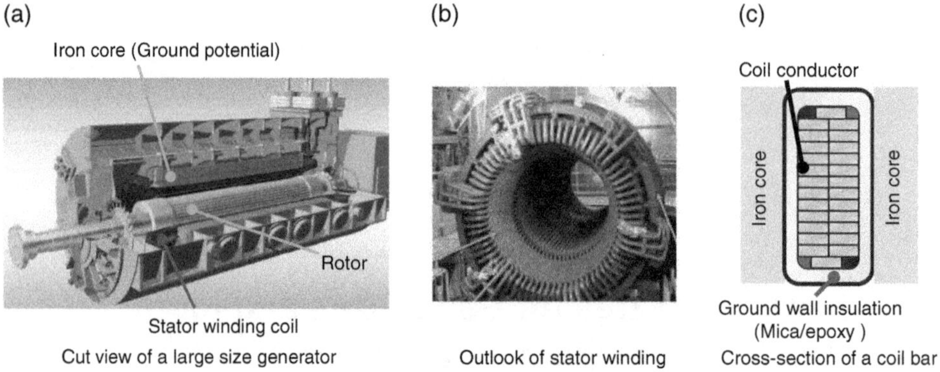

Figure 10.1 A large-sized generator and its electrical insulation structure.

order to realize thinner insulating material, it is necessary to increase the electrical breakdown voltage or lifetime of the electrical insulation material.

10.2 Polymer Composite in High-Voltage Rotating Machines

In high-voltage rotating machines, two types of polymer composites with different functions are in use. One is for the ground wall insulation and the other is electric field stress grading in coil end termination.

The ground wall insulation material is a polymer composite which is composed of mica, which is inorganic silica (SiO_2), and organic polymer materials. Although for smaller motors and traction motors, polyester, polyesterimide, and silicone resins are in use, epoxy systems dominate for a large rotating machine for power utilities. This basic material configuration for the insulation of rotating machines has long been used since the 1940s and now it is still in use [1].

In the 1990s, a challenge that to tries improve the thermal conductivity of the ground wall insulation has begun [2]. In addition to the electrical insulation performance, the thermal conductivity of the ground wall insulation of a rotating machine is another key factor that determines various performances of the machine such as loss efficiency, size, and power output. By adding inorganic micrometer-sized filler like boron nitride (BN), aluminum oxide (Al_2O_3), an increase of the thermal conductivity was shown [2]. The high thermal conductivity technology for the insulation of rotating machine was installed in the actual machine and commercialized by several manufacturers [3, 4] This insulation technology increased the maximum output power of hydrogen indirect cooled turbine generator from less than 500 up to 900 MVA [5–8]. At this stage, the ground wall insulation has come to be a polymer composite composed of epoxy polymer and two kinds of inorganic materials: mica flake for electrical insulation and inorganic micro filler for thermal conductivity.

The stress grading material is a composite of a polymer and small grain of conductive ceramics like silicon carbide (SiC) and zinc oxide (ZnO). The function of stress-grading material is smoothing the electric field to prevent partial discharges and flashover breakdowns when they withstand voltage tests. The nonlinear resistance property of the composite allows uniform electric field distribution and hence lower maximum electric field.

10.3 Ground Wall Insulation

10.3.1 Mica/Epoxy Insulation

In the insulation system of the stator winding of a rotating machine rated higher than around 2 kV, a mica tape is wound around a coil conductor bar as shown in Figure 10.2a. A low-resistance tape is wound over mica tape to maintain electrical contact with the ground potential of the stator core to prevent partial discharges caused by the separation of the insulator surface from the stator core.

The mica tape which is composed of inorganic silica flake and a glass fiber backing material is impregnated with epoxy resin in a vacuum. Figure 10.2b shows the cross section of ground wall insulation of rotating machines. The insulation has a layered structure of mica tapes. Half of the mica tape width is overlapped with each other. The insulation thickness after impregnated and cured is typically several mm depending on the rated voltage class of the machine. Each layer of the mica tape is composed of a stack of thin large area mica flakes on the glass cloth as shown in Figure 10.2c.

Figure 10.3 shows an SEM (Scanning Electron Micrograph) picture of the cross- section of mica/epoxy insulation. In the example shown in Figure 10.3a, three layers of mica sheet are stacked and impregnated with epoxy resin. Three groups of many dots between mica flake layers are cross- sections of glass cloth fiber of the backing material. Each layer of the mica tape is composed of many stacks of thin large-area mica flakes on the glass cloth. Although mica flakes are just weakly held together by Van der Waals (molecular-level electrostatic) forces, they are bonded tightly by the impregnated epoxy resins after the curing process. Thus, the epoxy resin is impregnated in between the mica tapes and between mica flakes as shown in Figure 10.3. Figure 10.4 shows an SEM picture of the top view of mica tape layer. The lateral shape of each mica flake can be seen. The lateral size of a single mica

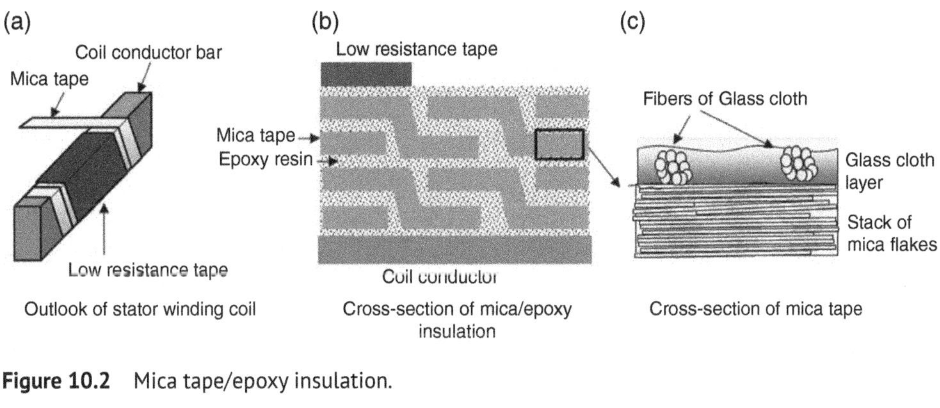

Figure 10.2 Mica tape/epoxy insulation.

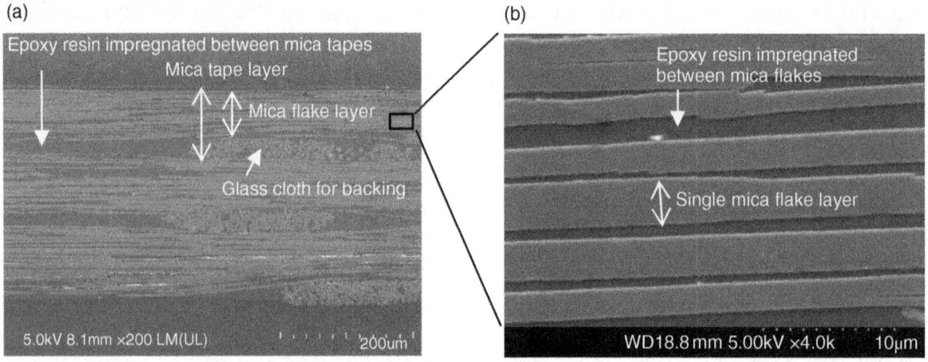

Figure 10.3 A SEM picture of a cross section of mica/epoxy insulation. (a) Mica tape layers (b) Mica flake layers in the mica tape.

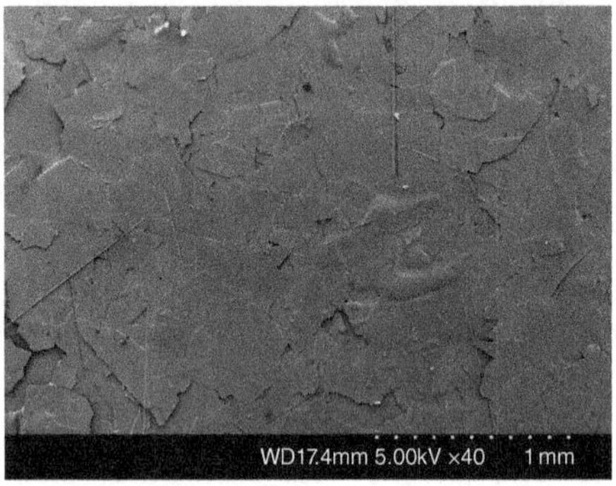

Figure 10.4 A SEM picture of the top view of the mica tape layer.

flake distributes from several tens to several hundred micrometers. During the vacuum pressure impregnation process, the epoxy resin must be impregnated well between the mica tapes and also between mica flakes inside the mica tape so that gas-filled void remains as few as possible.

10.3.2 Electrical Defect in the Insulation of Rotating Machines and Degradation Mechanism

Generally, electrical insulation defect in ground wall insulation of rotating machines is categorized into three types as illustrated in Figure 10.5: ①enclosed void, ②metallic particle or protrusion, and ③periphery of the conductor. The metallic particle (②) can be removed in the proper manufacturing process. The periphery of conductors (③) is treated by stress-grading material to be explained later in this chapter. However, the void defect (①) tends to remain in the insulation. Hence, the major cause that determines the electrical insulation property of rotating machine are void defects created inside the mica/epoxy insulation.

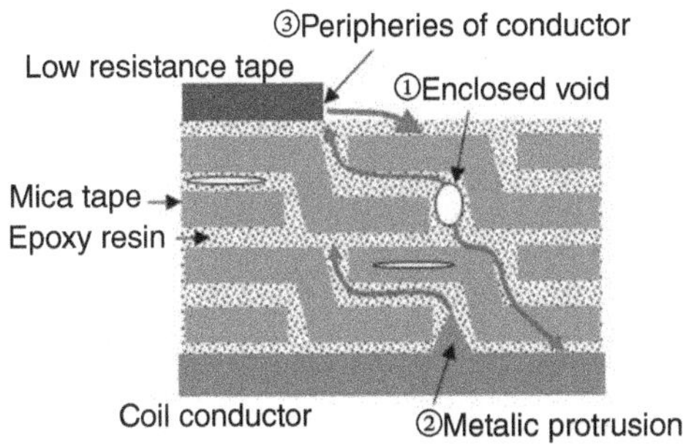

Figure 10.5 Mica/epoxy insulation structure and possible defects.

If there exists a void defect, initially gas discharge in the void starts followed by an electrical tree initiation at the internal surface of the void into surrounding epoxy resin between mica tapes. Then an electrical tree proceeds along two adjacent mica tapes and also inside the mica tape toward the opposite conductor resulting in a final breakdown at a long time later. Because the mica is a very strong physical barrier against partial discharge, the electrical discharges can't penetrate the mica layers. But, rather, go around them in a zigzag fashion resulting in a long pathway to the final breakdown. Multiple mica tape layers can act as multiple barriers against partial discharge. This electrical tree proceeds very slowly due to two reasons. One is mica itself which is very resistant against partial discharge. Inorganic material like silica is resistant to ion bombardment due to discharges compared to polymer materials. It is hard for the electrical discharges to penetrate the mica layer and forced to progress in the creeping direction along the mica layer surface resulting in a long pathway. Another reason is that the electric field magnitude toward the progressing direction is very low. The typical operating electric field that is perpendicular to the mica layer is 3 kV/mm. [9] Since the creeping direction along the mica surface is perpendicular to the applied electric field, the electric field in parallel to the mica layer is far lower than the applied electric field. That is why the PDs take a long time to traverse the insulation more than 30–40 years to final breakdown.

10.3.3 Insulation Design and V-t Curve

In the electrical design of a rotating machine, the insulation lifetime of the main insulation is the key specification that determines the thickness of the ground wall insulation. The rotating machine is designed so that the lifetime of the insulation at an operating electric field exceeds a specified period of time which is typically more than 30 years. Since the actual test with a period of 30 years is impossible, accelerated tests are performed by applying higher electric fields and/or higher frequencies. Thus, the lifetime curve, i.e. voltage endurance curve is obtained and the lifetime at operating electric field is extrapolated from the accelerated tests.

In Figure 10.6, the typical image of the lifetime curve of the mica/epoxy insulation system is shown as V-t (E-t) curve. The curve is conventionally approximated by the following expression,

$$t = \mathrm{k}\,\mathrm{E}^{-n}$$

$$\log t = \log \mathrm{k} - n \log \mathrm{E}$$

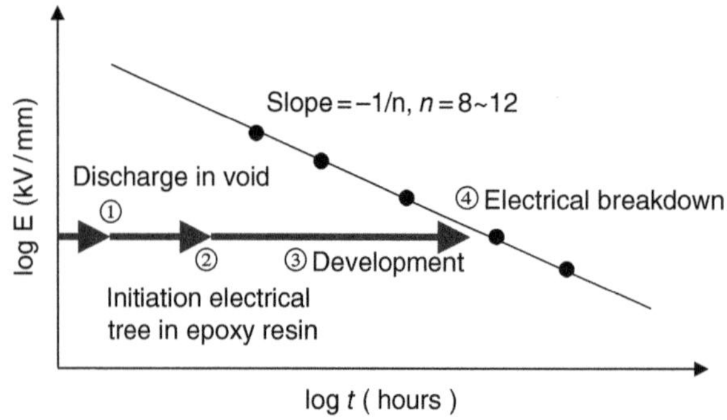

Figure 10.6 V-t curve and steps of partial discharge to a final electrical breakdown.

Here, t is breakdown time, k is a constant. The constant n is the inverse of the slope of the log t versus log E plot as shown in Figure 10.6. The number n is called "lifetime index" and typically ranges from 8 to 12 depending on the electric field level [10, 11]. When a new insulation system is developed, the V-t curve is compared to the existing systems and if the new one is comparable or better than the conventional one, the new system can be accepted.

10.4 Polymer Nanocomposite for Rotating Machine

Polymer nanocomposite is a polymer with small amounts of nanometer-sized inorganic fillers homogeneously dispersed. Research activities on polymer nanocomposite for electrical insulation emerged in the early 2000s [12]. Since then, various insulation-related properties such as PD resistance, treeing lifetime, dielectric breakdown strength, space-charge density, tracking resistance, permittivity, and DC conductivity have been studied and published. It was found that PD resistance is enormously enhanced by the addition of a small amount of nanofillers to polymer matrices. Although the positive effects of some of the properties are not clear yet, the superiority of polymer nanocomposite in terms of partial discharge property has been widely accepted. In the review paper published in 2013 [12], it was declared, "There is no doubt that PD resistance, the treeing lifetime can be significantly improved by loading the matrices with appropriate nanofillers."

In general, electrical insulation in high-voltage apparatus is subjected to partial discharges (PDs). Therefore, PD resistance is a long-time main concern for high-voltage researchers in academic institutions and engineers in manufacturers. Needless to say, this is the case for the people involved in the developments of high-voltage rotating machines. As mentioned in the previous section, the major electrical degradation process in the ground wall insulation is the partial discharge in gas-filled void defect and subsequent electrical tree development into epoxy resin. Therefore, if polymer nanocomposite technology is transferred into rotating machines, efficiency improvement, and compactness will be realized.

The insulation design of the rotating machine has a direct link to the V-t curve of the insulation. This leads to the idea that the most suitable application of polymer nanocomposite among high-voltage apparatus would be rotating machines. Other high-voltage apparatus like epoxy spacers of gas-insulated switchgear, mold transformers, low-voltage motors are designed so that they are PD-free. Therefore, although lifetime extension against PD definitely improves the insulation lifetime of those machines, it affects little its insulation (thickness) design of those machines except high-voltage rotating machine.

R&D steps for a novel insulation material of rotating machine are divided into four steps: (i) insulation material research, (ii) small insulation model with typical electrical defect, (iii) miniature model coil bars, and (iv) original stator bars of coil windings.

In the following, firstly, the key basic studies revealed mainly academic institutions briefly reviewed from the viewpoint of application to rotating machine insulation. Secondly, voltage endurance test in model defect of rotating machine insulation is shown. Thirdly, voltage endurance tests in model coil bars that simulate actual coil winding bar are mentioned.

10.4.1 Partial Discharge Resistance and a Treeing Lifetime of Nanocomposite as Material Property

10.4.1.1 PD Resistance

Figure 10.7 shows PD resistance of epoxy nanocomposites measured by the depth of surface erosion due to air-gap discharges between a rod and nanocomposite sheet in a rod-plate electrode system [13]. The erosion depths of TiO_2 and SiO_2 nanofiller-filled epoxy resins ([a]–[d]) are far shallower than the neat epoxy sample (f) showing the superior PD resistance of the nanocomposites.

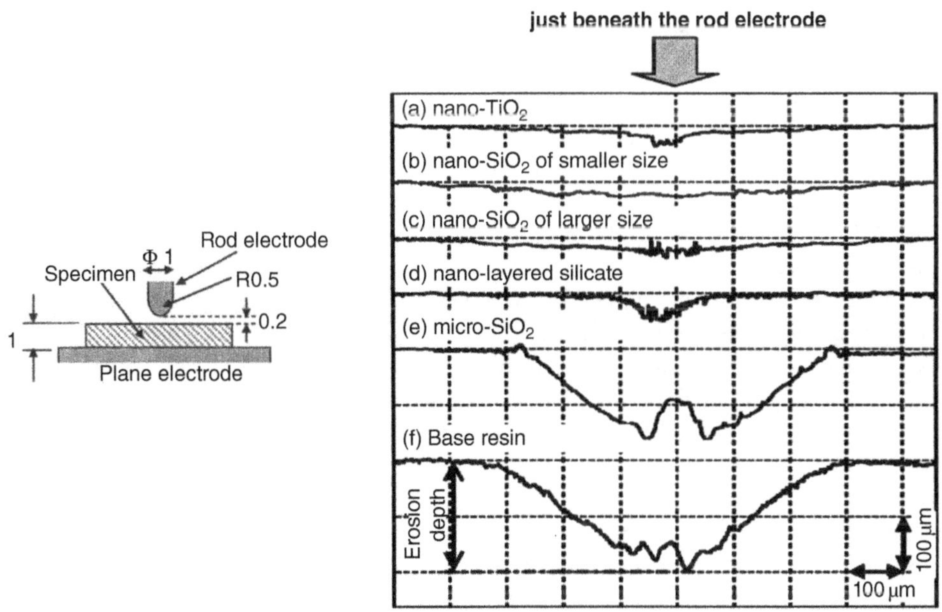

Figure 10.7 Surface profiles of eroded areas due to PDs in the specimens containing nano-TiO_2 (a), nano-SiO_2 of smaller size (b), nano-SiO_2 of larger size (c), layered silicate (d), micro-SiO_2 (e), and without fillers (f) after 120 hours aging at 720 Hz.[13]

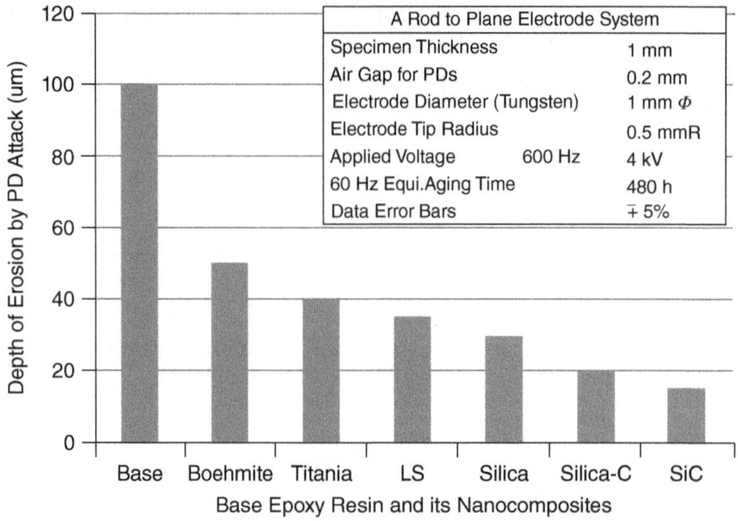

Figure 10.8 PD erosion depth of various epoxy nanocomposites.[14]

Figure 10.8 shows the PD resistance of nanocomposites filled with various kinds of nanoparticles under the condition of the same filler loading (5 wt%) [14]. Regardless of the material kinds of the fillers, inorganic particles dispersed in the epoxy polymer matrix enhance the PD resistance. Possible mechanisms are proposed: (i) fine partitioning of polymer matrix by nanofillers (segmentation effect), (ii) forming of interfaces and interphases exuding out of nanofillers onto nanocomposite surfaces (pile-up effect), coupling agent effect, permittivity difference effect [12]. These conceptual mechanisms may provide guidelines to modify insulating polymers to obtain highly PD-resistant nanocomposites. The dominant mechanism would be different depending on the filler specification and polymer matrices. The pile-up effect could be dominant for highly loaded nanocomposite. Segmentation would be effective for smaller filler size and so on.

10.4.1.2 Electrical Treeing Lifetime

As mentioned in the previous section, degradation of the insulation system of the rotating machine starting from discharges in a void defect possibly exists in imperfection in coil windings. After some duration of the discharges in the void, a small protrusion called "pit" is generated at the bottom of the void. At the tip of the pit, electric field enhancement occurs resulting in the initiation of an electrical tree into the epoxy bulk as reviewed by Tanaka [14]. Thus, the electrical trees start as the second long-term degradation process. This process can be stimulated by electrical tree inception and development in a needle-plane electrode system.

Longer electrical tree lifetime of the nanocomposite was first reported by Nelson [15] in TiO_2 particle-filled epoxy nanocomposite. Subsequently, Imai and Iizuka reported a detailed study on TiO_2 and SiO_2 epoxy nanocomposite and published various parameter studies such as filler kinds, size, and effects of coupling agent [16–21].

Figure 10.9 compares the average insulation breakdown time due to an electrical tree in the needle-plate electrode for two types of SiO_2 nanocomposites (layered silicate and silica)

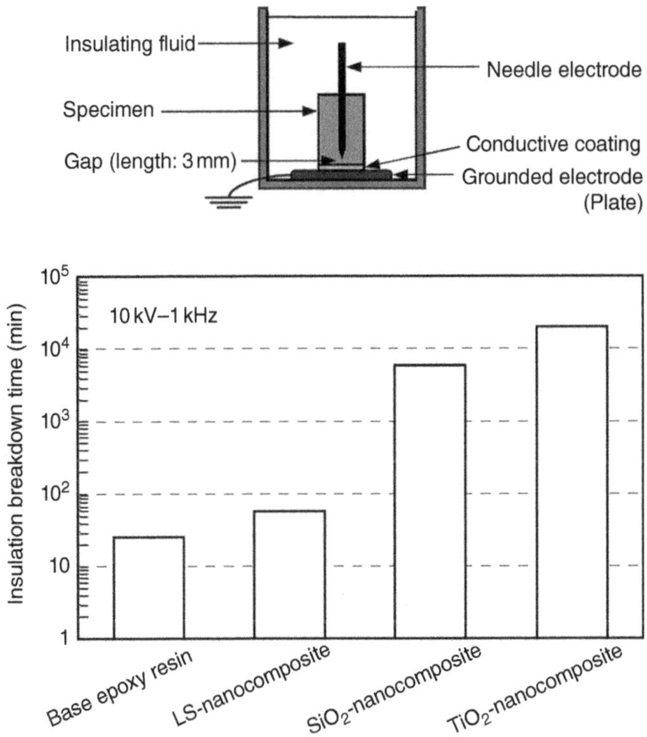

Figure 10.9 Comparison of insulation breakdown time due to electrical tree under divergent electric field (10 kV–1 kHz).[16]

and TiO_2 nanocomposite [16]. All three kinds of nanocomposites show longer breakdown times than the base epoxy resin. The LS-nanocomposite has a breakdown time of 57 minutes, whereas the base epoxy resin has a breakdown time of 25 minutes. In particular, the SiO_2- and TiO_2-nanocomposites have breakdown times of approximately 6000 and 20 000 minutes, respectively.

Figure 10.10 shows the dependence of breakdown time on the amount of TiO_2 nanofiller [17]. In the horizontal axis, the amount of 10 phr corresponds to 5 wt%. The breakdown time does not depend on the filler amount in the range from 10 to 20 phr. In a lower amount than 10 phr, it could depend on the amount as indicated in the reference [18].

Figure 10.11 shows the effect of the coupling agent on the treeing lifetime [19]. The nanocomposite with coupling agent (12 nm C) has a longer lifetime than that without the coupling agent. The reason is estimated to be detachments of the filler surface from the epoxy matrix, which was observed for the nanocomposite without the coupling agent.

Figure 10.12 shows experimental results of breakdown time for the various filler sizes from 12 up to 300 nm under the same weight % of SiO_2 filler loading [21]. It is clearly recognized that tree breakdown time is the longest for the nanocomposite with 12 nm fillers and decreases as the size becomes larger. Apparently, this finding is very important for the selection of filler size. Considering the actual fabrication processes of impregnation, the smaller size does not necessarily mean suitable since a higher viscosity of the epoxy resin

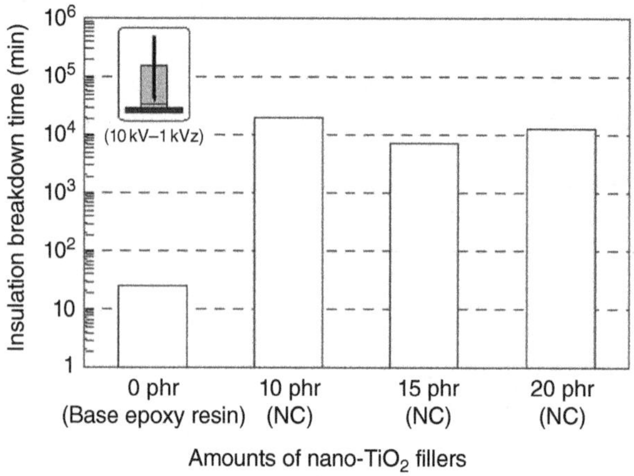

Figure 10.10 Dependence of insulation breakdown time of epoxy nanocomposites on the amount of TiO_2 filler.[17]

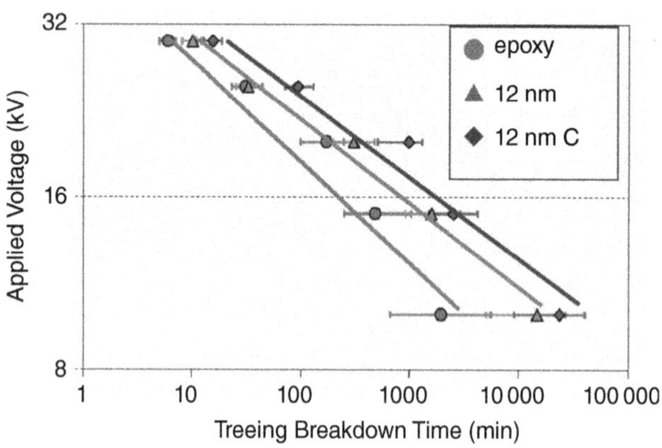

Figure 10.11 V-t characteristics of silica-filled epoxy nanocomposite and effect of coupling agent. The sample 12 nm C is the sample with a coupling agent.[19] © IEEJ.

increases the possibility of void defects due to an imperfect impregnation. The filler size would be in relation to a trade-off between the tree lifetime and appropriate viscosity of the nanocomposite resin for processing.

10.4.2 Breakdown Lifetime Properties of Realistic Insulation Defect in Rotating Machine

Positive results of surface discharge erosion and electrical tree breakdown experiments intensively done by academic researchers were encouraging for the application of polymer nanocomposite to the rotating machine. However, to proceed to industrial-scale development, it is necessary to confirm the effect of those nanocomposites in realistic insulation defects and structures. They are void defects in epoxy nanocomposite and mica/epoxy nanocomposite/mica-layered structure.

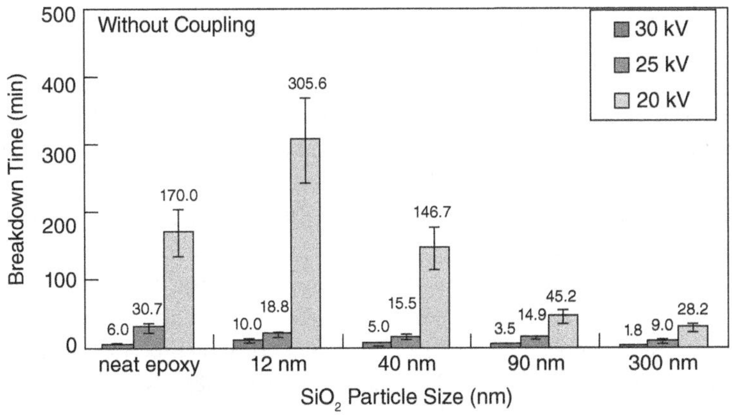

Figure 10.12 Treeing breakdown time for four kinds of filler size of epoxy/silica nanocomposites.[21]

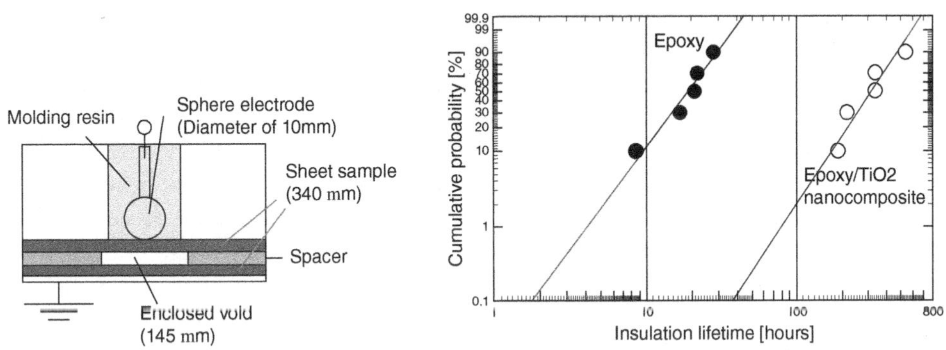

Figure 10.13 2-parameter Weibull distribution of the insulation breakdown lifetime at $14\,kV_{rms}/mm$.[22]

10.4.2.1 Voltage Endurance Test of Void Defect

Figure 10.13 shows a sample structure simulating the void defect in the main ground wall insulation in a rotating machine [22]. Between two epoxy nanocomposite sheets, an air gap with a thickness of 145 um is enclosed, which simulate void defect in solid. The sphere electrode is molded by epoxy resin to prevent undesired PD around it. The actual void size in a machine is far smaller like several tens of micron meters to hundreds of micron meters, the sample with the configuration is used to allow continuous stable gas discharges in the air gap. The degradation of this sample progresses in consecutive two steps similar to the actual defect: gas discharges in the void and tree propagation in epoxy bulk to a final breakdown.

Figure 10.13 shows insulation lifetime characteristics at $14\,kV_{rms}/mm$ for epoxy and epoxy/TiO_2 nanocomposite samples with a content of 5 vol%. Although the average diameter of a single TiO_2 filler is approximately 35 nm, observation of the cross section by SEM reveals the nanocomposite contains some agglomerates of the maximum size of 5 μm. The mean time to failure (MTTF), i.e. 63.2% breakdown lifetimes, of the nanocomposite is 16 times longer than that of the epoxy sample. The epoxy/TiO_2 nanocomposite clearly shows the superior insulating lifetime properties compared to the epoxy without nanofillers in the void defect model sample as well as the material property sample.

As mentioned, degradation in the existing void defect progresses in consecutive two steps to a final breakdown: firstly, gas discharge and pit formation in the void; secondly, electrical tree inception and propagation in the epoxy bulk. It is still not clear which process, surface erosion inside the void, or tree propagation inside bulk is more effectively retarded by the nanofillers. A more detailed study of those individual steps will contribute to the optimization of nanofiller specification.

In order to optimize the insulation properties of polymer nanocomposites, it is important to clarify the effect of particle dispersion state. Nanofillers are easy to form agglomerates to a greater or lesser extent during the mixing process. In the actual fabrication process, the inclusion of agglomerates is considered to be unavoidable. The formed agglomerates could deteriorate their insulation properties. Although the importance of agglomerate has long been pointed out, there have been few reports which handle the effect of the agglomerates from the right in the front. Kurimoto made an approach to this issue using his original centrifugation technique [23] by which agglomerates larger than a specified weight or size can be removed. The size of removed fillers is controlled by rotation speed and duration of the centrifugation.

Figure 10.14 shows the dependence of the partial discharge lifetime of the TiO_2 nanocomposite on the maximum agglomerate size [24]. The maximum size of the agglomerate is changed from 5 to 0.05 um as shown in Table 10.1. Sample #0 h with 5.6 vol% TiO_2 filler shows the longest lifetime such as approximately 17 times longer than the neat epoxy. As the maximum agglomerate size decreases, the lifetime becomes shorter. In centrifuging processing, the volume fraction of the nanofiller also decreases inherently as the maximum agglomerate size decreases. Since the decrease of the lifetime of nanocomposites with lower vol% is well-known property, the global tendency in Figure 10.14 is what is expected.

Even though, in Figure 10.14, another distinctive feature of the effect of agglomerate size can be seen. That is the distinctive increase of the lifetime of the sample #1 h compared to the #2 h. The difference of the filler content between the two samples is the volume fraction of

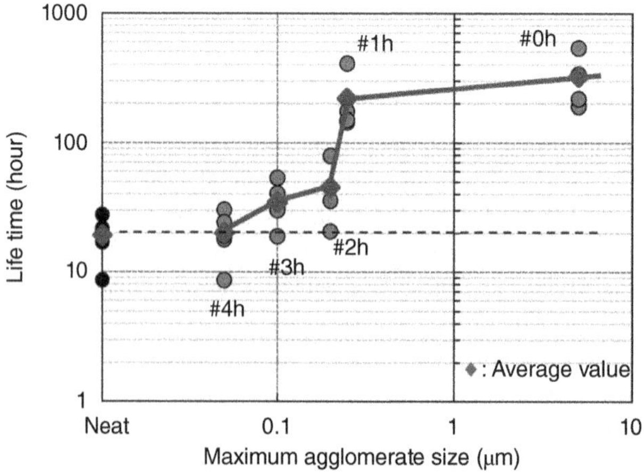

Figure 10.14 Effect of agglomerates on PD lifetime in an enclosed void sample of TiO_2-filled nanocomposite.[24]

Table 10.1 Maximum agglomerate size and volume % of TiO_2 nanofiller of the samples in Figure 10.14 [24].

Sample #	Centrifuging Time (Hours)	Maximum Particle Size	Total Volume %	vol% of Each Particle Size of TiO_2. 0–0.05 um	0.05–0.2 um	0.2–0.25 um	0.25–5 um	
#0 h	0	5 um	5.6	<0.1	0.8	1.3	1.4	2.0
#1 h	1	0.25 um	3.6	<0.1	0.8	1.3	1.4	
#2 h	2	0.20 um	2.2	<0.1	0.8	1.3	None	
#3 h	3	0.10 um	0.9	<0.1	0.8			
#4 h	4	0.05 um	<0.1	<0.1				

1.4% of the agglomerates whose sizes range between 0.2 and 0.25 um as shown in Table 10.1. Thus, it can be pointed out that the relatively large agglomerates of those sizes can contribute distinctively to the suppression of partial discharge developments. It has long been pointed out that good and homogeneous dispersion of nanofiller in a polymer is a requisite for better insulation properties. In regard to tree propagation property, agglomerates could play an important role in the tree lifetime extension. If agglomeration is allowed in the fabrication process, it is good news for industrial applications since the process margin would expand.

10.4.2.2 Voltage Endurance Test in Mica/Epoxy Nanocomposite-Layered Structure

Electrical trees propagate in zig-zag fashion in epoxy nanocomposite between mica layers. There are two directions for tree propagation: one is the direction parallel to the mica layer, the other is the direction perpendicular to the mica layer. If the discharge is strong enough, penetration through the mica flake layer can also take place. The lifetime for both directions in mica/epoxy nanocomposite structure is an important property when considering the degradation of the actual insulation structure of rotating machines.

Mitsubishi Electric Co. investigated the insulation lifetime extensions with mica/epoxy nanocomposite/mica-layered sample structure under the condition that the electric field is applied in the direction of perpendicular to the mica layer. The direction of the electric field is the same as that for the actual main wall insulation of the rotating machine. Two layers of mica sheet were stacked and the epoxy resin filled with nanofillers was impregnated and tested in a rod-plane electrode configuration.

Figure 10.15 shows V-t curves of nanocomposites with four kinds of nanofillers (TiO_2, SiO_2, MgO, $MgOH_2$). It can be clearly seen that except MgOH, all the other nanocomposites (TiO_2, SiO_2, MgO) show a better lifetime than the neat epoxy resin sample. However, the slopes of the V-t curves among the three nanocomposites are different. The slope of TiO_2 is smaller resulting in the best lifetime at the extrapolated low electric field. The curve for SiO_2 and $MgOH_2$ shifts parallel toward a longer lifetime.

TMEIC investigated the insulation lifetime extensions with mica/epoxy nanocomposite/mica-layered sample structure under the condition that the electric field is in the direction of parallel to the mica layer [26]. A sheet-form epoxy nanocomposite/mica sample shown in Figure 10.16 is prepared to investigate the effect of nanofillers and their weight contents on the insulation lifetimes. In this sample, a thin metal electrode with a thickness of

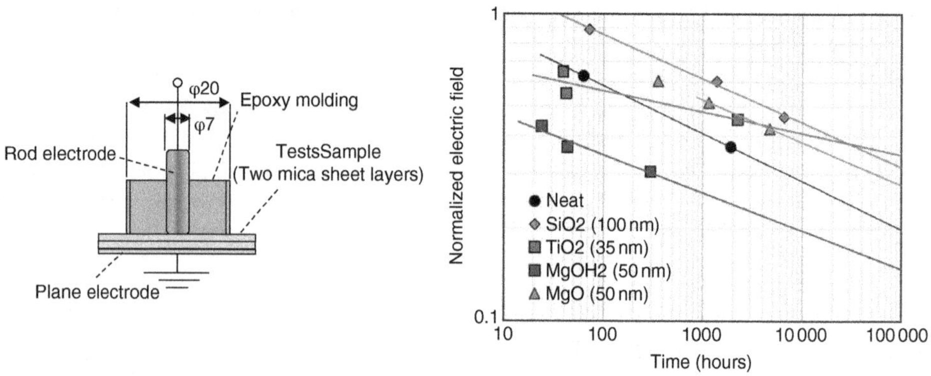

Figure 10.15 V-t curves of nanocomposites with various nanofiller materials. The electric field in a rod-pane electrode configuration was applied to two mica sheet layers impregnated with epoxy nanocomposite. The vertical axis is the normalized electric field by the short-term breakdown voltage of the neat sample.[25]

0.1 mm and a tip radius of 2 μm was sandwiched with mica layers to simulate the one unit layer in actual mica/epoxy insulation. In this configuration, the electrical tree initiates at the tip of the thin electrode. Then it propagates toward the counter-grounded electrode. Although the direction of the electric field is parallel to the mica layer and different from the actual machine, this sample structure can test the lifetime due to electrical trees which propagate in narrow two-dimensional epoxy space sandwiched by mica layers.

Figure 10.16 shows a 2-parameter Weibull distribution of the insulation breakdown times of the nanocomposites. Silica nanofillers with an average diameter of 12 nm were filled and the weight contents are 5 and 10 wt%, respectively. It can be clearly seen that the nanocomposites showed superior insulation breakdown times than the neat epoxy sample. Moreover, the nanocomposite with 5 wt% showed the best performance among these samples. This result suggests the optimal wt% exists to obtain the best performance in the insulation lifetime characteristics.

10.4.2.3 V-t Curves in Coil Bar Model with Mica/Epoxy Nanocomposite Insulation

As shown in the previous sections, polymer nanocomposites do have the potential to improve the PD resistance, tree lifetime characteristics. These properties can improve the efficiency of high-voltage rotating machines by making the main wall insulation thinner. In the insulation-manufacturing method called Vacuum Pressure Impregnation (VPI) systems, mica tapes are wound around stator bars or coils which are subsequently impregnated with a low-viscosity impregnating resin and thermally cured. For the new insulation system, nanofiller-mixed epoxy resin is used in place of the conventional epoxy resin. Two examples of the insulation properties of the nanocomposites in the coil bar structure are introduced below.

Voith Hydro Ltda and Siemens AG have published successful results on the insulation lifetime extension effect of stator winding bars [27]. The rated voltage and the length of the bars used for the voltage endurance test were 13.8 kV and 1.6 m, respectively. For the endurance test, sets of original bars, impregnated with pure epoxy resin and with 20wt% SiO_2 nanoparticle-loaded epoxy resin, were prepared by the VPI-manufacturing process. The diameter of the SiO_2 nanoparticles was 10–20 nm. Figure 10.17 compares the results of the

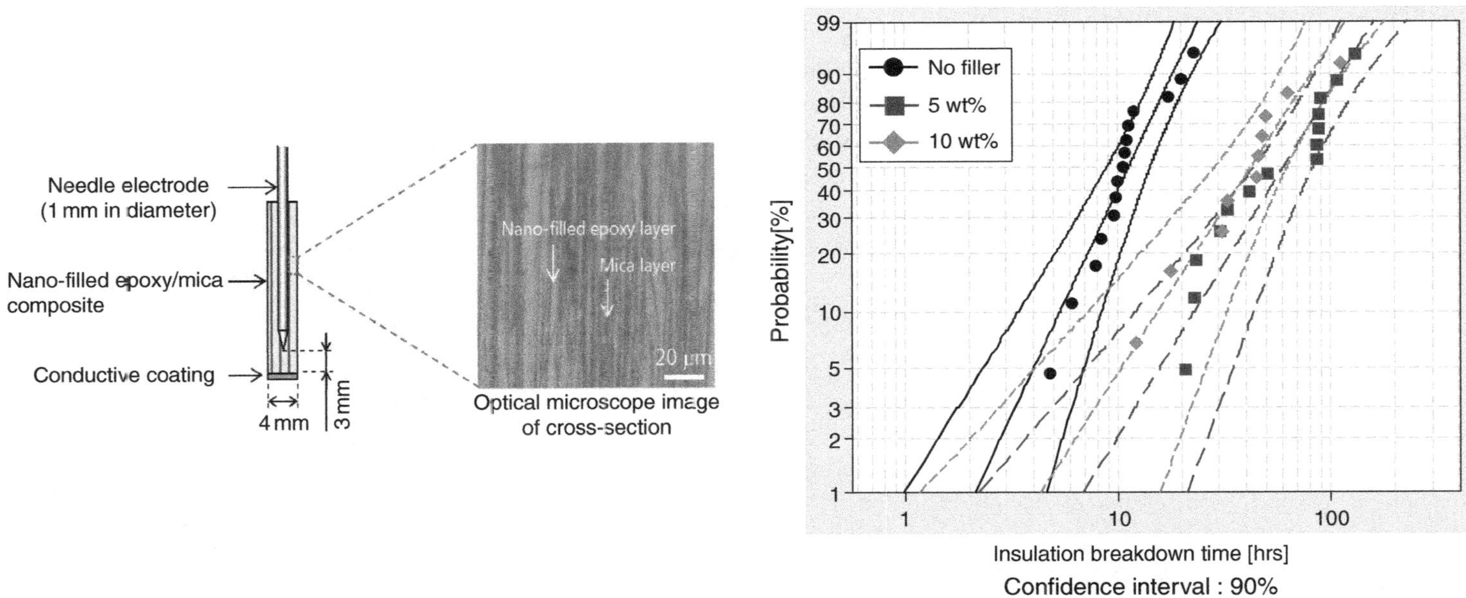

Figure 10.16 Weibull distribution of breakdown time of mica/epoxy nanocomposite.[26] © IEEJ.

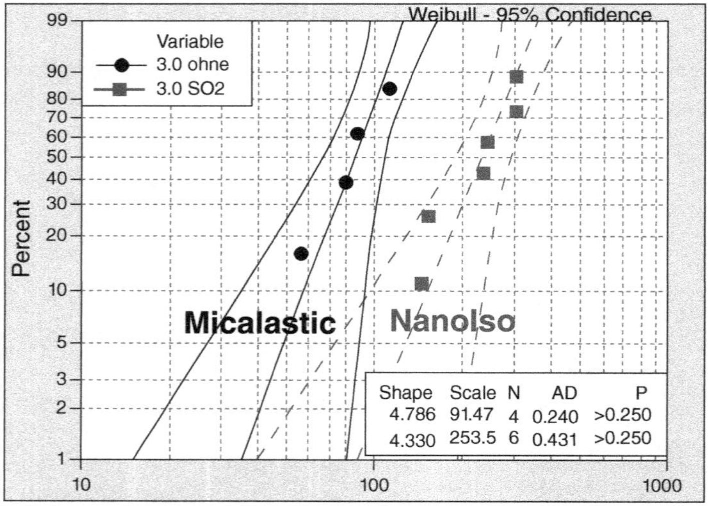

Test Voltage $U_P = 3U_n = 41.4$ kV

Mean electrical lifetime (63% scale)

Micalastic: 91.5 h

NanoIso: 254 h

→ **Improvement Factor VEI = 2.8**

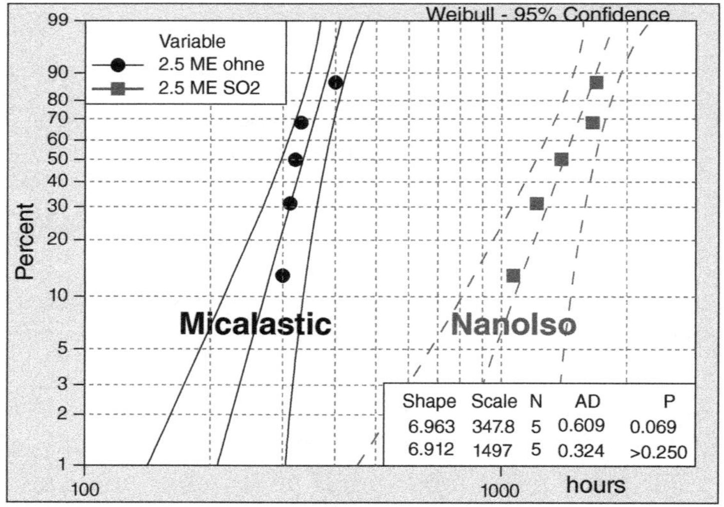

Test Voltage $U_P = 2.5U_n = 34.5$ kV

Mean electrical lifetime (63% scale)

Micalastic: 348 h

NanoIso: 1497 h

→ **Improvement Factor VEI = 4.3**

Figure 10.17 Weibull probability plot of voltage endurance tests at RT with standard insulation system MICALASTIC and new nanocomposite insulation system at 3-times and 2.5-times rated voltage.[27]

voltage endurance test at different voltage levels (41.4 and 34.5 kV) as Weibull probability distribution plots. These results indicate that the electrical lifetime significantly increased by introducing SiO_2 nanoparticles. Furthermore, the improvement of the insulation lifetime was more significant under the lower voltage (34.5 kV), which suggests that the insulation degradation was effectively suppressed by SiO_2 nanoparticles and the drastic lifetime extension was expected under the operating voltage.

It is described in the reference [28] that they were now focused on functional testing of the original stator winding bars before the last step of prototype production. They expect that the new nanocomposite insulation will increase the overall generator performance (MVA) between 10 and 15%.

Mitsubishi Electric Co. investigated the appropriate types of nanoparticles and their sizes for the insulation lifetime extensions with coil bar structures. Figure 10.18 shows the V-t characteristics of a miniature-sized bar coil model with nanocomposite insulation in which an aluminum metal bar with a rectangular cross section was wound by a mica tape and nanoparticle-filled epoxy resin was impregnated by the VPI process [25]. In this minibar model, the typical thickness of the mica/epoxy insulation was approximately 1 mm. As filler materials, TiO_2 and SiO_2 were selected since those fillers showed better V-t properties than the others in a laboratory-scale test (Figure 10.15).

As shown in Figure 10.18, the TiO_2 nanocomposite didn't show an improvement in the insulation lifetime compared to the neat epoxy sample. For the SiO_2 nanocomposites, the samples with 100 nm SiO_2 nanoparticles show superior insulation lifetime characteristics compared to those with 10 nm. This tendency of the particle size dependence agrees with the results of the insulation lifetime test with a void defect (Figure 10.14), and this lifetime extension was achieved since relatively large nanoparticles contribute to the suppression of the PD propagation.

It should be noted that there are still some inconsistencies in the results of the insulation lifetimes between the laboratory-scale fundamental model test and the coil or bar-form test. This is because not only controllable factors, e.g. the material type of nanoparticles,

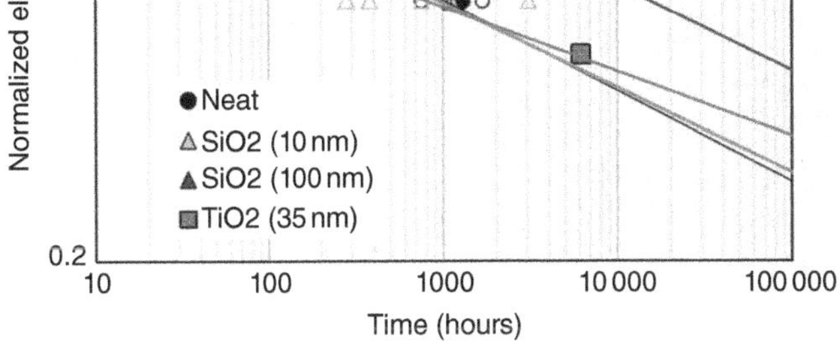

Figure 10.18 V-t curves of epoxy nanocomposites with two types of fillers and different sizes.[25]

the volume content and surface treatment, but also uncontrollable ones, e.g. the dispersion state of fillers and contaminated insulation defects, could have a strong influence on the lifetime properties. For the superior nanocomposite materials with higher insulation reliability, systematic experiment on the above factors as parameters will be required further.

10.5 Stress-Grading System of Rotating Machines

10.5.1 Silicon Carbide Particle-Loaded Nonlinear-Resistive Materials

In rotating machines, another type of polymer composites play an important role at stator end windings of high-voltage rotating machines over decades. Silicon carbide (SiC) particle-loaded polymer with electric-field-dependent conductivity has been used to control surface potential distribution and prevent partial discharges (PDs). The function is called "stress grading" (SG). As the stress-grading material, polyester and epoxy polymers are typically used as matrix resin. As fillers, SiC particles with a mean particle size of a few micrometers to few tens of micrometers are mixed into the above matrix resin, then the mixture of the matrix resin and SiC particles is cured at a prescribed temperature. Figure 10.19 shows a typical scanning electron microscope (SEM) image of SG material [29]. The size of the SiC particles has direct influence on the electrical properties of the SG materials. Figure 10.20 compares typical current (I)–voltage (V) characteristics of SG materials with different SiC particle size. In this figure, the average size of SiC particles of sample A is smaller than that of sample B. At low voltage, both materials show ohmic (linear-resistive) properties. However, the current of sample B with larger SiC particle increases rapidly and nonlinearly. It should be noted that the transition from linear to nonlinear resistivity of sample A with smaller SiC particles occurs at higher voltage application than that of sample B with larger particles.

By using the experimentally obtained I–V characteristics with various SiC particle size and a rough assumption that shape of a SiC particle is spherical, and the sample has primitive orthorhombic structure, a relationship of voltage across one interface between SiC particles (V) and the current flowing through the interface (I) can be calculated as shown

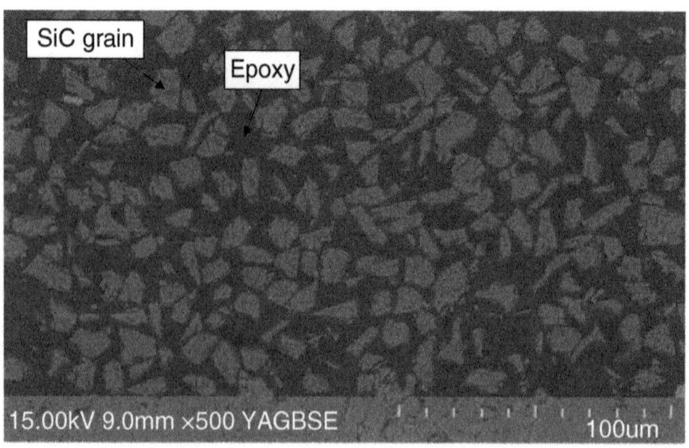

Figure 10.19 A typical SEM image of stress-grading material.[29]

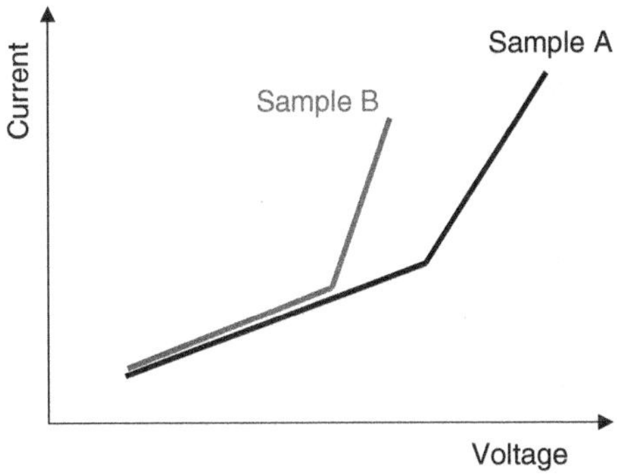

Figure 10.20 Current-voltage characteristics of stress-grading materials with different SiC particle size (sample A: smaller SiC particles, sample B: larger SiC particles).

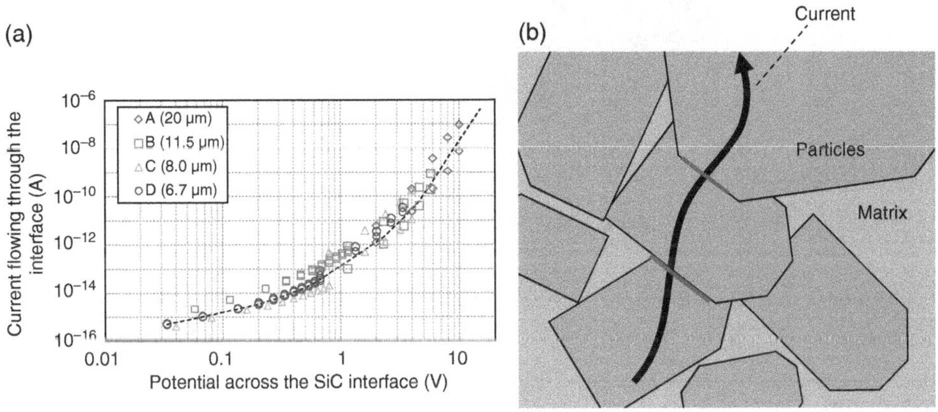

Figure 10.21 I–V characteristics across an interface between SiC particles and a schematic of SiC microstructure. (a) I–V characteristics at SiC/SiC interface [29]. (b) Schematic of SiC microstructure.

in Figure 10.21a [29]. The I–V characteristic shows a single-voltage-dependent conductive current property, regardless of the average SiC particle size. This result clearly indicates that the nonlinear conductivity of the SG material is determined by a SiC/SiC interfacial conduction phenomenon as in Figure 10.21b. Tunneling by field emission at the interface is the dominant conduction mechanism.

10.5.2 End-turn Stress-Grading System of High-Voltage Rotating Machines

Figure 10.22 shows a schematic of a turbine generator and typical insulation structure of the end-turn SG system [29], and Figure 10.23 shows its cross-section [30]. Carbon-powder-loaded semi-conductive slot layer (conductive armor tape, CAT) is wound over the ground-wall insulation (epoxy/mica insulation) within slot of the stator core, and it continues for some distance beyond where the stator coil exits the slot. The conductive slot layer can reduce the potential difference between outer of the stator coils and the stator core and can prevent PDs within the slot. The SiC-based nonlinear-resistive

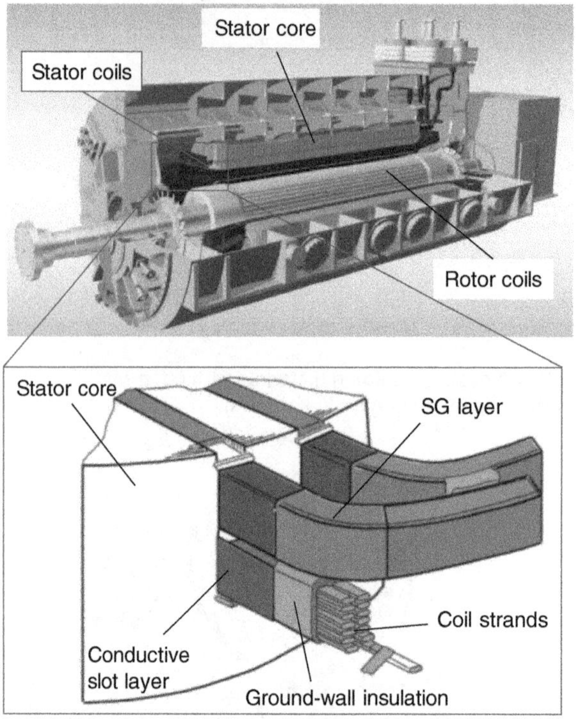

Figure 10.22 Schematic of a turbogenerator and typical insulation structure of the end-turn SG system.[29]

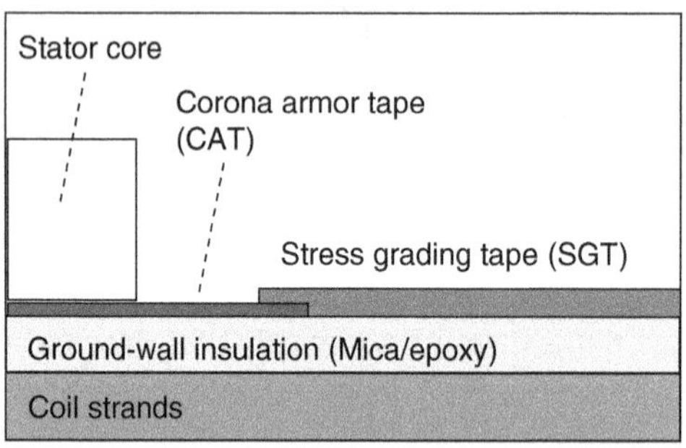

Figure 10.23 Cross-section of the end-turn SG system [30].

stress-grading layer (stress-grading tape, SGT or paint, SGP) is also wound so as to cover the end of the conductive slot layer.

There are typical two kinds of manufacturing processes of the SG system, depending on manufacturers of the rotating machines. In one method, the above mixture in the liquid state is painted over stator end windings, then is cured. In the other method, the semi-cured mixture is processed into tape (prepreg tape), and the tape is wound over stator windings. Then, prepreg tape is thermally cured to form the SG system.

Potential distribution behaviors along the SG system were experimentally evaluated by Pockels surface potential and field sensors [31, 32]. For instance, potential distribution along the SG system under 30 kV_p and 50 Hz sinusoidal voltage is shown in Figure 10.24.

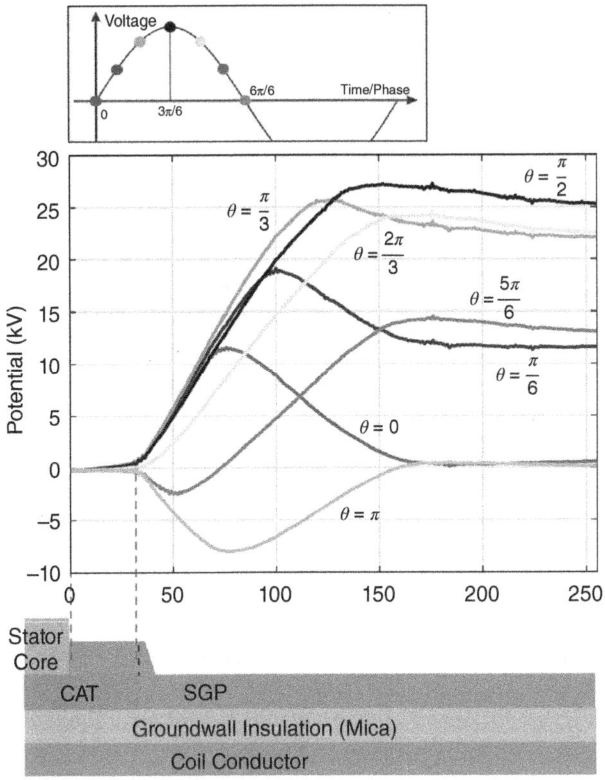

Figure 10.24 Potential distribution along the SG system under 30 kV_p 50 Hz voltage [31].

It can be seen that the potential of the CAT is maintained at ground potential under sinusoidal voltage. Furthermore, tangential electric field along the system (the slope of the curve) is kept at approx. 0.30–0.35 kV/mm, regardless of instantaneous voltage (phase of the applied voltage), due to the nonlinear-resistive characteristics of the SG layer.

References

1 Stone, G.C. and Miller, G.H. (2013). Progress in rotating machine insulation systems and processing. *IEEE Electrical Insulation Magazine* 29 (4): 45–51.

2 Brammer, R., Karlsson, T., Rudolfsson, D., and Lutz, A. (1994). Insulation mica tapes with increased thermal conductivity, 15-202, CIGRE (August 1994).

3 Miller, M.L. (1999). Turbine generator stator slot thermal conductivity. *Proceedings of Electrical Insulation Conference and Electrical Manufacturing and Coil Winding Conference* (August 1999), 65–69.

4 Tari, M., Yoshida, K., Sekito, S. et al. (2001). HTC insulation technology drives rapid progress of indirect-cooled turbo generator unit capacity. IEEE PES Summer Meeting.

5 Suzuka, S., Hatano, H., and Kato, Y. (2012). 820 MVA-50Hz large-capacity indirectly hydrogen-cooled turbine generator. *Toshiba Review* 67 (7): 42–45.

6 Harakawa, T., Sawa, F., Okamoto, T. et al. (2014). Current progress in high thermal conducing insulation technology for large capacity turbine generator. 17th International Conference on Electrical Machines and Systems (ICEMS), 2409–2412.

7 Satake, Y., Takahashi, K., Wake, T. et al. (2015). Development of large capacity turbine generators for thermal power plants. *Technical Report of Mitsubishi Heavy Industries* 52 (2): 44–51.

8 Sako, H., Koga, K., Maeda, H. et al. (2016). Development of larger output indirectly hydrogen-cooled turbine generator with high heat transfer main insulation. *CIGRE* 2016: A1–A116.

9 Fréchette, M., Convenor, A., Vaughan, A. et al. (2016). Functional nanomaterials for electric power industry. Cigre WG D1.40, TB-661.

10 Wichmann, A. (1983). Two decades of experience and progress in epoxy mica insulation system for large rotating machines. *IEEE Transactions on Power Apparatus and Systems* PAS-102 (1): 74–82.

11 Tsukui, T., Takahashi, G., and Isogai, T. (1976). Voltage endurance of mica insulation systems for rotating machines. *IEEJ-A* 96 (10): 463–470.

12 Tanaka, T. and Imai, T. (2013). Advances in Nanodielectric Materials Over the Past 50 Years. *IEEE Electrical Insulation Magazine* 29 (1): 10–23.

13 Kozako, M., Kuge, S., Imai, T. et al. (2005). Surface erosion due to partial discharges on several kinds of epoxy nanocomposites. Annual Report IEEE-CEIDP, 162–165.

14 Tanaka, T. and Iizuka, T. (2010). Generic PD resistance characteristics of polymer nanocomposites. Annual Report IEEE-CEIDP.

15 Nelson, J.K. and Hu, Y. (2004). The impact of nanocomposite formulation on electrical voltage endurance. International Conference on Solid Dielectrics, Toulouse.

16 Imai, T., Sawa, F., Ozaki, T. et al. (2006). Comparison of insulation breakdown properties of epoxy nanocomposites under homogeneous and divergent electric fields. Annual Report IEEE-CEIDP, 306–309.

17 Imai, T., Sawa, F., Ozaki, T. et al. (2007). Roles of fillers on properties of Nano-TiO_2 and Micro-SiO_2 filler mixed composites. International Conference on Solid Dielectrics, 407–410.

18 Kodama, N., Yamada, D., Kawashima, T. et al. (2019). Treeing breakdown property in epoxy/TiO_2 nanocomposites. International Conference on Solid Dielectrics.

19 Iizuka, T. and Tanaka, T. (2010). Effects of nano silica filler size on treeing breakdown of epoxy nanocompositees. *IEEJ Transactions on Fundamentals and Materials* 130 (9): 837–842. (in Japanese).

20 Iizuka, T., Uchida, K., Tanaka, T. (2007). Different voltage endurance characteristics of epoxy/silica nanocomposites prepared by two kinds of dispersion methods. Annual Report IEEE-CEIDP, 236–239.

21 Iizuka, T. and Tanaka, T. (2009). Effects of nano silica filler size on treeing breakdown lifetime of epoxy nanocomposites. Proceedings of the 9th International Conference on Properties and Applications of Dielectric Materials, 733–736.

22 Umemoto, T., Otake, Y., Muto, H. et al. (2019). Insulation lifetime characteristics of epoxy/TiO_2 nanocomposite with internal void. Annual Report IEEE-CEIDP.

23 Kurimoto, M., Yoshida, S., Umemoto, T., and Muto, H. (2019). Impulse breakdown strength of epoxy/TiO_2 nanocomposite enhanced by agglomerate removal. Annual Report IEEE-CEIDP.

24 Muto, H., Yoshida, S., Umemoto, T. et al. (2020). Critical agglomerate sizes to improve the insulation characteristics of epoxy/TiO_2 nanocomposites. Annual Report IEEE-CEIDP.

25 Muto, H., Yoshida, S., Umemoto, T. et al. (2020). Dielectric breakdown and partial discharge properties of nanocomposites for epoxy/mica insulation system in large-size rotating machines. Conference proceedings of ISEIM.

26 Hirai, H., Imai, T., Murata, K. et al. (2019). Fundamental consideration on application of nanocomposite insulation to electrical rotating machine. *IEEJ Transactions on Fundamentals and Materials* 139 (9): 374–379.

27 Hildinger, T. and Weidner, J.R. (2017). Progress in development of a nanocomposite stator winding insulation system for improved generator performance. IEEE Electrical Insulation Conference (EIC), 139–142.

28 Hildinger, T., Brockschmidt, M., Groppel, P. et al. (2016). Improved generator performance with a nanocomposite high voltage insulation system for stator windings – a status report. Cigre, A1–109.

29 Umemoto, T., Otake, Y., Yoshimura, M. et al. (2020). Optimization of double-layer stress grading system for high voltage rotating electrical machines by electric field and thermal coupled analysis. *IEEE Transactions on Dielectrics and Electrical Insulation* 27 (3): 971–979.

30 Nakamura, T., Karasawa, K., and Umemoto, T. (2020). Fundamental and carrier frequency-dependence on electric field and power dissipation for inverter-fed high voltage rotating machine. IEEE Electrical Insulation Conference(EIC), 265–268.

31 Hirakawa, T., Kumada, A., Hidaka, K. et al. (2019). Surface potential measurement of model stator bar with stress grading system by field sensor. *IEEE Transactions on Dielectrics and Electrical Insulation* 26 (4): 1146–1153.

32 Nakamura, T., Kumada, A., Ikeda, H. et al. (2019). Potential distribution on the stress grading system of high voltage rotating machines – part II transient measurements and simulation profiles. *IEEE Transactions on Dielectrics and Electrical Insulation* 26 (5): 1495–1502.

11

Polymer Composite Conductors and Lightning Damage

Xueling Yao

School of Electrical Engineering, Xi'an Jiaotong University, Xi'an, China

11.1 Lightning Environment and Lightning Damage Threat to Composite-Based Aircraft

11.1.1 The Lightning Environment

11.1.1.1 Formation of Lightning

To accurately assess the interaction between lightning and aircraft, some physics knowledge about lightning phenomena must be introduced, which will help one understand the conditions under which aircraft are struck by lightning.

Lightning flashes usually originate from charge centers in a cloud, particularly the cumulonimbus cloud, although they can occur in other atmospheric conditions. Typically, most positive charges accumulate at the top of the cumulonimbus clouds, leaving the lower regions negative, although there may be a small positive region near the base. The typical structure of cumulonimbus clouds is shown in Figure 11.1. The electric charges in clouds create an electric field. The electric field with sufficient intensity can ionize the air and produce electrical sparks which can develop into lightning flashes. These lightning flashes may be of three types, namely:

1) Flashes between regions of opposite polarity within a cloud (intra-cloud discharges),
2) Flashes between regions of opposite polarity in different clouds (inter-cloud charges), and
3) Flashes between clouds and ground. These may originate from clouds and propagate to the ground or originate from tall objects on the ground and propagate to clouds.

The typical cloud-to-ground discharge is initiated by the formation of the leader. When the charge accumulated in the charge center of the cloud reaches a certain amount, the surrounding electric field will cause the strong ionization of the air, and the leader will be emitted from the area with high electric field intensity (greater than 500 kV/m), and produce a bright spark-conducting channel.

Polymer Composites for Electrical Engineering, First Edition. Edited by Xingyi Huang and Toshikatsu Tanaka.

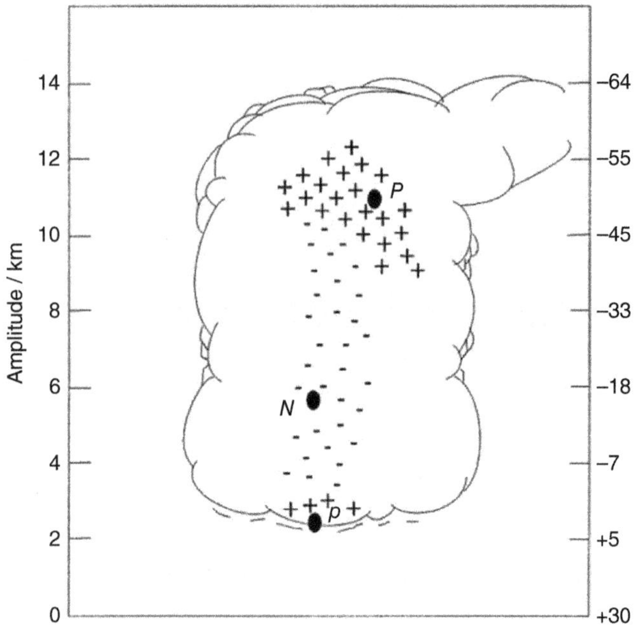

Figure 11.1 Distribution of electrical charge in a typical cumulonimbus cloud.

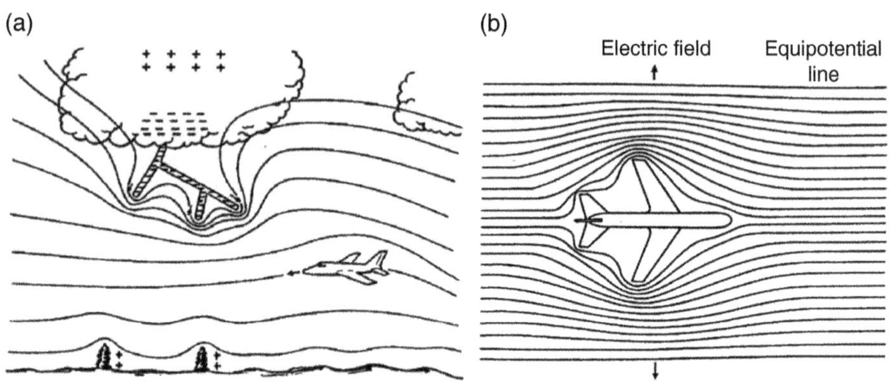

Figure 11.2 The process of aircraft triggered a lightning strike. (a) Steeped leader approaches an aircraft. (b) Compression of the electric field around an aircraft.

About 90% of lightning strikes are triggered by the aircraft itself. The conductive shell of aircraft can be regarded as an equipotential body. The aircraft can divert and compress the adjacent equipotential surfaces in a thunderstorm environment, as shown in Figure 11.2. The electric field strength of the compressed electric field increases as the distance between the aircraft and the leader decreases, which attracts subsequent leader propagation toward the aircraft. In addition, the electric field lines are compressed closest together around the extremities of the airframe, such as nose, wing, tail, and small protrusions (such as antennas and probes). When the leader advances to the point where the electric field adjacent to

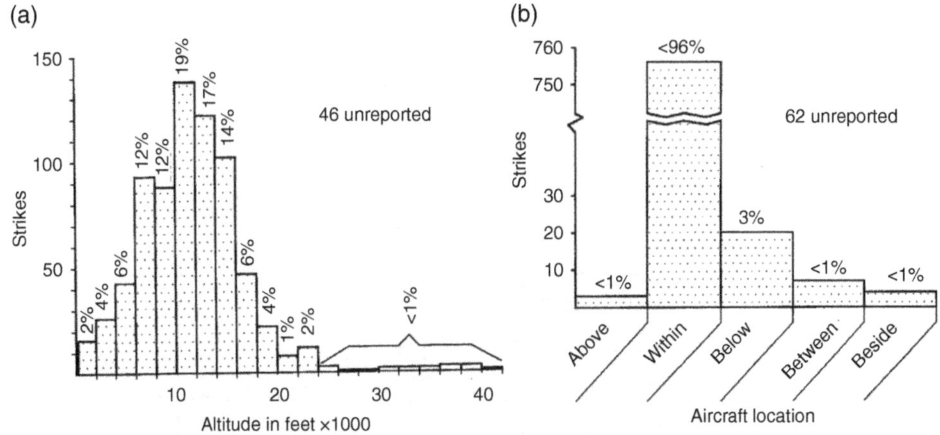

Figure 11.3 Statistics of aircraft lightning strike accidents. (a) Aircraft lightning accident at different flight altitudes. (b) Position of aircraft relative to cloud while lightning strikes.

the aircraft extremity is increased to approximately 30 kV/cm, air will ionize, and spark discharge will occur at the tips of the aircraft and extending along the direction of the oncoming leader [1, 2].

The probability of aircraft being struck by lightning is related to climate conditions, aircraft type, flight height, and other factors. Generally speaking, an average of one strike can be expected for every 3000 hours' flight for most commercial transport aircraft [3–6]. According to statistical analysis, there are more lightning strikes being experienced at intermediate altitudes (3.5 km), which are generally thought to be the region for cloud distribution, than at higher or lower altitudes, as shown in Figure 11.3. More than 54% of lightning accidents occur during the takeoff and landing stages of aircraft passing through clouds [7].

The hazards of lightning strikes to aircraft can be divided into two categories: one is direct damage effects, including the combustion, erosion, and explosion of airframe materials caused by the strong current, strong magnetic field, and strong shock wave accompanying lightning, which will lead to structural distortion and strength reduction of aircraft body [1, 8]; and the other is the indirect damage effects, which mainly refers to the interference and damage of the induced voltage and current to the internal electronic equipment of the aircraft. In a severe aircraft lightning accident, there is the risk of aircraft disintegration and personal injuries, which may cause serious economic losses and have a huge social impact.

11.1.2 Lightning Test Environment of Aircrafts

Lightning is a natural discharge with high voltage and large current, which directly threatens the safe operation of aircraft. In order to simulate the actual situation of multiple lightning discharges during lightning strikes, the American Society of Automotive Engineers (SAE-APR 5412 [9]), the American Military Standard (MIL-STD-464A [10]), and the European Organization for Civil Aviation Equipment (EUROCAE ED-14E [11]) have formulated detailed test specifications and requirements for aircraft. In addition, the surface

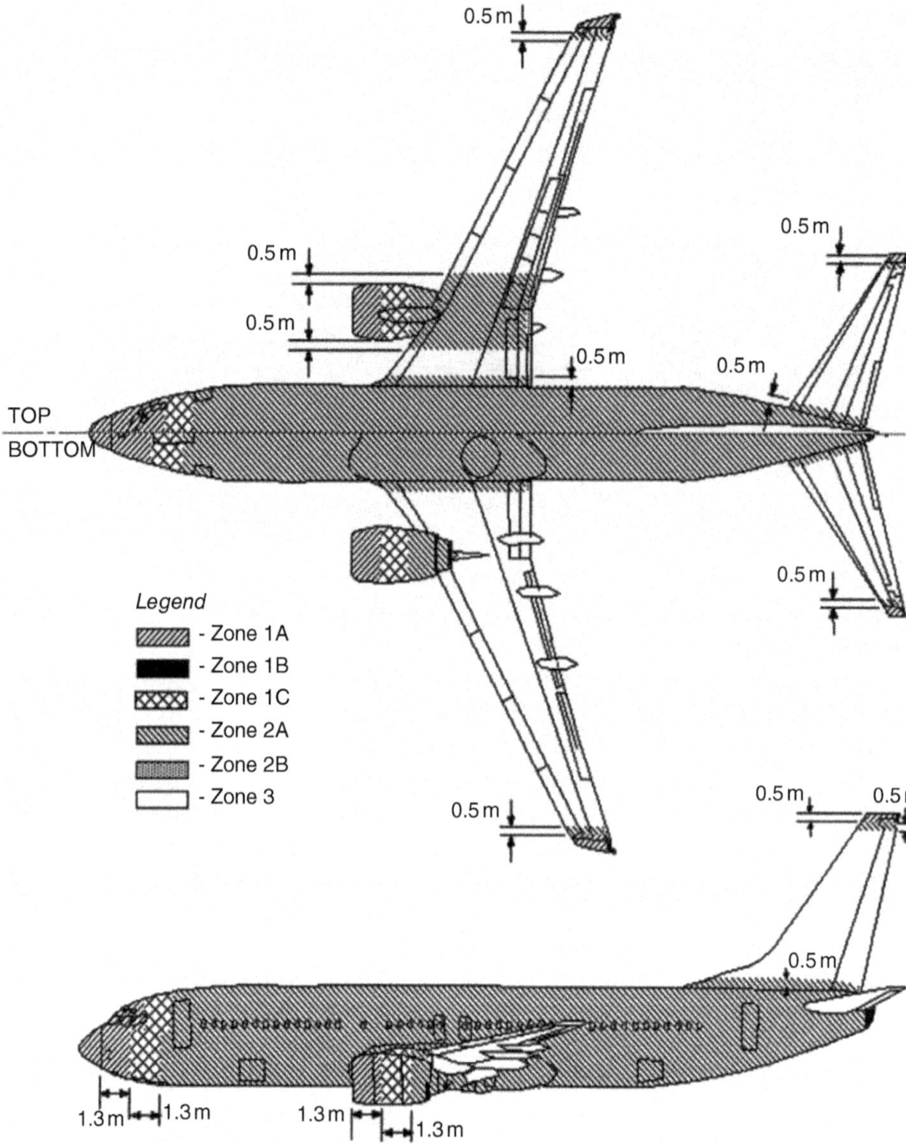

Figure 11.4 Lightning strikes zone of transport aircraft.

of an aircraft can be divided into a set of regions called lightning strike zones, which represent the areas likely to experience the various types of lightning currents and, consequently [12], as shown in Figure 11.4.

11.1.2.1 Zone 1

- Zone 1A – First return strike zone: All the areas of the aircraft surfaces where a first return strike is likely during lightning channel attachment with a low expectation of flash hang on.

- Zone 1B – First return strike zone with long hang on: All the areas of the aircraft surfaces where a first return strike is likely during lightning channel attachment with a high expectation of flash hang on.
- Zone 1C – Transition zone for first return strike: All the areas of the aircraft surfaces where a first return strike of reduced amplitude is likely during lightning channel attachment with a low expectation of flash hang on.

11.1.2.2 Zone 2

- Zone 2A – Swept strike zone: All the areas of the aircraft surfaces where subsequent return strike is likely to be swept with a low expectation of flash hang on.
- Zone 2B – Swept strike zone with long hang on: All the areas of the aircraft surfaces into which a lightning channel carrying a subsequent return strike is likely to be swept with a high expectation of flash hang on.

11.1.2.3 Zone 3

Those surfaces not in Zones 1A, 1B, 1C, 2A, or 2B, where any attachment of the lightning channel is unlikely, and those portions of the aircraft that lie beneath or between the other zones and/or conduct a substantial amount of electrical current between direct or swept strike attachment points.

Simulated lightning current testing is a common method to evaluate the effect of direct lightning strikes on aircraft. It can be used to test and evaluate the lightning resistance performance of CFRP composite materials and aircraft structures in the laboratory. According to SAE-APR 5412 standard, the external lightning environment for evaluating direct effects is comprised of current components A, B, C, and D, as shown in Figure 11.5 [9]. SAE-APR 5414 [12] standard described the application of the lightning environment

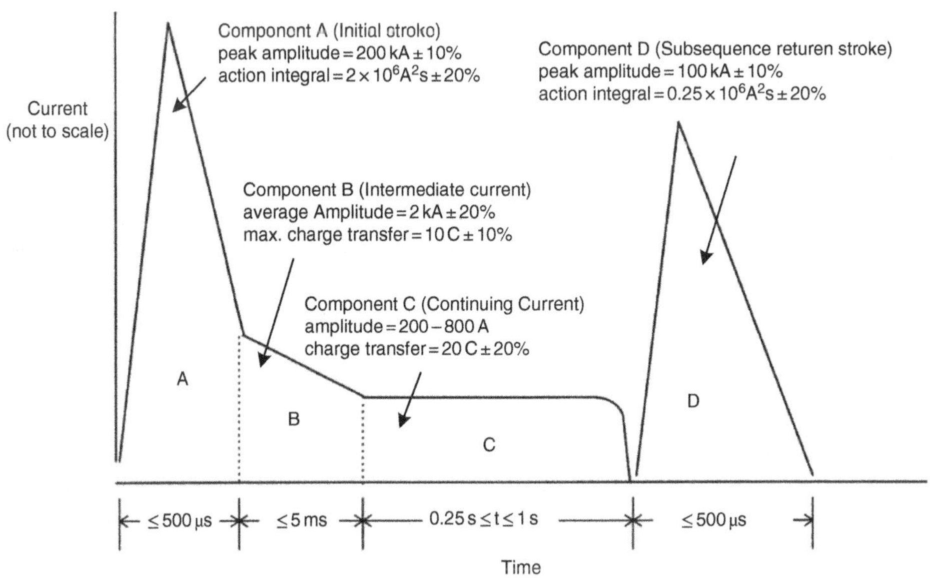

Figure 11.5 Lightning current components A through D for direct effect testing.

waveforms and components to specific lightning zones. The current components are defined in the following sections.

11.1.2.4 Current Component A – First Return Strike

This waveform combines the severe parameters of both the negative and the positive first return strikes. It occurs most frequently to aircraft flying at lower altitudes. The doubled exponential waveform shown in Figure 11.6a shall be applied for analysis purposes.

For direct effect testing purposes, component A can be simulated by an oscillatory (Figure 11.6b) or unidirectional (Figure 11.6c) waveform. The current must have an amplitude of 200 kA (±10%) with a rise time of up to 50 μs (the time between 10 and 90% of peak amplitude). The action integral has to be 2×10^6 A^2s (±20%) and the total time to 1% of peak value shall not exceed 500 μs.

11.1.2.5 Current Component Ah – Transition Zone First Return Strike

The amplitude and waveform of the first return strikes, which might hit an aircraft, depend on the flight altitude. In general, lower amplitudes and action integrals can be expected at higher altitudes. This waveform is applicable in the transition zone 1C and represents the estimated shape of the first return strike (Component A) at higher altitudes, as shown in Figure 11.7.

For direct effects testing, component Ah can also be simulated by an oscillatory or unidirectional waveform. The current must have an amplitude of 150 kA (±10%) with a rise time of up to 37.5 μs (the time between 10 and 90% peak amplitude). The action integral has to be 0.8×10^6 A^2s (±20%), and the total time for the current to decay to 1% of peak value shall not exceed 500 μs.

11.1.2.6 Current Component B – Intermediate Current

This component represents mainly the intermediate currents following some of the negative initial return strikes and/or restrikes. For analysis purposes, the current waveform could be used as presented in Figure 11.8.

11.1.2.7 Current Component C – Continuing Current

This current component represents the lightning environment that might be caused by the long-duration currents which may follow some restrikes of the negative cloud to ground lightning strikes and also the return strike of the positive cloud to ground lightning flashes.

For direct effects testing, Component C should have a current amplitude between 200 and 800 A, a time duration between 0.25 and 1.0 s and a transfer charge of 200 coulombs (±20%). This waveform should be unidirectional; e.g. rectangular, exponential, or linearly decaying. The current waveform could be used as presented in Figure 11.9.

11.1.2.8 Component C* – Modified Component C

This component represents the portion of component C that flows into an attachment point in Zone 1A or 2A if the dwell time at that point exceeds 5 ms. Component C* is primarily used for evaluating the melt-through of metal skins. Component C* is a current averaging not less than 400 A for a period equal to the dwell time minus the 5 ms duration of component B.

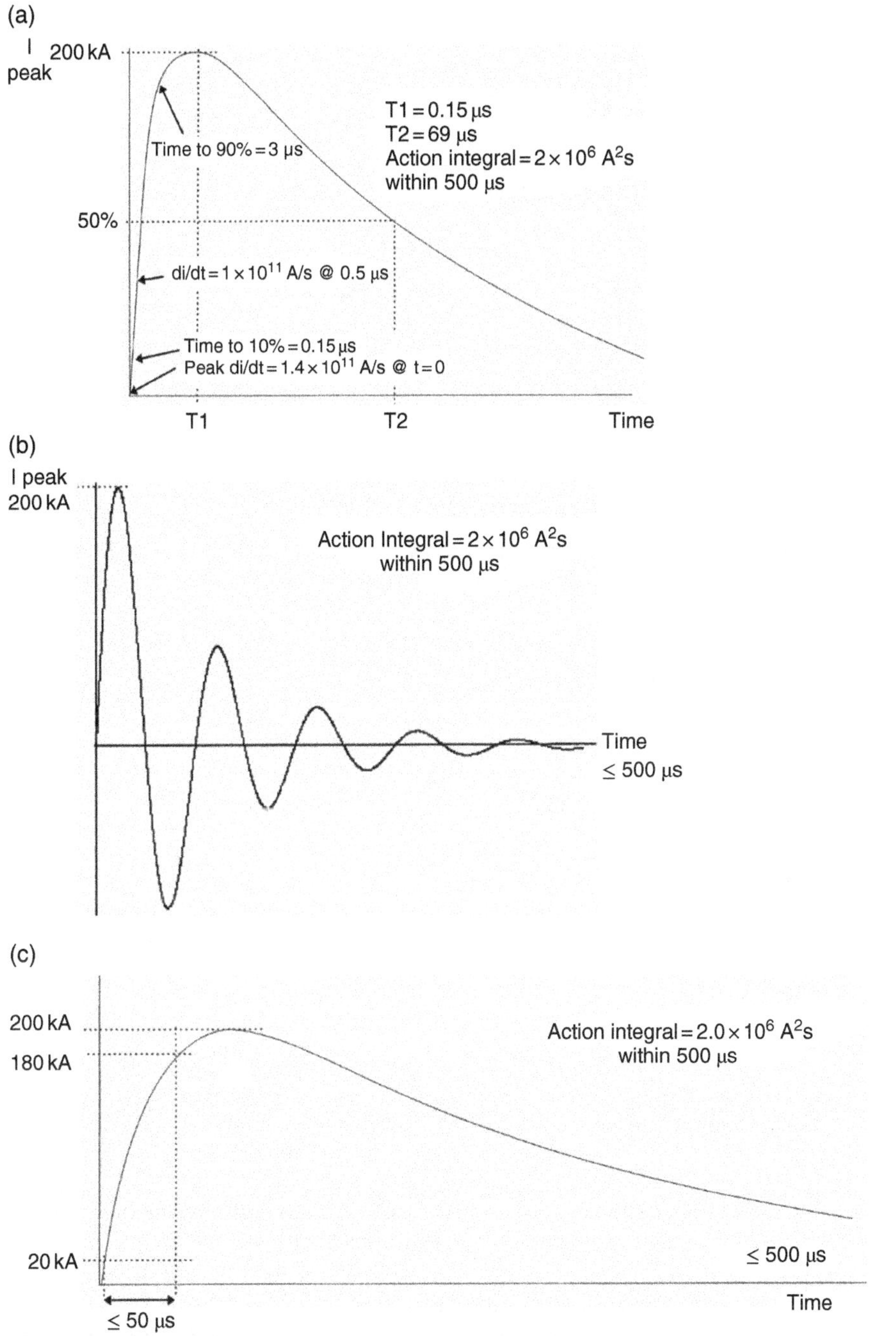

Figure 11.6 Current component A for direct effect testing. (a) Theoretical lightning current component A. (b) Damped sinusoidal current. (c) Unipolar lightning current component A.

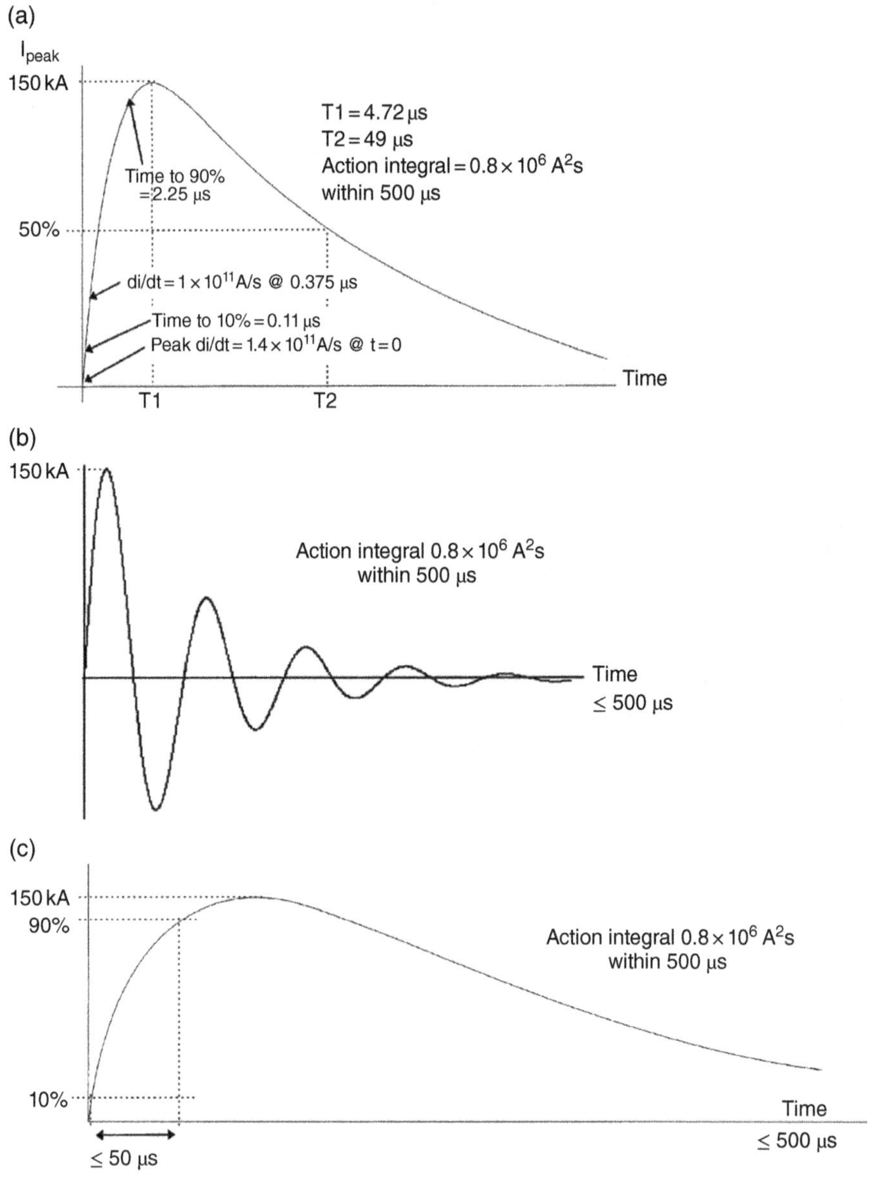

Figure 11.7 Current component A_h for direct effect testing. (a) Theoretical lightning current A_h component. (b) Damped sinusoidal current. (c) Unipolar lightning current A_h component.

11.1.2.9 Current Component D – Subsequent Strike Current

For direct effect assessments, current Component D represents a subsequent strike. For direct effect testing, Component D can be simulated by either oscillatory or unidirectional waveforms with total time duration to 1% peak value of 500 μs. The amplitude shall be 100 kA (±10%), the rise time shall not exceed 25 μs (time between 10 and 90% of the amplitude). The action integral is $0.25 \times 10^6\ A^2s$ (±20%). The current waveforms could be used as presented in Figure 11.10.

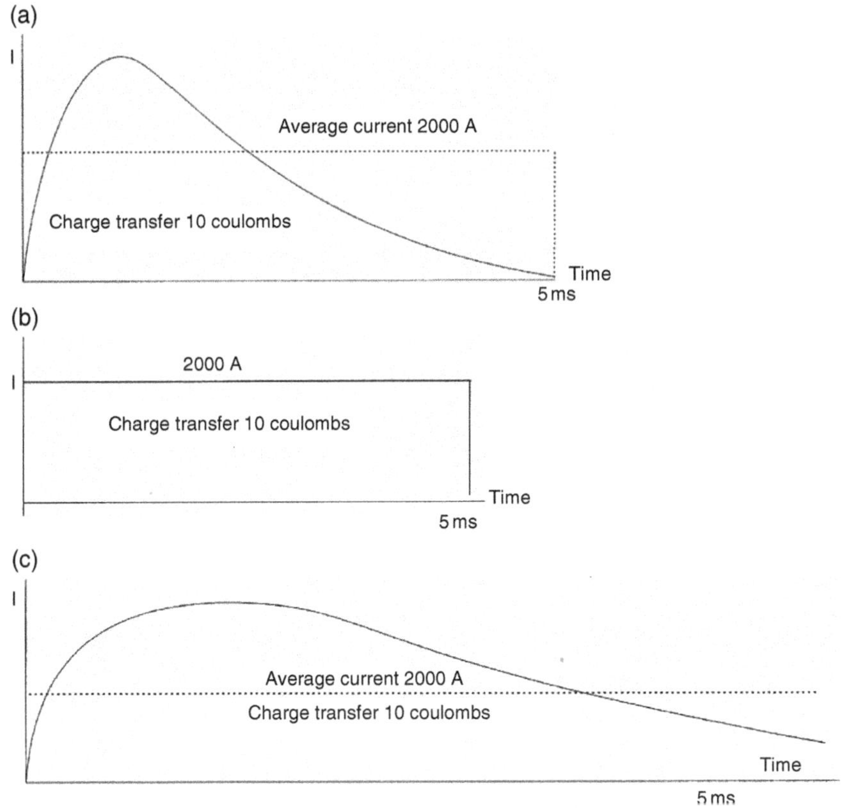

Figure 11.8 Current component B for lightning direct effect testing. (a) Theoretical current component B. (b) Rectangular current component B for direct effect testing. (c) Exponential current component B for direct effect testing.

Figure 11.9 Current component C for direct effect testing.

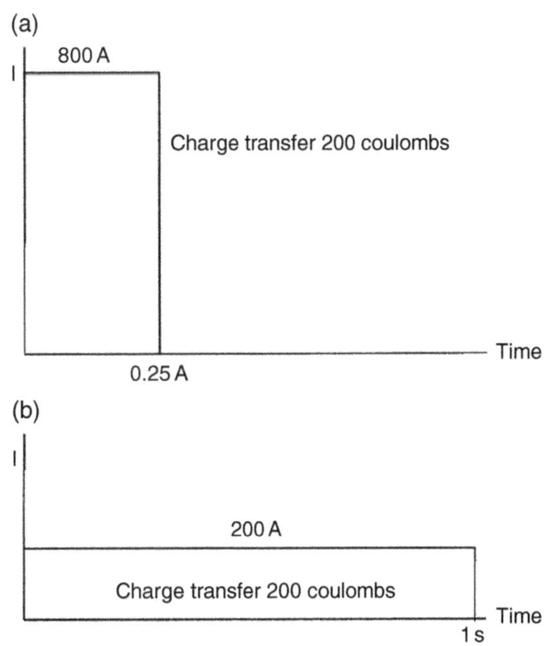

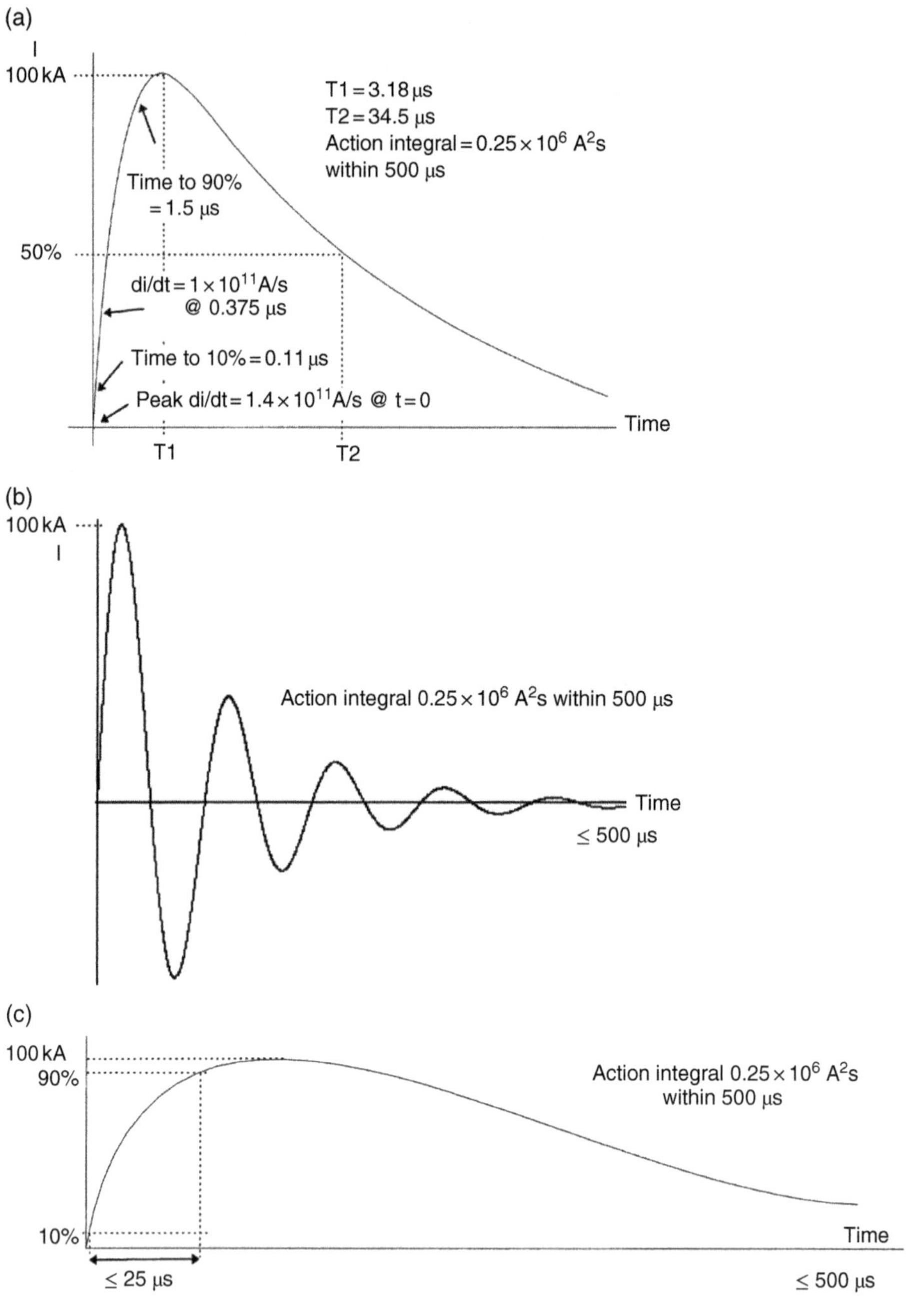

Figure 11.10 Current component D for direct effect testing. (a) Theoretical lightning current D component. (b) Damped sinusoidal current. (c) Unipolar lightning current D component.

Table 11.1 Application of lightning environment to aircraft zones.

Aircraft Zone	Voltage Waveforms	Current Component
1A	A, B, D	A, B, C*, H
1B	A, B, D	A, B, C, D, H
1C	A	A_h, B, C*, D, H
2A	A	D, B, C*, H
2B	A	D, B, C, H
3(Conducted)	—	A, B, C, D, H
3(Direct attachment)	A	A/5, B, C*

Source: Based on [12, 13]. © John Wiley & Sons.

11.1.3 Waveform Combination in Different Lightning Zones for Lightning Direct Effect Testing

The application of the lightning environment waveforms and components to specific zones as described in SAE-ARP 5414 [12] is shown in Table 11.1.

It can be concluded from Table 11.1 that for different aircraft zones, the specified combinations of lightning current components are different, but the common point is that the test lightning current components of each partition should exist in a combined form, which suggests that multiple lightning current components should be continuously applied to the tested object in one discharge process, rather than applied separately in multiple discharge processes [12, 13].

11.1.4 Application of CFRP Composites in Aircraft

CFRP composites have been widely used in defense, transportation, wind power, sports, and other industries due to their excellent mechanical properties at low density, excellent workability, and high fatigue and chemical corrosion resistance. With the improvement of aircraft design and the progress of CFRP, the usage of polymers in commercial aircraft, military aircraft, UAV, stealth aircraft is growing. In 1960, the amount of composites on McDonnell Douglas DC-9 was less than 1%; by the end of 2011, the main wing, tail wing, fuselage, and floor of Boeing 787 Dreamliner are basically made from CFRP composites which represent 50% of aircraft weight, and the proportion of composites on Airbus A350XWA in 2014 reached 53 wt% [14–16], as shown in Figure 11.11.

Compared with metal materials such as aluminum, steel, and titanium alloy traditionally used in aircrafts, the obvious defect of CFRP composites is its poor electrical conductivity. The resistivity of carbon fiber is three orders of magnitude higher than that of metal materials. The resistivity of typical carbon fiber is 10^{-5} Ω·m, while the polymer matrix is an insulating matcrial, and its typical resistivity is 10^{11}–10^{14} Ω·M. Generally speaking, the resistivity of CFRP laminates along the fiber orientation is 10^{-4} Ω·m, and the transverse resistivity is 10^{-2} Ω·m, while the in-thickness resistivity of the laminate can reach 10^{4} Ω·m [17–19].

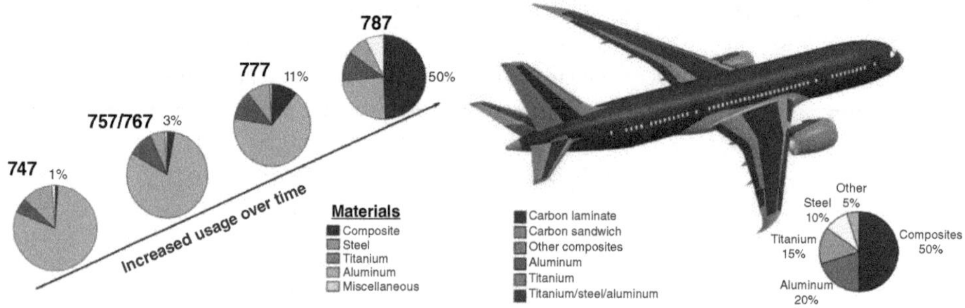

Figure 11.11 Application of CFRP in aircraft.

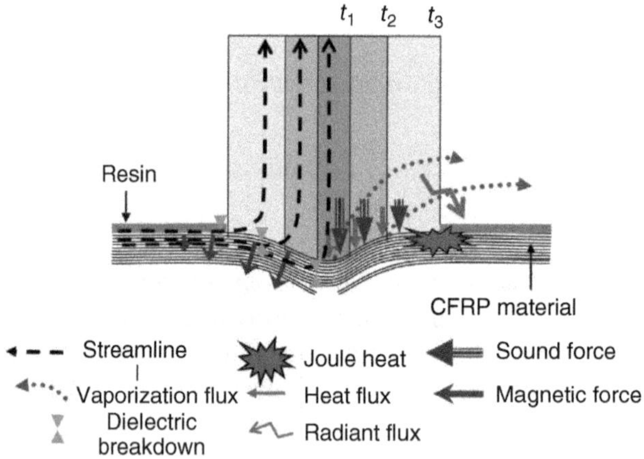

Figure 11.12 Schematic diagram of different damage effects at the lightning attachment point.

The occurrence of a lightning accident will seriously threaten the operation safety of aircraft, and also make CFRP composites, which are widely used in aircrafts, face severe challenges. The lightning strike process is a complex transient physical process involving strong lightning impact force, electromagnetic force, and thermal effect, as shown in Figure 11.12 [20].

The lightning arc is essentially a discharge plasma channel. The surface layers of CFRP laminate will undergo transient ablation with the attachment of high-temperature and high-pressure plasma. After the establishment of the conduction path to the composite surface, the high-intensity lightning current will be injected into the CFRP laminate, and the Joule heat generated by conducting current will cause the internal temperature of CFRP material to rise. However, the accumulated charge and heat in CFRP material can't be transferred or diffused rapidly in a short time due to its poor thermal and electrical conductivity. The accumulated heat energy will make the internal temperature of CFRP laminates increase sharply, and eventually lead to serious damage to the CFRP, such as fiber fracture, resin pyrolysis, and deep delamination.

In addition, the high-intensity electric field can cause an electric breakdown of the interlayer resin matrix, and the strong electromagnetic field of the lightning channel will also produce strong electromagnetic force between the current-carrying fiber bundles in CFRP. The direction of the Lorentz force is perpendicular to the fiber orientation. When the electromagnetic force exceeds certain strength, mechanical damage tends to destroy the complete structure of CFRP.

Moreover, the air around the lightning arc channel is compressed rapidly and vibrates violently during the lightning current injection. The acoustic shock wave will produce impact pressure on the CFRP surface and can cause mechanical damage.

11.2 The Dynamic Conductive Characteristics of CFRP

11.2.1 A Review of the Research on the Conductivity of CFRP

In the actual lightning strike process, the interaction between the lightning current and CFRP material is a complex multi-physics process with multiple physical fields, involving the formation of the lightning leader, the attachment of the lightning arc, the conducting of the lightning current (both on the surface of and inside CFRPs), thermoelectric ablation, and heat dissipation. The lightning damage effect accompanied by multi-physics lightning strikes includes the ablation effect of the high-temperature lightning arc, the shockwave impact caused by the thermal expansion of the surrounding air, the resin pyrolysis caused by the Joule heating effect of the lightning current, and the strong electromagnetic effect during the current conduction process. The conduction path and the distribution of the lightning current inside the CFRP laminate, as well as the intensity of the lightning current Joule heating effect, which will directly affect the CFRP lightning damage behavior, are determined by the conductive properties of the CFRP laminate. Specifically, CFRP laminates, composed of high-conductivity carbon fibers and insulating epoxy resin, have complex anisotropic electric properties [18, 19, 21]. The anisotropic static electrical conductivity (10^{-4}–10^{4} S/m) and thermal conductivity (10^{-1}–10^{0} W/m °C) of CFRP composites are much lower than those of conventional metals or alloys, such as aluminum and titanium alloys (10^{6}–10^{8} S/m and 10^{2} W/m °C) [22–24]. Thus, the large amounts of charge and heat associated with a lightning strike cannot be transferred and dissipated effectively in CFRP laminates, causing a sharp increase in the local temperature and serious lightning damage, such as fiber breakage, resin pyrolysis, and severe delamination [8]. Therefore, the measurement and modification of the conductive properties of CFRP materials are particularly critical for the application of this new advanced material in aerospace industries.

Regarding studies on the conductive properties of CFRP materials, Todoroki et al. [18] studied the static DC conductance in the longitudinal and transverse direction of CFRP laminates with different carbon fiber volume fractions based on the conductive principle of mixed materials. Cheng et al. [17] addressed the anisotropic bulk electrical conductivity of CFRP laminates by the eddy current method. They found that a cross-ply laminate has much better electrical conductance than a unidirectional plate and the bulk conductivity depends mainly on the number of current paths produced by the overlapping fibers at the interfaces between two angled layers. Haider et al. [19] investigated the anisotropy and nonlinear behavior in the electrical response of CFRP composites with different laminated structures. Park et al. [25] analyzed the electrical conduction behavior based on electrical percolation theory and calculated the electrical anisotropy of CFRP materials using the Kirchhoff criterion and Monte Carlo method. These studies indicated that the electrical properties of a CFRP depend on the fiber network that acts as the conduction pathway in the laminate. However, few studies have concentrated on the dynamic conduction model of CFRPs under lightning impulses. Because of the absence of research on the dynamic impedance characteristics of CFRP, in most coupled thermoelectric simulation studies, CFRP laminates are generally considered to be an anisotropic resistance network with constant values, in which the nonlinear equivalent conductance and the inductance effect under lightning impulse are ignored [26]. Additionally, in coupled thermoelectric simulation models, the setting conductivities of CFRP laminates are based on their DC static conductance parameters [27–31], but whether the DC conduction model is suitable for the dynamic conduction process of lightning impulses inside CFRPs remains to be further studied.

To analyze the conductive properties of CFRP materials regarding direct lightning effects, a nondestructive lightning impulse experimental platform for testing CFRP impedance was established based on the test waveforms of the direct lightning effect tests for aircraft stipulated by the Society of Automotive Engineers (SAE) and the European Organization for Civil Aviation Equipment (EUROCAE) [9, 11]. The dynamic conductivities of CFRP laminates subjected to lightning current impulses with different waveform parameters were measured. Then, the experimental results were listed and compared to analyze the factors affecting the dynamic conductivity of CFRPs, especially the nonlinear conductivity and inductance effect of CFRP material. Based on the study of the dynamic impedance properties of CFRP laminates, the intrinsic conduction mode of CFRPs under lightning current conduction was explained. In addition, the dynamic impedance of CFRPs can be used in lightning damage simulation studies to reduce the deviation between calculation and experimental results and improve simulation accuracy. This research improves the understanding of the electrical conductance processes inside CFRP laminates, and provides experimental support and a theoretical basis for future studies on the mechanism of CFRP lightning damage.

11.2.2 The Testing Methods

11.2.2.1 Specimens

Carbon woven fabric/epoxy and unidirectional CFRP laminates were selected in the dynamic conductivity test. The specific parameters of the specimens are shown in Table 11.2. A carbon woven fabric/epoxy laminate with a total of 16 plies was fabricated using epoxy resin 5228A developed by the AECC Beijing Institute of Aeronautical Materials and carbon fabric CCF300 manufactured by Weihai Expanded Fiber Co. Ltd. The ply orientation angle was [(0/90F)]. Epoxy resin 3021 supplied by Shanghai Duodi Polymer Material Co. Ltd and carbon fabric T300 from Toray Industries (China) Co. Ltd. were used as the matrix and reinforcement material to produce the unidirectional CFRP laminates, respectively. The properties of the fibers can be found in [32]. Dry carbon fabrics were placed on a mold with a stacking sequence of $[0]_{24}$. To study the conductive properties of CFRP laminates with different layup angles, the unidirectional laminates were cut into two different structures of $[0]_{24}$ and $[45]_{24}$. The specimens used in this experiment are shown in Figure 11.13.

Table 11.2 The parameters of CFRP specimens.

Type	Size/mm	Material Type	Fiber Volume Fraction	Stacking Sequence
A1	150 × 100 × 4	CCF300/5228A	55%	$[(0/90F)]_{16}$
A2	100 × 100 × 3.6	T300/3021	60%	$[0]_{24}$
A3	100 × 100 × 3.6	T300/3021	60%	$[45]_{24}$

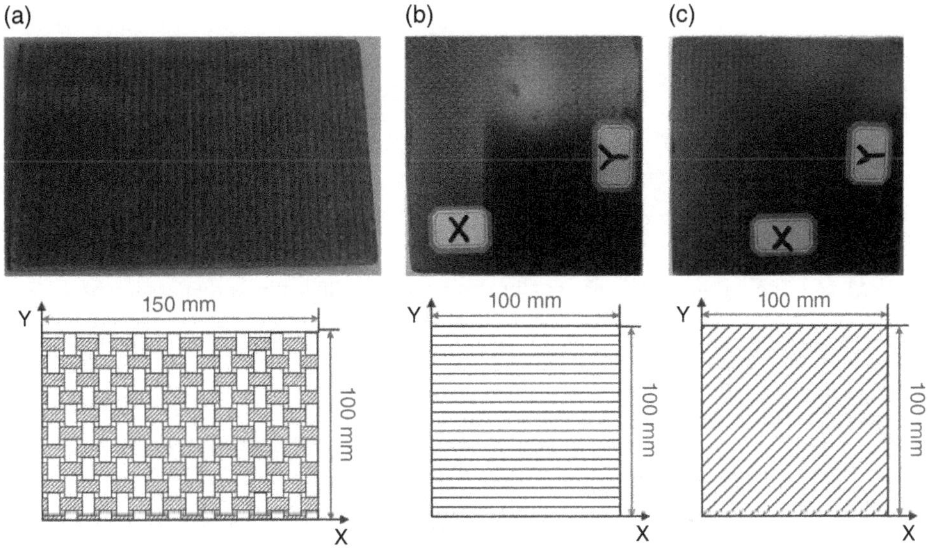

Figure 11.13 Tested CFRP specimens: (a) carbon woven fabric/epoxy laminate; (b) unidirectional carbon fabric/epoxy laminate, $[0]_{24}$; (c) unidirectional carbon fabric/epoxy laminate, $[45]_{24}$. *Source:* Wang et al. [43]. © Elsevier.

11.2.2.2 The Test Fixture

The CFRP specimens (Type A1–A3) were clamped along the length direction of the plane (X-direction), the width direction of the plane (Y-direction) and the thickness direction to measure the anisotropic dynamic conductivities of the CFRP laminates. The specimen fixture for measuring the in-thickness conductivities is shown in Figure 11.14a. The CFRP laminate was sandwiched between the two copper plates which were used as positive and negative electrodes. The fixture in Figure 11.14b was used to measure the longitudinal and transverse conductivity by clamping specimens along its X-direction and Y-direction. The two symmetrical copper bars were positive and negative electrodes. The lightning current was injected into the specimen through the positive electrode. After conducting through the specimen, the lightning current flowed back to the low-voltage terminal of the lightning impulse generator through the negative electrode. To ensure the reliability of the experimental results, the surface of the copper electrode was polished with sandpaper before each test to reduce the measurement error caused by the contact resistance. In addition, multilayer aluminum foils with high conductivity and soft texture were inserted

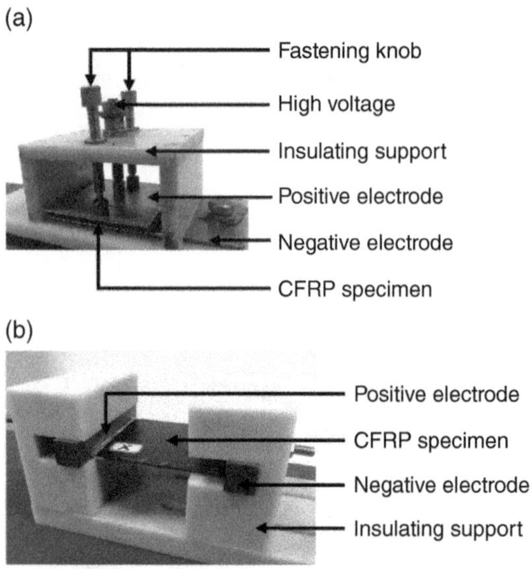

Figure 11.14 The specimen fixtures used for clamping the CFRP laminate to measure the (a) in-thickness conductivity; and (b) in-plane conductivity.

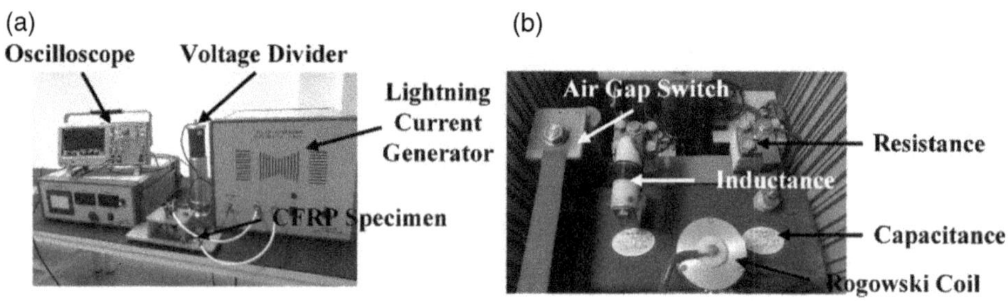

Figure 11.15 Lightning current impulse test platform.

between the electrode and the specimen which can effectively fill the small air gaps distributed on the contact surface between the electrode and the specimen and weaken the interference caused by the repeated breakdown occurred in the small gaps to the voltage waveform.

11.2.2.3 Lightning Impulse Generator and Lightning Waveforms

A lightning current impulse generator and test platform were designed to study the dynamic conduction properties of CFRP laminates. The lightning current impulse generator consisted of a high-voltage capacitance C, waveform forming components (including the inductance L and the resistance R), triggered air gap switch, grounding system, and tested specimen, as shown in Figure 11.15.

The capacitance C, inductance L, and resistance R were connected in series. The film capacitance bank could be switched between 14 and 28 μF, and the waveform forming resistance and inductance could be continuously adjusted in the range of 0–5 Ω and

Table 11.3 The parameters of the modulated components corresponding to different lightning current waveforms.

Waveforms	C/μF	R/Ω	L/μH
10/30 μs	14	1.0	9.9
20/60 μs	14	2.0	39.5
40/120 μs	28	2.0	79.1

0–100 μH, respectively. This generator was capable of providing a lightning current impulse with amplitudes of 10 kV and 1.5 kA. The lightning current impulse could generally be defined by the waveform parameters of the front time (T_f) and tail time (T_t). The waveform parameters of 20/60 and 10/30 μs current impulses, including the front time, the electrical action integral and the duration, satisfy the requirements of the standard lightning components A and D, respectively. Moreover, the 40/120 μs current impulse was selected in this experiment to further analyze the variation law of conductive characteristics with the waveform parameters. These current impulses with different waveform parameters can be generated by adjusting the modulated resistance and inductance in the *RLC* series circuit. The parameters of the modulated components corresponding to these lightning current waveforms are listed in Table 11.3. Figure 11.16 shows the waveforms of 10/30, 20/60, and 40/120 μs lightning current impulses.

The Rogowski coil, voltage impulse divider, and Tektronix MDO3012 oscilloscope were used to measure the lightning current impulse flowing through the tested CFRP specimens and the corresponding applied voltage impulse. The dynamic conductivities of the CFRP laminates clamped in different directions subjected to lightning current impulses with different amplitudes and parameters were measured. Then, the influences of the stacking sequence, clamping direction, and applied lightning current parameters on the dynamic impedance properties of CFRP laminates could be analyzed. In addition, to verify the reliability of the test data, the conductive property of the CFRP specimen was tested three times at each current amplitude and the average was adopted to be the conductivity result. The conductivity of another CFRP laminate with the same structural parameters as the specimen in this experiment was measured under the same test conditions to analyze the dispersion of specimens.

11.2.3 The Experimental Results of the Dynamic Impedance of CFRP

11.2.3.1 The Nondestructive Lightning Current Test

Via repeated measurements, it can be tested and verified that lightning current impulses with lower amplitudes do not cause irreversible damage to the conductive properties of CFRP specimens. When the lightning current conducts inside the CFRP laminate, the specimen is heated because of the generated Joule heat, which can be calculated by $\int U \cdot Idt$ based on the experimentally measured lightning current and voltage impulses.

To quantitatively analyze the temperature increase of CFRP specimens under lightning current impulses, the temperature-rising model for CFRP material was established using

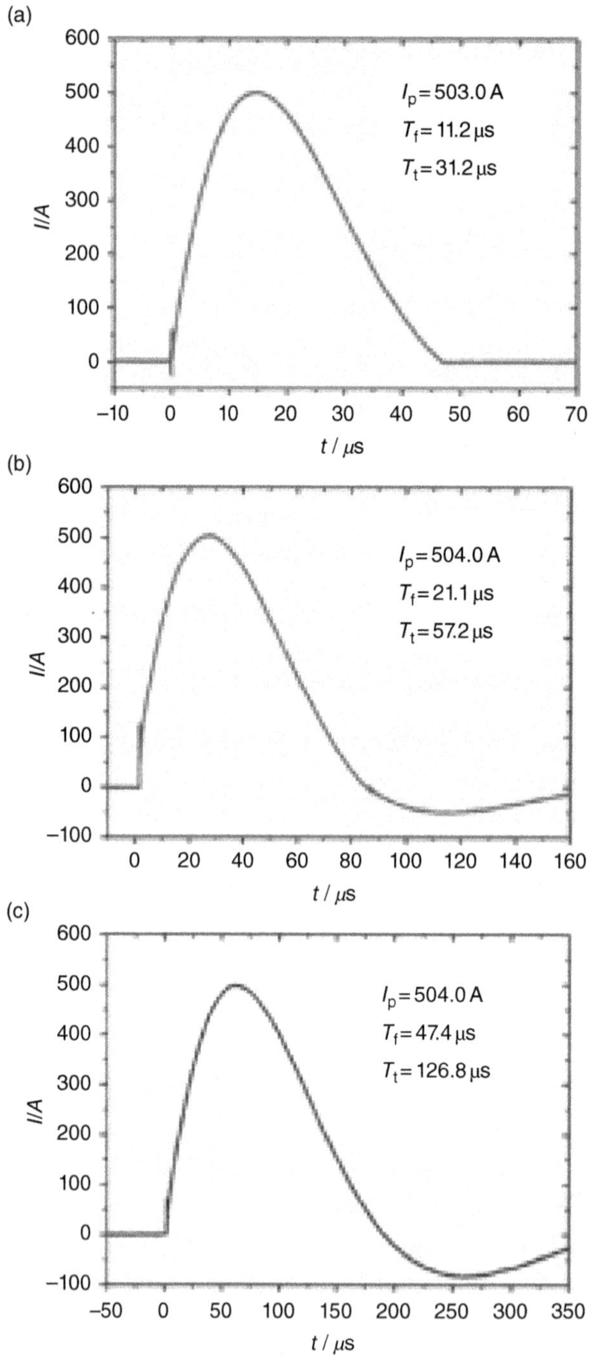

Figure 11.16 Lightning impulse waveforms and parameters: lightning impulses of (a) 10/30 μs; (b) 20/60 μs; (c) 40/120 μs; I_p: peak value; T_f: front time; T_t: duration time.

the coupled thermoelectric analysis module in ABAQUS. The geometry and boundary conditions of the finite element (FE) model were set in accordance with the laminate structures and experimental setup. The interaction between the CFRP specimen and the surrounding environment includes thermal convection and thermal radiation effects. Details of the simulation model for temperature rise in CFRP laminates are described in a later section. Under the maximum lightning current impulse of 1.5 kA, the simulated surface temperature on CFRP specimen is shown in Figure 11.17.

It can be seen from the simulated results that the temperature of the CFRP specimen increased from the initial value 25–31.7 °C (t = 165 μs) under a lightning impulse of

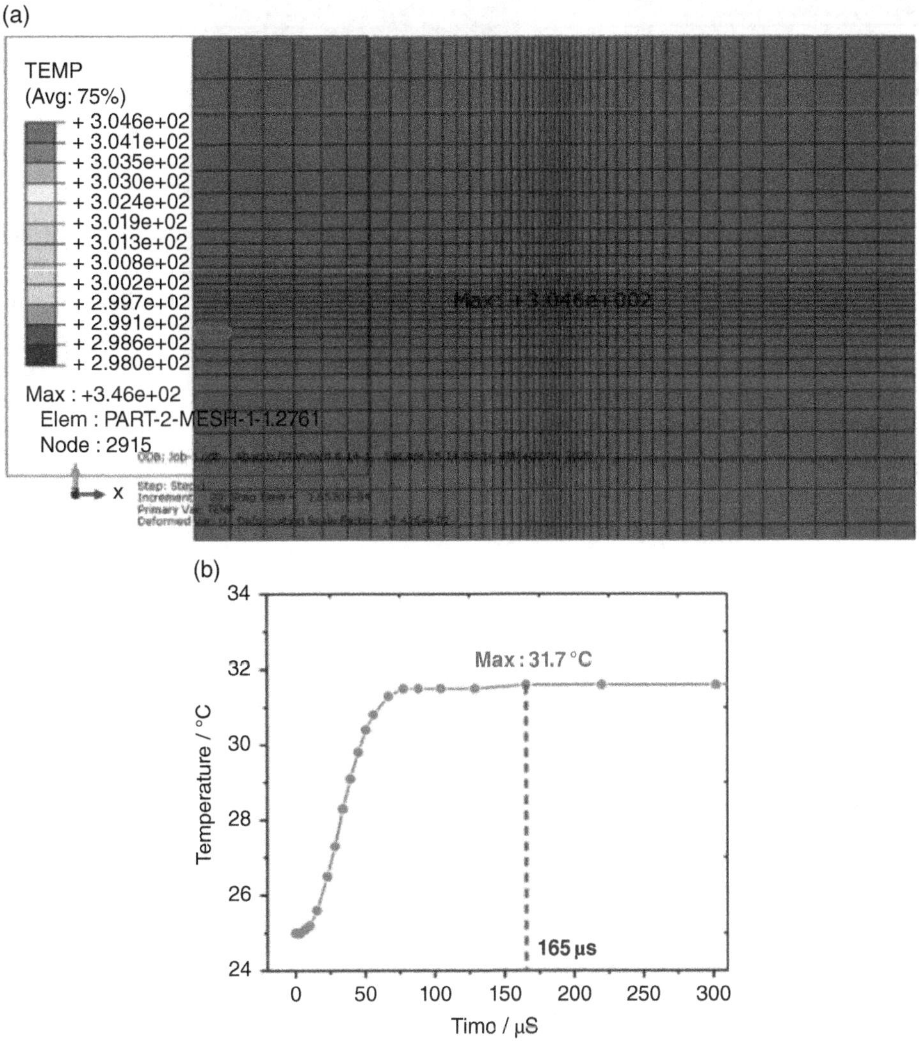

Figure 11.17 The surface temperature on the specimen after lightning current conduction (20/60 μs lightning impulse, I_{peak} = 1.5 kA): (a) surface temperature distribution (t = 165 μs); (b) maximum surface temperature varies with the lightning current injection time t.

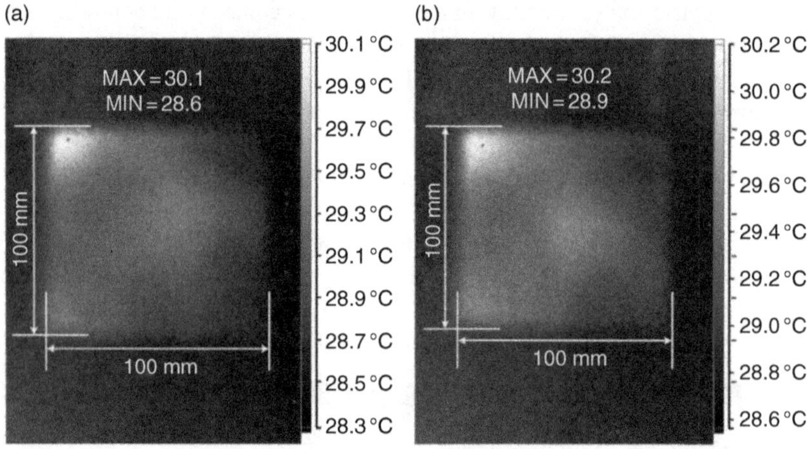

Figure 11.18 Temperature distributions: (a) before the lightning test; (b) after 10 consecutive lightning tests.

1.5 kA. Generally, the resin matrix pyrolyzes when the temperature exceeds 200 °C (aerobic conditions) and the carbon fibers sublime at temperatures approaching 3300 °C [33]. Therefore, the temperature increase of the CFRP laminate subjected to the low-intensity lightning current impulse can be neglected and does not cause irreversible damage. Additionally, the surface temperature distribution of specimen A2 before the lightning test and after 10 consecutive lightning tests was measured using a handheld infrared radiation thermometer (Fluke Ti125, measuring range −20~350 °C) to verify the estimation of the temperature increase, as shown in Figure 11.18. It can be seen that the measured maximum temperature on CFRP specimen is 30.2 °C, which means no observable temperature increase after the lightning current strikes.

11.2.3.2 The Applied Lightning Current Impulse and the Response Voltage Impulse

The dynamic conductive properties are distinctly affected by the applied lightning waveform parameters, the laminated structures, and the specimen clamped directions, but there are some similarities in the relationship between the lightning current and the response voltage impulse. For instance, when specimen A1 is struck by a 20/60 μs lightning impulse, it can be seen that the voltage impulse waveforms are similar in shapes to the current waveforms for different specimen clamping methods, as shown in Figure 11.19. In addition, all the current impulses flowing through the CFRP laminate lag behind the response voltage applied on the specimen, which means that when the voltage impulse reaches its peak value, the lightning current impulse is still in its increasing stage and will reach the peak after a period of time.

The dynamic volt-ampere characteristics of CFRP laminates can be obtained by analyzing the single lightning conducting process. Figure 11.20 shows obvious noncoincidence during the increasing and decreasing stages of the lightning current impulse. Specifically, the voltage on the CFRP laminate during the current decreasing stage is significantly smaller than that in the current increasing stage. In addition, in the initial increasing stage of the lightning current, when the current flowing through the CFRP specimen increases

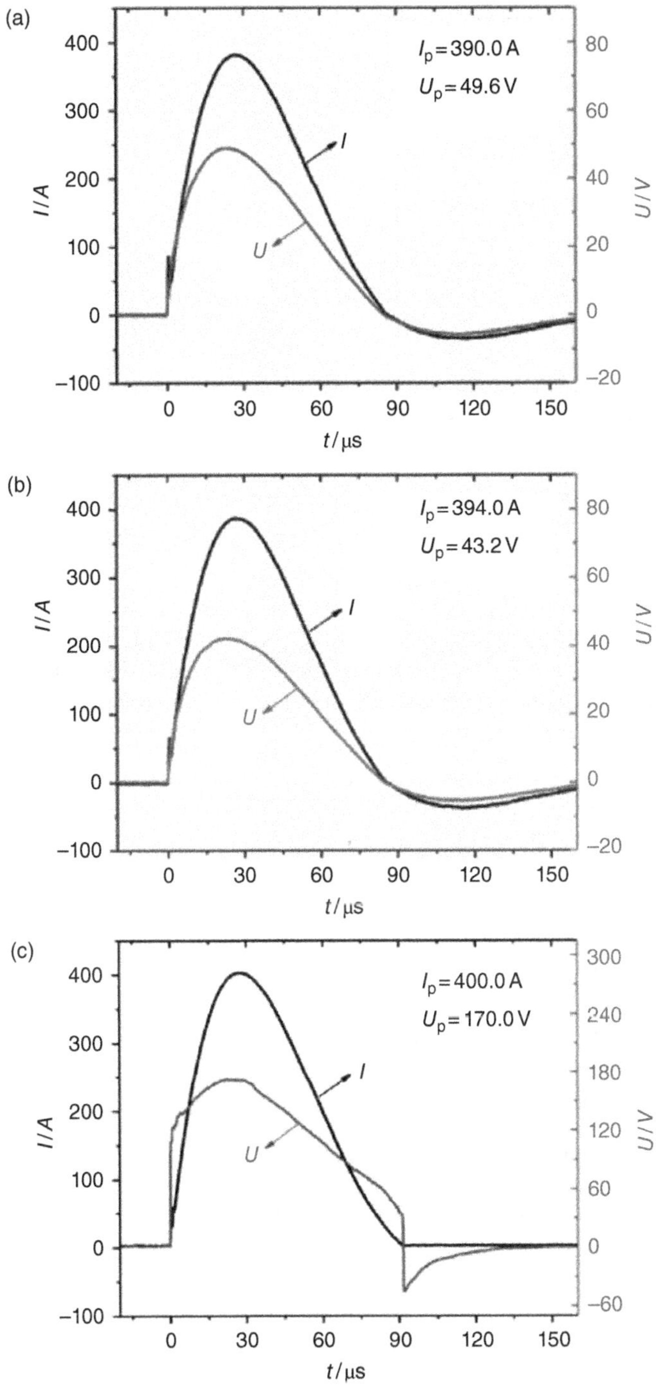

Figure 11.19 Current and voltage impulse waveforms of specimen A1 with different clamping methods: (a) in the X-direction of the plane; (b) in the Y-direction of the plane; (c) in the thickness direction. I_p: peak value of the current impulse; U_p: peak value of the voltage impulse.

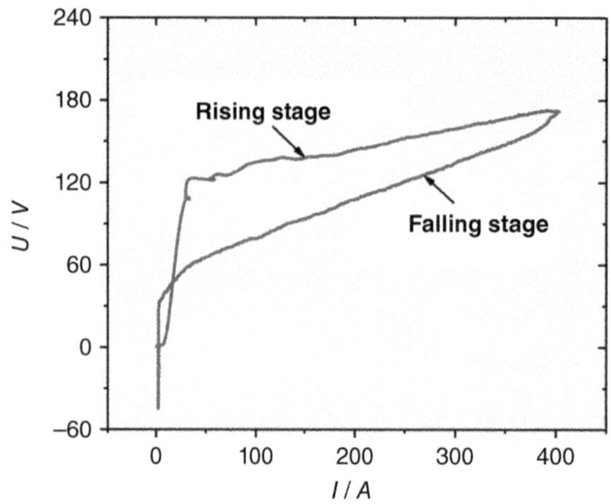

Figure 11.20 The dynamic volt-ampere characteristics of specimen A1 in the thickness direction under 20/60 μs lightning current impulse.

to 30 A, the voltage on the laminate abruptly changes from 0 to greater than 110 V. In comparison, as the current increases from 30 to 400 A, the voltage only increases by approximately 60 V. It can be inferred that the CFRP laminate has a large initial impedance in the thickness direction before it can conduct stably and the impedance decreases when the voltage applied on the laminate reaches a critical value.

11.2.3.3 Equivalent Conductivity of CFRP Laminates Under Different Lightning Impulses

To quantitatively compare the conductivities of specimens subjected to lightning impulses with different waveform parameters and intensities, the equivalent conductance $G = I_{peak}/U_{peak}$ was defined and calculated to represent the conductive properties of the CFRP laminate, temporarily ignoring the phase difference between the current and voltage impulses. Then, the anisotropic conductivities $\sigma_{(x, y, z)}$ (with a unit of S/mm) can be calculated based on the geometric parameters of specimens by the equation $\sigma_{(x, y, z)}=G_{(x, y, z)}\cdot l/S$. The relationship between $\sigma_{(X, Y, Z)}$ and the amplitude of the applied lightning impulse is shown in Figure 11.21.

From the experimental results in Figure 11.21, it can be seen that:

1) With an increase in the peak value of the lightning impulse, the equivalent conductivity gradually increases, indicating significant nonlinear conductive behavior.
2) The longitudinal and transverse conductivities of the CFRP laminates are three orders of magnitude larger than the in-thickness conductivity, which can be explained by the inference that the in-plane current inside the laminate is mainly conducted through the highly conductive carbon fibers and the in-thickness conductivity is weakened because of the insulating interlaminar resin.
3) The equivalent conductivities of the CFRP laminates are different under the lightning current impulses with different waveform parameters. Specifically, the electrical conductivity of the CFRP laminate increases with the wavefront time and duration of the applied lightning pulse.

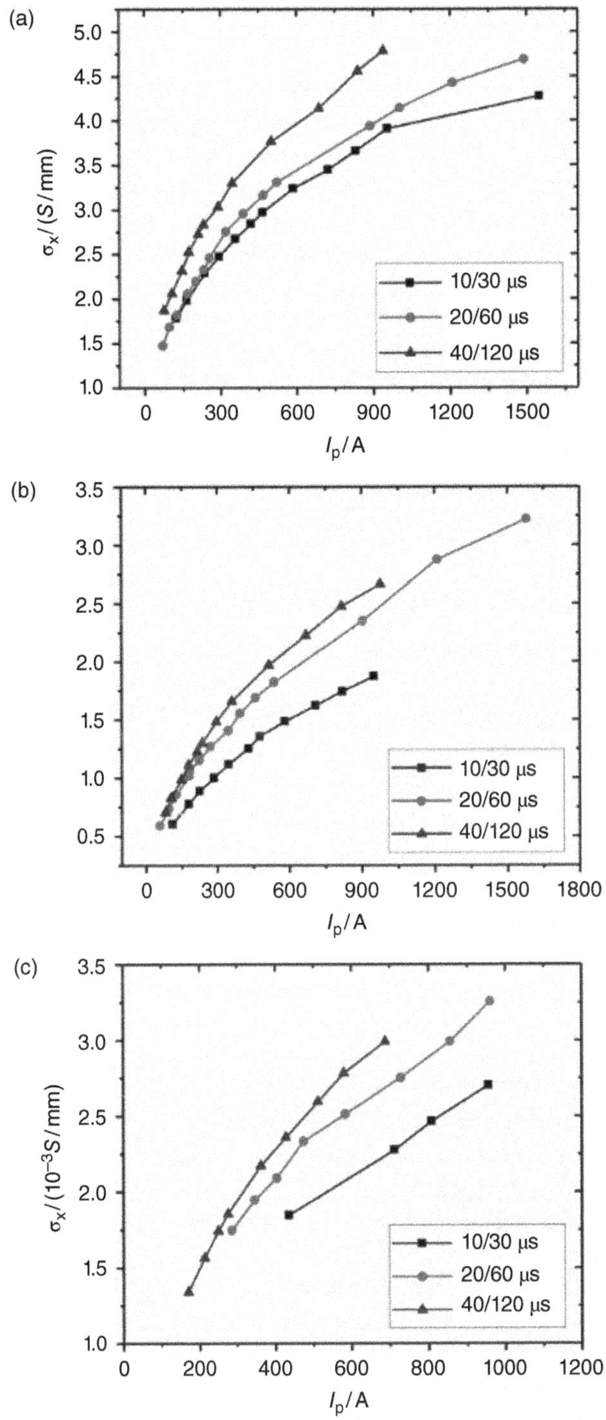

Figure 11.21 The equivalent conductivity of specimen A1 under the different lightning current impulses: (a) longitudinal conductivity; (b) transverse conductivity; (c) in-thickness conductivity.

11.2.3.4 Equivalent Conductivity of CFRP Laminates with Different Laminated Structures

The CFRP laminates demonstrated different responses to lightning impulses due to their different laminated structures. For instance, for specimen A2 with a stacking sequence of $[0]_{24}$, when it is clamped along its length direction (X-direction), the conductivities of the 0° laminated structure can be measured. The measured results for the same specimen A2 clamped in the width direction (Y-direction) correspond to the electrical conductivities of the 90° laminated structure. Similarly, the conductivities of the 45° laminated structure can be obtained from the in-plane conductivities of specimen A3. Figure 11.22 shows the equivalent conductivities of the 0, 90, and 45° laminated structures subjected to different lightning impulses.

The test results in Figure 11.22 indicate that the CFRP laminate has the best electrical conductivity along the longitudinal direction. The transverse conductivities (90° laminated structure) are at the lowest level. However, it is worth noting that the in-plane conductivities in different directions are on the same order of magnitude, but they are three orders of magnitude higher than the in-thickness conductivity.

11.2.4 The Discussion of the Dynamic Conductive Characteristics of CFRP

11.2.4.1 The Conduction Path of the CFRP Laminate Under a Lightning Current Impulse

From the experimental results in Section 11.2.3, the in-thickness conductivity of the CFRP laminate is found to be much smaller than the in-plane conductivities. The conductive properties along different directions are closely related to the internal layup structure. CFRP laminates are usually cured from carbon fiber prepregs. First, the carbon fiber prepregs should be cut to a certain size. Then, according to the designed layup structure, the carbon fiber prepregs are laid in the mold frame-coated with the release agent. Finally, the upper and lower mold frames are combined and placed in a flat vulcanizing machine for curing. The fiber monofilament inside the CFRP laminate is not completely straight, which causes random contacts between curved fibers, as shown in Figure 11.23 [18].

In the same layer inside the CFRP laminate, the carbon fiber monofilaments are placed in two different ways. The fibers may directly contact other fibers, or they may be separated by a thin insulating matrix layer. The in-plane conduction of current mainly depends on the conductive network composed of fiber monofilaments, but the conductive paths are different in different laminated structures [34, 35].

Since the current can be conducted along the relatively intact carbon fiber, the longitudinal conductivity is the highest. The electrical conductivity can be calculated according to the volume density of the carbon fibers in the CFRP laminate [18, 36]. When lightning current is injected into laminates with other in-plane directions, such as 45° or 90°, the conductive network connected by part of the carbon fibers and the random contact points between fibers is discontinuous. The thin resin layer can connect the fiber monofilaments that are distributed parallel, and the carbon fibers, thin resin layers, and direct contact points between fibers can form a continuous conductive path. The schematic of the current

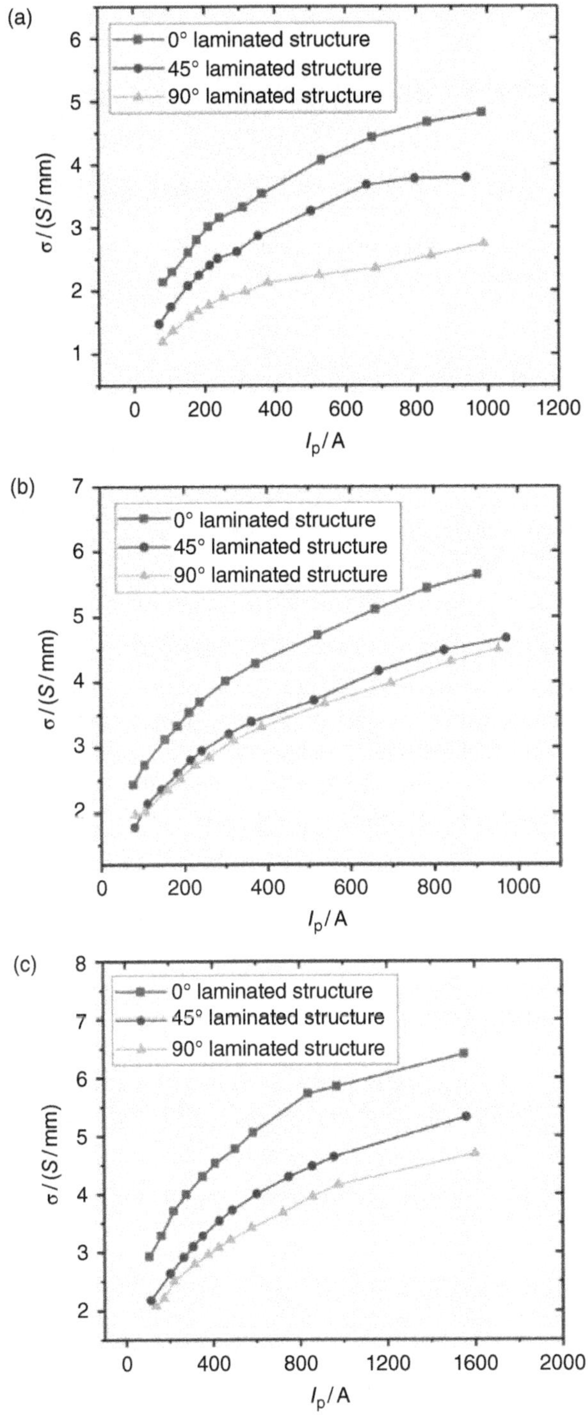

Figure 11.22 The equivalent conductivities of the 0, 45, and 90° laminated CFRP structures subjected to lightning impulses of (a) 10/30 μs; (b) 20/60 μs; (c) 40/120 μs.

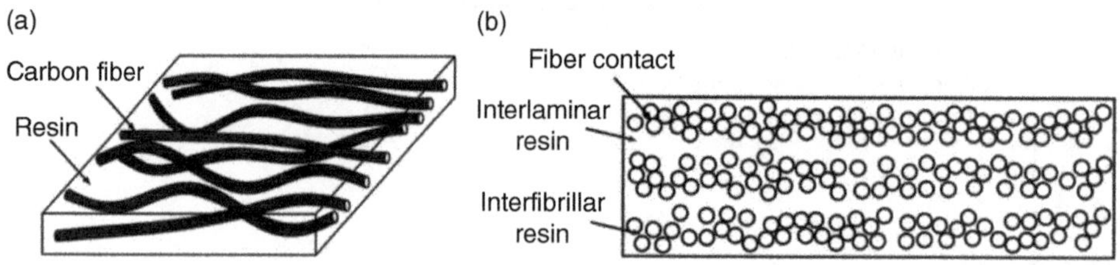

Figure 11.23 Schematic model of the actual carbon fiber network inside the CFRP laminate: (a) in the single layer; (b) side view of the CFRP laminate.

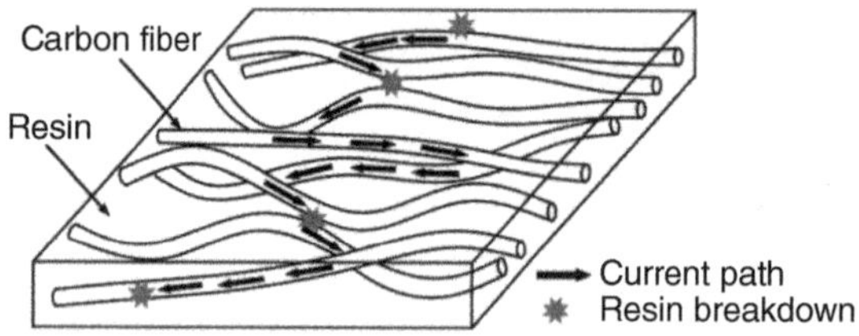

Figure 11.24 Schematic model of the current conduction path in a 90° laminated structure.

conduction path in the 90° laminated structure is shown in Figure 11.24, and a similar pattern for the conduction path also exists in the 45° laminated structure. The in-plane conductivities in different directions differ only in the length of the conductive paths. Thus, they are on the same order of magnitude as the conductivity of the 0° laminated structure. In comparison, the in-thickness conductivity is approximately three orders of magnitude lower than the in-plane conductivity due to the thick interlaminar resin layer with poor conductivity.

11.2.4.2 Dynamic Conductance of CFRP Laminate

During the conducting process of the 20/60 μs lightning current impulse, the in-thickness conductance of CFRP specimen A1 is shown in Figure 11.25.

The in-thickness conductance increases first and then decreases during the single lightning strike process and the trend is basically consistent with the lightning impulse waveform. The in-thickness conductive path can be simplified to a series circuit of the carbon filament network and epoxy resin layers. Thus, the total equivalent conductance (G_l) of the CFRP laminate can be expressed as:

$$G_l = \frac{G_c G_r}{G_c + G_r} \tag{11.1}$$

where G_c is the conductance of the carbon fibers connected in the conductive path and G_r is the equivalent conductance of the resin layers. There is a significant difference between the conductivities of carbon fibers (10^3 S/mm) and insulating resin (10^{-9} to 10^{-13} S/mm), namely $G_c >> G_r$. Then, Eq. (11.1) can be simplified as $G_l \approx G_r$. Therefore, the conductive properties of the resin layer greatly affect the dynamic variation rule of the CFRP conductance G_l with the applied voltage impulse.

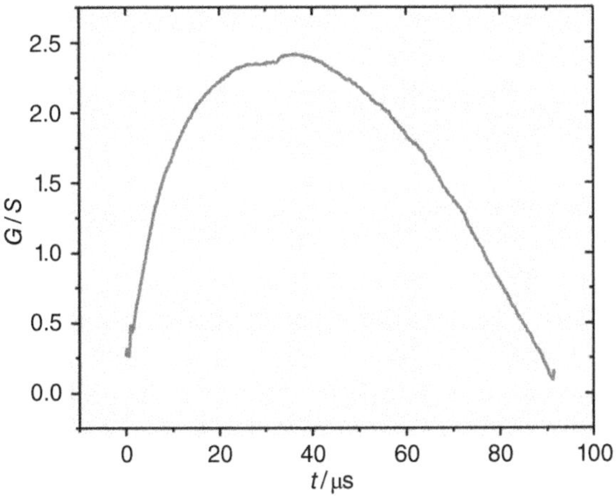

Figure 11.25 The dynamic in-thickness conductance of CFRP specimen A1.

The resin material is an insulating material accompanied by a low electrical conductivity. In the initial stage of lightning impulse, the resin layer remains in a high-resistance state. Thus, most of the discharge voltage, which is on the order from 10^2 to 10^3 V, is distributed on the specimen, and the current flowing through the specimen is at a very low value. The loaded voltage on the CFRP specimen may induce interior electrical carriers inside the laminate that can participate in the current flowing process and lead to an improvement in the conductivity of resin layers. With the continuous application of voltage impulses and an increase in electrical carriers injected from the external circuit, the conductivity of the resin layer increases gradually, and then the overall conductivity increases. In the voltage decrease stage, when the electric field intensity decreases to a value that is insufficient for maintaining conductive carriers inside the resin material, electrical carriers such as free electrons and electric doublets steadily dissipate, and the material conductivity decreases and returns to the original state. Therefore, it can be inferred that the varying regulation of the conductivity of the resin material with the applied impulse intensity is one of the key factors that cause nonlinear behavior in the conductivities of CFRPs subjected to lightning impulses.

At the initial moment of the lightning current generation, the voltage applied on the CFRP specimen can reach 1~4 kV according to the test conditions. However, in the static DC conductivity test, the test voltage used is only on the order of 10^0–10^1 V [19]. Therefore, there is a distinct difference between the dynamic conductive properties of CFRP laminates subjected to lightning current impulses and their DC static conductivity properties. According to the experimental results of Section 11.2.3.3, the electrical conductivities of 0 and 90° laminated structures are both on the order of 10^0 S/mm, and the in-thickness conductivity is on the order of 10^{-3} S/mm. The conductivities increase as the lightning current intensity increases. In comparison, in the DC conductance tests of the CFRP laminates, the measured static DC conductivities along the 90° laminated direction and in-thickness direction are both poor since the inter-fibrillar resin and the resin layers between adjacent carbon fiber layers are in an insulating state under the applied low-magnitude DC voltage. Specifically, the longitudinal and transverse DC conductivities of the CFRP laminate are on the order of 10^1 and 10^{-4}~10^{-3} S/mm, respectively, and the in-thickness DC conductivity is 10^{-7}~10^{-6} S/mm [27–29, 37]. Therefore, the use of the DC conductive properties of CFRP materials in the simulated calculation of the coupled thermoelectrical model for the evaluation of CFRP lightning damage behavior may cause distortion in the simulation results.

11.2.4.3 The Inductive Properties of CFRP Laminates

According to the experimental results in Section 11.2.3.2, the phase of the voltage impulse applied on the CFRP laminate is ahead of the phase of the current impulse, and the dynamic volt-ampere curves of CFRPs do not coincide during the increasing and decreasing processes of the lightning current impulse. In addition, the lower the increase in rate and the longer the duration of the lightning current impulse, the greater the equivalent conductivity of the CFRP laminates. These test results can be explained as follows.

The expression for the lightning current phase can be written as:

$$\dot{I} = \dot{U} / \left(R_m + jX_m\right) \tag{11.2}$$

where $\dot{I}$ and $\dot{U}$ are the lightning current and voltage phasor flowing through and applied on the specimen, respectively, and $Z_m = R_m + jX_m$ is the equivalent impedance of the CFRP laminate. The phase angles of the voltage phasor $\dot{U}$ and current phasor $\dot{I}$ are assumed to be φ_1 and φ_2, respectively. Based on the experimental results, the voltage impulse is ahead of the current impulse in the phase space, which means $\varphi_1 > \varphi_2$. The phase angle of the impedance, which is defined as $Z_m = R_m + jX_m$, can be calculated as $\varphi_3 = \arctan(X_m/R_m) = \varphi_1 - \varphi_2 > 0$. Therefore, the dynamic electrical properties of the CFRP laminate are inferred to be inductive properties according to the positive phase φ_3 of the impedance. At the macroscopic level, the equivalent conduction circuit of the CFRP laminate can be regarded as a series connection of equivalent resistance (R) and equivalent inductance (L). The applied voltage on the CFRP specimen can be written as:

$$U = iR + L\left(\mathrm{d}i / \mathrm{d}t\right) \tag{11.3}$$

where R and L represent the equivalent resistance and inductance of the CFRP laminates. The applied voltage values during the increasing (U_1) and decreasing (U_2) processes of the lightning current impulse can be expressed as follows:

$$\begin{cases} U_1(t) = i_1 R + L\left(\mathrm{d}i_1 / \mathrm{d}t\right) \\ U_2(t) = i_2 R + L\left(\mathrm{d}i_2 / \mathrm{d}t\right) \end{cases} \tag{11.4}$$

In Eq. (11.5), i_1 and i_2 are the current values in the increasing and decreasing processes of the lightning impulse, respectively. The deviation of the increasing and decreasing voltages is

$$U_1(t) - U_2(t) = \left(i_1 - i_2\right)R + L\left(\mathrm{d}i_1 / \mathrm{d}t - \mathrm{d}i_2 / \mathrm{d}t\right) \tag{11.5}$$

When i_1 and i_2 are the same, $i_1 = i_2$. Considering that $\mathrm{d}i_1/\mathrm{d}t > 0$ during the current increasing stage and $\mathrm{d}i_2/\mathrm{d}t < 0$ in the decreasing stage, it can be concluded that $U_1 - U_2 > 0$. In other words, with the same current value, the voltage in the current increasing stage is larger than that in the decreasing stage, resulting in the noncoincidence phenomenon in the dynamic volt-ampere curves (Figure 11.20).

In addition, since the lightning current lags behind the voltage impulse, when the voltage reaches the peak value, the current is still in its increasing stage. Therefore, for lightning impulses with different waveform parameters (10/30, 20/60, and 40/120 μs), the values of di/dt are all positive at the peak of the voltage impulse. When the current amplitude is at the same value, the lower the increasing rate and the longer the duration of the lightning impulse are, the smaller the value of di/dt. Then, the applied voltage on the CFRP specimen obtained according to Eq. (11.4) decreases, and the equivalent conductivity of the CFRP specimen increases. Additionally, it should be noted that as the lightning current continues flowing through the CFRP laminate, the injection of electrical carriers from the external circuit can improve the conductivities of the resin layers inside the CFRP laminates. Therefore, the equivalent conductivity of the CFRP material increases as the duration of the lightning current impulse increases, that is,

$$\sigma_{40/120\mu s} > \sigma_{20/60\mu s} > \sigma_{10/30\mu s} \tag{11.6}$$

The equivalent conductivity of the CFRP laminates under a lightning current impulse of 40/120 μs is the highest, followed by that under a 20/60 μs lightning impulse, and the conductive ability of CFRPs under a 10/30 μs lightning impulse is the worst.

As shown in Section 11.2.4.3, the current conduction path inside a single plane layer is mainly formed by the interconnection of carbon fiber filaments, although thin layers of resin may be interposed between the carbon filaments. Therefore, the inductive properties of CFRP laminates may be closely related to the electrical characteristics of the carbon fibers. The carbon fiber can be simply regarded as a straight conductive wire, and the equivalent inductance of the straight wire can be calculated as:

$$L = \frac{\mu_0 l}{2\pi}\left(\ln\frac{2l}{r} - 0.75\right) \tag{11.7}$$

where l is the length of the carbon fiber, r_c is the radius of the carbon fiber (approximately 3.5 μm), and $\mu_0 = 4\pi \times 10^{-7}$ H/m is the vacuum permeability. According to Eq. (11.7), the inductance of a carbon fiber monofilament with a length of 1 m is approximately 2.5 μH. When the current flows in the 90° laminated structure CFRP laminate with a size of 100 × 100 × 3.6 mm, it can be assumed that the length of the in-plane conductive path is approximately 2 × 100 mm. Thus, the equivalent inductance of the single conductive path formed by a carbon monofilament is approximately 2.5 μH/m × 2 × 100 mm = 0.5 μH.

The current conduction path inside a single layer is a parallel circuit formed by multiple carbon fiber conductive channels. However, not all carbon fiber filaments can participate in the lightning conducting process. Assuming that only 50% of the carbon fibers in the laminate are involved in the construction of the conductive path, the number of conductive fibers connected in the parallel circuit inside a CFRP laminate with a fiber volume fraction (f_v) of 60% can be estimated as

$$\begin{aligned} &\frac{w \times h \times f_c}{2r_c \times 2r_c} \times 50\% \\ &= \frac{100\,\text{mm} \times 3.6\,\text{mm} \times 60\%}{2 \times 3.5\,\mu\text{m} \times 2 \times 3.5\,\mu\text{m}} \times 50\% \approx 2.2 \times 10^6 \end{aligned} \tag{11.8}$$

In Eq. (11.8), w and h are the width and height of the CFRP specimen, respectively, and it is assumed that the cross section of the space that the carbon fiber occupies is a square with a side length of $2r_c$. When multiple carbon fiber wires are connected in parallel, the equivalent inductance of the circuit network is reduced to some extent. However, the equivalent inductance of the carbon fiber parallel circuit does not strictly follow the inverse-proportion law to the number of parallel filaments. According to engineering experience regarding the inductance of parallel wire circuits, an increase in the number of parallel wires can only reduce the total inductance to approximately 60% of a single wire's inductance due to the existence of mutual inductance between wires. It can be estimated that the equivalent inductance of the carbon fiber conductive network in the plane direction of the carbon fiber laminate is approximately $0.5 \times 60\% \approx 0.3\,\mu\text{H}$. When the lightning current impulse of 20/60 μs with the peak value of 384 A is applied on CFRP specimen A1 (90° laminated structure), the voltage change due to the inductive effect of the carbon fiber conductive network can be calculated as:

$$\Delta U = L\frac{di}{dt} = 0.3\,\mu\text{H} \times \frac{384\,\text{A}}{10\,\mu\text{s}} = 11.34\,\text{V} \tag{11.9}$$

The total voltage on the CFRP laminate is measured to be approximately 35.2 V. The voltage decreases due to the inductive properties of the carbon fiber accounts for 32.2% of the total voltage on the tested CFRP specimen, which indicates that the inductive effect of the carbon fiber network has a significant influence on the dynamic conductive properties of the CFRP laminate.

11.2.4.4 Equivalent Conductivity of CFRP Laminates Subjected to Lightning Current Impulses with Higher Intensity

Based on the former experiment on the dynamic conductive properties of CFRP laminates subjected to nondestructive lightning current impulses, it can be found that the dynamic conductive properties of CFRP laminates are different under lightning impulses with different waveform parameters and amplitudes. For the destructive lightning direct effect test, because of the distinct difference in the waveform parameters of lightning current components stipulated in lightning test standards [9–11], the equivalent conductivity of the CFRP material should be measured and calculated according to the specific lightning test environment. For instance, when lightning component A is applied, the dynamic equivalent conductivity of the CFRP laminate can be estimated based on its conduction properties under a lightning impulse of 20/60 μs, which has similar front time and duration parameters as lightning component A. From the test results of the equivalent conductivity of the $[0]_{24}$ unidirectional CFRP laminate in Figure 11.22, the conductivity along the fiber direction was found to increase with an increase in lightning current peak value according to the power function law via the curve fitting method, as shown in Figure 11.26.

The relationship between the anisotropic equivalent conductivity of the unidirectional CFRP laminate and the lightning current amplitude can be calculated as follows:

$$\begin{cases} \sigma_x = 0.58198 \times I_{\text{peak}}^{0.33507} \\ \sigma_y = 0.37959 \times I_{\text{peak}}^{0.36098} \\ \sigma_z = 0.0323 \times I_{\text{peak}}^{0.65571} \end{cases} \tag{11.10}$$

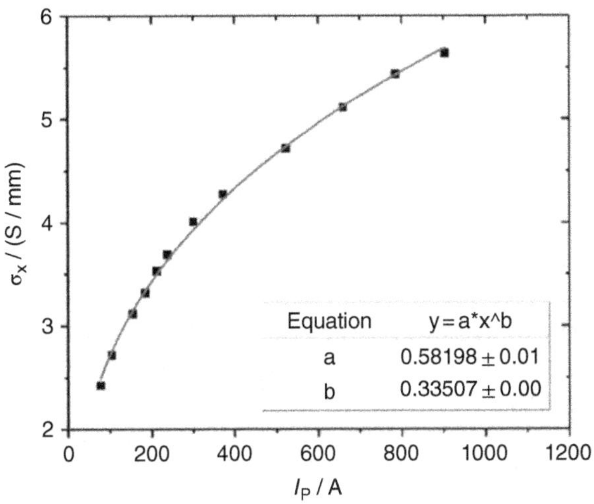

Figure 11.26 Relationship between the longitudinal conductivity of specimen A1 and the amplitudes of the 20/60 μs lightning impulse.

As for the applied ranges of Eq. (11.10), first, it has been verified to be valid under the experimental conditions of current amplitude 50~1000 A and discharge voltage 100 V~4 kV. Additionally, Eq. (11.10) can be used to estimate the equivalent conductivities of CFRPs subjected to lightning current impulses with higher intensity if there is no resin pyrolysis, delamination, and other damage inside CFRP material. In the destructive lightning test, the change in conductivities caused by material heating and resin pyrolysis should be fully considered. With low-intensity voltage applied, the resin layer distributed in the conduction path is an insulating material accompanied by a low electrical conductivity which is generally on the order of 10^{-9}–10^{-13} S/mm. When the applied voltage is less than 50 V, the in-thickness resistance of the CFRP specimen was found to be so large that the electrical energy can only be slowly released in the form of low-amplitude current, and cannot generate the current impulse with the expected waveform parameter. Therefore, it can be inferred that the conductivity of CFRP material does not conform to the law in Eq. (11.10) when the order of magnitude of the current intensity is lower than 10^0 A.

11.3 The Lightning Strike-Induced Damage of CFRP Strike

11.3.1 Introduction of the Lightning Damage of CFRP

With the wide application of CFRP in aircraft, the lightning damage behavior of CFRPs has attracted increasing interest. The experimental study is the most direct method to obtain the interaction effect between lightning and carbon fiber polymer-matrix composites. P. Ferraboli and Y. Hirano began to conduct experimental researches in 2007 and published relevant research results [37–40]. The lightning damage effects of lightning current component D with peak values of 10, 30, and 50 kA on CFRP laminates with and without fasteners were experimentally studied. The mechanical damage degree of CFRP was evaluated by residual tensile strength and residual compressive strength. The results showed that for CFRP laminates without fasteners, the lightning damage area is limited to the area near the lightning attachment point, and the damage depth is as deep as 2–4 fiber layers; however, for CFRP laminates with fasteners, the lightning damage is more concentrated and serious. The experimental results showed that the lightning damage depth of CFRP laminates is determined by the peak value of the lightning current component, while the lightning damage area of CFRP is determined by charge transfer amount and action integral.

The literatures [31, 41, 42] carried out an experimental study on the lightning damage of CFRP laminates under the lightning current component. The area and depth damage of carbon fiber laminates were detected by the ultrasonic imaging system, and the micromorphology of lightning damage was observed by electron microscope. The relationship between lightning damage and lightning current parameters of CFRP specimens was obtained, and the obtained correlation is basically consistent with the results of the literature [37]. Lightning strike experiments are also an important means to evaluate the lightning performance of the composite protective layer. The lightning protections of CFRP composites include mesh foil, aluminum coating, carbon nanotube doping, etc. The lightning protection performance of aluminum spraying layer in reference [43] showed that the metal protection layer can effectively prevent the damage of CFRP laminates from the lightning strike, and the influences of coating thickness and spraying method on lightning protection performance were studied. In recent years, lightning protection materials of CFRP laminates have become a new research hotspot. In addition to covering metal mesh, metal foil and other metal conductive layers on CFRP laminates to reduce lightning-induced damage [44, 45], various new lightning protection materials, and protection structures have been developed [46–54]. Many scholars have proposed new lightning protection methods including using a lightweight nonmetallic conductive protection layer (e.g. carbon nanotube paper) [46–48], doping conductive material (graphene, nickel-coated single-walled carbon nanotubes and other conductive nanoparticles) [49–51], and material modification [52–54] to improve the conductivities and strengthen the lightning protective ability of CFRP laminates.

11.3.2 Single Lightning Strike-Induced Damage

11.3.2.1 Experimental Setup for Single Lightning Strike Test

The platform for the single lightning strike test was developed by the High-Voltage and High-Current Measurement Technology and Equipment Engineering Laboratory of Xi'an Jiaotong University. A circuit schematic and an image of the lightning current generator are shown in Figure 11.27. A crowbar circuit, consisting of high-voltage film capacitance C (30 μF), waveform forming inductance L and resistance R, a triggered air-gap switch G_A, and a triggered vacuum switch G_V, was adopted to reduce the footprint and energy consumption.

Lightning component A, with peak values of 10–200 kA, a peak time of 20 μs, and a duration of 200 μs, can be generated as stipulated in SAE ARP5412 [9]. The lightning impulse was measured and recorded by a Rogowski coil (1 kA/V) and a Tektronix oscilloscope (MDO3012). In stage I in Figure 11.27c, the electrical energy stored in the capacitance C is released through G_A, L, and R_L, forming the wave front portion of the lightning impulse. In stage II, switch G_V

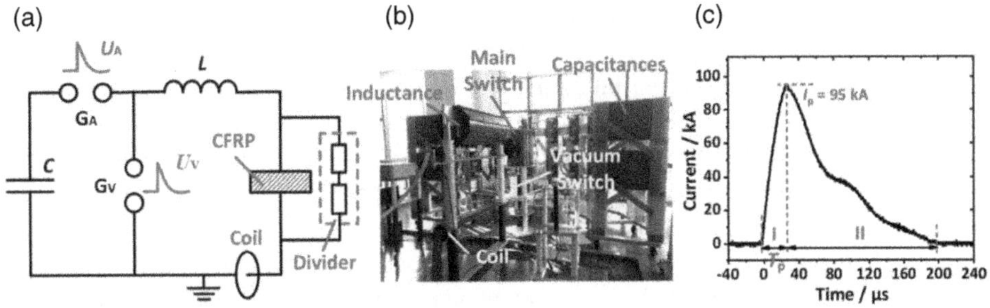

Figure 11.27 Experimental platform for lightning current component A: (a) circuit schematic; (b) generator; and (c) current waveform.

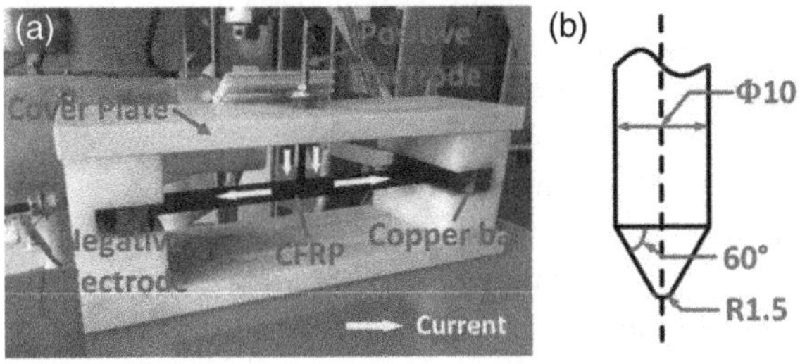

Figure 11.28 Test fixture for CFRP laminates: (a) image of the fixture; (b) conical head of the positive electrode.

is closed, while switch G_A is opened. Thus, the magnetic energy stored in L is consumed by the resistor R_L and CFRP specimen to form the wave tail of the current impulse.

The composite material system used in this experiment was a unidirectional T300/3021 carbon fiber/epoxy prepreg with a fiber volume fraction of 60% (Shanghai Duodi Polymer Material Co. Ltd.). The prepregs were placed on a mold with a 24-ply quasi-isotropic layup $[45/-45/0/90]_{3S}$ with a specimen dimension of 150 × 100 × 3.6 mm. The properties of T300 can be found in [32] from Toray Industries Co. Ltd. To simulate the spread and penetration of the lightning current along the in-plane and in-depth directions of CFRPs, a specimen fixture was designed, as shown in Figure 11.28.

In Figure 11.28a, two insulation plates are installed on the pedestal. The right plate can be adjusted on the guide rail according to the specimen size. Two symmetrical copper bars are utilized as negative electrodes and were connected to the ground. A 4-mm-wide groove is machined on the copper bar. The CFRP specimen is placed in the grooves and tightly embedded between the copper bars under the pressing force of a pair of horizontal springs, thus ensuring that the short edge surfaces of the CFRP specimen are fixed at their initial positions even under the force of the lightning impact. The positive electrode, which is fixed on the insulation cover by an insulating sleeve, is composed of a copper-lead screw with a conical head, as shown in Figure 11.28b. The discharge gap between the electrode and specimen is adjusted to 2 mm in this test. The symmetrical copper bars are used as negative electrodes and connect to the ground. The conduction path of the lightning current along the laminate is indicated by the yellow arrows in Figure 11.28a.

11.3.2.2 Experimental Results of Single Lightning Strike-Induced Damage

11.3.2.2.1 Lightning Strike Process The lightning strike process on CFRP composite can be divided into different stages, including the development of a lightning leader, the formation of a conductive channel, current injection, burning and cooling. A high-speed digital camera (Phantom MiroM310) was used to record real-time images of the lightning strike process. The frame rate and the recording resolution of the high-speed camera were 200 kfps and 126 × 126 pixels, respectively.

The formation of a lightning leader originates from air ionization caused by a high-intensity electric field near the electrode. When one of the leader branches approaches the specimen, it will trigger a connection leader from the CFRP surface, as indicated by the arc jets generated from both the electrode and CFRP laminate (Figure 11.29a1). After the moving leaders approach each other, a conductive channel is formed through the entire gap (Figure 11.29a2). The lightning current flows through this conductive channel with an emission of a dazzling white light (Figure 11.29a3). The temperature of the laminate increases sharply, resulting in firing and burning after the lightning current injection process, as exhibited by the red flame in Figure 11.29a4.

An infrared thermal imager (FLIR T460) with a frame rate of 30 Hz and a recording resolution of 320 × 240 pixels was utilized to obtain the surface temperature distribution of the CFRP specimens. Because the duration of lightning impulse does not exceed 400 μs, which is less than the recording interval of the infrared thermal imager (0.033 s), the infrared thermal imager can only be utilized to capture the surface temperature distribution of the CFRP specimens in the cooling process. The infrared temperature distributions show that the high-temperature region is concentrated around the lightning attachment point. Approximately 0.1 s after the lightning current injection, the maximum surface temperature exceeds 670.1 °C. The burst fiber fragments splash outward from the damaged area. At $t \approx 1.233$ s, high-temperature broken fiber bundles can also be observed on the laminate

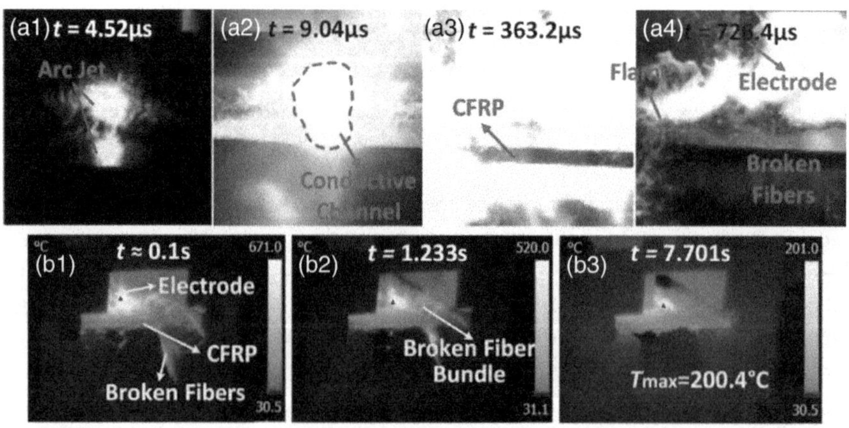

Figure 11.29 Lightning strike process recorded by a high-speed camera at (a1) $t = 4.52\,\mu s$; (a2) $t = 9.04\,\mu s$; (a3) $t = 363.2\,\mu s$; and (a4) $t = 726.4\,\mu s$; surface temperature distribution obtained from an infrared thermal imager during the cooling process at (b1) $t \approx 0.1$ s; (b2) $t \approx 1.233$ s; and (b3) $t \approx 7.701$ s.

surface, which match the actual damage shape. The maximum surface temperature at $t \approx 7.701$ s still exceeded 200.4 °C, that is, the resin matrix inside the damaged laminate continues to undergo pyrolysis for at least 7.6 s after the lightning strike, causing further expansion of lightning damage.

11.3.2.2.2 Surface Damage of the CFRP Laminates The lightning damage test was first performed on the quasi-isotropic specimens under lightning impulse A from 20 to 100 kA. The damage morphology of the CFRP specimens is illustrated in Figure 11.30.

Under a lightning current impulse of 20 kA, the fiber bundles at the lightning attachment point break and bulge outward along the fiber orientation in the top layer (45°). A series of fiber breakpoints is distributed along the ply direction of the second layer (−45°). Broken fiber bundles arising from the pressure of the internal pyrolysis gas are scattered on the laminate surface, as shown in Figure 11.30a. As the impulse intensity increases, additional typical damage modes can be observed on the specimen surface, including fiber sublimation, fiber fracture, resin pyrolysis, ply-lift, etc., the distribution area of the damage becomes larger, and damage depth at the lightning injection point increases. Under a lightning impulse of 100 kA, a deep ablation pit is formed in the lightning attachment zone, which damages several layers of the specimen. The ply-lift and outward warping of fiber bundles in the surface layup direction become more significant, which can also be observed in deeper layers along their layup directions (−45°, 0°, etc.). Moreover, the resin pyrolysis area increases gradually and expands symmetrically outward perpendicular to the surface fiber orientation with an increasing peak of lightning current, as shown from Figure 11.30a to d.

The damage morphology in Figure 11.30d illustrates uneven damage degrees in different zones. The overall view of surface damage can also be observed by ultrasonic C-scanning

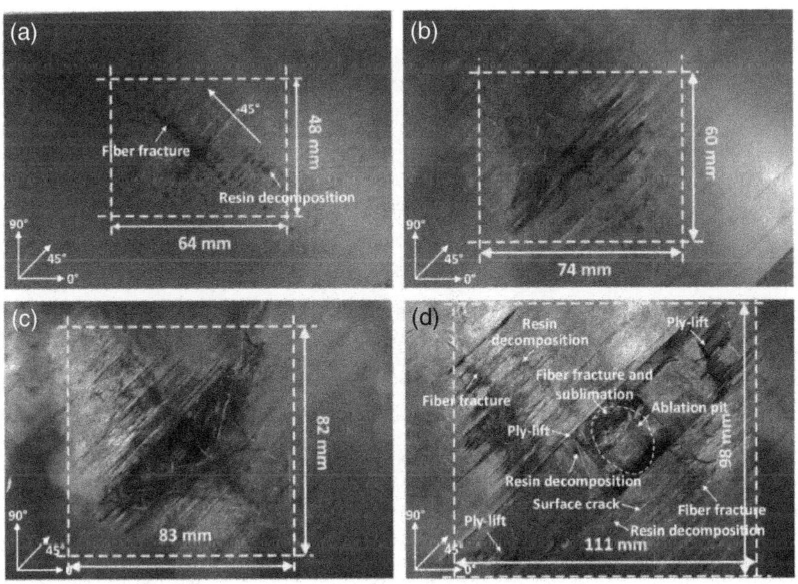

Figure 11.30 Images of CFRP specimens under lightning component A with different intensities: (a) I_p = 20 kA; (b) I_p = 40 kA; (c) I_p = 60 kA; and (d) I_p = 100 kA.

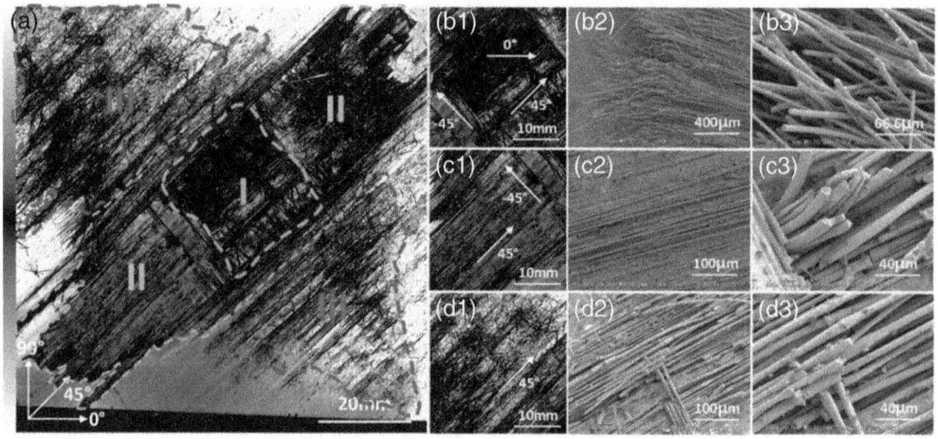

Figure 11.31 Surface damage observed by ultrasonic scanning: (a) overall view; (b1) Zone I; (c1) Zone II; and (d1) Zone III; SEM images: (b2) and (b3) Zone I; (c2) and (c3) Zone II; and (d2) and (d3) Zone III.

(Sonoscan, D9500). The damage of CFRP laminate can be divided into three different zones according to its damage morphology and properties, as shown in Figure 11.31a. A field emission scanning electron microscope (FESEM, HITACHI SU-8010) was used to further analyze the damage behavior of CFRP by observing typical microstructures in different damage zones.

11.3.2.2.2.1 Zone I Zone I exhibits the most severe damage, mainly including resin pyrolysis, fiber fracture, and sublimation. The lightning current arc directly contacts the tested laminate in this zone. In the ionization channel of the lightning arc, the instantaneous pressure and temperature exceed 5×10^6 Pa and 3×10^4 °C, respectively, and the electron movement speed can reach 5000 m/s with a transported energy of 39.55×10^3 J/Ω ($I_p \approx$ 40 kA) [4]. The pyrolysis temperature of resin generally ranges from 200 to 600 °C, and the carbon fiber sublimation temperature is approximately 3000 °C. Therefore, the fibers and resin in this zone may sublimate and pyrolyze within a very short time. In addition to the resin degradation, the resin pyrolysis gas will cause explosive rupturing near the lightning attachment zone, with at least three layers of fibers (45°, –45°, and 0°) being completely broken in this area (Figure 11.31b1). Due to the extreme temperature and pressure of the lightning arc, the carbon fiber bundles in the lightning attachment damage zone are ablated and scattered (Figure 11.31b2). The inter-fiber resin undergoes severe pyrolysis, and most of the surrounding matrix disappears (Figure 11.31b3).

11.3.2.2.2.2 Zone II Zone II, which contains the damage modes of ply-lift and fiber fractures, is a strip-shaped damage area starting from the lightning attachment area and is distributed along the surface fiber orientation (45°). The top fiber layer is completely peeled off from the laminate. To obtain better observations of the damage morphology, the lifted top fiber bundles in the ply-lift area were carefully trimmed off. Although slight ablation traces can be observed along the fiber orientations of the top layer (45°) and second layer

(−45°) on the exposed second layer, the overall arrangement of the fibers is relatively neat and the underneath layer is intact without large-scale degradation, as shown in Figure 11.31c1, suggesting that the ply-lift damage in this zone can be ascribed to inter-laminar cracking due to mechanical failure. Figure 11.31c2 to c3 presents the microstructure of the exposed second layer, which demonstrates milder damage compared to Zone I. The fiber arrangement is well ordered in this damage zone and the resin structure is relatively complete (Figure 11.31c2). From the enlarged view of the broken fiber fractures in Figure 11.31c3, a large number of flat fractures indicating brittle rupture can be seen instead of deformed fiber fractures.

11.3.2.2.2.3 Zone III Damage image of Zone III, which is a fan-shaped area symmetrically distributed adjacent to Zone II, illustrates more serious fiber fractures and resin decomposition compared to Zone II (Figure 11.31a). The broken fiber fractures and ply-lifts along the top fiber orientation (45°) are distributed over Zone III. The resin matrix around the fibers pyrolyzed as the temperature rises during the lightning strike, and the fiber network with a higher sublimation temperature remains on the top layer, resulting in a large number of cracks along the direction of the top fiber layup. The typical microstructures of the damaged fibers and matrix in the resin pyrolysis damage zone are shown in Figure 11.31d2 and d3. The resin is more severely pyrolyzed than that in Zone II. Obvious ablation pits appear on the filling matrix, and the number of exposed fibers is significantly increased. In contrast to the brittle fiber fractures in Zone II, the damaged fibers in this zone are more likely to become coarsened near their broken tips (Figure 11.31d3), which is caused by the deorientation of the crystalline carbon chain during pyrolysis.

11.3.2.2.3 Internal Damage Behavior To understand the failure process of CFRP laminate subjected to a lightning strike, the internal damage was further studied using cross section CT images obtained via a 3D X-ray imaging system (YXLON Y. Cheetah). A taper-shaped hole can be observed in the center of the cross section, which belongs to damage Zone I. Several delaminated plies bend outward (Figure 11.32a), indicating a certain force from the inner layer acting on the delaminated layers. During the lightning strike, a large proportion of the lightning current flows into the surface fiber layer due to the larger in-plane conductivity of the CFRP laminate than its in-thickness conductivity coupled with the ablation effect of the high-temperature lightning arc, resulting in the rapidly rising temperature at the lightning attachment point and its surrounding area. The resin matrix is thermally decomposed, releasing CO_2 and water vapor, due to the high temperature. The extreme heating area expands from the attachment point, which implies that the damage process can be traced from the central ablation point. Figure 11.32b to e presents the probable presumptions regarding the delamination damage process based on the analysis of the lightning strike and damage characteristics in different damage zones.

The laminate peeled-off damage observed in Zone II is thought to be related to the pyrolysis gas generated in the lightning attachment area, as presented in Figure 11.32b. The lightning energy density in the lightning attachment area is the largest. Thus, a large amount of pyrolysis gas can accumulate in a very short time during the initial lightning strike process. When the pyrolysis gas is trapped in the inter-laminar space due to the obstruction of upper laminate layers, it causes large internal pressure perpendicular to the

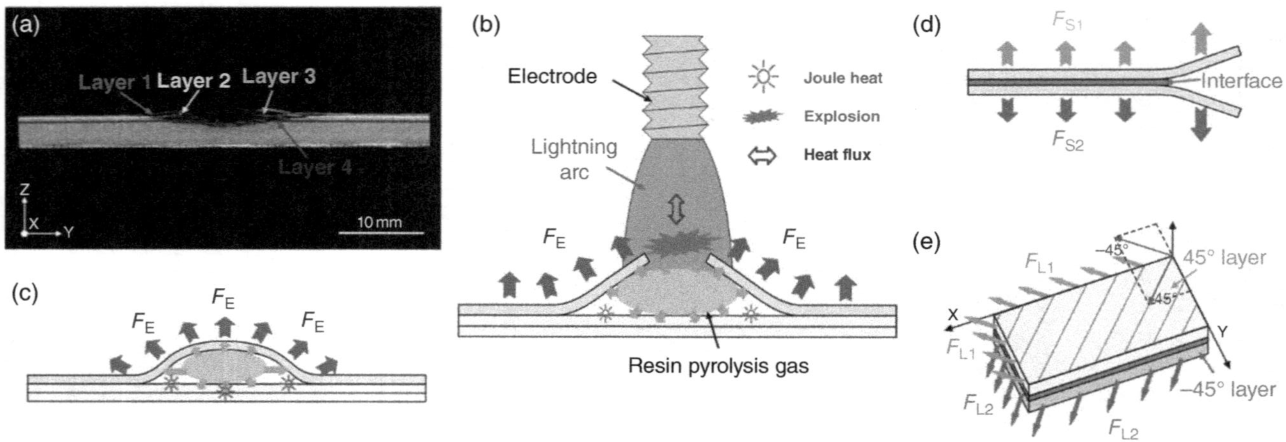

Figure 11.32 Internal damage process: (a) cross section CT image; (b) pyrolysis gas explosion; (c) pyrolysis gas expansion; (d) thermal stress: Mode I; and (e) thermal stress: Mode II.

laminate plane. Once the upward pressure exceeds the restriction tension of the upper sub-laminates, the fiber bundles break into fragments and form many brittle ruptures in the fibers (Figure 11.31c3). The top laminate is torn and peeled off under the explosive gas impact (F_E) and the inertial effect. As a result, the delamination section is drastically expanded to the strip-shaped torn damage region in Zone II. Moreover, the sub-laminate beneath the laminate peeled-off damage area, the structures of the fibers and the matrix remain intact, suggesting that this part of the damaged area is more due to the mechanical impact and is less exposed to thermal damage.

The injected impulse results in thermal degradation of the resin material over a large range when the shunt current flows into the laminate layers along different layup directions. The current density flowing along the transverse direction of the fiber orientation is smaller than that in the longitudinal direction of the top layer. Therefore, the increase in the material temperature and resin pyrolysis is slower than in the lightning central area and its vicinity. Compared with the laminate peeled-off damage in Zone II, damage in Zone III can be presumed to originate from this milder resin pyrolysis process. The weak interlaminar mechanical strength, which is generally within 100 MPa, leads to the poor ability of CFRP laminates to withstand mechanical and thermal impacts [29]. The initial delamination cracks occur in the interlaminar spacing due to the presence of pyrolysis gas. Therefore, the internal pressure of the accumulated pyrolysis gas, which is not sufficient to explode the fiber bundles, leads to a decline in the interlaminar bonding force and delamination propagation, as shown in Figure 11.32c. The delamination process is ascribed to the expanding pyrolysis gas, which can be traced from the reduction in resin matrix that survived the process compared with Zone II (Figure 11.31d2). Overall, the different delamination properties in Zone II and Zone III can be explained by the different intensities of the resin pyrolysis process.

The thermal decomposition of the resin matrix starts within a temperature range of approximately 200–600 °C. Apart from the extension of pyrolysis gas, even if the pyrolysis gas has not been generated, the thermal expansion stress generated during the material-heating process can also be regarded as an accelerating factor for the extension of delamination damage. Figure 11.32d and e exhibits two different thermal stress modes in adjacent layers. In Mode I, the tensile stresses from the upper and lower layers are both normal to the plane of the laminates. The tensile stresses in the opposite directions will exacerbate the delamination damage and cause interface failure. On the contrary, the thermal expansion of layers along the in-plane direction also contributes to the delamination damage. Tearing failure is thought to be generated by the mutually perpendicular stresses due to the different fiber orientations of adjacent layers in Mode II. Especially when the matrix material is partially pyrolyzed, the interlayer bonding force will be reduced, and the slipping damage may be induced by the in-plane thermal stress along perpendicular directions. Therefore, it can be speculated that the delamination process is also affected by the mixed thermal stress of Mode I and Mode II.

11.3.2.3 Evaluation for Single Lightning Strike-Induced Damage

The internal damage of the CFRP specimens was detected by ultrasonic T-scanning (Sonoscan, D9500), which can superimpose the damage areas of different layers. The damage area presented in the T-scan image is symmetrically distributed along the fiber

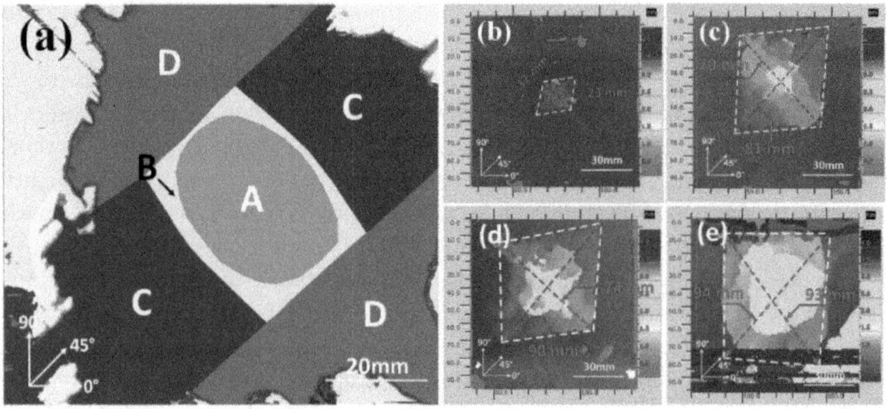

Figure 11.33 Evaluation for lightning damage: (a) damage region division based on the T-scan result and in-depth damage scanning of specimens under (b) $I_p = 20\,kA$; (c) $I_p = 40\,kA$; (d) $I_p = 60\,kA$; and (e) $I_p = 100\,kA$.

orientation of the top layer. In addition to this anisotropic behavior, the total damage region is not represented by a single uniform area. Combining the surface morphology and microstructural changes, the damage area presented in the T-scan image can be divided into different regions according to the damage characteristics and damage mechanism, as shown in Figure 11.33a.

The central damage area marked in green (Region A) corresponds approximately to Zone I in Figure 11.31a. The direct contact with the lightning arc accompanied by the concentrated current Joule heating and heat flux results in a severe thermal impact in this region. On the exposed second layer in Zone II, the region near the lightning attachment area exhibits many scratches in the −45° direction, which demonstrate the thermal ablation propagating along the fiber orientation of the second layer. In contrast, in the remainder of Zone II, which is far away from the lightning center, instead of −45° scratches, ablation traces parallel to the fiber direction of the top layer (45°) can be observed (Figure 11.31c1). It can be inferred that the laminate peeled-off damage is formed by the mechanical impact of pyrolysis gas explosion instead of the thermal damage effect, as shown in Figure 11.32b. Moreover, the blast gas is more likely to be generated by the rapid resin pyrolysis in the lightning attachment area rather than the pyrolysis of resin in this region, as demonstrated by the relatively complete microstructures of the fiber arrangement and matrix structure in Region C (Figure 11.31c2). Blue and yellow marks are used to distinguish damage Regions B and C in Figure 11.33a. The red region representing Region D is approximately equal to damage Zone III. The extended delamination damage is presumed to be induced by the expanding internal pressure of the pyrolysis gas, as shown in Figure 11.32c. The origination of the pyrolysis gas can be inferred from the large proportion of in-situ resin deterioration in Figure 11.31d2.

The projected total damage area S_Z in the T-scan image can be divided into two parts: (i) the thermal damage area S_P, consisting of the Regions A, B, and D in Figure 11.33a, which can be attributed to the thermal damage effect of the lightning strike, and (ii) the laminate peeled-off damage area S_B marked in yellow, which is not directly related to lightning thermal effects and can be regarded as a type of mechanical damage. Thus, the total damage S_Z can be written as:

$$S_Z = S_B + S_P \tag{11.11}$$

Figure 11.33b to e illustrates the damage depth images obtained from the ultrasonic scanning. The experimental results of the damage areas (S_Z, S_B, and S_P) and damage depth D_d are shown in Table 11.4 and Figure 11.34.

The thermal damage effect of the current impulse, including arc ablation and current Joule heating effects, increases as the intensity of the lightning current increases. At the same time, a larger current amplitude is accompanied by a higher lightning voltage, suggesting that the penetration depth of the lightning current will increase due to inter-laminar breakdown. The expansion of the potential resin pyrolysis region toward deeper layers indicates the possibility of more pyrolysis gas accumulating in the lightning attachment area. Therefore, the increasing current intensity is deduced to be corresponding to a stronger explosive impact force of pyrolysis gas, which enhances the laminate peeled-off damage. The experimental results (Figure 11.34) show consistent characteristics with this inference: the damage area, the damage depth and the proportion of damage area S_B relative to the total damage area S_z increase with increasing intensity of the applied lightning impulse.

Table 11.4 Waveform parameters and the corresponding lightning damage of specimens.

I_p (kA)	di/dt (kA/μs)	Q_C (C)	W_A (A²s)	S_Z (cm²)	S_B (cm²)	S_P (cm²)	D_d (mm)
21.3	1.224	1.53	19208	5.24	1.26	3.98	0.70
41.9	2.341	4.52	108475	16.5	7.31	9.16	1.02
56.3	3.217	7.74	260205	29.2	13.7	15.5	1.27
94.8	3.646	11.17	628393	61.2	35.7	25.5	1.58

I_p: peak value; T_f: front time; T_t: half-peak time; di/dt: current rising rate; Q_C: transfer charge; and W_A: electrical action integration.

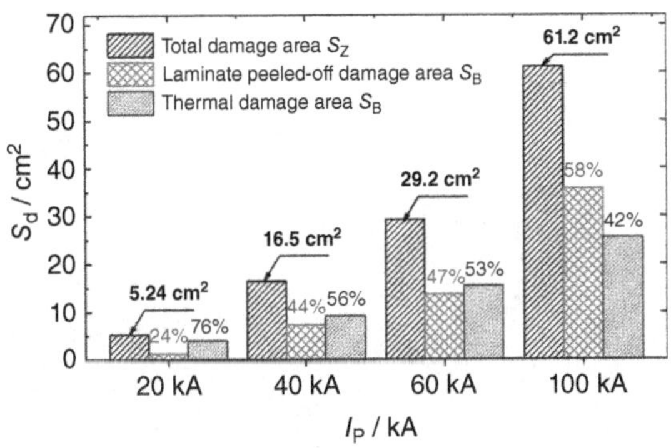

Figure 11.34 Experimental lightning damage areas of CFRP specimens.

11.3.3 Multiple Lightning Strikes-Induced Damage

In fact, lightning is a high-voltage and high-current discharge process including multiple consecutive lightning current components. To simulate the lightning damage conditions in CFRP laminates for a real lightning process, the lightning current components should be applied in a combination within the given test shot, rather than as individual, distinct shots. As the lightning test method recommended in SAE ARP5412 and ARP5416 [9, 12], the stipulated lightning test on CFRPs used in aircraft should be conducted with multiple consecutive lightning waveforms consisting of four components including initial stroke, intermediate current, continuing stroke, and subsequent return stroke, as shown in Figure 11.5.

11.3.3.1 Experimental Method for Multiple Consecutive Lightning Strike Tests

11.3.3.1.1 Test Specimens Carbon fiber/epoxy (T300/3021) composite laminate, in which the volume fraction of carbon fibers is approximately 60%, was selected as the test specimens in multiple lightning damage experiment of CFRP laminates. The size of the specimens is 300 × 45 × 3.6 mm with a 24-ply quasi-isotropic layup ($[-45/45/0/90_2/-45/0/45/0/90/45/-45]_S$). The thickness of each layer is approximately 0.15 mm. The material properties at room temperature of 25 °C, including the anisotropy electrical conductivity, thermal conductivity, and specific heat, are given in Table 11.5.

11.3.3.1.2 Experimental Setup A multiple continuous sequential lightning direct effect test system (Figure 11.35) developed by the High-voltage and High-current Measurement Technology and Equipment Engineering Laboratory (Xi'an Jiaotong University) was used.

Table 11.5 The material properties of test CFRP laminates.

Density (kg/mm^3)	Specific Heat (J/kg °C)	Electrical Conductivity (S/m)			Thermal Conductivity (W/mm °C)	
		Longitudinal	Transverse	In-thickness	Longitudinal	Transverse
1.52×10^{-6}	1065	35293	1.764	0.2822	0.008	0.00067

Source: Based on [55]. © John Wiley & Sons.

Figure 11.35 The multiple lightning direct effect test system for CFRP composite.

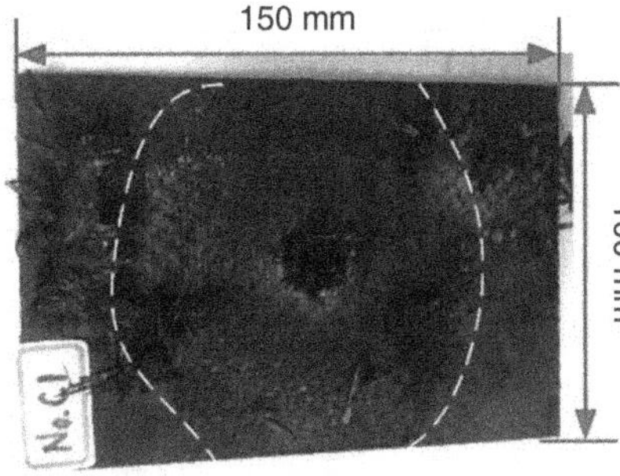

Figure 11.36 The lightning damage of a CFRP laminate under lightning component A of 100 kA.

The test system can generate multiple continuous sequential lightning current components, including lightning components A, B, C, and D, as stipulated in SAE ARP 5412 [9]. Lightning current components A and D are unipolar current impulses generated by a crowbar circuit. The lumped parameters chain consisting of capacitances and inductances is used to generate lightning current component B. Lightning component C is the continuing current generated by a three-phase controllable rectifier circuit.

It is worth noting that the lightning components shown in Figure 11.5 are generally used for the lightning direct effect test for complete aircraft or its parts in a large scale. However, under the effect of lightning component A with an amplitude of 100 kA, the CFRP laminates of limited size suffer severe lightning damage, as shown in Figure 11.36.

From Figure 11.36, it can be observed that the surface fiber was fractured and warped significantly and the damage area on the test CFRP laminate extended to the edge of the specimen laminate. If the subsequent lightning components are continuously applied under this damage condition, the damage area of the test CFRP laminate will not increase reasonably due to the limitation in the size of the test specimens, which will adversely affect the validity and accuracy of the experimental results. Therefore, lightning impulse components with the same intensity reduction were used in this section. Specifically, the experimental lightning waveforms differ from the standardized lightning waveforms only in intensity and they meet the requirements of the standard in their other parameters, including rise time and duration, as shown in Figure 11.37.

Since the intensities of the test lightning current components are reduced by the same proportion, the research results in this section are valuable for future studies of the contribution of higher intensities of each lightning component to the lightning damage effect of CFRP laminates.

11.3.3.1.3 Test Method for Multiple Lightning Damage Test Because the intensity and frequency of lightning strikes on different parts and structures differ, the aircraft is divided into several zones [12]; the corresponding test waveforms of different lightning strike zones have been shown in Table 11.1.

In this experiment, 15 tested specimens of the same type were divided into five groups with different test modes, which were selected based on the stipulated test mode [13], as shown in Table 11.6.

The lightning components in every test mode should be applied continuously in one shot, rather than as individual, distinct test shots. For example, in test mode "ABCD,"

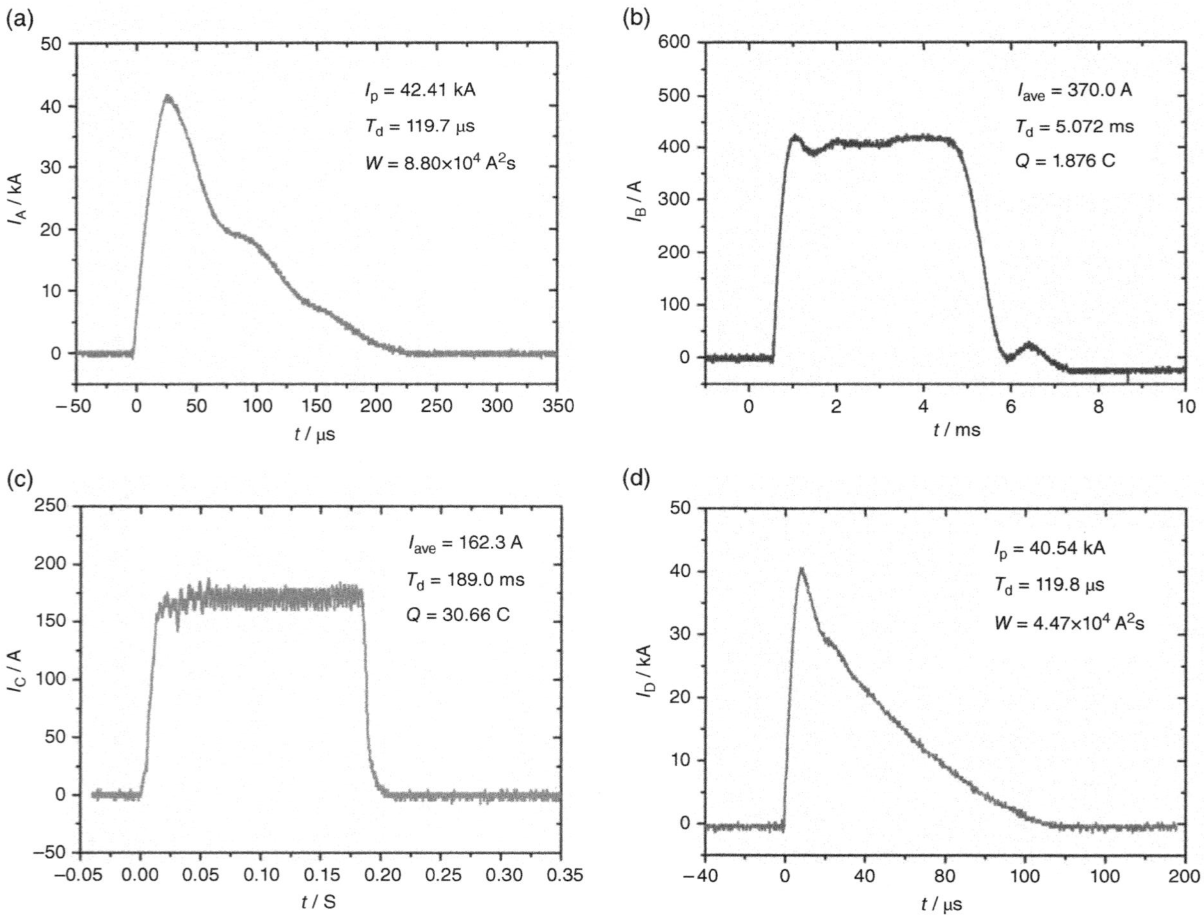

Figure 11.37 The waveforms and parameters of the lightning current components: (a) lightning component A, (b) lightning component B, (c) lightning component C, and (d) lightning component D. I_p: amplitude; I_{ave}: average current; T_d: duration; W: electrical action integration; and Q: transfer charge.

Table 11.6 The applied lightning components in different test modes and corresponding specimens.

Specimen	B1	B2	B3	B4	B5
Test mode	"A"	"AB"	"DBC"	"ABC"	"ABCD"

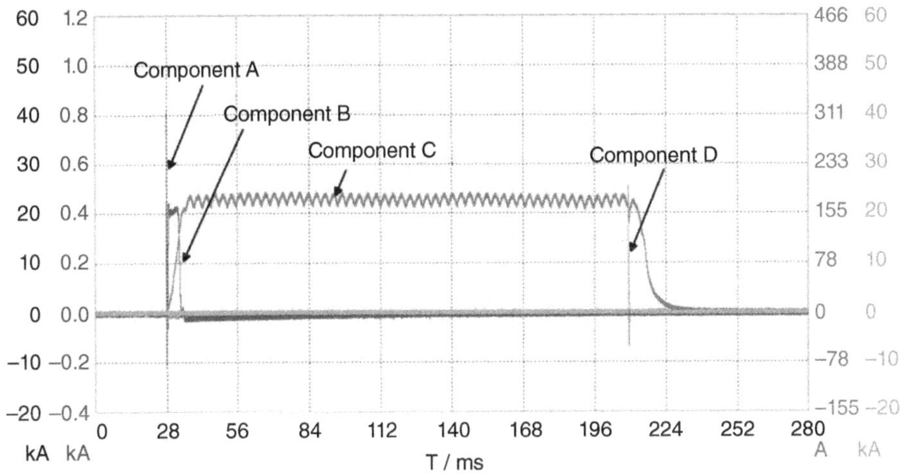

Figure 11.38 Continuous lightning test waveform for the "ABCD" test mode.

lightning component B is injected within 500 μs after lightning component A; lightning component C is injected into the CFRP specimens within 5–10 ms after lightning component B. Lightning component D is applied before the end of lightning component C, as shown in Figure 11.38, and the test fixture for CFRP laminates is the same as that used in the single lightning test (Figure 11.28).

11.3.3.2 Experimental Results of Multiple Lightning Damage

11.3.3.2.1 Visual Inspection of Lightning Damage The lightning process is recorded by the high-speed camera. During the lightning strike, a dazzling white light was emitted by the arc, a red flame was generated from the burning laminate, and burst fibers were observed on the CFRP specimen. After the lightning process, the black smoke and a pungent smell were generated from the degraded resin after the lightning strike. The parameters of the applied lightning current components in different combination and images of the CFRP specimens after a lightning strike are shown in Table 11.7 and Figure 11.39.

From the lightning damage shown in Figure 11.39, the lightning damage results are summarized as follows:

1) The tested specimen B1 which is subjected to the lightning component A (peak current $I_p \approx 40$ kA; transfer charge $Q \approx 3$C) suffered from the obvious bulge and break of carbon fibers along the surface fiber layer direction (−45° direction), especially at the lightning attachment point. At the same time, bundles of carbon fibers swelled up and were

Table 11.7 The parameters of lightning component waveforms in different test modes.

	Component A			Component B			Component C			Component D		
	I_p/kA	T_d/μs	W/A²s	I_{ave}/A	T_d/ms	Q/C	I_{ave}/A	T_d/ms	Q/C	I_p/kA	T_d/μs	W/A²s
B1	41.67	26.76	79225	—	—	—	—	—	—	—	—	—
B2	41.28	25.75	81659	374.0	5.036	1.883	—	—	—	—	—	—
B3	—	—	—	362.9	5.070	1.840	163.0	187.8	30.61	35.21	14.00	47968
B4	41.11	25.50	78359	364.3	5.060	1.843	158.4	189.5	30.00	—	—	—
B5	40.81	25.50	76353	375.4	5.054	1.897	170.3	191.3	32.57	19.89	14.50	18804

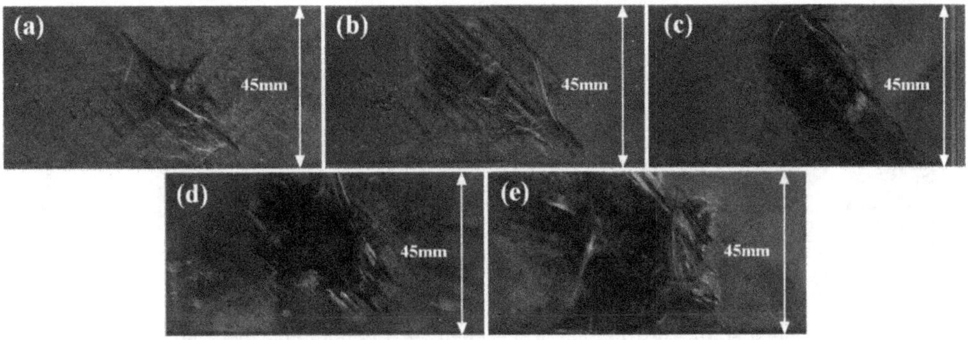

Figure 11.39 Images of the CFRP laminates after a lightning strike test: (a) B1, (b) B2, (c) B3, (d) B4, and (e) B5.

delaminated in the directions of 45° and −45°, which are the ply directions of the first and the second layers. The symmetrical delamination area expanded outward along the axis of the fracture fiber. Some local damage points caused by fiber fracture and resin pyrolysis scattered on the surface of CFRP laminates, as shown in Figure 11.39a. The damage in specimen B2 is similar to specimen B1, but the damage degree was more serious (Figure 11.39b).

2) For specimens B3, B4, and B5, the injection of the lightning component C led to an obvious near-circular erosion pit at the lightning attachment point. A large area of severe resin pyrolysis, fiber breakage and delamination for several layers were observed around the lightning attachment point. Several piles of flaky black carbon black powder and residual melt material were deposited on the specimen surface, which were resulted from sublimation of carbon fibers and pyrolytic carbonation due to the high temperature of the lightning arc. In specimen B5, the damage area extended to the edge of the specimen laminate.

11.3.3.2.2 Temperature Properties of CFRP Laminates After Multiple Lightning Strike When a CFRP laminate is struck by lightning current components, the temperature near the lightning attachment point will rise rapidly. Considering the thermal physical properties of resin and carbon fibers, the resin matrix will pyrolyze when the temperature exceeds 200 °C (aerobic conditions) and the carbon fibers will sublime at the temperature approaching 3300 °C [27, 30]. The temperature distribution of specimen B4 decreased after the injection of lighting components in different combinations was measured using an infrared radiation thermometer (Fluke Ti125), as shown in Figure 11.40.

As shown in Figure 11.40, the high-temperature region is concentrated around the lightning attachment point and forms a near-circular erosion pit. After the lightning strike, for 13 s, the maximum temperature of specimen B4 is 142 °C. Considering the heat exchange with the outside environment, the CFRP laminate surface temperature drops exponentially after the lightning strike, as shown in Figure 11.41.

Based on the temperature measured after the lightning strike test and considering both the thermal physical properties of CFRP and the temperature characteristics of arc, which have been measured to be on the order of 10^3~10^4 K in some researches [56, 57], the curve

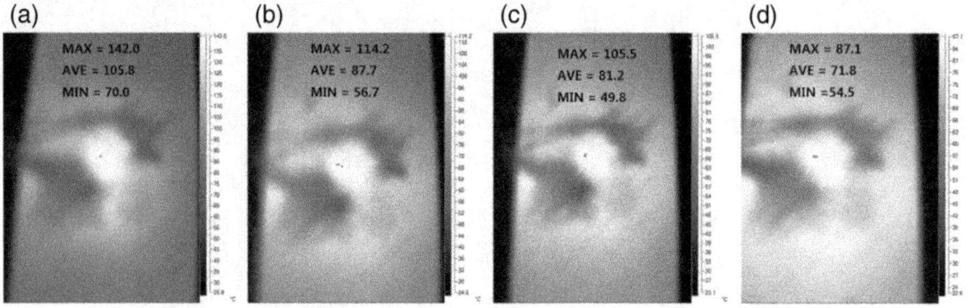

Figure 11.40 Temperature distribution of specimen B4 after lightning strike: (a) $t = 13$ s, $T_m = 142.0$ °C; (b) $t = 36$ s, $T_m = 114.2$ °C; (c) $t = 50$ s, $T_m = 105.5$ °C; (d) $t = 87$ s, $T_m = 87.1$ °C.

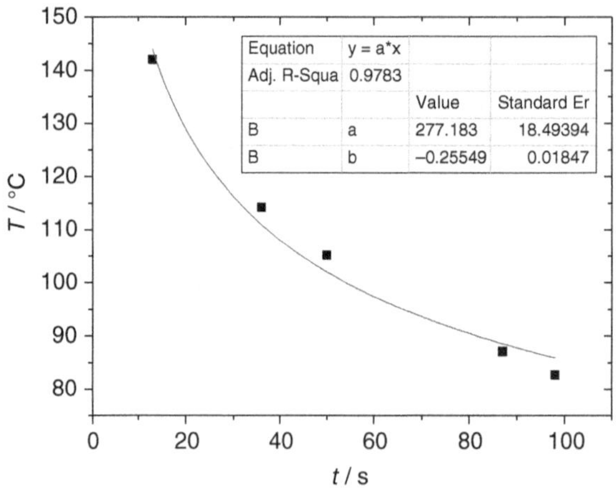

Figure 11.41 Descending trend of the CFRP's surface temperature (specimen B4).

fitting formula $T = 277.183 \times t^{-0.25549}$ and the certain value of $t = 1$ ms are selected to evaluate the surface temperature of CFRP laminates. Therefore, the instantaneous temperature of the CFRP laminate surface after a lightning strike ($t = 1$ ms) can be estimated as 1,600° C. Considering the heat exchange between CFRP laminates and the ambient environment during the loading process of lightning components, which can last for more than 200 ms, the maximum instantaneous temperature of CFRP laminates could be higher. The thermal effect of the lightning current components can cause the temperature to rise in a large area of CFRP laminates, which results in the sublimation of carbon fiber and the pyrolysis of the resin layer.

Comparing the instantaneous temperature characteristics ($t = 1$ ms) of the CFRP specimens with different lightning test modes, as shown in Figure 11.42, the surface temperature of the CFRP laminates due to a high-energy lightning strike increases with the number of applied lightning components. Correspondingly, the lightning damage of CFRP specimens in Figure 11.39 increases as the temperature of the lightning injection point increases. Moreover, the lightning instantaneous temperature of specimens B3 and B4, which have

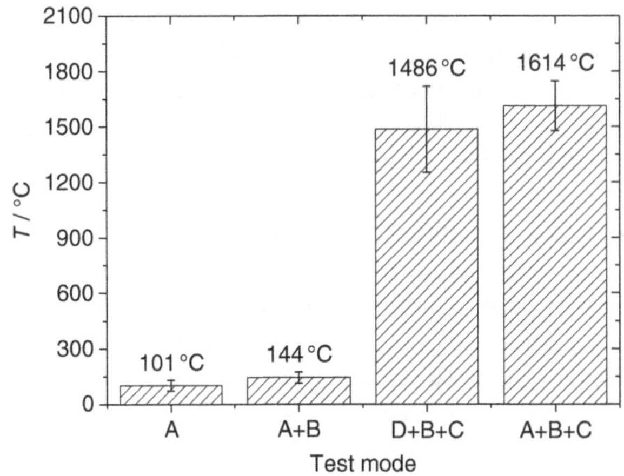

Figure 11.42 Instantaneous temperature of CFRP specimens subjected to different lightning test modes.

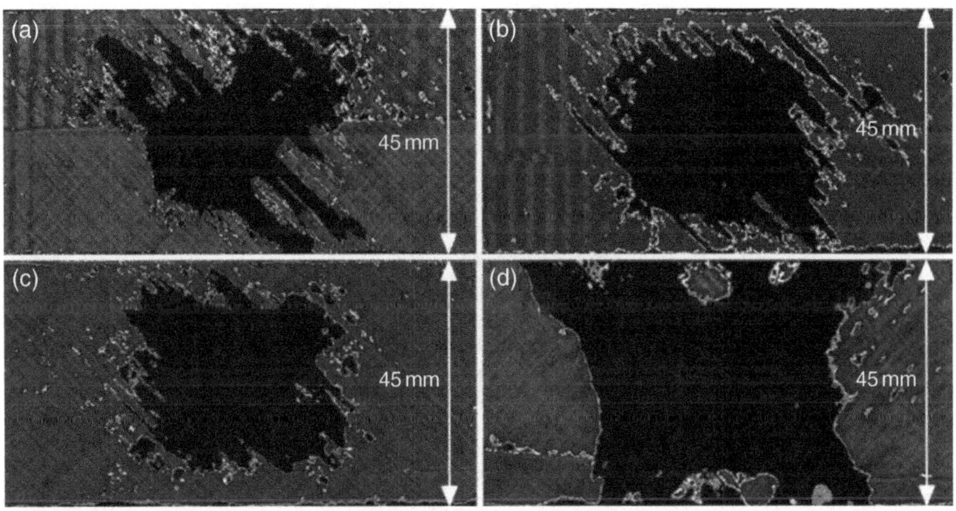

Figure 11.43 Ultrasonic T-scan images of the lightning damage areas in CFRP specimens: (a) B1, (b) B2, (c) B3, and (d) B5.

endured the injection of the lightning component C, increases significantly. The evaluated temperature results indicate that the thermal effect of continuing current component C is more significant than that of the impulse current components A and B when flowing in CFRP laminates.

11.3.3.2.3 Internal Damage Images The ultrasonic T-scan and C-scan were used to detect the multiple lightning damage area and damage depth of CFRP laminates. The ultrasonic test results for the post-lightning test are shown in Figures 11.43 and 11.44.

After the lightning strike, the smaller the thickness of the damage area was, the greater the damage depth. That is, the damage is more serious when the thickness is smaller in Figure 11.44. From the scan results, the lighting damage areas are found to develop mainly in the direction of −45° and 45°. Particularly, the lightning damage area of specimen B5 extended to the edge of specimen B5 in the "ABCD" lightning test mode.

11.3.3.3 Multiple Lightning Damage Areas and Depths of CFRP Laminates

The lighting damage types of CFRP laminates mainly include fiber breakage, delamination, resin pyrolysis, and ablation, and the corresponding damage degrees are difficult to define and observe. Therefore, to quantitatively discuss the damage degree of the tested CFRP laminate, both the damage areas and depths were calculated as a function of different test modes using the pixel statistics method from the scan images, as shown in Table 11.8.

At the macroscopic level of the lightning effect, the damage area and damage depth represent different interaction processes of multiple continuous sequential lightning strikes. To be specific, the damage depth may be related to the shock wave and the overpressure effect, electromagnetic force effect and the local thermal effect which is caused by contact

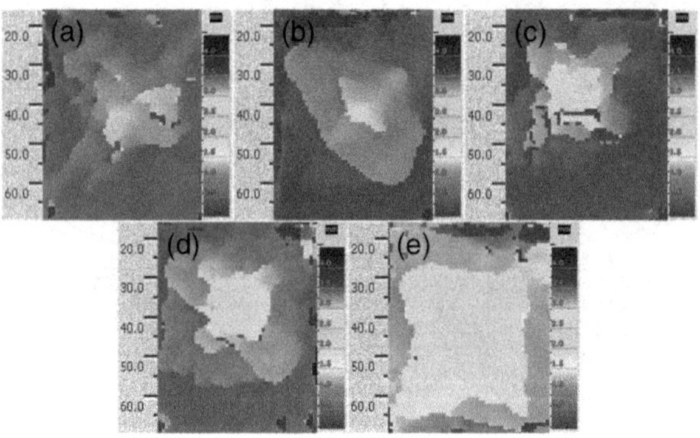

Figure 11.44 Ultrasonic C-scan images of the lightning damage depth in the CFRP specimens: (a) B1, (b) B2, (c) B3, (d) B4, and (e) B5. (The legend represents the thickness of the tested CFRP laminate).

Table 11.8 Lightning damage areas (S_d) and damage depths (D_d) of the CFRP specimens in different test modes.

Specimen	Test Mode	Damage Depth/mm	Damage Area/mm^2
B1	A	0.513	1104
B2	AB	0.810	1323
B3	DBC	0.900	1453
B4	ABC	0.930	2400
B5	ABCD	1.280	2790

with the high-temperature arcs, and the expansion of the damage area should result from the Joule-heating effect generated when the current conducts in the CFRP laminates.

Since the lightning components contained in one test mode are applied sequentially, compared to the lightning damage effect in the test mode with fewer lightning components, the increase in the damage area and damage depth can be regarded as a contribution of the newly added lightning current component. The contributions of individual lightning components to damage depth and damage area are shown in Figure 11.45.

The contribution of the lightning current component to the damage depth and damage area in Figure 11.45 can be summarized as follows:

1) Both the damage area and depth of CFRP specimens increase with the increasing numbers of applied multiple lightning current components. The damage area was as large as 27.9 cm^2 and more than 8 fiber layers (1.28 mm) were destroyed when the lightning components A, B, C, and D were applied continuously on specimen B5 in a single shot.
2) The damage depth is affected significantly by lightning current components A, B, and D, especially components A and D, rather than the low-amplitude, long-duration component C. The damage depth is 0.5 mm (specimen B1) subjected to the single lightning current component A. The damage depth increases by 0.3 mm when lightning component B is applied consecutively after lightning component A. However, when lightning component C is applied after lightning components A and B, the damage depth increases slightly by only 0.12 mm (specimen B4). Moreover, the final strike of lightning component D ($I_p = 20$ kA) results in the damage depth continuously increased by 0.35 mm in specimen B5.
3) Lightning component C with a large transfer charge and a long duration seems to have a greater influence on the damage area than the other lightning current components. The damage areas of specimens B2 and B5 increased by 219 and 390 mm^2, respectively, due to the effects of lightning current components B (transfer charge of 1.88 C) and D (transfer charge of 2.07 C). Lightning component C led the damage area to significantly increase by 1077 mm^2 in specimen B4, when it was applied continuously after lightning components A and B.
4) Although lightning current A contributes most to the damage depth and damage area (0.513 and 1104 mm^2, respectively), considering that lightning component A is applied on an intact CFRP specimen B1, the damage is mainly distributed on the upper surface of the test specimen. More than 80% of the damage area is concentrated in several upper layers, and the damage area with a damage depth greater than 0.5 mm not exceeding 150 mm^2.

11.3.3.4 Analysis for Multiple Lightning Damage of CFRP Laminates

The characteristic parameters of each lightning current component contained in the multiple continuous lightning combinations are very different. For example, there is a difference of two orders of magnitude in the current amplitudes between lightning current C and lightning current A, and the difference in the current rising rate is close to four orders of magnitude. The transfer charge of the lightning component C is approximately 20 times those of lightning components A and D. Therefore, the lightning strike process and damage effect of each lightning current component on CFRP laminate are also different.

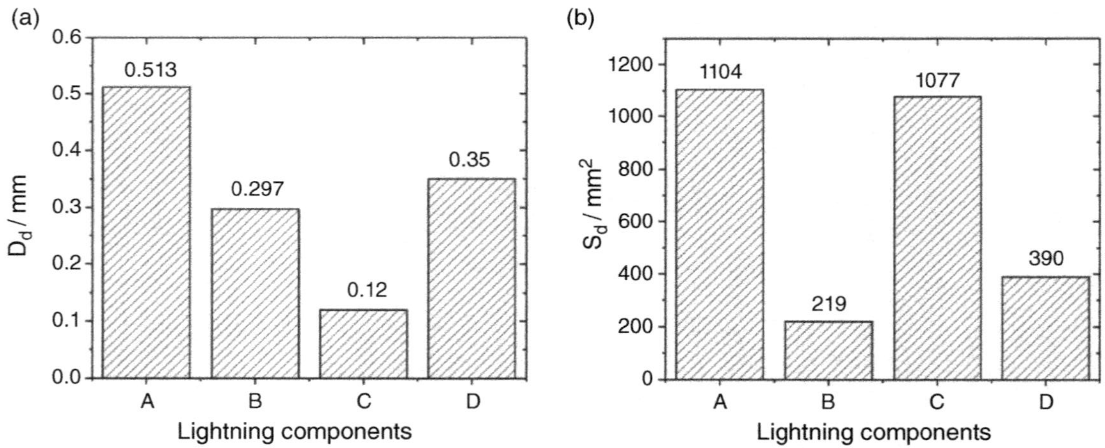

Figure 11.45 The contributions of individual lightning current components to (a) damage depth (D_d); and (b) damage areas (S_d).

11.3.3.4.1 Lightning Damage Depth When lightning current components A and D, which have a high current rise rate and high amplitude, are applied to the CFRP laminates, the huge amount of energy will be instantaneously released through the narrow discharge channel, leading to the rapid compression and violent vibration in the surrounding air. The air shock will impose tremendous pressure on the CFRP laminates, and the intensity increases for higher-magnitude impulse current and higher energy release rate. The impact effects and dielectric breakdown caused by the high-intensity electric field result in instantaneous rupture and exposure at the lightning arc attachment point. In addition, the lightning damage depth of CFRP laminates increases due to the extremely high temperature of the impulse arc. Therefore, the carbon fibers at the lightning attachment point melt and break, leading to exposure of the inner layer of the CFRP. Additionally, the current flowing through the inner carbon fiber layers further increases the damage depth. On the basis of the aforementioned analysis, we conclude that the lightning damage depth of the CFRP laminates may be closely related to the amplitude and rise rate of the lightning impulse current.

11.3.3.4.2 Lightning Damage Area When a lightning strike is applied on the CFRP laminates, the lightning current will flow along the fiber orientation inside the CFRP laminates. Joule heating is generated along the current conduction path, resulting in pyrolysis of the resin and erosion or sublimation of the carbon fibers. The lightning damage expands outward along the axis of the fracture fiber and leads to a symmetrical fan-shaped delamination damage area in CFRP laminates. In addition, the consistency between the damage area and surface temperature of CFRP specimens also indicated that the lightning damage area of a CFRP laminate should be closely related to the thermal effect of lightning components.

It should be noted that the lightning direct effect is a comprehensive effect that includes the shock wave and the overpressure effect, the thermal effect, the electromagnetic effect, and the dielectric breakdown effect, which means that the expansion of damage is not determined by a single lightning component but affected by multiple factors.

11.3.3.4.3 Comparison of the Lightning Damage of Different Test Modes of Lightning Strike

11.3.3.4.3.1 The Comparison of Lightning Damage Effect Between Single Lightning Component and Multiple Sequential Lightning Strike

In the multiple continuous sequential lightning strike experiments, since the application of the pre-pulses will influence the thermal, electrical, and mechanical properties of test CFRP laminate, the lightning damage caused by a lightning component in multiple continuous impulse sequence differs greatly from the damaging effect of the single lightning component.

For example, when lightning component D, with an amplitude of 40 kA, whose action integral is approximately half of the action integral of the lightning current A component applied to specimen B1, was applied to the CFRP laminate alone. The experimental results are shown in Figure 11.46. Visual inspection of the test specimen and ultrasonic scan results show that the lightning damage of the CFRP laminate is concentrated on the upper surface, with 1–2 layers, and the damage area is approximately 900 mm^2. It is worth noting that the type of lightning damage caused by lightning component D is similar to that of specimen B1 due to the similarity between lightning components A and D.

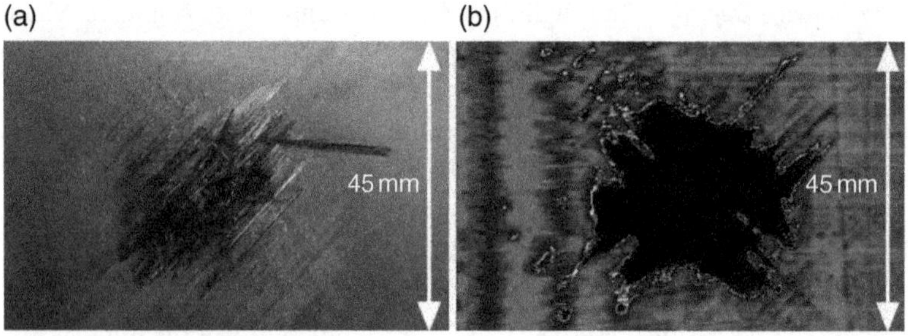

Figure 11.46 The lightning damage of the CFRP laminate under lightning component D: (a) visual inspection, (b) ultrasonic scan.

However, the damage effect caused by lightning component D in "ABCD" test mode differs greatly from the test result in Figure 11.46. From the experimental results shown in Figure 11.44d and e, it can be seen that with the continuous injection of the lightning current A, B, and C components, the damage area at depths greater than 1.5 mm is less than 100 mm^2, and most of the damage is distributed in the upper surface of the laminate, at depths less than 0.5 mm. In comparison, with the additional application of lightning component D to specimen B5, the newly generated damage area is only 390 mm^2, while the damage area at depths greater than 1.5 mm increases to more than 1000 mm^2, which means that the expansion effect of lightning component D on the damage area with deeper damage depth is quite significant. It can be explained by the fact that after the interaction of lightning components A, B, and C, most of the damage area is in an extremely high-temperature state, which can cause melting and sublimation of the carbon fibers. Additionally, the ply delamination occurs due to the internal pressure generated by pyrolysis of the resin, which will weaken the cohesive force and cause a significant decrease in the mechanical properties of the CFRP laminate. The subsequent application of lightning component D, which is accompanied by a large shockwave, overpressure effects, and electromagnetic effects, leads to an obvious expansion of the damage area at greater depths.

11.3.3.4.3.2 The Comparison of Lightning Damage Effect in "ABC" and "DBC" Test Modes The pre-pulses will influence the conducting process of the subsequent lightning current components on the test specimens. Therefore, the application sequence of multiple continuous lightning components will greatly affect the lightning strike process and damage results of CFRP laminates.

Despite the similarity of the lightning components A and D, a large difference exists in the damage degree for "ABC" and "DBC" test mode. Specifically, in test modes "DBC" and "ABC," the damage distributions of specimens B3 and B4 (Figure 11.44c and d) show that at greater depths, the damage is concentrated around the lightning current injection point in both "DBC" and "ABC" test modes. The obvious difference is that the shallow damage zone of test specimen B3 is mainly distributed in the –45° and 45° directions, while the shallow damage in test specimen B4 has no obvious directionality, which means that the damage develops in the –45°, 45°, 0°, and 90° directions.

Considering the layup of the test CFRP laminate ($[-45/45/0/90_2/-45/0/45/0/90/45/-45]_S$), it can be found that the damage distributions of test CFRP specimens in different test modes show significant differences between test specimens B3 and B4, resulting from differences in the damage depths caused by lightning components A and D. As shown by the damage of specimen B1 under lightning component A of 40 kA, the damage depth at the lightning injection point is close to the depth of the fourth carbon fiber layer with layup directions of –45°, 45°, 0° and 90° ($D_d = 0.513$ mm, the thickness of each layer is approximately 0.15 mm). Therefore, although the subsequently applied lightning component C with a low voltage does not easily penetrate the resin layer in the thickness direction, lightning component C can conduct in the carbon fiber directions of –45°, 45°, 0°, and 90° and results in a nearly circular distribution of the lightning damage area at the lightning attachment point. However, in the test mode of "DBC," both the amplitude and the energy of the lightning component D are smaller than component A, resulting in a smaller damage depth under the effect of lightning component D. Thus, lightning components B and C can only conduct in the direction of the first and second plies of the test CFRP laminate, which leads the corresponding damage area to extend along the –45° and 45° directions.

11.3.3.4.3.3 The Influence of Material Properties on Lightning Damage of Multiple Sequential Lightning Strikes The lightning damage degree of test CFRP specimens is affected by the electrical conductivity of CFRP laminates, which is the key parameter to determine the Joule heat generated during the current conducting process. For instance, the action integral of lightning component D with the amplitude of 20 kA is approximately 2×10^4 A^2s, while the action integral of lightning component C with the average current of 200 A and duration of 200 ms is only 5×10^3 A^2s. However, the thermal effect, which can be represented by the surface temperature, of lightning component C is significantly higher than that of the lightning component D. According to the calculation, the contribution of lightning current component D to the lightning damage area of specimen B5 is 13.9% with the action integral of lightning component D representing 18.55% of the total action integral of the four lightning components. Correspondingly, the contribution of the lightning current component C to the lightning damage area is approximately 39%, while the action integral of lightning component C merely accounts for 5.4% of the total amount.

The reason for the inconsistency between the action integral and damage effect of lightning components is the nonlinear characteristics of the dynamic impedance of CFRP laminates [26]. When the high-amplitude lightning component A or D is applied, the equivalent impedance of CFRP specimen is smaller, while the dynamic impedance will increase under the effect of the low-amplitude lightning component C. Therefore, the Joule heat (I^2R) generated by the lightning current component C is greater and the thermal effect is more remarkable, which lead to the significant increase in surface temperature and lightning damage area of CFRP laminate after a lightning strike.

11.3.3.5 Evaluation for Multiple Lightning Damage of CFRP Laminates

11.3.3.5.1 Evaluation for the Lightning Damage Depth The impact effects and dielectric breakdown caused by the lightning current impulses result in instantaneous rupture and exposure of both carbon fibers and resin at the lightning arc attachment point. When lightning current components A and D, which have a high current rise rate and high amplitude, are applied to the CFRP laminates, a huge amount of energy will be instantaneously released through the narrow discharge channel.

This energy release leads to rapid compression and violent vibrations in the surrounding air. The air shock will impose tremendous pressure on the CFRP surface, and the intensity will increase for higher-magnitude impulse currents and higher energy release rates. In addition, the lightning damage depth increases due to the local thermal effect of the lightning components. The mechanical properties of the CFRP laminates are seriously damaged as a result of the extremely high temperature of the impulse arc. Therefore, the carbon fibers at the lightning attachment point melt and break, leading to exposure of the inner layer of the CFRP. Additionally, the current flowing through the inner carbon fiber layers further increases the damage depth.

On the basis of the aforementioned analysis, we conclude that the lightning damage depth of the CFRP laminates may be closely related to the amplitude and rise rate of the lightning impulse current.

11.3.3.5.1.1 Influence of the Current Amplitude on the Lightning Damage Depth According to Tables 11.7 and 11.8, the experimental data of lightning current amplitude and damage depth can be obtained, as shown in Table 11.9.

Assuming that the relationship between the lightning damage depth in the CFRP specimen and the current amplitudes:

$$D_d = k_{1DP} \times I_{pA} + k_{2DP} \times I_{pB} + k_{3DP} \times I_{pC} + k_{4DP} \times I_{pD}, \tag{11.12}$$

where k_{1DP}, k_{2DP}, k_{3DP}, and k_{4DP} are the damage depth coefficients affected by the amplitudes of lightning current components A, B, C, and D, respectively; these parameters can be obtained from experimental data for specimens B1, B2, B3, and B4 in Table 11.9, as shown in formula (11.13):

$$\begin{cases} k_{1DP} = 1.231 \times 10^{-2} \left(\text{mm} \cdot \text{kA}^{-1}\right) \\ k_{2DP} = 0.8070 \left(\text{mm} \cdot \text{kA}^{-1}\right) \\ k_{3DP} = 0.8204 \left(\text{mm} \cdot \text{kA}^{-1}\right) \\ k_{4DP} = 1.345 \times 10^{-2} \left(\text{mm} \cdot \text{kA}^{-1}\right) \end{cases}. \tag{11.13}$$

Table 11.9 The current amplitude I_p and the corresponding damage depths D_d.

Specimen	I_{pA} [a] (kA)	I_{pB} [b] (A)	I_{pC} [b] (A)	I_{pD} [a] (kA)	D_d/mm
B1	41.67	—	—	—	0.513
B2	41.28	374.0	—	—	0.810
B3	—	362.9	163.0	35.21	0.900
B4	41.11	364.3	158.4	—	0.930
B5	40.81	375.4	170.3	19.89	1.280

[a] I_{pA}, I_{pD}: amplitude of current components A and D.
[b] I_{pB}, I_{pC}: average values of current components B and C.

Thus, the estimated result of the lightning damage depth in specimen B5 can be obtained from formula (11.12). The deviation between the calculated result (1.212 mm) and the test result (1.28 mm) is approximately 5.27%.

11.3.3.5.1.2 Influence of the Rise Rate on the Lightning Damage Depth The relationship between the rise rate of the lightning components and the damage depth of the CFRP specimens can be obtained using a method similar to that described above.

$$D_{\mathrm{d}} = k_{1\mathrm{DR}} \times \frac{\mathrm{d}i(\mathrm{A})}{\mathrm{d}t} + k_{2\mathrm{DR}} \times \frac{\mathrm{d}i(\mathrm{B})}{\mathrm{d}t} + k_{3\mathrm{DR}} \times \frac{\mathrm{d}i(\mathrm{C})}{\mathrm{d}t} + k_{4\mathrm{DR}} \times \frac{\mathrm{d}i(\mathrm{D})}{\mathrm{d}t}, \tag{11.14}$$

where, the damage depth coefficients affected by the rise rate are

$$\begin{cases} k_{1\mathrm{DR}} = 3.295 \times 10^{-10}\left(\mathrm{mm \cdot s \cdot A^{-1}}\right) \\ k_{2\mathrm{DR}} = 1.835 \times 10^{-7}\left(\mathrm{mm \cdot s \cdot A^{-1}}\right) \\ k_{3\mathrm{DR}} = 4.442 \times 10^{-6}\left(\mathrm{mm \cdot s \cdot A^{-1}}\right) \\ k_{4\mathrm{DR}} = 1.985 \times 10^{-10}\left(\mathrm{mm \cdot s \cdot A^{-1}}\right) \end{cases}. \tag{11.15}$$

On the basis of formulas (11.14) and (11.15), the calculated result for the lightning damage depth for specimen B5 is 1.197 mm, which deviates 6.48% from the measured value.

The contributing percentage of each lightning component to the lightning damage depth, as estimated using formulas (11.12) and (11.14), is shown in Table 11.10.

The lightning damage depth is affected by lightning current components A, B, and D, which are characterized by high amplitude and a high-rise slope, but not by lightning component C. Furthermore, the two sets of estimated results determined using the different evaluation parameters are consistent. Therefore, the lightning damage depth for CFRP laminates can be evaluated using both the amplitudes and raise rates of the applied lightning components.

11.3.3.5.1.3 Evaluation for the Lightning Damage Area When a lightning strike is applied to the CFRP laminates, the lightning current will flow along the carbon fibers inside the CFRP laminates. Joule heating is generated along the current conduction path, resulting in pyrolysis of the resin and erosion or sublimation of the carbon fibers. The lightning damage expands outward along the transverse fiber direction and leads to a symmetrical fan-shaped delamination damage area in CFRP laminates. Therefore, the lightning damage area of a CFRP laminate should be closely related to the waveform parameters of the

Table 11.10 The contributing percentages of lightning components to the lightning damage depth.

Lightning Components	A	B	C	D
Using formula (11.12)	41.45%	24.99%	11.52%	22.05%
Using formula (11.14)	44.05%	25.69%	9.03%	21.23%

Table 11.11 The current electrical action integral W and the corresponding damage area S_d.

Specimen	W_A [a] (A^2s)	W_B [a] (A^2s)	W_C [a] (A^2s)	W_D [a] (A^2s)	S_d (mm^2)
B1	79225	—	—	—	1104
B2	81659	573.9	—	—	1323
B3	—	638.7	4989.7	47968	1453
B4	78359	643.7	4757.7	—	2400
B5	76353	683.5	5548.1	18804	2790

[a] W_A, W_B, W_C, W_D: electrical action integral of current components A, B, C, and D, respectively.

electrical action integral and the transfer charge, which can be used to determine the thermal effect of the lightning current components.

11.3.3.5.1.4 The Influence of the Electrical Action Integral on the Lightning Damage Area
Table 11.11 shows electrical action integral experimental data used to evaluate the lightning damage area in the CFRP specimens.

Here, we assume that the relationship between the lightning damage area of the CFRP specimen and the electrical action integral of the lightning current components is:

$$S_d = k_{1SW} \times W_A + k_{2SW} \times W_B + k_{3SW} \times W_C + k_{4SW} \times W_D. \quad (11.16)$$

The damage area coefficients k_{1SW}, k_{2SW}, k_{3SW}, and k_{4SW} can be calculated:

$$\begin{cases} k_{1SW} = 1.3935 \times 10^{-2} \left(mm^2 A^{-2} s^{-1}\right) \\ k_{2SW} = 3.2248 \times 10^{-1} \left(mm^2 A^{-2} s^{-1}\right) \\ k_{3SW} = 2.3131 \times 10^{-1} \left(mm^2 A^{-2} s^{-1}\right) \\ k_{4SW} = 1.9358 \times 10^{-3} \left(mm^2 A^{-2} s^{-1}\right) \end{cases}. \quad (11.17)$$

The deviation between the calculated result (2604 mm^2) and the test result (2790 mm^2) is approximately 6.66%. Notably, the electrical action integral of lightning component D with an amplitude of 20 kA is approximately 2×10^4 A^2s, whereas the electrical action integral of lightning component C with an average current of 200 A and a duration of 200 ms is only 5×10^3 A^2s. However, the thermal effect, which can be represented by the resin delamination area of lightning component C, is more significant than that of lightning component D. According to the fitting formula (11.16), the contribution of lightning component C to the damage area of specimen B5 is approximately 49.28%, whereas the electrical action integral of lightning component C only accounts for 5.47% of the total electrical action integral for the four lightning components. The contribution of lightning current component D to the damage area is merely 1.40%, although the electrical action integral of lightning component D represents 18.55% of the total electrical action integral.

The reason for the inconsistency between the electrical action integral and damage effect of lightning components is the nonlinear characteristics in the dynamic impedance of the CFRP laminates [26]. When high-amplitude lightning components A and D are applied, the equivalent impedance of the CFRP specimen is smaller, whereas the dynamic impedance increases due to the effect of low-amplitude lightning component C. Therefore, the Joule heating (I^2R) induced by lightning current component C is greater, and the thermal effect is more significant, leading to a substantial increase in the lightning damage area of the CFRP laminate in specimens B3, B4, and B5.

11.3.3.5.1.5 Influence of the Transfer Charge on the Lightning Damage Area On the basis of the nonlinear dynamic impedance of the CFRP material, the transfer charge of lightning current component C ($Q_C \approx 30$ C), which is far greater than that of lightning current components A ($Q_A \approx 3$ C), B ($Q_B \approx 1.8$ C), and D ($Q_D \approx 2$ C), should be considered. The lightning damage area is also affected by the transfer charge:

$$S_d = k_{1SQ} \times Q_A + k_{2SQ} \times Q_B + k_{3SQ} \times Q_C + k_{4SQ} \times Q_D. \tag{11.18}$$

The damage area coefficients k_{1SW}, k_{2SW}, k_{3SW}, and k_{4SW} are calculated as follows:

$$\begin{cases} k_{1SQ} = 339.9\left(\text{mm}^2\text{C}^{-1}\right) \\ k_{2SQ} = 120.3\left(\text{mm}^2\text{C}^{-1}\right) \\ k_{3SQ} = 37.28\left(\text{mm}^2\text{C}^{-1}\right) \\ k_{4SQ} = 38.69\left(\text{mm}^2\text{C}^{-1}\right) \end{cases}. \tag{11.19}$$

On the basis of formulas (11.18) and (11.19), the deviation between the calculated result (2563 mm^2) and the test result (2790 mm^2) for the lightning damage area of specimen B5 was approximately 8.15%.

The contributing percentages for each lightning component to the lightning damage area were calculated by the estimation formula. The results are shown in Table 11.12.

As shown in Table 11.12, the lightning damage area in the CFRP laminate is affected mainly by lightning current component C. The estimated results obtained separately from the electrical action integral and transfer charge exhibit great agreement.

The increase in the lightning damage area of the CFRP laminate is related to both the electrical action integral and the transfer charge from the applied current impulse. However, for a short-duration and high-amplitude lightning current impulse, such as lightning components A and D, the action integral is the key parameter for assessing the ability to increase the damage area. Correspondingly, the amount of transfer charge can be used to evaluate the lightning damage area when a lightning component that has a long duration and low average amplitude is applied. Therefore, both the electrical action integral and the transfer charge should be comprehensively considered when estimating the lightning damage area in a CFRP laminate, especially in experiments involving multiple continuous sequential lightning strikes.

Table 11.12 The contributing percentages of the lightning components to the lightning damage area.

Lightning Components	A	B	C	D
Using formula (11.16)	40.86%	8.46%	49.28%	1.40%
Using formula (11.18)	40.59%	8.91%	47.38%	3.12%

In addition, the calculation results presented in Table 11.12 show that in the "ABCD" test mode, the damage area of CFRP is mainly influenced by lightning components A and C, whereas component D has little effect. Because the lightning damage area is calculated as the sum of the area with defects in CFRP laminates, the increase in the damage area under multiple continuous sequential lightning strikes only represents the in-plane extension of the damage area to the undamaged area under the influence of the additional lightning component. However, the lightning damage effects in CFRP laminates involve three-dimensional damage expansion in both the in-plane and in-depth directions.

Therefore, the surface damage effect caused by lightning component D in the "ABCD" test mode can be comprehensively analyzed:

1) According to Table 11.12, the contribution of lightning component A to the damage area accounts for approximately 40%, but it can be seen from the damage distribution in Figure 11.44a that more than 80% of the damage area caused by lightning component A is concentrated in the first and second layers, and the damage area with a damage depth greater than 0.5 mm is approximately 150 mm^2.
2) In the "ABCD" test mode, the damage depth reaches 0.8 mm under the lightning damage effect of pre-sequence lightning components A and B. However, when lightning component C is injected, the applied voltage in the thickness direction is low due to the low amplitude of lightning component C, which leads to its limited penetration ability in the thickness direction. Therefore, a larger proportion of the injected current will be conducted along the in-plane fiber sheet. The Joule heat generated during current in-plane dispersion will cause interlaminar resin pyrolysis and the expansion of the damage area in the upper surface (depth less than 0.8 mm). In the "ABC" test mode, the damage area with a damage depth greater than 1.5 mm is less than 100 mm^2.
3) When lightning component D is applied, both the discharge voltage and the current amplitude are high, so the potential gradient generated in the thickness direction is large. On the one hand, under the effect of higher-amplitude voltage and current, the resistance in the depth direction of the laminated plate is smaller, which makes it easier for lightning component D to shunt into the depth direction, causing an increase in the damage depth. On the other hand, the high potential gradient will cause the breakdown of the interlayer resin, which allows lightning component D to be conducted in deeper lamination, resulting in the expansion of the damage area in deep layers. It is worth noting that after the interaction of lightning components A, B, and C, most of the damage area exists in an extremely high-temperature state, causing melting and sublimation of carbon fibers. The ply delamination caused by the internal pressure generated by resin pyrolysis will weaken the cohesive force and the mechanical properties of CFRP laminates. Lightning component D is accompanied by a very large shockwave, as well as overpressure and electromagnetic impacts, which lead to an obvious expansion of the damage area at greater depths.

The deep-layer damage caused by lightning component D is mostly distributed beneath the damage area of the previously applied lightning components. Although the newly generated damage area is limited, Figure 11.44e shows that with the addition of lightning component D, the damage area with a damage depth greater than 1.5 mm increases to more than 1000 mm^2.

Analysis of the damage area can deepen our understanding of lightning damage effects from one side, but it cannot completely assess the lightning damage degree for CFRP laminates without considering the lightning damage depth. Especially for specimens that have suffered from the pre-sequence impulse, damage extension at different depths is also an important factor for evaluating the damage effect of the newly added lightning component.

11.3.3.5.2 Summary of the Evaluation Methods The damage depth of the test CFRP laminates can be evaluated using the applied lightning waveform parameters of current amplitudes and raise rates. The amount of transfer charge and the action integral represent the thermal effects that are used to assess the damage area. Using the waveform parameters as the evaluation factors, the damage depth coefficients and damage area coefficients correspond to lightning current components can be calculated, and the calculation formulas for the lightning damage depth and area were deduced. The deviations between the calculated and experimental values for lightning damage depth and area were less than 7 and 9%, respectively.

However, the multiple-factor lightning direct effect is a comprehensive process that includes the shockwave and overpressure effect, the thermal effect, the electromagnetic effect, and dielectric breakdown effect. On the one hand, the expansion of damage is affected by multiple factors. For instance, the damage area is not only related to the Joule heat generated, which is, in turn, related to the transfer charge and action integral, during the current conduction process; it is also affected by the shockwave, overpressure effect, and electromagnetic force, which correlate with the current amplitude and raise rate. On the other hand, the lightning waveform parameters are inherently interrelated. For example, the amount of transfer charge and the action integral increase as the current amplitude increases.

Although in this section, only typical parameters of the lightning waveforms were selected to assess the lightning damage degree, a lightning damage evaluation method with higher accuracy and universality, which incorporates more comprehensive waveform parameters with different weights, should be developed in future studies.

11.4 The Simulation of Lightning Strike-Induced Damage of CFRP

11.4.1 Overview of Lightning Damage Simulation Researches

The lightning experiment is the most direct and effective method to study the lightning damage of CFRP composites. However, experimental investigations are limited by the realization of the extremely high-intensity and transient duration of lightning current in lightning tests, especially in multiple consecutive lightning strike tests. Moreover, the absence of a unified standard for the direct effect test method of CFRP composites reduces the comparability between different experimental studies. Simulation research became a fast, convenient, and effective way to carry out the lightning damage research of CFRP composite.

Finite element analysis (FEA) has been widely adopted to quantitatively assess lightning damage with enhanced effectiveness. In general, prior simulation studies regarding the lightning damage failure of CFRP composite have been carried out via thermoelectrical analysis based on the current Joule heating effect. Metwally et al. [58] first introduced the thermoelectrical FE model to assess the temperature rise of CFRP under the current Joule heating effect, although the constant material parameters used in the simulation model resulted in large deviations between the calculations and experimental results. Abdelal and Murphy [59] proposed introducing a temperature-dependent material parameter system into the coupled thermoelectrical model and discussed the lightning damage distribution in CFRPs. However, the deviation between the transient decomposition of resin and the resin ablation in thermal gravimetric analysis resulted in inaccurate simulation results. Ogasawara et al. [27] introduced a temperature criterion to evaluate lightning damage. Guo et al. [60] evaluated and compared lightning in-plane and in-depth damage to CFRPs using temperature- and pyrolysis-dependent models. The pyrolysis damage area was found to be smaller than in the experimental results, indicating the limitation of the pyrolysis criterion for lightning damage area assessment.

The electrical conductivity of CFRP material is the key parameter affecting the evolution and distribution of lightning damage in the coupled thermoelectric simulation model. Nevertheless, the conductive property of CFRP composite under an applied lightning current cannot be simply equated with its DC conductivity. The utilization of DC conductive properties could result in unreliable simulation results, especially in terms of the in-depth conductivity; the DC conductivity is 3–4 orders of magnitude smaller than the equivalent conductivity of CFRP under a lightning impulse [19, 26], which greatly suppresses the penetration of lightning energy in the depth direction of the laminate.

In addition to the current Joule heating effect, the acoustic shockwave, lightning arc ablation, electromagnetic force, and thermal stress also exacerbate the propagation of lightning damage in CFRP composite. In the simulation study of Millen and Murphy [61], the output temperature profile from the thermoelectric simulation was passed to a thermomechanical model, which aimed to understand the delamination behavior considering strain and heating rates and to assess the damage with different boundary conditions. Kamiyama, et al. [62] used a coupled thermoelectric model with cohesive elements and thermal decomposition for thermal stress analysis to simulate the delamination behavior. Fu et al. [63] developed a dynamic FE model with a plasma expansion-induced pressure model to predict the intralaminar damage and interlaminar delamination of CFRP composite. Moreover, simulations have been used to study the dynamic motion characteristics and thermal ablation effects of the attached lightning arc. Millen et al. [64] combined a magnetohydrodynamics (MHD) FE multiphysics model employed to simulate the lightning arc and an FE thermal-electric modeling approach, applying the MHD-predicted current density to the thermal-electric damage model. In the work of Chen et al. [65], seven interpolation methods for coupling were tested to transfer the plasma information on the fluid-solid interface in a thermal-electric model.

The lightning damage of CFRP composite can't be simply equated with the superposition of one or several damage modes. Under the complex lightning strike effects, the lightning damage in the same damage zone of a CFRP laminate often contains a variety of damage modes, such as fiber breakage, sublimation, resin pyrolysis, and ply-lift. The difference between calculated values and experimental results should be attributed to the damage types other than the damage induced by Joule heating. However, this difference has not been considered in previous simulation studies. Most numerical models, however, presume lightning failure under specific conditions without fully incorporating the actual complex lightning conditions. Moreover, the proposed failure mechanism should be demonstrated by experimental evidence, but the lack of quantitative analysis of lightning damage induced from different effects has become a bottleneck restricting the future development of lightning damage mechanisms.

In Section 11.3.2, the lightning damage zones were classified by inspecting the damage morphology and microstructural characteristics, and the lightning damage was quantitatively investigated based on an analysis of the internal damage behavior of CFRP during a lightning strike. In this section, a coupled thermoelectric model based on the dynamic conductive properties of the composite material was established to simulate the lightning damage. Then, a reasonable application range of the simulation model was evaluated to verify the rationality of the damage criterion and the accuracy of the calculated results. The simulation study could provide an enhanced understanding of the direct lightning damage mechanism and the improvement to coupled multi-damage effect studies.

11.4.2 Establishment of the Coupled Thermal-Electrical Model

11.4.2.1 Finite Element Model

The lightning damage model for CFRP composite was established in Abaqus using coupled thermal-electrical-structural analysis. The subroutine USDFLD is utilized to calculate the changing material status based on the simulated temperature field. The material parameters are updated according to pyrolysis degree, which contributes to the realization of the coupled thermal-electrical analysis. A finite element (FE) model, with the dimension of 150 × 100 × 3.6 mm and quasi-isotropic layup angles of $[45/-45/0/90]_{3S}$, is defined in local coordinates. The boundary conditions of the model are set in accordance with the experimental setup shown in Figure 11.47.

The short edge surfaces of the laminate model are completely fixed, complying with the fixed boundary condition (ENCASTRE). The electric potential is presumed to be 0 V at the short-edge surfaces. Joule heat dissipates from all laminate surfaces into the atmosphere by convection and radiation with the heat transfer coefficient of $1\,W/(m^2 \cdot s \cdot K)$ and thermal emissivity of 0.9. The ambient temperature and the initial temperature of the model are set to 25 °C.

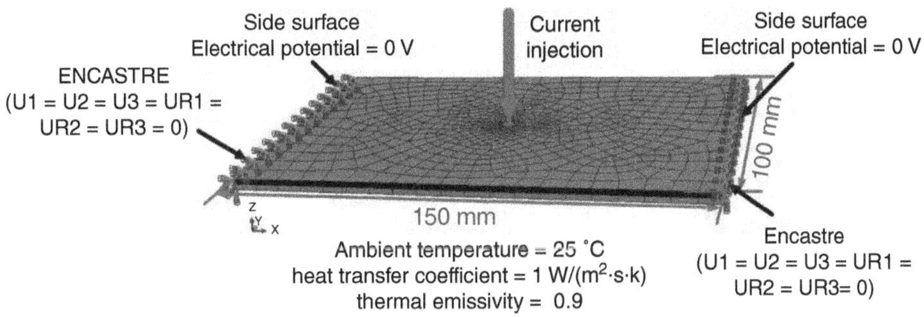

Figure 11.47 Coupled thermal-electrical lightning damage simulation model for CFRP composite.

11.4.2.2 Simulated Lightning Component A

The current load comes from the discretized impulse waveform measured in experiments. Instead of a concentrated current, which makes the temperature profiles of the central area depend greatly on the meshing method [59], a surface current was adopted to better simulate real-life lightning conditions. The radius of the lightning arc column is generally 1–10 mm [66, 67]. A straight cylindrical channel was proposed by Braginskii [68]:

$$R(t) = \alpha \rho_0^{-1/6} \left[I(t) \right]^{1/3} t^{1/2} \tag{11.20}$$

Where, $R(t)$ is the channel radius; $\rho_0 = 1.29\,\text{kg/m}^3$ is the air density; $\alpha = 0.294$ is a constant; $I(t)$ is the instant current; and t is time. Based on a series of simplifications, we assume that the lightning current is injected into the model through a central circular area with a radius of

$$R(t) \approx 4.34 \times 10^{-4} \times I_{\text{peak}}^{1/3} \tag{11.21}$$

Within a cylindrical channel, current density distributes according to a Gaussian-shaped profile [69] as follows:

$$J(r,t) = I(t) / \left(\int_0^{2\pi} \int_0^{R(t)} r e^{cr^2} \mathrm{d}r \mathrm{d}\theta \right) \cdot e^{cr^2}, \qquad r \le R(t) \tag{11.22}$$

where $J(r, t)$ is the lightning current density, and $c = \ln(0.1)/(0.55 \cdot R(t))^2$ is a constant.

11.4.2.3 Pyrolysis Degree Calculation

Thermal degradation of the resin is essentially a reaction with chemical kinetic characteristics, which can be described by an n-order chemical kinetic equation [31, 70]. Because the phase transition in the resin is a non-isothermal process, the numerical expression for the pyrolysis degree C can only be derived based on the Scheil additivity rule, which discretizes the continuous temperature variation process into a stepped isothermal process:

$$\begin{cases} C_1 = 1 - \left[(n-1) A \exp(-Q / \mathrm{R} T_1) \Delta t + C_0 \right]^{1/(1-n)} \\ C_i = 1 - \left[(1 - C_{i-1})^{1-n} + (n-1) A \exp(-Q / \mathrm{R} T_i) \Delta t \right]^{1/(1-n)} & i \ge 2 \end{cases} \tag{11.23}$$

11.4.2.4 Dynamic Conductive Properties

Through our previous studies, the conductive characteristics in the longitudinal, transverse, and in-thickness directions depend highly on the parameters of the applied lightning current [26]. Therefore, the equivalent conductivity of CFRP composite subjected to lightning component A should be calculated by extrapolating its conductivities under lightning component A at low intensity. Via numerically fitting of experimental results, the anisotropic conductivities increase with increasing lightning current peak according to the following power-law function (Eq. 11.24):

$$\begin{cases} \sigma_{xx} = 0.58198 \times I_{\text{peak}}^{0.33507} \\ \sigma_{yy} = 0.37959 \times I_{\text{peak}}^{0.36098} \\ \sigma_{zz} = 0.03024 \times I_{\text{peak}}^{0.66395} \end{cases} \tag{11.24}$$

For the first time, the dynamic conductivities were introduced into the coupled thermal-electrical model to simulate the real-life conductive conditions of CFRP composite under a lightning current impulse.

11.4.2.5 Pyrolysis-Dependent Material Parameters

The pyrolysis of the resin matrix, a chemical reaction process in which the molecular structure changes, is an important index for characterizing the variation in material properties. The change in material properties depends highly on the pyrolysis degree, which should be taken into account. The parameters of CFRP, including the electrical conductivity $\sigma^E = \sigma_0{}^E [1-C(T, t)] + \sigma_1{}^E$, specific heat $c_p = c_{p0}[1-C(T, t)] + c_{p1}$, thermal conductivity $k^E = k_0{}^E [1-C(T, t)] + k_1{}^E$, and density $\rho = \rho_0[1-C(T, t)] + \rho_1$, are assumed to change with the pyrolysis degree $C(T, t)$ according to the mixed parameter rule. In the above equations, $\sigma_0{}^E$, c_{p0}, $k_0{}^E$, and ρ_0 are material parameters when the pyrolysis degree $C = 0$, whereas $\sigma_1{}^E$, c_{p1}, $k_1{}^E$, and ρ_1 are material parameters when $C = 1$. The material parameters, including the thermal, electrical, and mechanical parameters, are updated according to the pyrolysis degree, which contributes to the realization of the coupled thermal-electrical analysis.

11.4.3 Simulation Physical Fields of Lightning Current on CFRP Laminates

11.4.3.1 Temperature and Pyrolysis Fields

The development of ablation damage is closely related to the physical field distributions inside the CFRP laminate. When a lightning impulse (40 kA) is applied, the calculated temperature and pyrolysis degree at the central node in different layers vary with the injection time of the lightning current, as shown in Figure 11.48.

As shown in Figure 11.48a, the highest temperature (approximately 3000 °C), which is close to the sublimation temperature of carbon fiber (3316 °C), appears in the first and second layers in the CFRP model. The simulation model focuses on the current injection process, while the temperature recorded by the infrared thermal imager represents the status of the specimen during the cooling process. Thus, the simulated maximum temperature (exceeding 3000 °C) corresponds to the highest value that the CFRP specimen can reach during the entire lightning process, which is thought to be much higher than the measured temperature (671.0 °C in Figure 11.29b) in the lightning test. At the peak time of the lightning impulse ($t = 16.9\,\mu s$), the temperatures in the upper seven layers all exceed 600 °C, and the resin matrix in these several layers pyrolyzes rapidly. Specifically, at $t = 16.9\,\mu s$, the pyrolysis degrees of the central nodes in the upper six layers all exceed 0.8. With the subsequent injection of lightning current, the resin in deeper layers pyrolyzes gradually to severe degrees, as shown in Figure 11.48b.

By defining the normalized temperature $T^* = T/T_{max}$, where T_{max} is the highest simulated temperature at that moment, the pyrolysis degree can be compared with the normalized temperature T^*. The comparison results in Figure 11.49 show that in the same layer, the pyrolysis degree lags behind the evolution of the normalized temperature. When the temperature of the lightning injection center in the ninth layer rises to 450 °C, the pyrolysis degree at this point is only 0.04. This calculated result indicates that the local temperature reaching the critical temperature is only a necessary requirement for resin pyrolysis. An increase in the pyrolysis degree is achieved based on a continuously high temperature, while a transient high temperature only has a limited contribution.

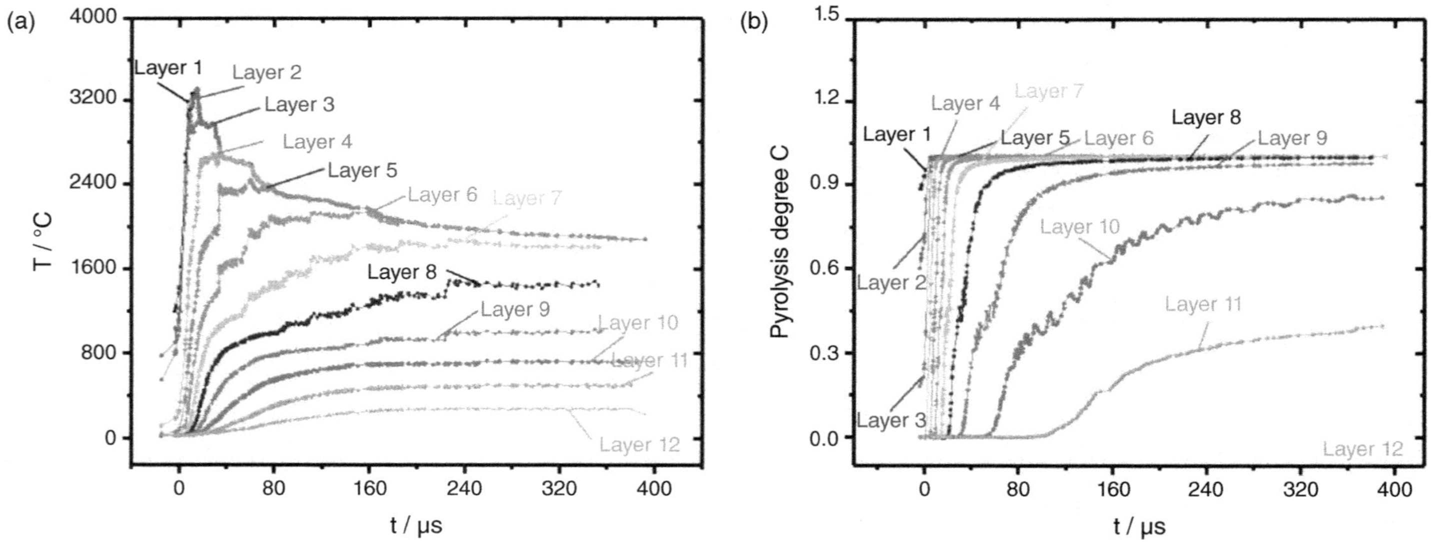

Figure 11.48 (a) Temperature; and (b) pyrolysis degree at the central node in different layers of the CFRP model.

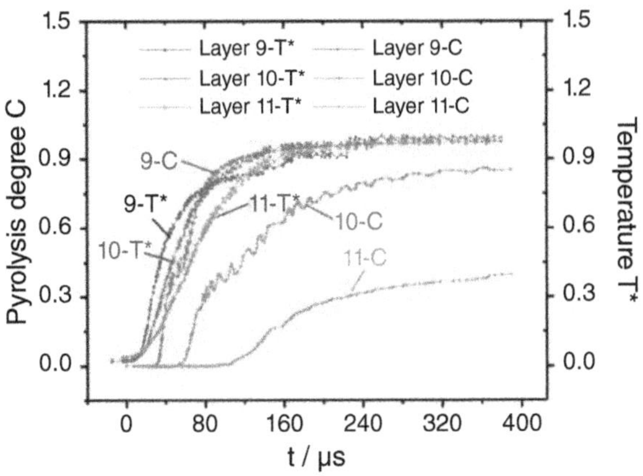

Figure 11.49 Pyrolysis degree *C* and normalized temperature T^* at the center node of the CFRP model.

11.4.3.2 Mechanical Analysis

The structure of carbon fibers is relatively stable at temperatures below 400 °C; thus, the thermal deformation characteristics of CFRP composites depend primarily on the thermal expansion behavior of the resin matrix. In the coupled thermal-electrical model, the temperature difference $\Delta T(x, y, z)$ occurs and causes structural thermal expansion. According to the basic theory of elastic mechanics of finite element analysis, the strain contours due to thermal expansion can be calculated in the simulation model according to the temperature distribution (Figure 11.50).

The thermal effect of the lightning current results in a rapid increase in thermal expansion and an expansion of the strain area extending from the lightning attachment point to the surrounding area. At the current peak time ($t = 16.9\,\mu s$), the maximum strain on the surface reaches 1.743×10^{-1} mm. In the decreasing stage of the lightning current, the growth of thermal expansion slows. From the strain distribution in the specimen cross section at $t = 20\,\mu s$ (Figure 11.50b), it can be found that the thermal expansion strain of the composite material occurs mainly in the E33 direction, suggesting the flexure of CFRP composite in the thickness direction due to the large lightning impact, which could cause delamination damage to the composite material.

11.4.4 Simulated Lightning Damage Results

11.4.4.1 Numerical Criterion for Lightning Damage

Because temperature characteristics of the material can, to some extent, reflect the degree of ablation damage, pyrolysis gas pressure, and thermal expansion stress, the numerical model established is based on the coupled thermal-electrical process, which concerns the damage behavior caused by the Joule heating effect of the lightning current. According to the characteristics of different damage regions, it is speculated that the generation and evolution of thermal damage in damage Regions A, B, and D, corresponding to damage area S_P, can be simulated by the coupled thermal-electrical model. The lightning damage area of Region C (damage area S_B) induced by the mechanical impact of pyrolysis gas,

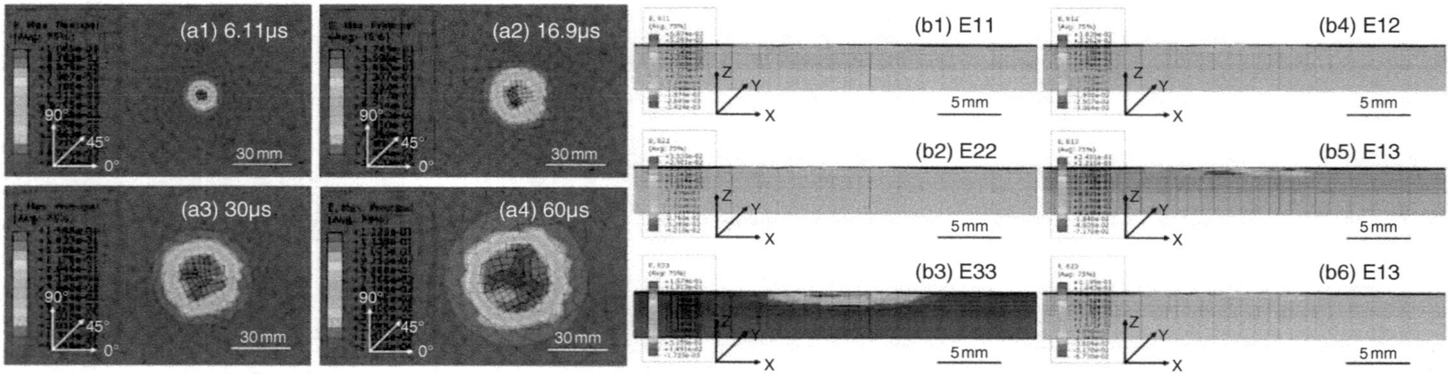

Figure 11.50 Strain contours (a) on the surface; and (b) in a specimen cross section during the lightning strike.

however, is not included in the simulation scope. Many previous studies have presented simulation results based on coupled thermal-electrical models to study lightning damage behaviors, where the temperature characteristics, including temperature and pyrolysis fields, have been commonly used to evaluate the damage distribution in thermal-electrical damage models [27, 31, 60, 71, 72]. The temperature and pyrolysis degree fields can be used to analyze the lightning damage of CFRPs.

11.4.4.1.1 Temperature Field Criterion Determining the initial pyrolysis temperature of the resin is the most critical step when using the temperature field to evaluate lightning damage of CFRPs. The recommended values for the estimated parameters in Eq. (11.23) are $n = 3.5$, $A = 3 \times 10^{15}\ s^{-1}$ and $Q = 180\,kJ/mol/K$ [73]. The calculated pyrolysis degree results reveal that the initial temperature of resin pyrolysis increases with increasing heating rate, namely the larger the heating rate is, the slower the resin pyrolysis. The rate of the temperature increase of CFRPs under a lightning strike is extremely high (10^7–10^8 °C/s); thus, the onset temperature of lightning ablation damage is larger than that in ordinary situations (200 °C). Additionally, the oxygen entering the lightning attachment region can promote resin pyrolysis, while carbon fibers will slow down the resin ablation process; thus, Eq. (11.23) requires proper modifications to simulate the lightning damage effect in CFRPs more reasonably.

11.4.4.1.2 Pyrolysis Degree Field Criterion The change in the pyrolysis degree is closely related to the dynamic heating process. For example, when the heating rate increases from 100 to 10^6 K/min, the pyrolysis degrees of the resin material at the same temperature are different, such as 0.98 and 0.26 at 450 °C. Using only the temperature field to define the damage region is insufficient for thoroughly analyzing the lightning ablation damage of CFRPs; thus, the pyrolysis field was also introduced as a supplement to lightning damage assessment in this numerical model, which is helpful for understanding the lightning damage mechanism in the time and space domains.

11.4.4.2 In-Plane Lightning Damage Evaluation

The in-plane damage in the temperature model is evaluated by the superposition of the 450 °C isotherm and the areas encircled by the $C = 0.9$ contour line. The simulated damage profiles based on the temperature and pyrolysis criteria are plotted in yellow and red dotted lines in Figure 11.51.

The in-plane damage areas are approximately elliptical in shape, without obvious anisotropy compared with the numerical results in existing studies [27, 29, 31, 59, 60, 65, 70–75], which can be ascribed to the similar split-flow ratios of the lightning current in the different in-plane directions due to the similarity in the dynamic conductivities in the longitudinal and transverse direction of the fiber layup [26]. The pyrolysis damage area is obviously smaller than the temperature damage area. To determine the appropriate damage criterion, the simulated in-plane damage obtained by the pixel statistics method using the temperature and pyrolysis criteria are compared with the experimental results in Figure 11.52.

The simulated damage areas obtained based on the temperature criterion exhibit good agreement with the thermal damage area S_P, with a maximum deviation of 11.13%, while they are quite different from the total damage area S_Z, which is highly consistent with the

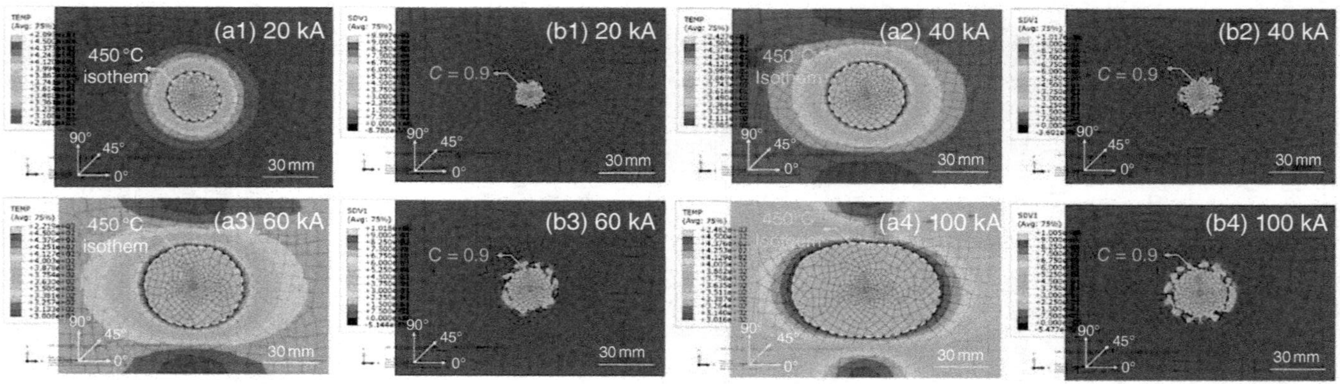

Figure 11.51 Simulated in-plane lightning damage evaluated by the temperature field criterion (450 °C, yellow dotted line): (a1) I_p = 20 kA, (a2) I_p = 40 kA, (a3) I_p = 60 kA, and (a4) I_p = 100 kA; pyrolysis degree criterion (C = 0.9, green dotted line): (b1) I_p = 20 kA, (b2) I_p = 40 kA, (b3) I_p = 60 kA, and (b4) I_p = 100 kA.

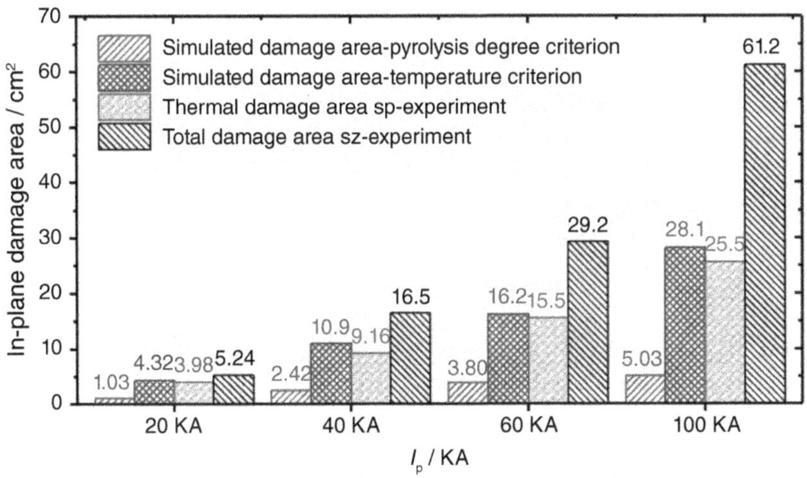

Figure 11.52 Comparison of experimental damage areas and simulated damage areas based on the temperature criterion and pyrolysis degree criterion.

simulation scope analyzed above. The temperature change is the foundation of thermal damages, including the massive ablation of both resin and fibers in the lightning-attached area, the generation of pyrolysis gas and the increased thermal expansion. Considering the great consistency in the simulated damage areas and experiment data in Figure 11.52, the temperature field can be regarded as a well-founded indicator to evaluate the in-plane lightning damage. On the contrary, the pyrolysis degree field criterion underestimates the in-plane damage, with the largest deviation of –80.28%. This result can be explained by the inference that material pyrolysis has not finished at the end of the lightning current injection process, as indicated by the surface temperature not falling below 200 °C until 7 s after current injection (Figure 11.29b3). In addition, there exist other probable causes for damage propagation, such as thermal expansion stress and extension of internal pyrolysis gas, suggesting that the pyrolysis area can only reflect a portion of the in-plane damage.

Considering that the partition of the damage region contains artificial errors, to verify the simulation scope of the coupled thermal-electrical model and the universality of the temperature criterion, woven fabric/epoxy laminates were selected for lightning damage tests, and the same temperature criterion (450 °C) was utilized, as shown in Figure 11.53. The specifications of the woven/fabric epoxy laminates are the same as described in Section 11.2.2.1.

The circular distribution of the damage area on the woven structure laminates is closely linked to their quasi-isotropic conductive properties. The lightning damage area can also be classified into three different damage zones similar to those in quasi-isotropic specimens. The damage near the lightning attachment point (Zone I) is mainly an ablation pit. The outermost yellow dotted line is the boundary where the first layer intersects the second layer. The second layer in damage Zones II and III is both damaged and exposed, while a certain difference exists between the damage properties in these two zones. The surface in Zone III is grayish black, whereas the apparent color of Zone II is relatively white. The outer ring part between the red dotted line and the yellow dotted line belongs to the

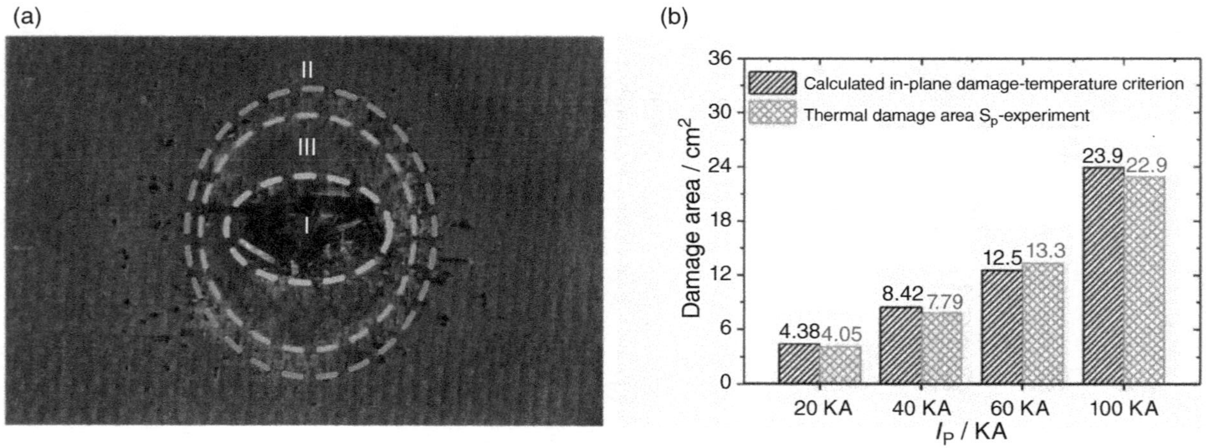

Figure 11.53 Lightning test results of woven fabric/epoxy laminates: (a) damage zones of the damaged specimen under lightning component A of 100 kA; (b) comparison of simulation and experimental results of the in-plane damage area (temperature criterion of 450 °C)

laminate peeled-off damage area (Zone II). Different from the quasi-isotropic CFRPs, the restricted laminate peeled-off damage results from the unique texture of the woven fabric/epoxy laminate: when it is struck by a lightning impulse, the movement of internal fibers is greatly hindered by adjacent fiber bundles. The inner small ring represents damage Zone III induced by the lightning thermal effect, and the external black powder is the product of resin carbonization. The comparison in Figure 11.53b indicates that the simulated lightning damage area exhibits great consistency with the experimental thermal damage area S_P, with a maximum deviation of 8.15%. The results in Figures 11.52 and 11.53 reveal that the established coupled thermal-electrical CFRP model with the temperature field criterion can well simulate the thermal damage.

11.4.4.3 In-Depth Lightning Damage Evaluation

With the same temperature criterion and pyrolysis degree criterion (450 °C and $C = 0.9$, respectively), the simulated in-depth lightning damage, as well as a comparison of the simulated and experimental damage depths, are presented in Figure 11.54.

The damage with the greatest depth is located directly below the lightning attachment point. The comparison profile shows that the in-depth damage based on the temperature criterion is larger than the test results, presenting greater errors than the pyrolysis degree model, and the simulated results using the pyrolysis degree criterion exhibit better consistency with the experimental results. The temperature of the specimen changes transiently and varies nonmonotonically, as shown in Figure 11.48a, which implies that the duration of the high-temperature state extending to deep layers is probably too short to induce detectable in-depth damage. In comparison, the deviations in the simulated damage depths using the pyrolysis criterion and test results are within an acceptable range. Because the increase in the resin pyrolysis degree is irreversible (Figure 11.48b), the appearance of resin pyrolysis necessarily means that the damage has developed to the depth of the layer. In addition, if the temperature has exceeded the critical pyrolysis temperature but no substantial pyrolysis has occurred, due to the high-rigidity, high-strength fiber layup, and the weak inter-laminar mechanical strength, the probable damage caused by the thermal expansion effect is more likely to induce in-plane damage rather than the in-depth damage. Overall, the overestimated damage suggests that the temperature criterion is not suitable to simulate in-depth damage propagation, and the pyrolysis degree can be regarded as a significant indicator to evaluate in-depth lightning damage.

References

1 Rupke, E. (2002). *Lightning Direct Effects Handbook*. Pittsfield, MA, United States: Lightning Technologies Inc.

2 Mazur, V. and Moreau, J.P. (1992). Aircraft-triggered lightning-processes following strike initiation that affect aircraft. *Journal of Aircraft* 29 (4): 575–580.

3 Gabrielson, B.C. (1988). *The Aerospace Engineer's Handbook of Lightning Protection*. Gainesville, VA, United States: Interference Control Technologies, Inc.

4 Uman, M.A. and Rakov, V.A. (2003). The interaction of lightning with airborne vehicles. *Progress in Aerospace Sciences* 39: 61–81.

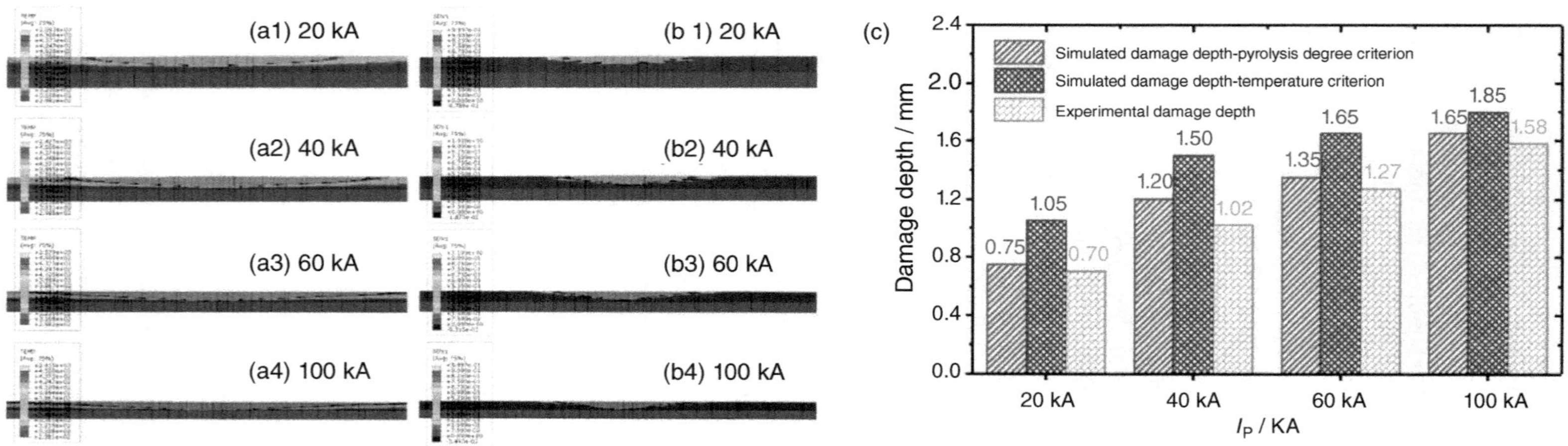

Figure 11.54 Simulated in-depth lightning damage evaluated by the temperature criterion: (a1) $I_p = 20$ kA, (a2) $I_p = 40$ kA, (a3) $I_p = 60$ kA, and (a4) $I_p = 100$ kA; pyrolysis degree criterion: (b1) $I_p = 20$ kA, (b2) $I_p = 40$ kA, (b3) $I_p = 60$ kA, and (b4) $I_p = 100$ kA; and (c) comparison of simulated and experimental damage depths.

5 Golding, W.L. (2005). Lightning strikes on commercial aircraft: how the airlines are coping. *The Journal of Aviation/Aerospace Education & Research* 15 (1): 40–50.

6 Mazur, V. (2015). Lightning threat to aircraft: do we know all we need to know? *Journal of Aircraft* 30 (1): 156–159.

7 Fisher, F.A., Plumer, J.A., and Perala, R.A. (1989). *Aircraft Lightning Protection Handbook*. Pittsfield, MA, United States: Lightning Technologies Inc.

8 Plumer, J.A. and Robb, J.D. (1982). The direct effects of lightning on aircraft. *IEEE Transactions on Electromagnetic Compatibility* 24 (2): 158–172.

9 SAE ARP 5412 (2005). *Aircraft Lightning Environment and Related Test Waveforms*. Warrendale, PA, United States: SAE International Group.

10 MIL-STD-464A (2002). *Electromagnetic Environmental Effects Requirements for Systems*. Washington DC, United States: Department of Defense Interface Standard.

11 EUROCAE ED-14E (2015). Environmental conditions and test procedures for airborne equipment, section 23: lightning direct effects. European Organization for Civil Aviation Equipment.

12 SAE ARP 5414 (2005). *Aircraft Lightning Zoning*. Warrendale, PA, United States: SAE International Group.

13 SAE ARP5416 (2005). *Aircraft Lightning Test Methods*. Warrendale, PA, United States: SAE International Group.

14 Roeseler, B., Sarh, B., and Kismarton, M. (2007). Composite structures-the first 100 years. 16th International Conference on Composite Materials, Kyoto, Japan (July 2007).

15 Marsch, G. (2010). Airbus A350 XWB update. *Reinforced Plastics* 45 (6): 20–24.

16 Chand, S. (2000). Review carbon fibers for composites. *Journal of Materials Science* 35 (6): 1303–1313.

17 Cheng, J., Ji, H., Qiu, J. et al. (2014). Role of interlaminar interface on bulk conductivity and electrical anisotropy of CFRP laminates measured by eddy current method. *NDT & E International* 68: 1–12.

18 Todoroki, A., Tanaka, M., and Shimamura, Y. (2002). Measurement of orthotropic electric conductance of CFRP laminates and analysis of the effect on delamination monitoring with an electric resistance change method. *Composites Science and Technology* 62: 619–628.

19 Haider, M.F., Majumdar, P.K., Angeloni, S. et al. (2018). Nonlinear anisotropic electrical response of carbon fiber-reinforced polymer composites. *Journal of Composite Materials* 52 (8): 1017–1032.

20 Chemartin, L., Lalande, P., Peyrou, B. et al. (2012). Direct effects of lightning on aircraft structure: analysis of the thermal, electrical and mechanical constraints. *AerospaceLab* 5: 1–15.

21 Corum, J.M., Battiste, R.L., Liu, K.C., and Ruggles, M.B. (2000). *Basic Properties of Reference Crossply Carbon-Fiber Composite*. Oak Ridge, TN, USA: Oak Ridge National Laboratory.

22 Abry, J.C., Bochard, S., Chateauminois, A. et al. (1999). In situ detection of damage in CFRP laminates by electrical resistance measurements. *Composites Science and Technology* 59: 925–935.

23 Yao, X., Sun, J., Chen, J., and Bai, D. (2016). Direct current resistance testing methods of carbon fibre reinforced polymer. *Materials Research Innovations* 19: 64–69.

24 Kawakami, H. (2011). *Lightning Strike Induced Damage Mechanisms of Carbon Fiber Composites*. Seattle, WA, USA: University of Washington.

25 Park, J.B., Hwang, T.K., Kim, H.G., and Doh, Y.D. (2007). Experimental and numerical study of the electrical anisotropy in unidirectional carbon-fiber-reinforced polymer composites. *Smart Materials and Structures* 16: 57–66.

26 Sun, J., Yao, X., Xu, W., and Chen, J. (2018). Dynamic characteristics of carbon fiber reinforced polymer under nondestructive lightning current. *Polymer Composites* 39: 1514–1521.

27 Ogasawara, T., Hirano, Y., and Yoshimura, A. (2010). Coupled thermal-electrical analysis for carbon/epoxy composites exposed to simulated lightning current. *Composites. Part A, Applied Science and Manufacturing* 41 (08): 973–981.

28 Yin, J., Li, S., Yao, X. et al. (2016). Lightning strike ablation damage characteristic analysis for carbon fiber/epoxy composite laminate with fastener. *Applied Composite Materials* 23 (4): 821–837.

29 Dong, Q., Guo, Y., Chen, J. et al. (2016). Influencing factor analysis based on electrical–thermal-pyrolytic simulation of carbon fiber composites lightning damage. *Composite Structures* 140: 1–10.

30 Abdelal, G. and Murphy, A. (2014). Nonlinear numerical modeling of lightning strike effect on composite panels with temperature dependent material properties. *Composite Structures* 109: 268–278.

31 Dong, Q., Guo, Y., Sun, X., and Jia, Y. (2015). Coupled electrical-thermal pyrolytic analysis of carbon fiber/epoxy composites subjected to lightning strike. *Polymer* 56: 385–394.

32 Toray (2008). TORAYCA T300 Data Sheet, Technical Data Sheet No. CFA-001. Toray Carbon Fibers, Inc., Tokyo, Japan.

33 Sun, J., Yao, X., Tian, X. et al. (2019). Damage characteristics of CFRP laminates subjected to multiple lightning current strike. *Applied Composite Materials* 26 (3): 745–762.

34 Taipalus, R., Harmia, T., Zhang, M.Q., and Friedrich, K. (2001). The electrical conductivity of carbon fibre reinforced polypropylene/polyaniline complex-blends: experimental characterization and modelling. *Composites Science and Technology* 61: 801–814.

35 Weber, M. and Kamal, M.R. (1997). Microstructure and volume resistivity of composites of isotactic polypropylene reinforced with electrically conductive fibers. *Polymer Composites* 18 (6): 726–740.

36 Wang, Y. and Zhupanska, O.I. (2016). Thermal ablation in fiber-reinforced composite laminates subjected to continuing lightning current. 57th Scitech/AIAA/ASCE/AHS/ASC Structures, Structural Dynamics, and Materials Conference, San Diego, California, USA.

37 Hirano, Y., Katsumata, S., Iwahori, Y. et al. (2010). Artificial lightning testing on graphite/epoxy composite laminate. *Composites Part A: Applied Science and Manufacturing* 41 (10): 1461–1470.

38 Feraboli, P. and Miller, M. (2009). Damage resistance and tolerance of carbon/epoxy composite coupons subjected to simulated lightning strike. *Composites: Part A* 40 (6–7): 954–967.

39 Feraboli, P. and Kawakami, H. (2010). Damage of carbon/epoxy composite plates subjected to mechanical impact and simulated lightning strike. *Journal of Aircraft* 47 (3): 999–1012.

40 Hirano, Y., Reurings, C., and Iwahori, Y. (2011). Damage resistance of graphite/epoxy laminates with a fastener subjected to artificial lightning. 18th International Conference on Composite Materials, Jeju Island, South Korea.

41 Li, S., Yin, J., Yao, X. et al. Damage analysis for carbon fiber/epoxy composite exposed to simulated lightning current. *Journal of Reinforced Plastics and Composites* 35 (15): 1201–1213.

42 Yin, J., Chang, F., Li, S. et al. (2017). Experimental and numerical simulation analysis of typical carbon woven fabric/epoxy laminates subjected to lightning strike. *Applied Composite Materials* 24 (6): 1353–1372.

43 Wang, F., Ji, Y., and Yu, X. (2016). Ablation damage assessment of aircraft carbon fiber/epoxy composite and its protection structures suffered from lightning strike. *Composite Structures* 145: 226–241.

44 Dhanya, T.M. and Sekher, Y.C. (2018). Lightning strike effect on carbon fiber reinforced composites-effect of copper mesh protection. *Materials Today Communications* 16: 124–134.

45 Che, H., Vo, P., and Yue, S. (2017). Metallization of carbon fibre reinforced polymers by cold spray. *Surface & Coatings Technology* 313: 236–247.

46 Khan, Z.U., Kausar, A., Ullah, H. et al. (2016). A review of graphene oxide, graphene buckypaper, and polymer/graphene composites: properties and fabrication techniques. *Journal of Plastic Film & Sheeting* 32 (4): 336–379.

47 Gou, J., Tang, Y., Liang, F. et al. (2010). Carbon nanofiber paper for lightning strike protection of composite materials. *Composites Part B: Engineering* 41 (2): 192–198.

48 Han, J.H., Zhang, H., and Chen, M.J. (2015). The combination of carbon nanotube buckypaper and insulating adhesive for lightning strike protection of carbon fiber/epoxy laminates. *Carbon* 94: 101–113.

49 Rehbein, J., Wierach, P., Gries, T. et al. (2017). Improved electrical conductivity of NCF-reinforced CFRP for higher damage resistance to lightning strike. *Composites Part A: Applied Science and Manufacturing* 100: 352–360.

50 Chakravathi, D.K., Khabashesku, V.N., and Vaidyanathan, R. (2011). Carbon fiber-bismaleimide composites filled with nickel-coated single-walled carbon nanotubes for lightning-strike protection. *Advanced Functional Materials* 21 (13): 2527–2533.

51 Wang, B., Duan, Y., Xin, Z. et al. (2018). Fabrication of an enriched graphene surface protection of carbon fiber/epoxy composites for lightning strike via a percolating-assisted resin film infusion method. *Composites Science and Technology* 158: 51–60.

52 Hirano, Y., Yokozeki, T., Ishida, Y. et al. (2016). Lightning damage suppression in a carbon fiber-reinforced polymer with a polyaniline-based conductive thermoset matrix. *Composites Science and Technology* 127: 1–7.

53 Katunin, A., Krukiewicz, K., Turczyn, R. et al. (2017). Synthesis and characterization of the electrically conductive polymeric composite for lightning strike protection of aircraft structures. *Composite Structures* 159: 773–783.

54 Tomohiro, Y., Teruya, G., Tatsuhiro, T. et al. (2015). Development and characterization of CFRP using a polyaniline-based conductive thermoset matrix. *Composites Science and Technology* 117: 277–281.

55 Yin, J.J., Chang, F., Li, S.L. et al. (2016). Lightning strike ablation damage influence factors analysis of carbon fiber/epoxy composite based on coupled electrical-thermal simulation. *Applied Composite Materials* 24 (5): 1–18.

56 Munoz, R., Delgado, S., Gonzalez, C. et al. (2014). Modeling lightning impact thermo-mechanical damage on composite materials. *Applied Composite Materials* 21: 149–164.

57 Semenov, S. and Cetegen, B. (2001). Spectroscopic temperature measurements in direct current arc plasma jets used in thermal spray processing of materials. *Journal of Thermal Spray Technology* 10 (2): 326–336.

58 Metwally, I.A., A-Rahim, A.A., Heidler, F. et al. (2006). Computation of transient-temperature profiles in objects exposed to simulated lightning currents. *International Journal of Thermal Sciences* 45 (7): 691–696.

59 Abdelal, G. and Murphy, A. (2014). Nonlinear numerical modelling of lightning strike effect on composite panels with temperature dependent material properties. *Composite Structures* 109: 268–278.

60 Guo, Y., Dong, Q., Chen, J. et al. (2017). Comparison between temperature and pyrolysis dependent models to evaluate the lightning strike damage of carbon fiber composite laminates. *Composites. Part A, Applied Science and Manufacturing* 97: 10–18.

61 Millen, S.L.J. and Murphy, A. (2020). Understanding the influence of test specimen boundary conditions on material failure resulting from artificial lightning strike. *Engineering Failure Analysis*: 104577.

62 Kamiyama, S., Hirano, Y., and Ogasawara, T. (2018). Delamination analysis of CFRP laminates exposed to lightning strike considering cooling process. *Composite Structures* 196: 55–62.

63 Fu, K. and Ye, L. (2019). Modelling of lightning-induced dynamic response and mechanical damage in CFRP composite laminates with protection. *Composite Structures* 218: 162–173.

64 Millen, S.L.J., Murphy, A., Abdelal, G. et al. (2020). Specimen representation on the prediction of artificial test lightning plasma, resulting specimen loading and subsequent composite material damage. *Composite Structures* 231: 111545.

65 Chen, H., Wang, F.S., Ma, X.T., and Yue, Z.F. (2018). The coupling mechanism and damage prediction of carbon fiber/epoxy composites exposed to lightning current. *Composite Structures* 203: 436–445.

66 Gallimberti, I. (1979). The mechanism of the long spark formation. *Le Journal de Physique Colloques* 40 (C7): 193–250.

67 Wang, Y. and Zhupanska, O.I. (2015). Lightning strike thermal damage model for glass fiber reinforced polymer matrix composites and its application to wind turbine blades. *Composite Structures* 132: 1182–1191.

68 Braginskii, S.I. (1958). Theory of the development of a spark channel. *English trans.: Soviet Phys, JEPT* 34: 1068–1074.

69 Perera, C., Rahman, M., Fernando, M. et al. (2012). The relationship between current and channel diameter of 30 cm long laboratory sparks. *Journal of Electrostatics* 70: 512–516.

70 Lee, J.H., Kim, K.S., and Kim, H. (2013). Determination of kinetic parameters during the thermal decomposition of epoxy/carbon fiber composite material. *Korean Journal of Chemical Engineering* 30 (4): 955–962.

71 Fu, K., Ye, L., Chang, L. et al. (2017). Modelling of lightning strike damage to CFRP composites with an advanced protection system. Part I: thermal-electrical transition. *Composite Structures* 165: 83–90.

72 Wan, G., Dong, Q., Zhi, J. et al. (2019). Analysis on electrical and thermal conduction of carbon fiber composites under lightning based on electrical-thermal-chemical coupling and arc heating models. *Composite Structures* 229: 111486.

73 Zaman, W., Li, K.Z., Ikram, S. et al. (2012). Morphology, thermal response and anti-ablation performance of 3D-four directional pitch-based carbon/carbon composites. *Corrosion Science* 61 (4): 134–142.

74 Dong, Q., Wan, G., Guo, Y. et al. (2019). Damage analysis of carbon fiber composites exposed to combined lightning current components D and C. *Composites Science and Technology* 179: 1–9.

75 Millen, S.L.J., Murphy, A., Abdelal, G., and Catalanotti, G. (2019). Sequential finite element modelling of lightning arc plasma and composite specimen thermal-electric damage. *Computers and Structures* 222: 48–62.

12

Polymer Composites for Switchgears

Takahiro Imai

Infrastructure Systems Research and Development Center, Toshiba Infrastructure Systems & Solutions Corporation, Tokyo, Japan

12.1 Introduction

The ever-increasing demand for electric power around the world emphasizes the importance of electric apparatuses such as generators, transformers, and switchgears. Electricity consumption in advanced countries has been maintained since 1990. However, the world average has doubled in that time. In particular, electricity consumption in China, India, and Korea has increased rapidly during the past three decades [1]. Thus, our interest in energy conservation toward a sustainable future has given a boost to the development of high-efficiency electric apparatuses for power generation and transmission.

Figure 12.1 shows the electric power distribution in Japan [2]. Electricity generated in thermal, hydroelectric, and nuclear plants is delivered to factories, buildings, and residences via substations and pole transformers. In substations, switchgear for current interruption and transformers for voltage conversion play important roles in providing stable electricity supplies. These apparatuses are arranged in distribution grids from power plants to customers to decrease voltages gradually, which have the advantage of diminishing power transmission losses due to Joule heat attributed to the flow of electricity.

In switchgear and transformers, insulation technologies to maintain voltages are very important. In particular, insulation materials are elemental technologies and the overall performance of apparatuses depends on the properties of insulation materials. Table 12.1 shows the insulation materials used in these apparatuses. These materials are divided into three categories: insulation medium, structural materials, and other materials. This chapter focuses on epoxy resins (casting resins) and fiber-reinforced plastics (FRP) as structural materials for switchgear. Most of these materials are polymer composites consisting of epoxy resin and inorganic fillers/fibers. Epoxy resins are excellent insulation materials. However, materials for switchgear need good mechanical and thermal properties as well as insulation properties. Reinforcement with inorganic fillers/fibers are effective approaches to achieving both these properties.

Polymer Composites for Electrical Engineering, First Edition. Edited by Xingyi Huang and Toshikatsu Tanaka.

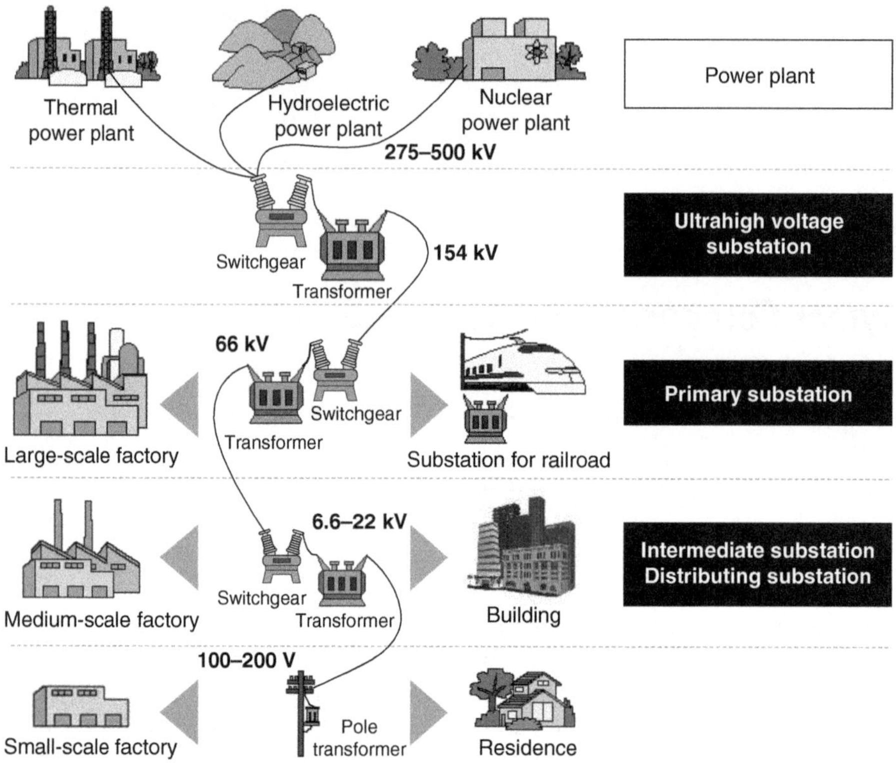

Figure 12.1 Electric power distribution in Japan.

Table 12.1 Insulation materials used in switchgear and transformers.

Insulation Materials	Switchgear	Transformer
Insulation gas (SF_6 gas)	Insulation medium	Insulation medium
Insulation oil	—	Insulation medium
Epoxy resin (Casting resin)	**Structural material**	Structural material
Fiber-reinforced plastic (FRP)	**Structural material**	Structural material
Rubber, Silicone rubber	Seal material	Seal material
Fluorine resin	Sliding material	Sliding material
Pressboard	—	Structural material
Creped paper	—	Turn insulation material

12.2 History of Switchgear

High-voltage power-receiving switchgears have changed from the open type to the housing type to the metal-enclosed type to accommodate social expectations, as shown in Figure 12.2. Engineering innovation enables downsizing as well as preventing pollution

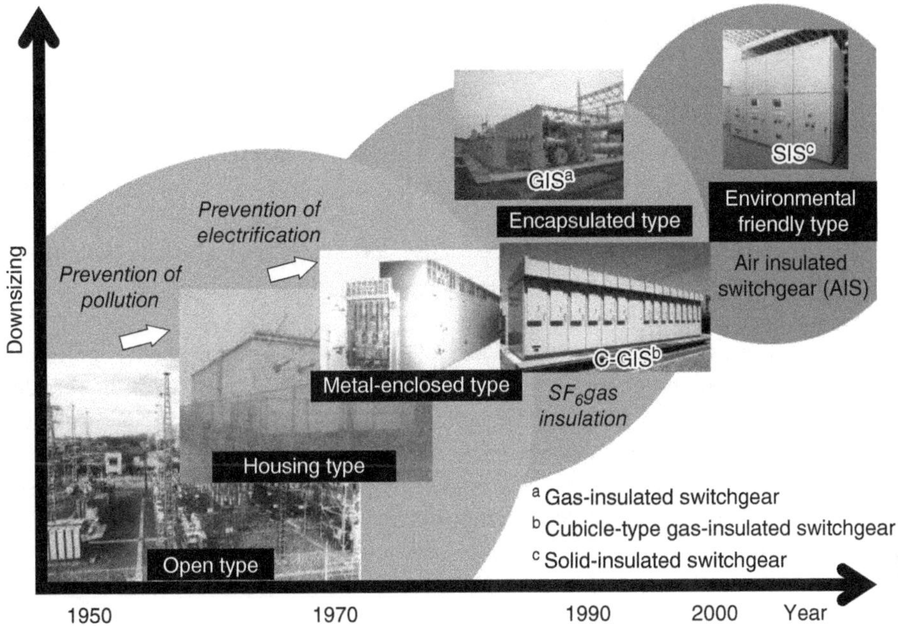

Figure 12.2 Transition of high-voltage power-receiving switchgears.

and achieving high voltage in switchgear. The transition of switchgear is briefly reviewed in the following.

In open-type switchgear, the main components such as circuit breakers, disconnectors, and instrument transformers were placed in an outdoor location. The main components and conductors between them were insulated with air. Thus, large premises were needed to ensure the required offset distance. In housing-type switchgear, the main components were placed in an indoor location. The housing protected the switchgear from pollution from the ambient environment. However, the main insulation systems were similar to the open type, which did not allow downsizing.

Since the 1970s, sulfur hexafluoride (SF_6) gas insulation, which has superior insulation properties to air insulation, has accelerated the downsizing of switchgear. Gas circuit breakers (GCBs) utilizing SF_6 gas and vacuum circuit breakers (VCB) utilizing vacuum state enabled the use of metal-enclosed switchgear. The main components were surrounded by a grounded metal housing. Unexposed electrically conductive parts prevented electrification and enhanced safety.

In currently used encapsulated switchgears, the main circuits insulated with SF_6 gas enable size reduction. Their structures, in which the main circuits are encapsulated in a metal housing filled with SF_6 gas, are completely isolated from the ambient environment, which dramatically decreases maintenance frequency. In gas-insulated switchgear (GIS), the main components such as circuit breaker and disconnector are placed in cylindrical tanks filled with high-pressure SF_6 gas (over 0.3 MPa). In cubicle-type gas-insulated switchgear (C-GIS), the main components are placed in metal housings filled with low-pressure SF_6 gas. Under divergent electrical fields, the insulation characteristics of SF_6 gas tend not to depend on gas pressure. Therefore, an appropriate insulation design of C-GIS enables the SF_6 gas pressure to be near atmospheric pressure.

In the late 1990s, environmental concerns such as global warming promoted the development of environment-friendly switchgear. SF_6 gas has approximately 23 000 times the global warming potential (GWP) as carbon dioxide (CO_2) gas [3]. Therefore, the 1997 Kyoto protocol on global warming designated SF_6 gas as a greenhouse gas. This protocol was issued in 2005 after Russia approved its ratification in 2004. SF_6 gas emissions into the atmosphere from GIS and C-GIS are strictly controlled at present. These switchgears still play important roles in power supplies. However, the electric power industry is positively working to reduce SF_6 gas use.

In environment-friendly switchgear, solid insulation systems and compressed dry air insulation systems are used instead of SF_6 gas insulation systems. The main circuits are molded of solid materials in solid-insulated switchgear (SIS) [4]. The main circuits are encapsulated in a metal housing filled with compressed dry air in air-insulated switchgear (AIS) [5]. Moreover, hybrid insulation systems using both solid insulation materials and compressed dry air are also utilized in AIS [6]. Most solid insulation materials in these switchgears are polymer composites. Thus, polymer composites have attracted increasing attention.

12.3 Typical Insulators in Switchgears

12.3.1 Epoxy-based Composite Insulators

As mentioned in the introduction, switchgears utilize various insulation materials. Epoxy resin (casting resin) and fiber-reinforced plastic (FRP) are important materials for switchgear. Most of these materials are polymer composites consisting of epoxy resin and inorganic fillers/fibers. Typical epoxy-based composite insulators are described in the following.

Figure 12.3 shows a molded vacuum interrupter in a solid-insulated switchgear. The main circuit conductor and vacuum interrupter to which high-voltage is applied are

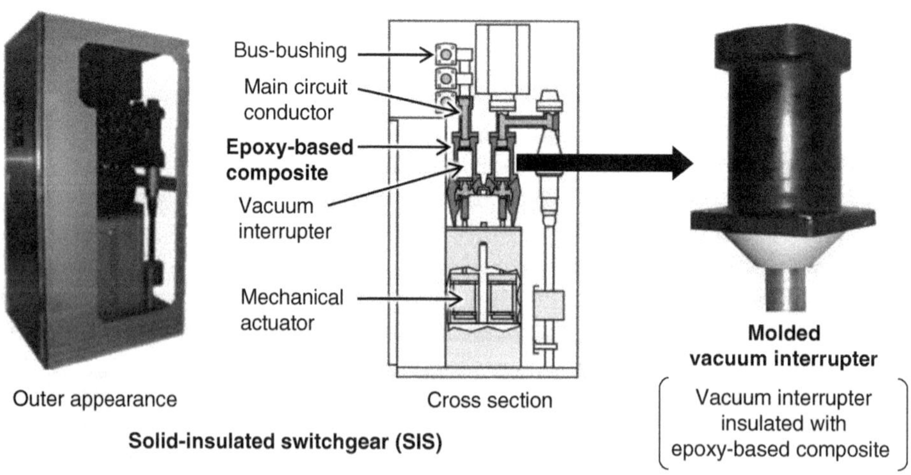

Figure 12.3 Molded vacuum interrupter in a solid-insulated switchgear. *Source:* Based on [7, 8].

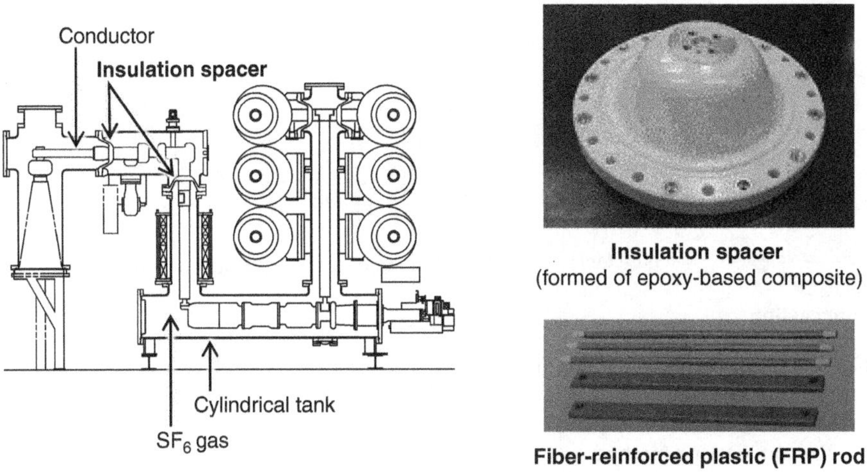

Figure 12.4 Insulation spacer and fiber-reinforced plastic (FRP) rod in SF_6-gas-insulated switchgear (GIS). *Source:* Takei and Hirano [9].

insulated with epoxy-based composite containing silica fillers. The molded interrupter is the most important component and, therefore, needs insulation reliability and availability. Generally, the insulation strength of an epoxy-based composite is approximately three times that of SF_6 gas. Thus, a solid insulation system using such a composite enables switchgear to have superior reliability and availability, and to be downsized.

Figure 12.4 shows an insulation spacer and fiber-reinforced plastic (FRP) rod in SF_6-gas-insulated switchgear [9]. The main circuit conductor and components to which high voltage is applied are insulated with SF_6 gas in cylindrical tanks. The insulation spacers hold the conductor up and work as a boundary to create gas compartments in the tanks. They are hat-shaped to ensure that the creepage distance does not cause a flashover on the surface of insulation spacers. The insulation spacers are formed of epoxy-based composites containing alumina or silica fillers.

Fiber-reinforced plastic (FRP) is used as an operation rod to drive the electrodes in the circuit breaker, which is one of the main components of the GIS. Therefore, the FRP rods need excellent insulation properties as well as high mechanical strength. They are composed of a combination of epoxy resin and glass fiber.

12.3.2 Insulator-Manufacturing Process

Inherent defects in insulators are unacceptable in switchgear, which need insulation reliability and availability. Remaining voids (very small cavities) and exfoliations between the resin and inserts such as vacuum interrupters and conductors in insulators would cause the generation of partial discharges. Such discharges deteriorate the insulators, leading to insulation breakdown. Therefore, vacuum casting is used in insulator manufacturing. The vacuum completely eliminates bubbles and air pockets, enabling the production of defect-free insulators. The following outlines the processes of manufacturing insulators and FRP rods.

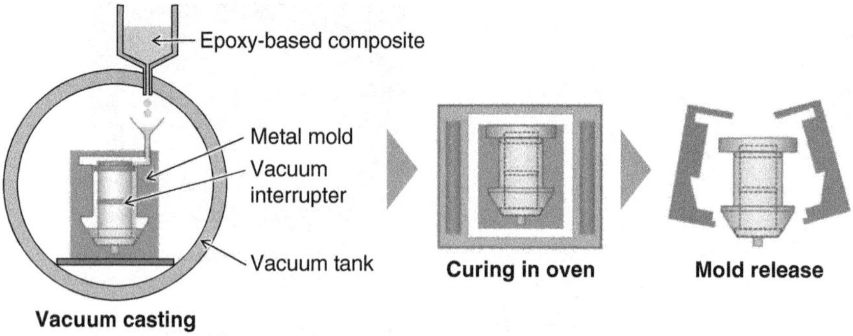

Figure 12.5 Vacuum casing process for insulator manufacturing.

12.3.2.1 Vacuum Casting Method

Vacuum casting is the most common method of manufacturing insulators such as molded vacuum interrupters and insulation spacers. Figure 12.5 depicts the vacuum casting process. Pre-degassed casting resin in the liquid state, which is epoxy-based composite containing inorganic fillers, is poured into a preheated metal mold inside a vacuum tank. Negative pressure is controlled in the tank. After casting, the resin is cured inside an oven under atmospheric pressure. Curing time (heating duration) is relatively long, e.g. 3–15 hours at temperatures below 100 °C because slow curing suppresses curing shrinkage of the cast resin, which would cause exfoliation between resin and inserts. The insulator is finally released from the mold. Although vacuum casting is a useful method for insulator manufacturing, it has the disadvantages that automatization of the process is difficult and turnover of the metal mold is not effective due to the long curing time.

Before curing, the viscosity and gelation time of the cast resin have an impact on the process and on insulator quality. Resin viscosity generally shows temperature and shear rate dependence. Low viscosity prevents bubbles and air pockets from occurring. So, there are no voids in the cured insulators. Moreover, gelation time means the amount of time it takes for the resin in the liquid state to transform into a gel state and generally shows temperature dependence. Resin in the gel state is a highly viscous material that can no longer flow. Therefore, appropriate gelation time is necessary for the casting process.

12.3.2.2 Automatic Pressure Gelation Method

Insulator manufacturing by the automatic pressure gelation (APG) method has increased recently. This method overcomes the disadvantages of the vacuum casting method and curing under pressure is its technological feature. Figure 12.6 depicts the APG process. Predegassed epoxy resin and hardener containing inorganic fillers are continuously mixed. The mixture is injected into the preheated metal mold inside a vacuum chamber. The temperature of the metal mold and the negative pressure in the chamber are computer-controlled. Resin curing is performed under pressure, which compensates for the shrinkage of resin curing. This prevents exfoliations between the resin and inserts due to internal strain attributed to resin shrinkage, and enables quicker resin curing at higher temperatures than in the vacuum casting method.

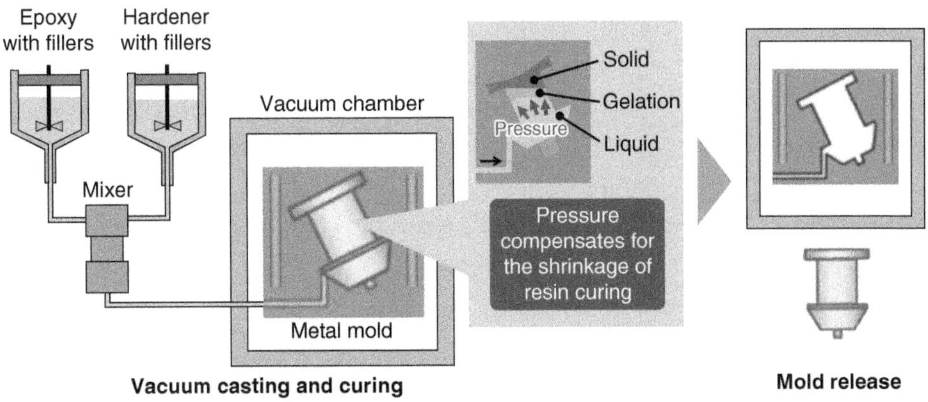

Figure 12.6 Automatic pressure gelation (APG) process of insulator manufacturing.

The APG method is mainly used for manufacturing small insulators. However, recent technological progress of APG casting equipment, such as size increase and improved performance, enable the manufacturing of large insulators. Moreover, improvement of the cast resin based on material design expands the range of APG application.

12.3.2.3 Vacuum Pressure Impregnation Method

There are various methods of manufacturing fiber-reinforced plastic (FRP), such as hand layup, filament winding, sheet winding, sheet molding compound (SMC) and bulk molding compound (BMC). FRP rods in switchgear need good insulation properties as well as good mechanical properties. In particular, the edgewise direction of glass fiber needs to endure the high electrical field in the FRP rods. Therefore, remaining voids are also unacceptable in FRP rods.

FRP rods of switchgear are generally manufactured by the vacuum pressure impregnation (VPI) method. Stacked glass fiber sheets are impregnated into the mixture of epoxy resin and hardener in a vacuum tank. After impregnation, the mixture is cured by heating under pressure. The VPI process can produce bulk FRP rods without any voids. The FRP rods are finally produced by machine processing.

12.4 Materials for Epoxy-based Composites

Epoxy-based composites for insulators in switchgears are composed of epoxy resin, hardener, and inorganic fillers/fibers. Additives such as silane coupling agents are also used to improve material properties. This section describes the materials for epoxy-based composites.

12.4.1 Epoxy Resins

Epoxy resin was first used in power apparatuses in the 1940s and is still commonly used. Bisphenol A epoxy resin is versatile and the most common. It is no exception in the

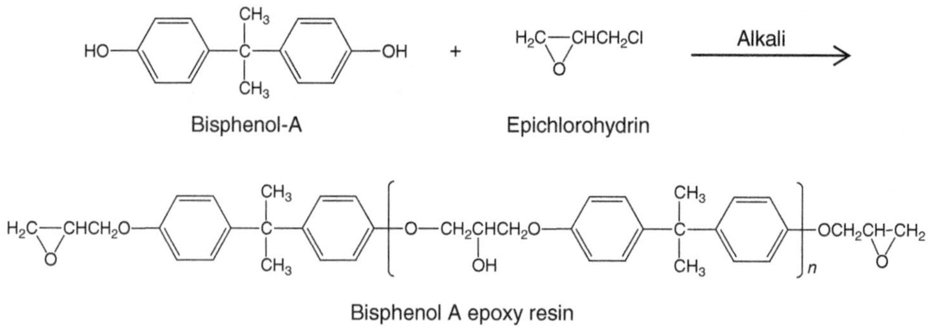

Figure 12.7 Bisphenol A epoxy resin.

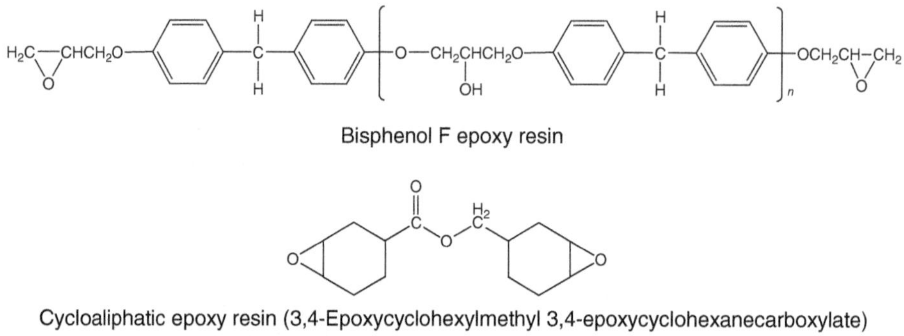

Figure 12.8 Epoxy resins with low viscosity.

epoxy-based composites for switchgears. The chemical reaction between bisphenol A and epichlorohydrin in the presence of alkali yields bisphenol A epoxy resin, as shown in Figure 12.7. Various epoxy resins depending on the repeat unit n are commercially available. In general, epoxy resins with n under 1 are in the liquid or semisolid state, and epoxy resins with n over 16 are in the solid state at room temperature. The most used epoxy resin has $n = 0.1–0.2$. Moreover, solid epoxy resins with large n reduce curing shrinkage due to decreasing cross-linked density in the curing process.

Epoxy resins with low viscosity are often used in the VPI method of FRP manufacture. Bisphenol F epoxy resin and cycloaliphatic epoxy resin are typical low-viscosity resins, and have chemical structures, as shown in Figure 12.8.

Bisphenol A (BPA) has been included in the Registration, Evaluation, Authorization, Restriction, and Chemicals (REACH) candidate list for chemical safety of Substances of Very High Concern (SVHC) since 2017. The remaining BPA in bisphenol A epoxy resin has a very low concentration of a few ppm. Although a concentration below 1000 ppm is unnecessary to notify REACH, appropriate treatments of bisphenol A epoxy resin are necessary for human and environmental safety.

12.4.2 Hardeners

Amine, polyamine, acid anhydride, and imidazole can be used as hardeners for a combination of epoxy resin. Hardener selection depends on the application. An acid

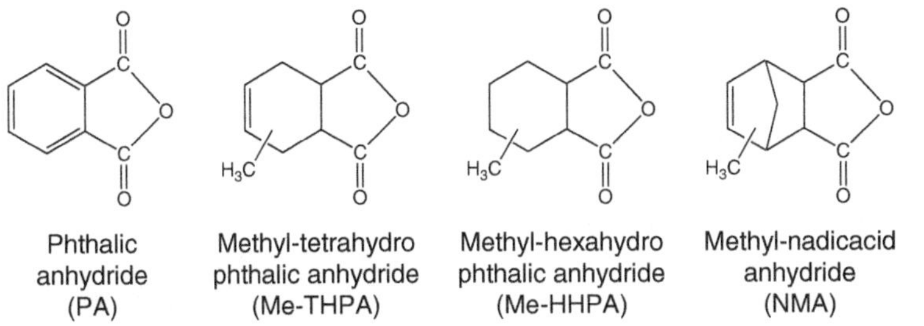

Figure 12.9 Acid anhydride hardeners.

anhydride is mainly used as a hardener in insulators of switchgears. Figure 12.9 shows the chemical structures of acid anhydrides. Each acid anhydride has the following characteristics: PA has low cost and sublimation; Me-THPA has low viscosity and is a versatile hardener; Me-HHPA has low viscosity, transparency and resistance to climatic conditions; and NMA has a heat resistance and long shelf time.

In acid anhydrides, it is relatively easy to generate free acid due to moisture absorption, which affects the properties of cured epoxy/hardener materials. However, acid anhydrides have the following advantages as hardeners for insulators.

1) Low viscosity of epoxy resin/acid anhydride mixture and long shelf life.
2) Small exotherm and curing shrinkage in the curing process.
3) Cured epoxy resin/acid anhydride mixture has excellent insulation characteristics, heat resistance, and mechanical properties.

Moreover, the curing reaction of the epoxy resin/acid anhydride system is slow compared to other hardener systems. Thus, curing accelerators (catalysts) are often used. Imidazole, metallic chelate, and ammonium salt are useful curing accelerators.

Me-HHPA has been included in the REACH candidate list of SVHC for chemical safety since 2012. Although it is not intended to prohibit the use of Me-HHPA, appropriate treatments are needed in its usage.

12.4.3 Inorganic Fillers and Fibers

Inorganic fillers and fibers play important roles in epoxy-based composites. They reinforce the epoxy matrix consisting of epoxy resin and hardener and enable the matrix to have sufficient properties for insulators of switchgears.

Most inorganic fillers used in casting resin for molded interrupters and insulation spacers are silica (SiO_2) and alumina (Al_2O_3). Table 12.2 summarizes their characteristics. Silica fillers show lower permittivity, thermal expansion, and density than alumina fillers. These are advantageous properties for insulators of switchgear. On the other hand, alumina fillers show superior thermal conductivity and resistance to SF_6 decomposition gas. They enable the insulator to improve the performance of switchgears. Therefore, filler selection depends on the application. Moreover, filler shape strongly affects the viscosity of the epoxy resin. Generally, it is known that viscosity of epoxy resin containing spherical fillers is 10% lower than that of epoxy resin containing irregular fillers.

Table 12.2 Characteristics of silica and alumina fillers.

Filler		Filler Shape	Properties
Silica (SiO_2)	Crystalline	Irregular	• Low permittivity
	Fused	Irregular or spherical	• Low thermal expansion • Low density
Alumina (Al_2O_3)	Sintered	Irregular	• High thermal conductivity
	Fused	Spherical	• Resistance to SF_6 decomposition gas

The viscosity of polymer-containing fillers is estimated from Mooney's viscosity equation [10]:

$$\eta = \eta_0 \exp\left[\frac{K_E \cdot \varphi}{1-(\varphi/\varphi_m)}\right]$$

where η is the viscosity of composite with fillers, η_0 is the viscosity of epoxy matrix, φ is the volume fraction of fillers, φ_{m} is the maximum filled volume fraction of fillers, and K_{E} is a coefficient.

In epoxy resin-containing fillers, the viscosity (η) increases with increasing volume fraction of fillers (φ). Besides, filler diameter has distribution. In the case of the highly filled epoxy resin, e.g. over 50 vol% filler loading, the distribution has an influence on resin viscosity. Mooney's viscosity equation indicates that the viscosity (η) decreases with increasing the maximum filled volume fraction of fillers (φ_{m}). The maximum filled volume fraction of fillers depends on the filler diameter distribution. The model of Horsfield's filling theory shows the closest packed structure [11]. Optimized wide filler diameter distribution composed of primary (large) fillers 20–30 μm in diameter, secondary (medium) fillers 10–20 μm in diameter, and tertiary (small) fillers 1–3 μm in diameter based on the theory provides the maximum filled volume fraction of fillers and decreases resin viscosity.

Glass fiber is used in fiber-reinforced plastic (FRP) for switchgear. There are several kinds of glass fibers such as E-glass and ECR-glass. E-glass fiber is the most common and consists of alumino-borosilicate glass with less than 1% alkali oxides. Moreover, E-glass is a favorable material for FRP rods for switchgears because it has excellent insulation properties.

12.4.4 Silane Coupling Agents

Silane coupling agents are often used for surface treatment of inorganic fillers, as part of the fabrication of epoxy-based composites, where they have the following roles:

1) Wettability improvement of fillers for epoxy resins.
2) Compatibility improvement of fillers and epoxy resins.
3) Hydrogen bonding between fillers and epoxy resins.
4) Chemical bonding between fillers and epoxy resins.

Generally, wettability, compatibility, and hydrogen bonding strength affect the dispersion of fillers in epoxy resins. Silane treatment is the most commonly used procedure to improve wettability and compatibility. Many kinds of silane coupling agents have been commercialized, and representative examples for epoxy resin are summarized in Table 12.3. Silane coupling agents have the general structure of R-Si(OR')$_3$, where the R group determines the nature of the interaction between the polymer and the OR' group, resulting in chemical bonding with the filler. Consequently, the silane coupling agent should be selected based on the relevant combination of polymer and filler.

Silane treatments are divided into two methods as follows:

1) Pre-treatment of nanofillers.
   - Wet method
   - Dry method
2) *In situ* treatment of nanofillers.

In the pretreatment method, the fillers are treated before compounding with the epoxy resin and two methods for this exist: the wet method and the dry method. In the wet method, the fillers are dispersed in a dilute silane solution and, after sufficient mixing, they are separated by filtration, decanting, or centrifugation. Drying by heating finally leads to a silane-treated filler. In the dry method, the silane coupling agent or the diluted silane solution is added to the filler and mixed at high speed. The separation and drying process is not needed in this method.

In the *in situ* treatment method, the fillers, silane coupling agent, and epoxy resin are compounded together. This method has the advantage of low cost and is widely used in industry, but the effectiveness of the method is inferior to that of pre-treatment. Also, the addition of a silane coupling agent such as γ aminopropyltriethoxysilane causes gelation of epoxy resins; the stability of the epoxy resin in the presence of the silane coupling agent is an important factor when considering *in situ* treatment.

Silane couplings are typical bonding states that combine epoxy resin with fillers. The bonding strength is medium in hydrogen bonds and becomes stronger when the chemical bonding is converted from a hydrogen bond to a covalent bond. Epoxy resin and fillers are immiscible with each other. So, coupling agents like silane combine them with the help of hydrogen bonding, as shown in Figure 12.10. Strong ionic bonding and weak van der Waals bonding are also effective in integrating them. "Wettability" is also an important factor and plays a functional role in enhancing affinity by which two kinds of substances are mutually adjacent to each other.

Finally, the silane coupling agents are also available to glass fiber for FRP. Pretreated fiber strengthens the interface between epoxy resin and fiber.

12.4.5 Fabrication of Epoxy-based Composites

By using the materials described above, the fabrication processes of epoxy-based composites are briefly explained herein. Figure 12.11 illustrates the simplified fabrication processes of epoxy-based composites.

The process shown in Figure 12.11a is mainly available to the composite for the vacuum casting method. Fillers are dispersed in the epoxy resin by mixing. Silane treatment on

Table 12.3 Silane coupling agents for epoxy resins.

Functional Groups	R	OR'	Chemical Structure	Chemical Name
Amino	$H_2N-CH_2CH_2CH_2-$	$-OC_2H_5$	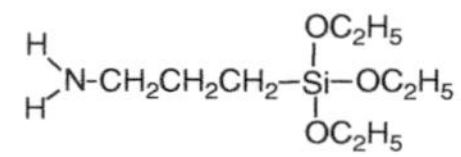	γ-Aminopropyl triethoxysilane
Epoxy	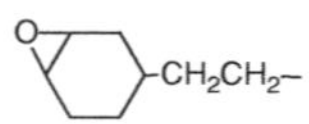	$-OCH_3$	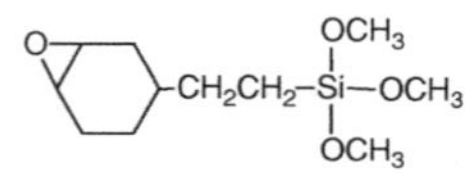	β-(3,4-Epoxycyclo hexyl)ethyl trimethoxysilane
	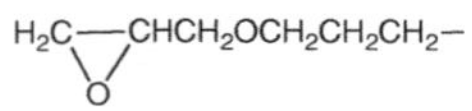	$-OCH_3$	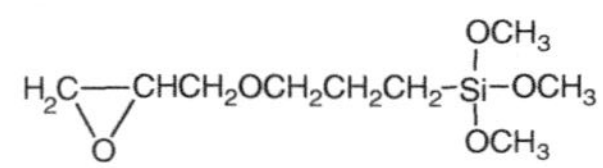	γ-Glycidoxypropyl trimethoxysilane
Laocyanate	$O{=}C{=}N-CH_2CH_2CH_2-$	$-OC_2H_5$	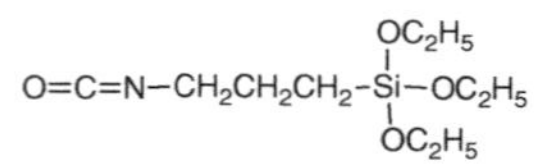	3-Isocyanatepropyl triethoxysilane
Mercapto	$HS-CH_2CH_2CH_2-$	$-OCH_3$	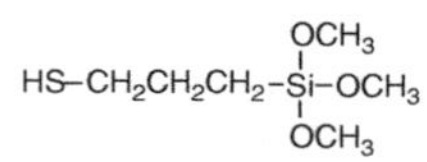	γ-Mercaptproyl trimethoxysilane

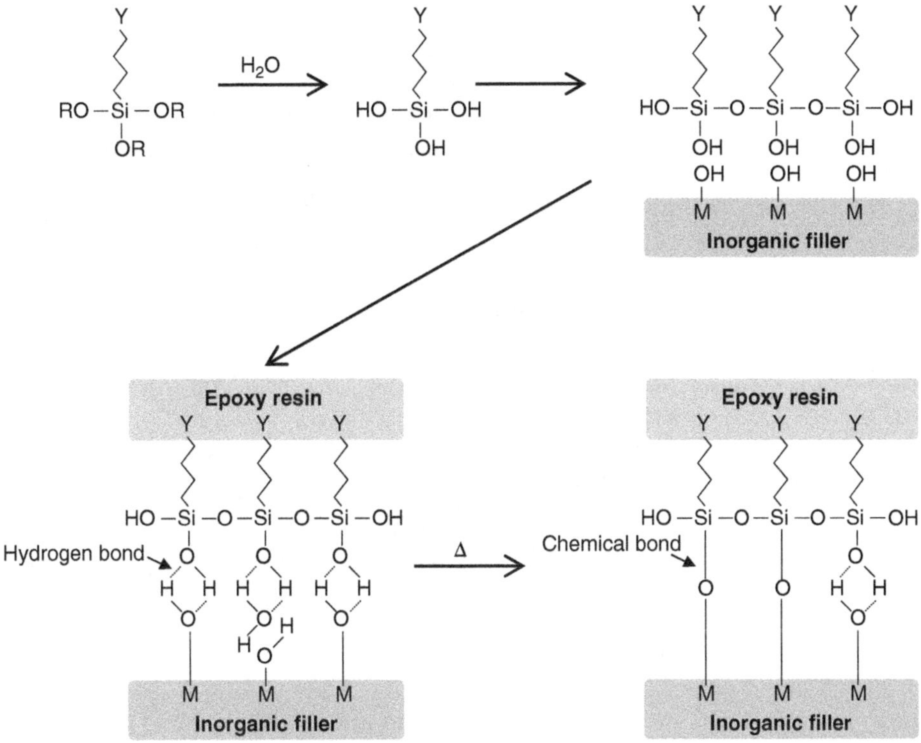

Figure 12.10 Chemical structures of silane coupling.

filler surfaces is performed by pretreatment or *in situ* methods. The mixture after the addition of the hardener is cast and degassed in a metal mold. The composite is obtained after curing in an oven.

Moreover, the process shown in Figure 12.11b is mainly available to the composite for the automatic pressure gelation (APG) method. Fillers are separately dispersed in the epoxy resin and hardener by mixing. Filler surfaces are treated with silane coupling. Prefilled epoxy resin and hardener are mixed and the composite is obtained in the same manner.

12.5 Properties of Epoxy-based Composites

Epoxy-based composites for insulators in switchgear need various properties and functions, as described in this section.

12.5.1 Necessary Properties of Epoxy-based Composites for Switchgears

The necessary properties of epoxy-based composites are divided into resistance to TEAM stresses, i.e. Thermal, Electrical, Ambient, and Mechanical stresses. Figure 12.12 summarizes the necessary properties for insulators in solid- and gas-insulated switchgears based on the categories of TEAM stresses [12].

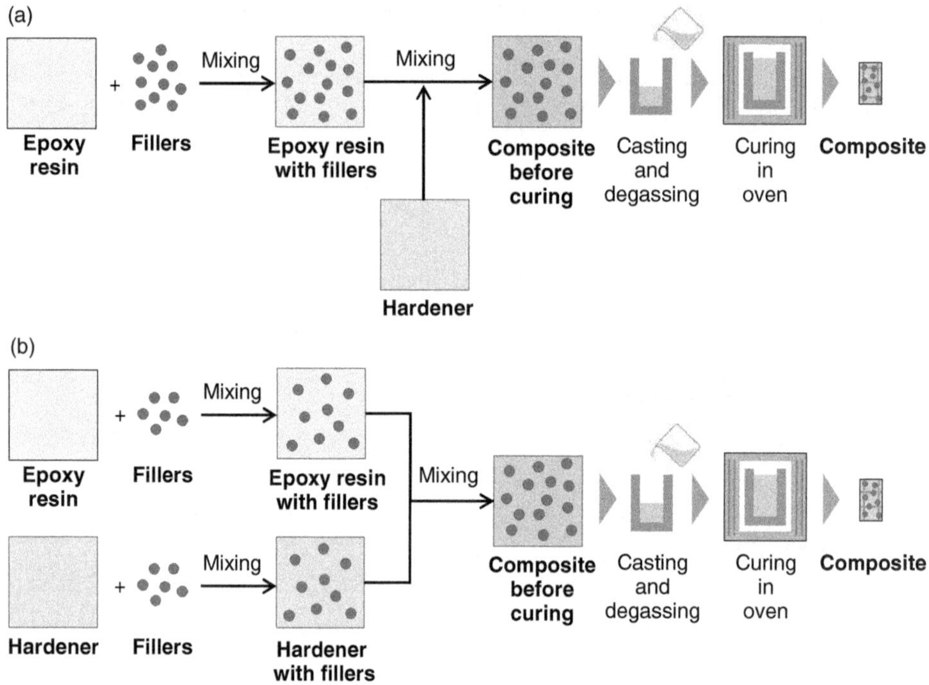

Figure 12.11 Fabrication processes of epoxy-based composites.

The composites need to have all these properties. For example, filler dispersion in epoxy resin improves flexural and tensile strength, but increases relative permittivity and decreases resistivity, which are unfavorable electrical properties. Therefore, material design is important for composites of insulators. Optimized performance of composites could dominate the performance of switchgears.

The following subsections describe representative properties that indicate resistances to TEAM stresses. The composites are epoxy-based and contain 50–70 wt% micro-sized inorganic fillers 1–50 μm in diameter.

12.5.2 Resistance to Thermal Stresses

12.5.2.1 Glass Transition Temperature

Insulator temperature in switchgear increases due to Joule heat attributed to the flow of electricity under operation. IEC international standard 62271-1:2017 specifies the limits of temperature and temperature-rise in switchgears [13]. The temperature-rise limits depend on the part (e.g. contacts and terminals for the connection to external conductors), materials (e.g. bare-copper and coated copper), and dielectric (e.g. in air, SF_6, and oil). Table 12.4 extracts the temperature-rise limits in the connection part: bolted or equivalent.

Glass transition temperature (T_g) above the maximum temperature is favorable in epoxy-based composites for insulators. Because T_g means the temperature of the border where the amorphous epoxy matrix changes from glassy state (rigid state) to rubbery state (more

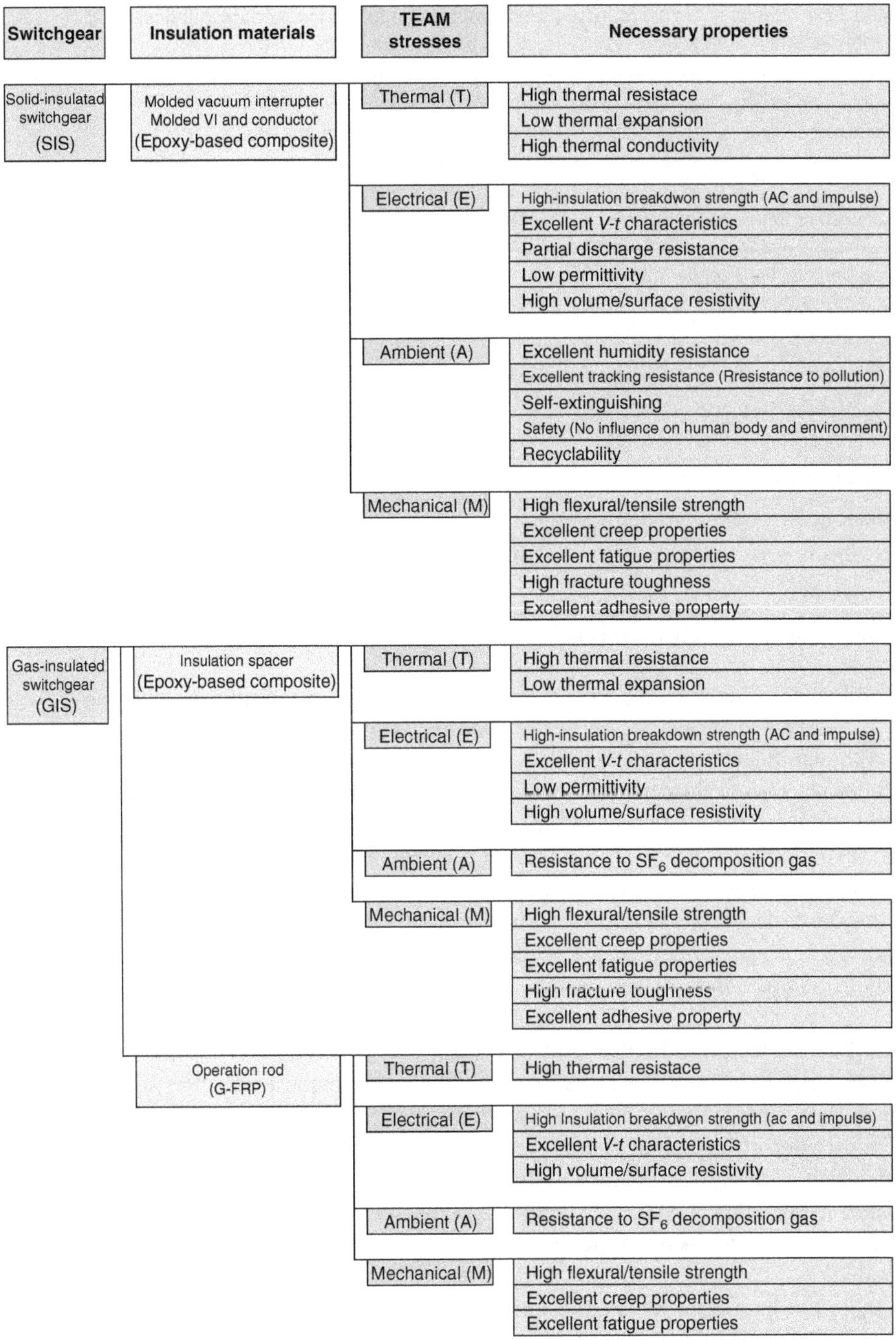

Figure 12.12 Necessary properties in epoxy-based composites in switchgears. *Source:* Investigating R&D Committee on Advanced Tailor-made Composite Insulation Materials [12].

Table 12.4 Limits of temperature and temperature rise for part, materials, and dielectrics of high-voltage switchgear.

Nature of the Part, of Material and the Dielectric		**Maximum Value [°C]**
Connection, Bolted or Equivalent	**Temperature**	**Temperature-Rise at Ambient Air Temperature not Exceeding 40 °C**
Bare-copper, bare-copper alloy, or bare-aluminum alloy		
– in air	90	50
– in SF_6	115	75
Silver-coated or nickel-coated		
– in air	115	75
– in SF_6	115	75
Tin-coated		
– in air	105	65
– in SF_6	105	65

Source: Based on IEC 62271-1:2017 [13].

flexible state). Therefore, the properties of the epoxy-based composites dramatically change at T_g. Most mechanical and electrical properties above T_g are inferior to those below T_g.

Although fillers, curing accelerators, and some additives affect T_g of epoxy-based composites, the chemical structure of an epoxy matrix formed from the combination of epoxy resin and hardener predominate. Table 12.5 compares T_g and heat distortion temperature (HDT) of epoxy matrices by the bisphenol A epoxy resin and typical acid anhydride hardeners [14]. HDT indicates the deflection temperature under specified load due to the transition from a glassy state to a rubbery state and has similar thermal properties to T_g. All T_g and HDT of epoxy matrices are higher than the maximum temperature, as shown in Table 12.4. Thus, these acid anhydride hardeners are used in insulators.

12.5.2.2 Coefficient of Thermal Expansion (CTE)

In solid insulation switchgear (SIS), the high-voltage metal conductors are molded of epoxy-based composites. In SF_6 gas-insulated switchgear (GIS), the conductors are held up at the center of insulation spacers made of composites. If the coefficients of thermal expansion of conductor and composite are different, exfoliation occurs at boundaries due to heat in operation, as shown in Figure 12.13. Therefore, high loading of inorganic microfillers is needed to decrease the thermal expansion of epoxy resin to the level of metal conductors.

Table 12.6 compares the linear coefficients of thermal expansion (CTE) in epoxy-based composites [15]. Linear, area, and volume coefficients of thermal expansion are calculated. Epoxy-based composites containing fillers are generally free to expand in any direction (isotropic expansion), and linear CTE is commonly used. The base epoxy resin without fillers has a CTE of 68 ppm/K, which is higher than that of the metal conductor (aluminum).

Table 12.5 Thermal properties of cured epoxy resin by typical acid anhydride hardener (without fillers).

Condition and Thermal Properties		Typical Acid Anhydride Hardener		
		Me-THPA	Me-HHPA	NMA
		Methyl-Tetrahydro phthalic Anhydride	Methyl-Hexahydro phthalic Anhydride	Methyl-Nadic Acid Anhydride
Compounding raion [phr[a]]	Bisphenol A epoxy resin[b]	100	100	100
	Hardener	87	85	80
	Curing accelerator[c]	1	1	1
Curing condition			120 °C in 5 hours + 150 °C in 15 hours	
Glass transition temperature (T_g) [°C]		137	151	150
Heat distortion temperature (HDT) [°C]		129	144	139

[a] Parts per hundred resin.
[b] Epoxy equivalent weight = 190.
[c] Tris(dimethylaminomethyl)phenol.*Source*: Ochi et al. [14].

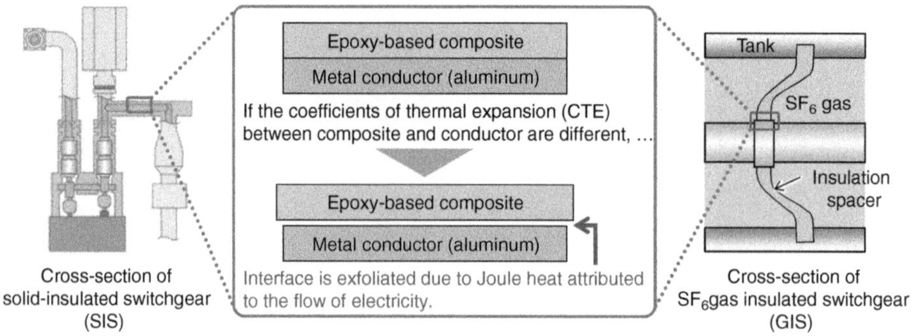

Figure 12.13 Exfoliation at the interface between composite and metal conductor.

Table 12.6 Coefficients of thermal expansion (CTE) of epoxy-based composites.

Epoxy-based Composites	**Coefficient of Thermal Expansion (α_1) [ppm/K]**
Base epoxy resin (without fillers)	68
Composite with 10 wt% SiO_2 nano-fillers (Nanocomposite)	65
Composite with 65 wt% SiO_2 micro-fillers (Microcomposite)	24
Metal conductor (Aluminum)	24

Source: Based on Imai et al. [15].

Dispersion of 10 wt% SiO_2 nano-fillers with 12 nm in average diameter slightly decreases the CTE of epoxy resin in the nanocomposite. However, the microcomposite, which contains 65 wt% SiO_2 micro-fillers with 17 μm in average diameter, has the same CTE of 24 ppm/K to the metal conductor. On the other hand, high-filled composites are prone to show high viscosity before curing. Therefore, optimized distribution of filler diameter is necessary to decrease the viscosity, as mentioned in subsection 12.4.3.

12.5.3 Resistances to Electrical Stresses

12.5.3.1 Short-term Insulation Breakdown

Good short-term insulation breakdown characteristics of epoxy-based composites are essential for insulators in switchgears. The size and shape of insulators depend on these characteristics.

Figure 12.14a shows the insulation breakdown strength under spherical and plate electrode configuration [16]. An AC ramp voltage with the rate of 50 Hz–1 kV/s is applied to the electrodes at 20 °C. The epoxy-based composite with 67 wt% irregular-shaped SiO_2 fillers of 6 μm average diameter shows lower breakdown strength than the base epoxy resin without fillers. The decline of breakdown strength seems to result from weak interfaces between epoxy matrix and fillers.

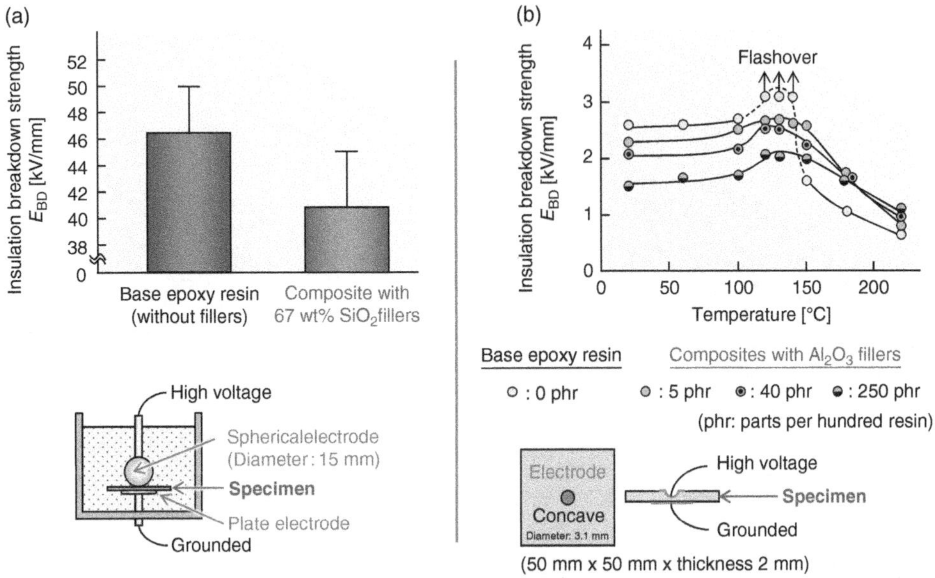

Figure 12.14 Insulation breakdown strength of epoxy-based composites. (a) Epoxy-based composite with SiO_2 fillers. (b) Epoxy-based composites with Al_2O_3 fillers. *Source:* Based on [16, 17].

Figure 12.14b shows the filler amount and temperature dependence of insulation breakdown strength under concave and plate electrode configurations [17]. An AC ramp voltage with a rate of 50 Hz–1 kV/s is applied to the electrodes at various temperatures. In epoxy-based composites with spherical-shaped Al_2O_3 fillers of 5 μm average diameter, increase in filler amounts decreases insulation breakdown strength. An increase in filler amounts expands the interfaces between epoxy matrix and fillers. These results prove that the reduction of breakdown strength is attributed to the interfaces. Moreover, the breakdown strengths of the base epoxy resin and all composites decline above the glass transition temperature (T_g) of approximately 130 °C.

Figure 12.15 illustrates the effects of spherical fillers and core-shell rubber fillers on insulation breakdown voltage [4, 7]. Figure 12.15a shows the microstructure of the epoxy-based composite with spherical SiO_2 micro-fillers of 10–15 μm average diameter and core-shell rubber filler of a few μm average diameter. The composite shows a higher breakdown voltage than any other composites when an AC stepwise voltage of 50 Hz is applied to the embedded columnar facing electrodes, as shown in Figure 12.15b. The electric field tends to concentrate more at the edges of irregular-shaped Al_2O_3 fillers with high relative permittivity than at the edges of the spherical-shaped SiO_2 fillers. However, the spherical-shaped SiO_2 fillers mitigate the local concentration of the electric field. Moreover, the core-shell rubber fillers decrease internal stresses thanks to the reduction in the elastic modulus of the epoxy matrix. Thus, dispersion of spherical fillers and core-shell rubber fillers in epoxy resin enhances the breakdown voltage in its composites.

12.5.3.2 Long-term Insulation Breakdown (*V-t* Characteristics)

Long-term insulation breakdown characteristics (*V-t* characteristics) of epoxy-based composites are important to guarantee the reliability and availability of insulators in

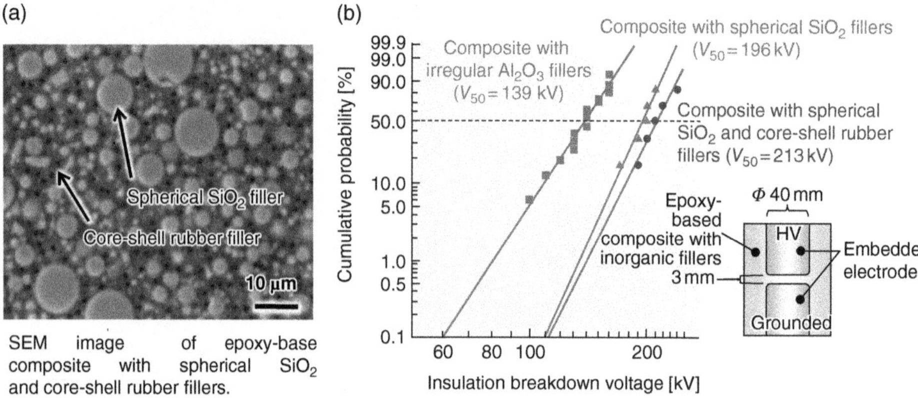

Figure 12.15 (a) SEM image of epoxy-base composite with spherical SiO_2 and core-shell rubber fillers. (b) Insulation breakdown voltage under homogeneous electric Field. Microstructure and insulation breakdown voltage of epoxy-based composite. *Source:* Sato et al. [7].

(a)

No.	Electrode		Filler	Gap [mm]	*n*
(1)	Half-ball rod / Plate	(50φ, 25R)	SiO_2	2.7	12
(2)	Plate / Plate	(67φ)	SiO_2	5	13.2
			Al_2O_3		12.3
(3)	Plate / Plate	(30φ)	Al_2O_3	15	15
(4)	Half-ball rod / Plate	(40φ, 20R)	Al_2O_3	3	16
	Plate / Plate	(40φ)			
(5)	Plate / Plate	(25φ)	SiO_2	2–4	10
	Coaxial cylinder	(22φ, 150L)		6–20	9.8
(6)	Ball / Plate	(7R)	SiO_2	2	12

(b)

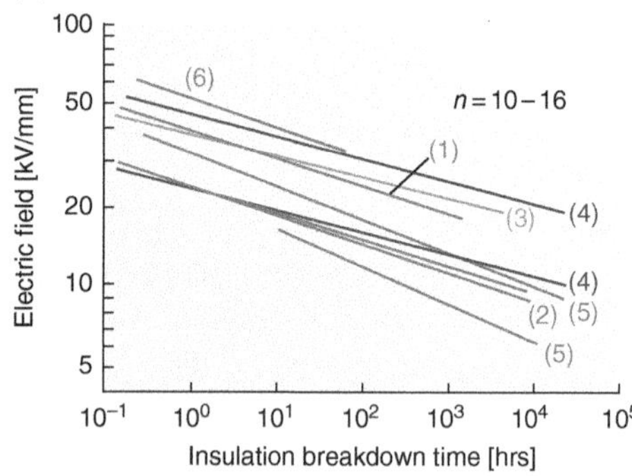

Figure 12.16 *V-t* characteristic of various epoxy-based composites. *Source:* Based on Aoyagi et al. [18].

switchgears. Figure 12.16 shows the *V-t* characteristics of various epoxy-based composites under several electrode configurations [18]. Insulation breakdown time *t* increases with decreasing applied AC electric field *E* according to $t \propto V^{-n}$ in each composite. Coefficient *n* indicates the electric field dependence of degradation in the composite. The figure shows that composites without voids or any other defects have *n* of 10–16. In these cases, degradation of the composites slowly progresses under the high-voltage application. Partial discharge and electron bombardment at very minute interspace between the composites and electrodes degrade the composites and the decline of resistance to the electric field due to the degradation causes insulation breakdown.

Switchgears are required to operate for over 20 years. *V-t* characteristics estimate the electric field of insulation breakdown after two decades. Insulation design based on *V-t* characteristics enables insulators to maintain electrical stability for a long time. Moreover, electrode area has an influence on the electric field of insulation breakdown. This is called the electrode area effect, which means that increase of electrode area causes decreases in the electric field of insulation breakdown up to a certain degree. Therefore, insulation design of insulators should take into consideration the electrode area effect, which implies the area effect related to the surface area of metal conductors in switchgear.

12.5.3.3 Relative Permittivity and Resistivity

The following equation calculates the relative permittivity of the epoxy-based composite where there is no interfacial polarization between fillers and adjacent resin.

$$\varepsilon_c^{\ k} = V_f \cdot \varepsilon_f^{\ k} + \left(1 - V_f\right) \cdot \varepsilon_e^{\ k}$$

where ε_c, ε_f, and ε_e are relative permittivity of composite, filler, and epoxy matrix, respectively, and V_f is the volume fraction of fillers. The equation is based on the capacitor model, which consists of two materials (filler and epoxy matrix). Therefore, the composition of the composite is reflected in k, which is determined on the basis of theoretical equations and experimental data. k ranges from −1 to 1. $K = 1$ maximizes and $k = -1$ minimizes the relative permittivity of the composite. SiO_2 filler, Al_2O_3 filler, and epoxy matrix have relative permittivities of 3.7–4.5, 9–10, and 3.0–4.5, respectively. Composites with low relative permittivity are favorable for insulators of switchgears.

Figure 12.17a shows the effects of filler content (volume fraction) on relative permittivities in epoxy-based composites with Al_2O_3 micro-fillers [19]. The relative permittivities of the composites increase with increasing filler content according to the above equation. Al_2O_3 filler has resistance to SF_6 decomposition gas as described in detail in the next subsection. Therefore, Al_2O_3 filler is often used for insulators in SF_6 gas-insulated switchgears (GIS). However, the relative permittivity of Al_2O_3 is higher than that of the epoxy matrix and SiO_2 fillers. Consequently, they show high permittivities. Alternative aluminum fluoride (AlF_3) filler, which has resistance to SF_6 decomposition gas, shows lower relative permittivity than Al_2O_3. Figure 12.17a shows that composites with AlF_3 fillers have low permittivities than composites with Al_2O_3 fillers. However, the mechanical properties of composites with AlF_3 fillers are inferior to those with Al_2O_3 fillers.

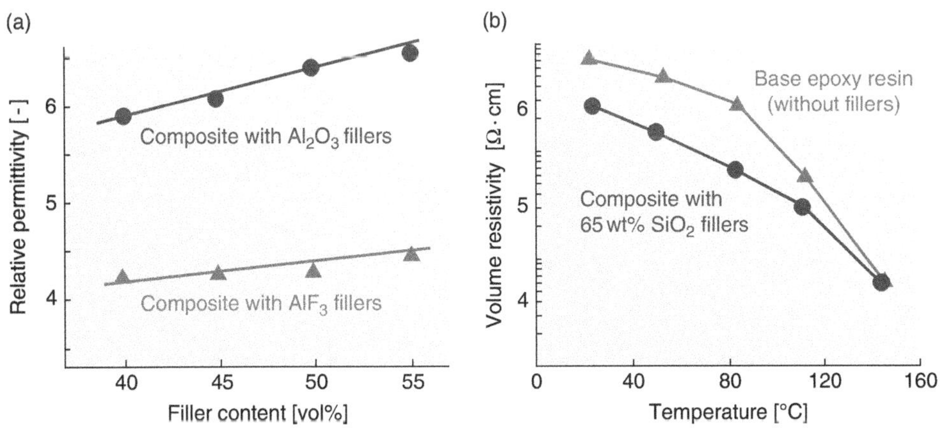

Figure 12.17 Relative permittivity and volume resistivity of epoxy-based composites. (a) Effects of filler content on relative permittivity. (b) Temperature dependence of volume resistivity. *Source:* Based on [19, 20].

Figure 12.17b shows the temperature dependence of volume resistivity in the base epoxy resin and a composite with 65 wt% SiO_2 micro-fillers 17 μm in average diameter [20]. Resistivities of composites are lower than that in base epoxy resin at every temperature. Ionic impurity due to SiO_2 filler loading causes a decline in the volume resistivities of the composites. Moreover, in both the base epoxy resin and the composite, the resistivities decrease with temperature increase. Increasing temperature activates the mobility of the cross-linking structure of the epoxy matrix and makes the ionic impurity attributed to contamination and curing accelerator move around the epoxy matrix easily.

12.5.4 Resistances to Ambient Stresses

12.5.4.1 Resistance to SF_6 Decomposition Gas

SF_6 gas-insulated switchgear (GIS) uses an insulation system based on SF_6 gas characteristics, which include excellent arc extinguishing characteristics. Although SF_6 gas is ordinarily very stable, the arc generated between electrodes at the time of current interruption produces activated decomposition gas SF_4, as shown in Figure 12.18a. The SF_4 reacts with H_2O in the presence of a slight amount of moisture and produces the SOF_2 and HF. These compounds adhere to the surfaces of insulators in the GIS tank and degrade SiO_2 and Al_2O_3 fillers in the epoxy-based composites. In particular, H_2SiF_6, which has deliquescent property, significantly decreases the surface resistivity.

Figure 12.18b compares the surface resistivities of epoxy-based composites with SiO_2 and Al_2O_3 fillers after 12 repetitions of 30 kA interruption test in a circuit breaker model [21]. The decline of surface resistivity due to the SF_6 decomposition gas in the Al_2O_3-filled composite is smaller than that of the SiO_2-filled composite. HF reacts with SiO_2 and Al_2O_3

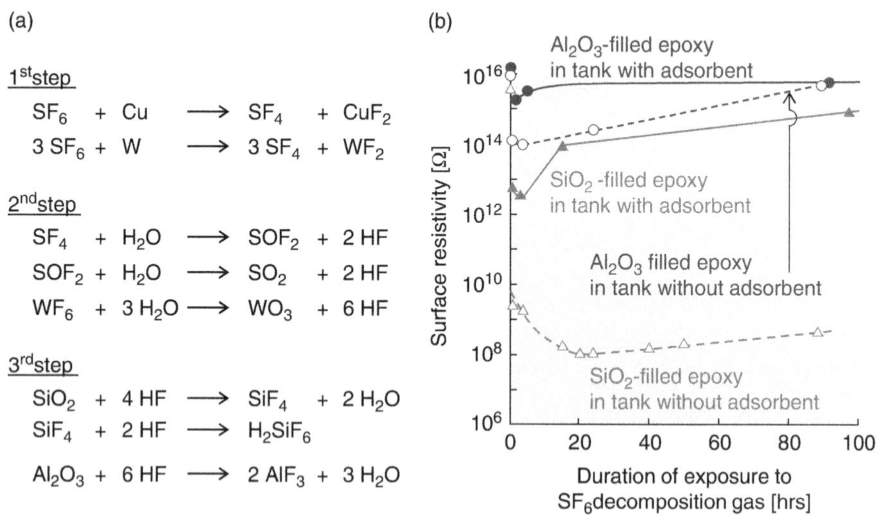

Figure 12.18 Degradation of SiO_2 and Al_2O_3 fillers in epoxy-based composites due to SF_6 decomposition gas. (a) Degradation of SiO_2 and Al_2O_3 fillers due to SF_6 decomposition gas. (b) Surface resistivities of epoxy-based composites after 12 repetitions of 30 kA interruption test in circuit breaker model tank. *Source:* Based on Tominaga et al. [21].

on the insulator surface. The surfaces of the SiO_2-filled composite are prone to moisture absorption, which boosts the production of HF and accelerates the reaction of SiO_2 and HF. However, the reaction of Al_2O_3 and HF progresses very slowly. In addition, AlF_3 shows low resolvability to absorbed moisture near the surface of the Al_2O_3-filled composite. Therefore, the Al_2O_3-filled composites show resistance to SF_6 decomposition gas.

Moreover, moisture absorption by adsorbent placed in the GIS tank is effective in suppressing the decline of surface resistivities in the SiO_2-filled composite as well as in the Al_2O_3 composite. The SiO_2-filled composite in the tank with adsorbent has much higher surface resistivity than that without adsorbent. With respect to the Al_2O_3-filled composites, the effects of moisture absorption on surface resistivity are confirmed in the same manner.

12.5.4.2 Water Absorption

Epoxy-based composites absorb ambient moisture. Absorbed water causes decline of some properties such as resistivity and flexural/tensile strength.

Figure 12.19 illustrates the influence of water absorption on relative permittivity in the epoxy-based composites with several kinds of SiO_2 fillers [22]. Figure 12.19a lists the specimens. Micrographs elucidate the differences among dispersed filler shapes in the composites and the irregular SiO_2 shows the angular shapes.

Figure 12.19b shows changes in water absorption in the composites as a function of immersion time. Immersion in water at 50 °C accelerates water absorption in the

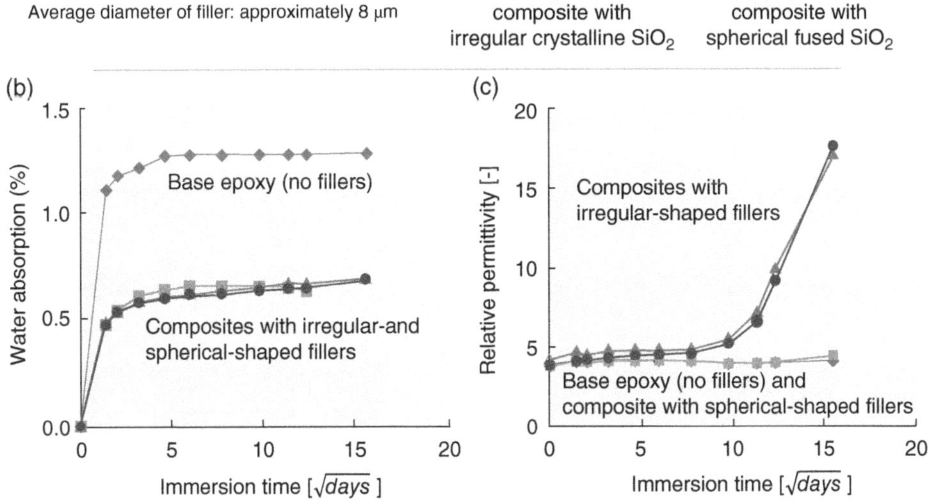

Figure 12.19 Properties of epoxy-based composites immersed in water at 50 °C. (a) Specimens (b) Changes of water absorption (c) Changes of relative permittivity. *Source:* Based on Ezoe et al. [22].

composites. The absorbed water is accumulated inside the epoxy matrix and at epoxy matrix/filler interfaces. Base epoxy resin without filler has a larger volume of the epoxy matrix than other composites. Thus, the water absorption of the base epoxy resin increases to double that of the other composites. On the other hand, significant differences do not appear in the water absorption of three kinds of composites.

Figure 12.19c shows changes of relative permittivity in the composites as a function of immersion time. Although water absorption of each composite saturates in a few days, the relative permittivities increase in the composites with irregular-shaped SiO_2 fillers after 10 days' immersion. When relative permittivities measured by using electrodes, the conductive path connecting the electrodes due to the absorbed water seems to form easily in the composites with irregular-shaped SiO_2 fillers compared to the composite with spherical-shaped SiO_2 fillers.

Generally, low relative permittivity is favorable for composites of insulators in switchgears. Permittivity increase due to water absorption should be checked in switchgears installed in high humidity areas.

12.5.5 Resistances to Mechanical Stresses

12.5.5.1 Flexural and Tensile Strength

Insulators work as structural materials as well as insulation materials in switchgears and are subjected to various mechanical stresses. Thus, epoxy-based composites for insulators need to have good mechanical properties.

Figure 12.20 shows the flexural and tensile strength of epoxy-based composites with irregular (angular) SiO_2 fillers at room temperature [23]. Increase in mean filler diameter decreases flexural and tensile strength in filler contents of both 55 and 64 wt%. In particular, dispersion of fillers over 30 μm in diameter reduces the flexural and tensile strengths of the composite to below those of the original base epoxy resin. Fillers with small diameters provide better flexural and tensile strengths. However, small fillers tend to increase the

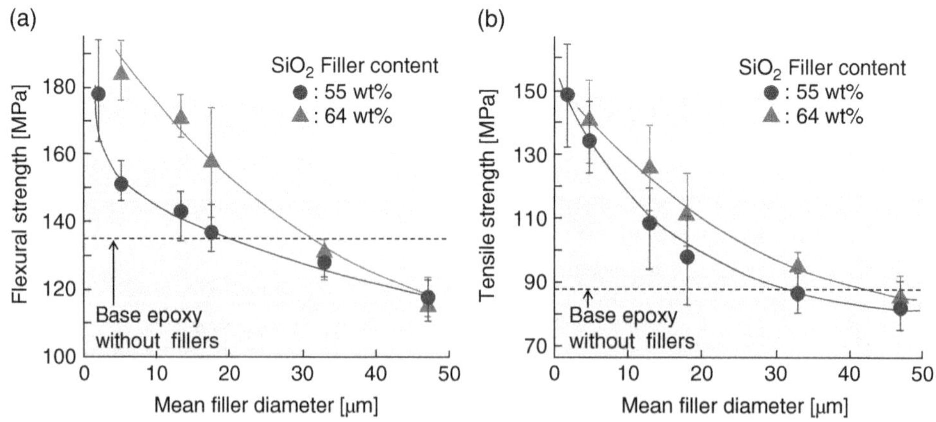

Figure 12.20 Flexural and tensile strength of epoxy-based composites with irregular-shaped SiO_2 fillers. (a) Filler diameter dependence of flexural strength. (b) Filler diameter dependence of tensile strength. *Source:* Based on Nakamura et al. [23].

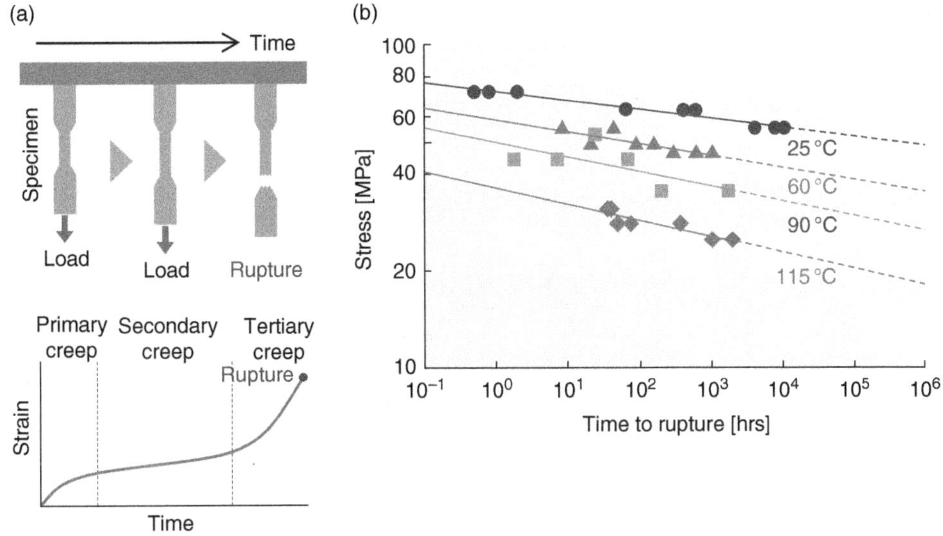

Figure 12.21 Creep properties of epoxy-based composite with SiO_2 fillers. (a) Typical creep behavior (b) Creep properties at various temperatures [24].

viscosity of composites before curing. Fillers approximately 10 μm in diameter satisfy both mechanical strength and viscosity.

12.5.5.2 Creep

Creep is a long-term mechanical property. Evaluation of creep of epoxy-based composites is needed to guarantee reliability and availability of insulators in switchgears for over 20 years.

Constant load lower than the short-term rupture strength gradually deforms the epoxy-based composite, as shown in Figure 12.21a. Creep is divided into three stages: primary, secondary, and tertiary. Secondary creep is based on the viscoelasticity of the composite.

Figure 12.21b shows the creep properties of epoxy-based composite with spherical SiO_2 micro-fillers at various temperatures [24]. Times to rupture shorten with increasing stress and temperature. In particular, temperature promotes the deformation of composites. Insulator temperatures increase due to Joule heat attributed to the flow of electricity during switchgear operation. Creep properties of composites must take account of the temperature rise of insulators.

12.5.6 International Standards for Evaluation of Composites

This section describes various properties of epoxy-based composites. Table 12.7 summarizes international standards for evaluating the properties of composites in relation to TEAM stresses. ISO (International Organization for Standardization), ASTM (American Society for Testing and Materials), and IEC (International Electrotechnical Commission) set and issued standards for materials including epoxy-based composites. Measurements and evaluations of composite properties according to the methods shown in international standards are strongly recommended.

Table 12.7 International standards for materials including epoxy-based composites.

TEAM Stresses	Properties	Standard	Title
Thermal stresses	Glass transition temperature	ISO 3146:2000	Plastics – determination of melting behavior (melting temperature or melting range) of semicrystalline polymers by capillary tube and polarizing-microscope methods
		ASTM D3418 – 15	Standard test method for transition temperatures and enthalpies of fusion and crystallization of polymers by differential scanning calorimetry
	Coefficient of thermal expansion	ASTM D696 – 16	Standard test method for the coefficient of linear thermal expansion of plastics between −30 and 30 °C with a vitreous silica dilatometer
Electrical stresses	Insulation breakdown strength	ASTM D149 – 20	Standard test method for dielectric breakdown voltage and dielectric strength of solid electrical insulating materials at commercial power frequencies
		IEC 60243-1:2013 IEC 60243-2:2013 IEC 60243-3:2013	Electric strength of insulating materials – test methods – parts 1, 2, and 3
	Permittivity	ASTM D150 – 18	Standard test methods for AC loss characteristics and permittivity (dielectric constant) of solid electrical insulation
		IEC 62631-2-1:2018	Dielectric and resistive properties of solid insulating materials 1 Part 2-1: relative permittivity and dissipation factor – technical frequencies (0,1 Hz–10 MHz) – AC methods
	Resistivity	ASTM D257 – 14	Standard test methods for DC resistance or conductance of insulating materials
		IEC 62631-3-1:2016 IEC62631-3-2:2015 IEC 62631-3-32015 IEC 62631-3-4:2019	Dielectric and resistive properties of solid insulating materials – parts 3-1, 3-2, 3-3, and 3-4: Determination of resistive properties (DC methods)
Ambient stress	Water absorption	ISO 62:2008	Plastics – determination of water absorption
Mechanical stresses	Flexural strength	ISO 178:2019	Plastics – determination of flexural properties
	Tensile strength	ISO 527-1:2019 ISO 527-2:2012	Plastics – determination of tensile properties – parts 1 and 2
		ASTM D638-14	Standard test method for tensile properties of plastics

12.6 Advances of Epoxy-based Composites for Switchgear

"Tailoring" is the keyword of recent research and development of epoxy-based composites. Precise material design to meet social demands as well as required material properties is essential to ensure that composites will enable high performance, downsizing, and energy conservation in switchgears. This section introduces recent examples of tailoring of composites.

12.6.1 Nanocomposites

Epoxy resins have superior insulation properties. However, if they contain defects such as voids, these properties are severely deteriorated due to degradation such as electrical treeing and partial discharge. Therefore, epoxy resins with superior insulation properties under high voltage are needed to expand the use of solid insulation systems.

Nano-filler dispersion is an effective approach to improving the insulation properties of epoxy resin. Nanocomposites containing a few wt% clay nano-filler have higher insulation breakdown time than the base epoxy resin without fillers under constant AC voltage using the needle-plate electrode configuration, as shown in Figure 12.22a [15].

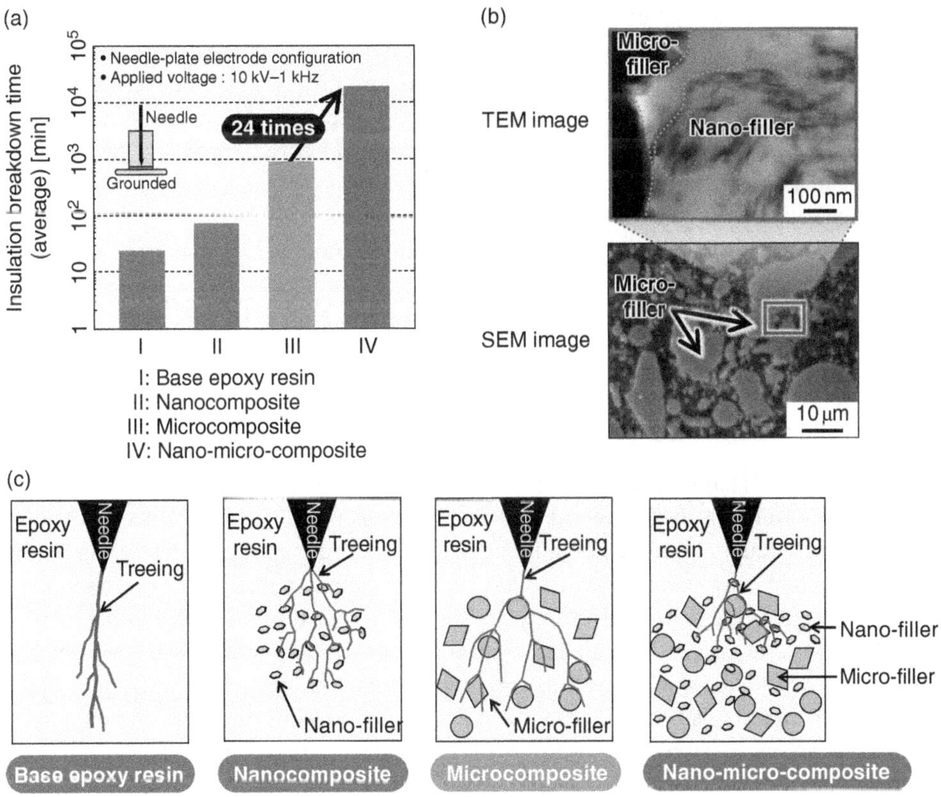

Figure 12.22 Improvement of insulation properties by nano-filler dispersion. (a) Comparison of insulation breakdown time. (b) Micrographs of nano-micro-composite. (c) Estimated mechanism of resistance to tree propagation. *Source:* Based on Imai et al. [15].

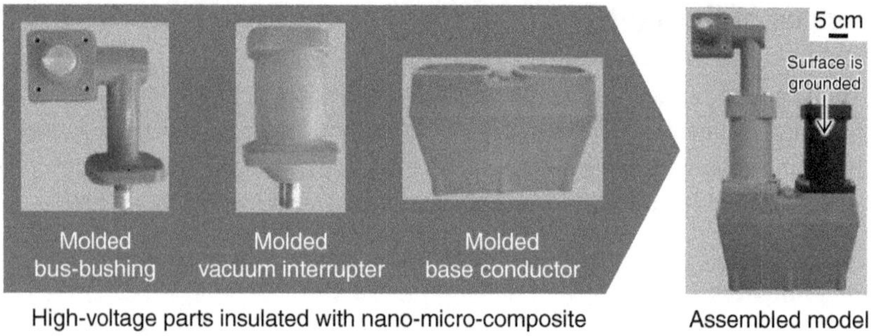

Figure 12.23 Component model of solid insulated switchgear (SIS) insulated with nano-micro-composite. *Source:* Imai et al. [25].

In solid insulation switchgear (SIS), high-voltage parts such as vacuum interrupters and metal conductors are molded from epoxy-based composite. In the conventional filled epoxy resin (microcomposite), SiO_2 fillers are used as inorganic micro-fillers, and filler content is 65 wt%. Nano-filler dispersion also improves the insulation properties of the microcomposite. The epoxy resin containing 64 wt% SiO_2 micro-fillers and a few wt% clay nano-fillers, which is called a nano-micro-composite, shows 24 times longer insulation breakdown time [20]. Figure 12.22b shows an electron micrograph of nano-micro-composite. The distance between fillers in a nano-micro-composite is one-fifth that of a conventional filled epoxy with the same amount of SiO_2 micro-fillers. The picture demonstrates that nano-mciro-composite has a densely packed filler structure. Moreover, Figure 12.22c estimates the mechanism of improvement in which the dispersed nano-fillers and micro-fillers seem to effectively prevent electrical tree propagation.

In addition, a component model of SIS composed of a connecting conductor, a vacuum interrupter and a base conductor insulated with nano-micro- composite was made for practical use, as shown in Figure 12.23 [25]. However, the nano-micro-composite is not applied to commercial SIS.

12.6.2 High Thermal Conductive Composites

The international standard for switchgear IEC 62271-1 specifies temperature-rise limits for various parts and materials, as mentioned in subsection 12.5.2. In bolt-connected external terminals, the standard limit temperatures rise to 65 K in tin-coated conductors and 50 K in bare conductors. Large current switchgear needs to suppress temperature rise due to Joule heat attributed to the flow of electricity. In SF_6 gas and air-insulated switchgears (C-GIS and AIS), gas and air convection enable the conductors to promote outward heat dispersion. In solid insulated switchgear (SIS), however, conductors are molded with epoxy-based composite containing silica (SiO_2) and heat dispersion depends on thermal conductivity. Therefore, an epoxy-based composite with high thermal conductivity is needed for large current SIS.

Boron nitride (BN), alumina (Al_2O_3), and magnesia (MgO) are typical thermally conductive fillers for epoxy-based composites. Their thermal conductivities are approximately 60, 36, and 42 W/(m·K), respectively. MgO fillers are most cost-effective, but have the

(a)

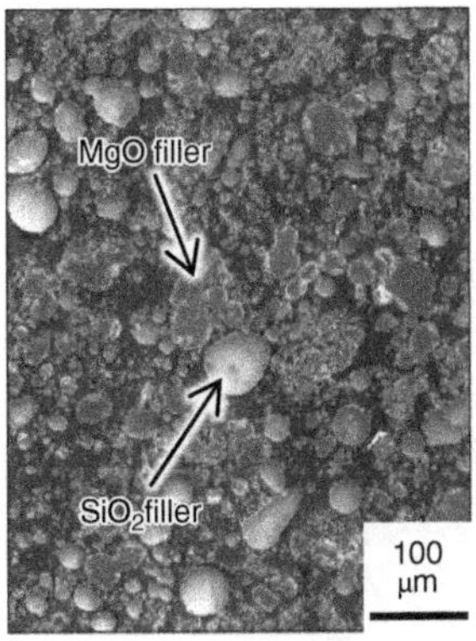

(b)

Properties	Measured value
Thermal conductivity	1.39 W/(m·K)
Coefficient of thermal expansion (α_1)	23 ppm/K
Insulation breakdown strength (at RT)	16.3 kV/mm
Dielectric constant (at 50 Hz and RT)	4.9
Volume resistivity (at RT)	6.0×10^{12} Ω·cm
Surface resistivity (at RT)	6.0×10^{13} Ω
Flexural strength (at RT)	122.6 MPa
Tensile strength (at RT)	74.0 MPa
Young's modulus (at RT)	18.2 GPa
Fracture toughness (at RT)	3.0 MPa·$m^{1/2}$
Glass transition temperature	115°C
Resin viscosity before curing (at 80°C)	2513 mPa·s

(RT: room temperature)

Figure 12.24 SEM image and properties of high thermal conductive composite. (a) SEM image. (b) Material characteristics. *Source:* Based on Komiya et al. [26].

disadvantage of large coefficient of thermal expansion (CTE). The CTE of BN, Al_2O_3, and MgO is approximately 1, 6, and 13 ppm/K respectively. Thus, the combination of an MgO filler with high thermal conductivity and an SiO_2 filler with a low CTE of 0.5 ppm/K is selected to have both properties in the developed epoxy-based composite as shown in electron micrograph (SEM) of Figure 12.24a [26]. The composite has a total filler loading of 51 vol% and the optimized ratio of 51: 49 vol% in MgO and SiO_2 filler loading enable the composite to have a high thermal conductivity of 1.39 W/(m·K) compared to approximately 0.5 W/(m·K) of the conventional composite with only SiO_2 fillers, as shown in Figure 12.24b. Moreover, the composite shows a low CTE of 23 ppm/K: the same as an aluminum metal conductor, and other properties are also sufficient for composites used in switchgears.

Figure 12.25 compares thermal analysis results by using CAE (computer-aided engineering) simulation in 40.5 kV/2500 A SIS applied with the thermal conductive composite and conventional composite. The results demonstrate that the thermally conductive composite suppresses temperature rise up to 10 K under large current flow.

12.6.3 Biomass Material-Based Composites

It is generally difficult to recycle thermo-setting resins compared to thermoplastic resins including epoxy resin due to the polymer structures with cross-linking. Used epoxy-based composites are generally burned and landfilled, which would cause increasing of carbon dioxide (CO_2) gas emission and a waste of petroleum resources. Therefore, biomass-based plastics have attracted attention based on carbon offset. Epoxidized vegetable oil such as soybean, linseed, palm, and castor oil are well known as biomass-based epoxy resins [27]. In power apparatuses including switchgears, biomass-based epoxy resins and their composites are developed as electrical insulation materials [28, 29].

Epoxidation of unsaturated bonds in linseed oil produces biomass-based epoxy resin. Table 12.8 compares the properties of epoxidized linseed-oil-based composites containing silica (SiO_2) fillers [28]. The improved biomass-based composite contains fused SiO_2 fillers

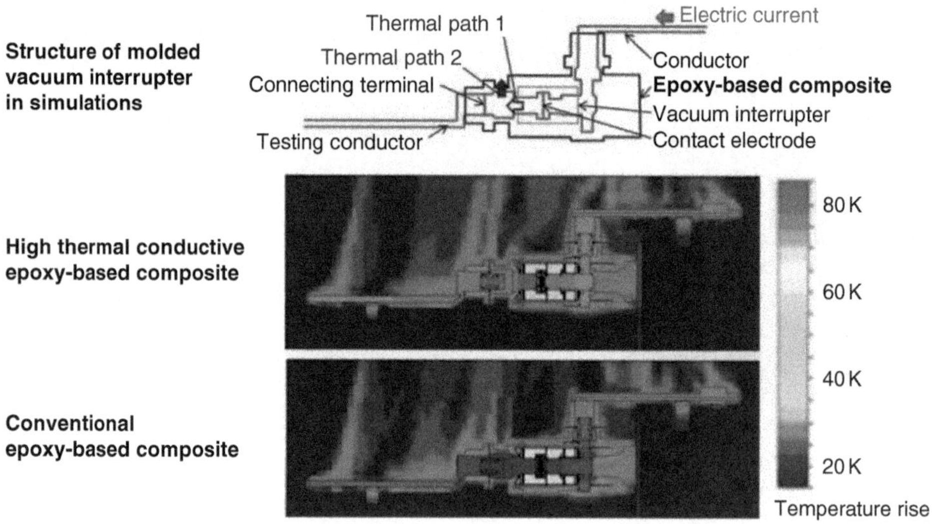

Figure 12.25 Comparison of temperature-rise analysis in 40.5 kV/2500 A solid insulated switchgear. *Source:* Komiya et al. [26].

instead of crystalline SiO_2. The fused SiO_2 enables the composite to have approximately 1.8 times higher filler loading than crystalline SiO_2. Mechanical, thermal, and insulation properties of the biomass-based and improved biomass-based composites are slightly inferior to those of existing epoxy-based composite. However, these biomass-based composites show superior resistance to thermal shock to existing epoxy-based composite. Therefore, although biomass-based composites have limited applications, they are applicable to insulation materials in power apparatuses.

12.6.4 Functionally Graded Materials

Functionally graded materials (FGM) are at the cutting edge of material technology and they will enable high efficiency and downsizing of apparatuses leading to energy conservation in the power sector [30–32]. An appropriate relative permittivity gradient of an insulation spacer in gas-insulated switchgear (GIS) decreases the diameter of its tank, as shown in Figure 12.26a. A reduction of 30% in GIS tank diameter is estimated based on the calculation of electric field distribution on the surface of the spacer. However, manufacturing is the core issue, and the distribution of epoxy-based composite materials (resins) with different permittivities dominate the permittivity gradient in spacers.

Two kinds of resins containing fillers with different relative permittivities have been developed to control the permittivity gradient. Mixtures with various mixing ratios between high-permittivity resin-containing strontium titanate ($SrTiO_3$) and low-permittivity resin-containing silica (SiO_2) fillers are sequentially injected into a metal mold to form an insulation spacer. This is called the "flexible mixture casting method" [33]. Computer-aided engineering (CAE) is available to preliminarily investigate concerns about the method in a full-scale insulation spacer. Computational fluid dynamics calculates resin

Table 12.8 Comparison of epoxidized linseed oil-based composites.

Properties[a]		**Existing Epoxy-based Composite (Petroleum-derived)**	**Biomass Composite based on Expoxidized Linseed Oil**	**Improved Biomass Composite based on Epoxidized Linseed Oil**
Filler	Type	Crystalline silica	Crystalline silica	Fused silica
	Ratio by weight	1	1	1.8
Thermal shock		Cracked by thermal shock at 0-100oC	Cracked after the end of curing	Cracked by thermal shock at -10 - 100oC
Mechanical property	Bending strength	1	0.8	0.8
	Bending elastic modulus	1	0.7	1.0
Thermal property	Glass transition temperature	1	0.7	0.7
Electrical property	Dielectric strength	1	1.1	—
	Dielectric constant	1	0.8	—

[a] Properties are normalized values based on the existing epoxy resin (petroleum-derived).
Source: Based on Ohtake et al. [28].

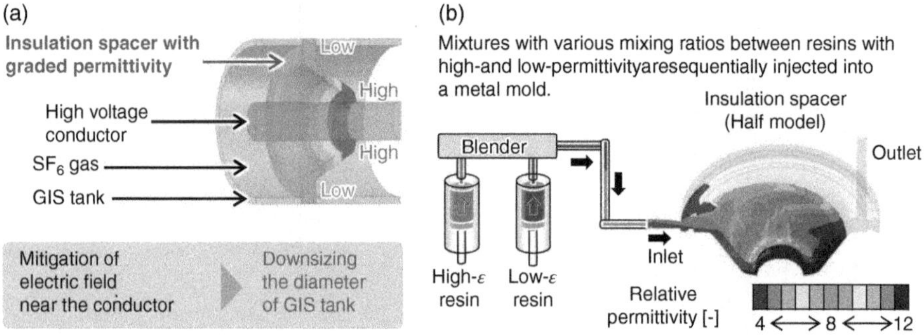

Figure 12.26 Insulation spacer with permittivity gradient and simulation of resin injection behavior in manufacturing process. (a) Cross section of insulation spacer with graded permittivity in a GIS tank. (b) Resin injection behavior in the process of manufacturing insulation spacer using computational fluid dynamic. *Source:* Imai et al. [35].

injection behavior and determines the distributions of resins with different permittivities, as shown in Figure 12.26b [34]. Moreover, simulation and experiment show similar tendencies in resin distribution and permittivity gradient. The comparison demonstrates that the simulation is valid [35].

Effects of insulation spacers with relative permittivity gradient on insulation breakdown characteristics are investigated in cone-shaped insulation spacer models with quarter full-scale spacers [36]. Figure 12.27a shows an insulation spacer model prepared using the flexible mixture casting method. The spacer model has the permittivity distribution shown in Figure 12.27b. Impulse insulation breakdown (negative lightning impulse) tests were conducted in the tank filled with 0.4 MPa-Gauge. Figure 12.27c compares breakdown voltages between the spacer model with permittivity gradient (ε-FGM) and a uniform permittivity of 4 (uniform). The ε-FGM spacer model shows a higher breakdown voltage than the uniform spacer model. Moreover, the experimental results almost coincide with estimations based on volume-time theory [37]. Mitigation of the electric field near a high-voltage conductor attributed to permittivity gradient seems to enhance breakdown voltage in ε-FGM.

12.6.5 Estimate of Remaining Life of Composites

The end of this section describes a unique development concerning polymer composite insulation materials. The remaining life estimation based on an evaluation of the condition in polymer composite insulators has recently emerged as a preventive measure against insulation breakdown in switchgears [38, 39]. Diagnosis of the insulators contributes to control of their operation and maintenance.

Surface resistivities of polymer composite insulation materials reflect their conditions. Material degradation attributed to electrical and thermal stress and pollution attributed to ambient stress such as atmosphere gas and powder dust cause a reduction of surface resistivities, as shown in Figure 12.28a. Therefore, evaluation of material condition (X_a), pollution condition (X_b), and temperature and humidity (X_c) elicit the following equation [39]:

$$Y = C + n_1X_1 + n_2X_2 \cdots\cdots n_iX_i$$

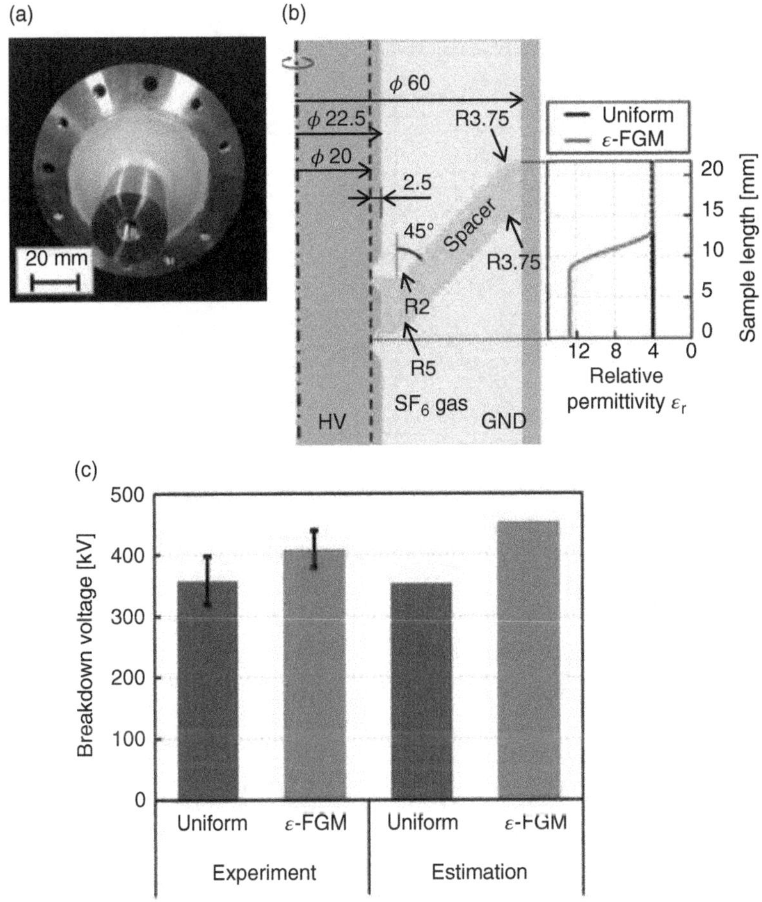

Figure 12.27 Improvement of breakdown voltage in insulation spacer with permittivity gradient. (a) ε-FGM spacer (convex surface). (b) Cross section and permittivity distribution. (c) Comparison of breakdown voltage. *Source:* Based on Miyazaki et al. [36].

where Y is estimated resistivity, n_1–n_i are weighting factors, X_1–X_i are measured values in each evaluation, and C is a coefficient.

An extended line connecting the dot of the initial value and the dot of the estimated resistivities Y_1 reaches the line of threshold resistivity, and its crossover point shows the remaining life T_1 of the polymer composite insulation materials, as shown in Figure 12.28b. Plural evaluations are needed to take into consideration the variability, which distributes the estimated resistivities in the range from Y_1 to Y_2. Thus, the shortest remaining life T_1 and the longest remaining life T_2 are estimated. Moreover, estimated resistivity Y_3 based on the evaluation under a high-humidity environment elicits a much shorter remaining life T_3 in similar degraded materials.

In addition, the advent of artificial intelligence (AI) and machine learning has accelerated proactive approaches toward cutting-edge diagnosis of the insulators.

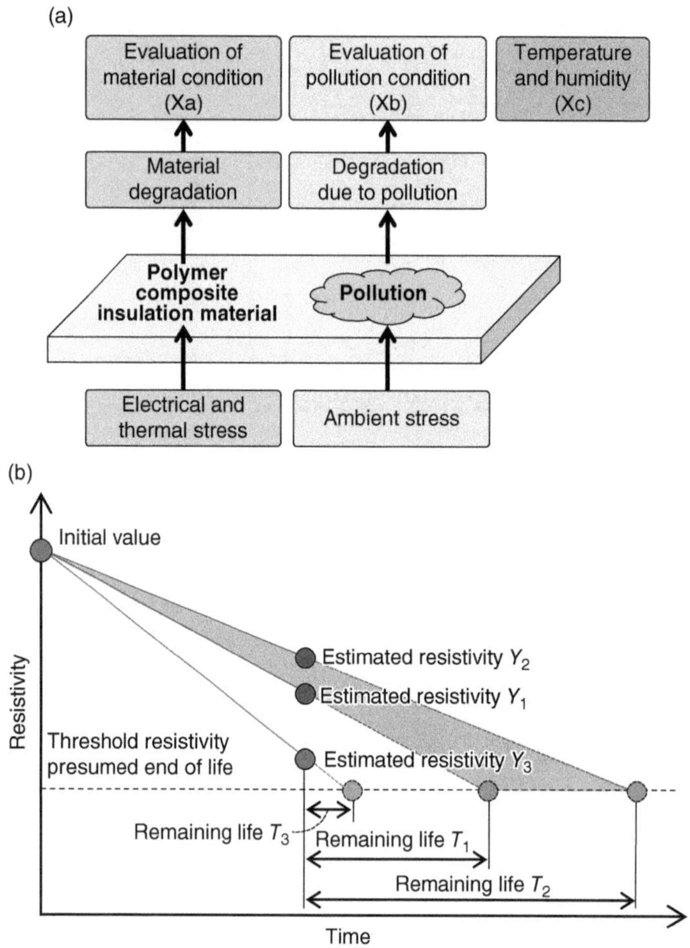

Figure 12.28 Evaluation of material degradation and estimate of the remaining life. (a) Evaluation criteria for material degradation. (b) Estimation method of remaining life. *Source:* Based on Murayama et al. [39].

12.7 Conclusion

The objective of this chapter is to describe polymer composites for switchgears. Related technologies covering a wide range from the history of switchgears, manufacturing of insulators, materials for composites, properties of composites, and emerging composites are treated. Epoxy-based composites are essential materials for insulators in switchgear. They need to have resistance to TEAM (Thermal, Electrical, Ambient, and Mechanical) stresses. However, most properties mutually interact: one property is improved but another property is sometimes deteriorated. Therefore, material design is very important, and an optimized combination of epoxy resin, hardener, and inorganic filler enables the composite to meet the material properties required in insulators.

The United Nations adopted "The 2030 Agenda for Sustainable Development" as our common vision for a better future in 2015. The agenda includes "The Sustainable Development Goals (SDGs)" which consists of 17 "Global Goals" with 169 targets. The 7th of these 17 global goals is "Ensure access to affordable, reliable, sustainable, and modern energy (Affordable and clean energy)." Switchgears have played important roles in electric energy distribution. The growing role of switchgears stimulates our interests in technological development toward a sustainable future.

Ever-improving technologies related to polymer composite insulation materials enable high efficiency, high performance, downsizing, long-life operation, energy conservation, and environmental friendliness in switchgears. Furthermore, reuse and recycle technologies for the insulation materials also matter in the development. I believe that these technologies will contribute to the achievement of the 7th goal. Recently, new approaches such as quantum chemical calculation and materials informatics are emerging in material developments. Interdisciplinary research and development by integration of new and conventional technologies are needed to achieve innovative polymer composite insulation materials that support affordable and clean energy.

References

1. The Federation of Electric Power Companies of Japan (2012). Nuclear and energy drawings, Information Library, 1-1-9, 1-1-10 (in Japanese).
2. The Federation of Electric Power Companies of Japan (2012). The path of electricity. http://www.fepc.or.jp/enterprise/souden/keiro/ (in Japanese).
3. The Japan Center for Climate Change Actions (2010). Properties of Green House Gases, 1–2 (in Japanese). http://www.jccca.org.
4. Shimizu, T. (2004). Material technology for Solid Insulated Switchgear. *EINA (Electrical Insulation News in Asia) Magazine*, pp. 41–42.
5. Yoshida, T., Yano, T., Tohya, N. et al. (2006). The development and practical use of a new 24 kV dry air insulated switchgear. *IEEJ Transactions on Power and Energy* 126 (3): 359–364. (in Japanese).
6. Yoshida, T., Koga, H., Harada, T. et al. (2011). Insulation technology in dry air and vacuum for a 72-kV low-pressure dry-air insulated switchgear. *Electrical Engineering in Japan* 175 (1): 18–24.
7. Sato, J., Sakaguchi, O., and Miyagawa, M. (2003). 24 kV solid-insulated switchgear conforming with environmental requirements. *Toshiba Review* 58 (12): 66–69. (in Japanese).
8. Fujii, S., Sakaguchi, O., and Sato, J. (2004). 24/36 kV solid-insulated switchgear. *Toshiba Review* 59 (12): 56–59. (in Japanese).
9. Takei, M. and Hirano, Y. (2008). *Denzai Journal* 598: 35–39. (in Japanese).
10. Mooney, M. (1951). The viscosity of a concentrated suspension of spherical particles. *Journal of Colloid Science* 6 (2): 162–170.
11. Horsfield, H.T. (1934). The strength of asphalt mixtures. *Journal of the Society of Chemical Industry* 53: 107–115.

12 Investigating R&D Committee on Advanced Tailor-made Composite Insulation Materials (2019). Advanced Tailor-made Composite Insulation Materials – Towards Innovative Composite Materials to Meet Demand in Electric Power and Electronics Sectors. Technical report No. 1455. The Institute of Electrical Engineers of Japan, 1–116. (in Japanese).

13 IEC 62271-1:2017 (2017). High-voltage switchgear and controlgear – part 1: common specifications for alternating current switchgear and controlgear.

14 Ochi, M., Kishi, H., and Fukui, T. (eds.) (2011). *The Latest Technology of Epoxy Resin for Electronic Devices II*. CMC Publishing Co., Ltd. (in Japanese).

15 Imai, T., Komiya, G., Murayama, K. et al. (2008). Improving epoxy-based insulating materials with nano-fillers toward practical application. *Proceedings IEEE International Symposium on Electrical Insulation*, 201–204.

16 Flandin, L., Vouyovitch, L., Beroual, A. et al. (2005). Influences of degree of curing and presence of inorganic fillers on the ultimate electrical properties of epoxy-based composites: experiment and simulation. *Journal of Physics D: Applied Physics* 38: 144–155.

17 Yoshimura, N., Fujita, S., Noto, F., and Tamura, M. (1982). AC short time breakdown characteristics of alumina filled epoxy resin. *IEEJ Transactions on Fundamentals and Materials* 102 (12): 673–677. (in Japanese).

18 Aoyagi, H., Koya, M., and Oakmoto, Y. (1981). *Convention record of four institutes of electrical engineers*, No. S13-4, 2-119-2-122 (in Japanese).

19 Miyagawa (1988). IEEJ the Papers of Technical Meeting, EIM-88-33 (in Japanese).

20 Imai, T., Sawa, F., Nakano, T. et al. (2006). Effects of nano- and micro-filler mixture on electrical insulation properties of epoxy based composites. *IEEE Transactions on Dielectrics and Electrical Insulation* 13 (2): 319–326.

21 Tominaga, S., Kuwahara, H., and Hirooka, K. (1979). Influence of arc-decomposed SF_6 gas on cast epoxy insulators for gas insulated switchgears. *IEEE Transactions on Power Apparatus and Systems* PAS-98 (6): 2107–2114.

22 Ezoe, M., Yamanaka, S., Tada, M., and Kawamura, H. (1999). The influence of filler shape and water absorption on dielectric properties of filled epoxy resin. *IEEJ Transactions on Fundamentals and Materials* 119 (7): 1011–1018. (in Japanese).

23 Nakamura, Y., Yamaguchi, M., Okubo, M., and Matsumoto, T. (1992). Effect of particle size on mechanical properties of epoxy resin filled with angular-shaped silica. *Journal of Applied Polymer Science* 44: 151–158.

24 Kinoshita, S., Shimizu, T., Komiya, G. et al. (2007). *Sokeizai* 12: 36–39. (in Japanese).

25 Imai, T., Yamazaki, K., Komiya, G. et al. (2010). Component models insulated with nanocomposite material for environmentally-conscious switchgear. *Proceedings IEEE International Symposium on Electrical Insulation*. 299–232.

26 Komiya, G., Imai, T., and Miyauchi, Y. (2018). Development of high thermal conductive epoxy composite for large current switchgear. *Electrical Engineering in Japan* 205: 9–15.

27 Mustapha, R., Rahmat, A.R., Majid, R.A., and Mustapha, S.N.H. (2019). Vegetable oil-based epoxy resins and their composites with bio-based hardener: a short review. *Polymer-Plastics Technology and Materials* 58 (12): 1311–1326.

28 Ohtake, M., Yamashita, T., Maruyama, M., and Otsuka, T. (2014). Improvement in the thermal shock of epoxy resin based from vegetable oil for the application to power electric mold euipment. *Annual Meeting Record IEE Japan* No. 2-010: 12–13 (in Japanese).

29 Ohtake, M., Yamashita, T., Ogawa, K. et al. (2015). Optimization of the curing process for the epoxy resin based on vegetable oil for electric mold equipment. *Annual Meeting Record IEE Japan* No. 2-019: 23–24 (in Japanese).

30 Kurimoto, M., Kato, K., Hanai, M. et al. (2010). Application of functionally graded material for reducing electric field on electrode and spacer interface. *IEEE Transactions on Dielectrics and Electrical Insulation* 17 (1): 256–263.

31 Ju, H.-J., Kim, B., and Ko, K.-C. (2011). Optimal design of an elliptically graded permittivity spacer configuration in gas insulated switchgear. *IEEE Transactions on Dielectrics and Electrical Insulation* 18 (4): 1268–1273.

32 Talaat, M., El-Zein, A., and Amin, M. (2018). Electric field simulation for uniform and FGM cone type spacer with adhering spherical conducting particle in GIS. *IEEE Transactions on Dielectrics and Electrical Insulation* 25 (1): 339–351.

33 Ochiai, K., Izu, A., Oishi, R. et al. (2018). Fabrication of permittivity graded materials (ε-FGM) by flexible mixture casting method. *Annual Report Conference on Electrical Insulation and Dielectric Phenomena* (CEIDP): 696–699.

34 Imai, T., Hoshina, Y., Kojima, H., and Hayakawa, N. (2019). Resin injection behavior in process of manufacturing insulation spacer with permittivity gradient using computational fluid dynamics. *IEEE Transactions on Dielectrics and Electrical Insulation* 26 (5): 1669–1677.

35 Imai, T., Takeuchi, M., Hoshina, Y. et al. (2019). Comparison of permittivity gradient in resin injection experiment and simulation for cone-shaped insulation spacer. *IEEE Transactions on Dielectrics and Electrical Insulation* 26 (5): 1678–1685.

36 Miyazaki, Y., Izu, A., Liang, Z. et al. (2020). Breakdwon characteristics of cone-type e-FGM (functionally graded materials) spacer for GIS. *Annual Meeting Record IEE Japan* No. 6 006: 9 (in Japanese).

37 Hayakawa, N., Miyaji, Y., Kojima, H., and Kato, K. (2018). Simulation on discharge inception voltage improvement of GIS spacer with permittivity graded materials (ε-FGM) using flexible mixture casting method. *IEEE Transactions on Dielectrics and Electrical Insulation* 25 (4): 1318–1323.

38 Miki, S., Hasegawa, T., Umemura, S. et al. (2008). Remaining service life diagnostic technology of phenol insulators for circuit breakers. *IEEE Transactions on Dielectrics and Electrical Insulation* 15 (2): 476–483.

39 Murayama, K., Tamura, T., and Mizuide, T. (2013). Degradation diagnosis technologies for power distribution facilities. *Toshiba Review* 68 (8): 52–55. (in Japanese).

13

Glass Fiber-Reinforced Polymer Composites for Power Equipment

Yu Chen

School of Electrical Engineering, Xi'an Jiaotong University, Xi'an, China

13.1 Overview

Neither fiber nor matrix can operate alone to achieve a combined property of two or more composite materials. Fiber-reinforced composites have been successfully used in a wide range of technological applications for decades. When it came to producing composite materials, the most frequently used kind was GFRP (glass fiber-reinforced polymeric). A wide range of resins was employed in the matrix, including organics, polyesters, thermostable vinyl ester, phenolic, and epoxy resins. Bisphenolic and ortho- or isophthalic polyester resins are the two types of resins available. Fiber strength and modulus, chemical stability, matrix strength, and the interface bonding between the fiber and matrix are all factors that influence the mechanical performance of a fiber-reinforced composite. GFRP composites were designed to be as strong as steel, stiffer than aluminum, and have a specific gravity of one-fourth that of steel, thanks to its fiber composition and orientation. Different GF reinforcements, such as long longitudinal woven mats, chopped fibers (distinct), and chopped mats, have been developed to improve the mechanical and tribological properties of composites. The fibers used in composite manufacturing determine the material's properties, such as strength, stiffness, and dimensional stability. Polymers were slow to gain traction in commercial settings due to their high cost. As a consequence, fillers improved composite properties while decreasing preparation and product costs overall. These composite materials may be used in a broad range of industries, from marine structures to pipes. Their high specific strength and stiffness, along with their exceptional environmental resilience, make them ideal. Lightweight decreased fatigue resistance in fasteners, and an increased component number made polymeric composites popular in the aviation industry, where they were utilized for rudders, elevators, fuselages, and landing gear doors. In maritime applications, polyester matrix-based composites have been widely used, and polymer composite deterioration in the marine environment necessitated water

Polymer Composites for Electrical Engineering, First Edition. Edited by Xingyi Huang and Toshikatsu Tanaka.

absorption. There were several mechanisms used to indicate material degradation, including initiation, propagation, branching, and termination. Epoxy resins have been widely used because of their chemical/corrosion resistance and low shrinkage during the curing process. Epoxy resin networks with high levels of cross-linking produced brittle material due to their ability to be treated under various conditions. Composites' ability to dissipate energy was essential when they were placed in a vibrating environment. Fiber volume, fiber orientation, matrix material, temperature, moisture, and other factors, such as lamina thickness and composite thickness, all affect the amount of energy that FRP composites dissipate. Depending on the temperature, the mechanical properties of polymeric composites change. A polymer matrix composites' dynamic stability at low and high temperatures was important to study, including the storage modulus and damping effects.

Glass fiber-reinforced polymer composites are composed of the resin matrix, reinforced fiber, and auxiliary materials. Its main molding processes include hand lay-up molding, compression molding, winding molding, and pultrusion molding. When it is used in the power grid, winding molding, and pultrusion molding are mostly used. GFRP is widely used in national defense, aviation, electric power, construction, automobile, and other industries because of its high specific strength, high specific modulus, low coefficient of linear expansion, corrosion resistance, good durability, and designability. In recent years, the State Grid Corporation of China (SGCC) and the China Southern Power Grid (CSG) have been increasingly demanding power supply reliability and environmental friendliness of power transmission. The excellent performance of GFRP materials has become the key choice to solve the above problems. This chapter mainly introduces the application of GFRP in the power equipment at substation, power transmission, and distribution network.

13.2 Glass Fiber-Reinforced Polymer Composites

13.2.1 Fibers

Glass fiber is a type of fiber-reinforced plastic in which the reinforcing material is made of glass fiber. The fact that glass fiber is also known as glass-reinforced plastic or glass-fiber-reinforced plastic suggests that it is for this reason. Glass fiber is usually woven into a fabric, although it can also be spread out randomly or made into a sheet. Depending on its intended use, glass fiber can be made from various types of glass. A variety of glass fiber and glass cloth varieties are shown in Figure 13.1.

Glass fiber has many advantages over other materials, including being lightweight, strong, and less brittle. Glass fiber's best characteristic is its ability to be molded into a wide variety of complex shapes. Thus, glass fiber may be found in a variety of products, including bathtubs, boats, aircraft, and roofs, among others.

13.2.1.1 Chemical Description

Glass is made from sand, an abundant natural resource. Except for possible transportation and melting, it's completely independent of petrochemicals. Chemical-resistant "C" glass made of soda-lime borosilicate and high-strength S-2 glass are the most common glass compositions, with grade "E" glass being the most common of them. During the pulling

(1) Conventional roving

(2) Woven roving

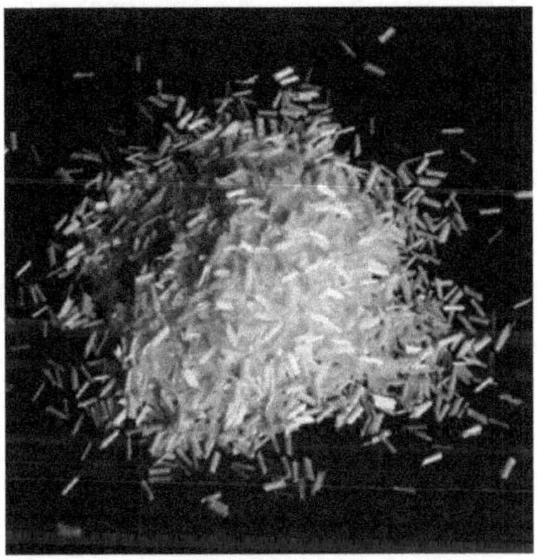

(3) Chopped strand

(4) Chopped strand mat

(5) Glass yarn

(6) Glass fabric cloth

Figure 13.1 Glass fiber and glass fabric. (1) Conventional roving; (2) Woven roving; (3) Chopped strand; (4) Chopped strand mat; (5) Glass yarn; (6) Glass fabric cloth.

process, filaments might have surface treatments (binders/sizing) applied to them instantly. Organic binders like starch oil are used in textile weaving to provide excellent weaving and strand preservation. They are then washed and thermally cleaned before being finished or size with coupling agents at the weaver, which improves the resin's compatibility, and then removed from textiles.

Untwisted glass-roving product yarns are commonly finished with the final coupling agents during the filament manufacturing process. As a result, the products will be easily identified by the product identifiers provided by the glass producer, negating the need for the usual desizing step associated with fabric "greige goods" forms. Heat-cleaned solutions are also available for products that are almost pure glass. Silicone laminates frequently employ these brittle materials that are prone to breakage. Using demineralized water after heat-cleaned product results in a different finish designation (neutral pH). Coupling agents utilized in structural applications are more frequent than special coupling agents for organic polymers. In the 1940s, Volan finishes became popular. Since then, a slew of modifications and enhancements have emerged, each with a unique company identity. Most people have heard about Volan A. This finish offers good wet and dry strength properties when used with polyester, epoxy, and phenolic resins. Before applying the finish, the cleaned glass is immersed in a solution of methacrylate chromic chloride, resulting in a chrome concentration of the finish ranging from 0.03% to 0.06%. In the curing process, this additive enhances resin wet-out. Silane coatings are frequently used with epoxy, although they are not the sole option. The bulk of them is aimed at improving the wet-out performance of laminates. Some additionally give great laminate clarity or remarkable composite properties in wet circumstances. Many more enhance high-pressure laminating while being impervious to chemicals or severe environmental conditions. As long as epoxy isn't the only finish utilized in conjunction with other matrix materials, the compositions are considered to be easily available to resin compounders (also known as "pre preggers") to determine compatibility and end-use reasons.

13.2.1.2 Classification of Glass Fibers

The major classification of glass fiber and the physical properties are shown in Table 13.1. Also, the chemical compositions of glass fiber in wt% are shown in Table 13.2. The physical and mechanical properties of glass fiber are shown in Table 13.3.

13.2.1.3 Properties of Glass Fiber

Some typical properties of glass fiber are shown in Table 13.4.

13.2.1.4 Glass Fabrics

Due to the high quantity of commercial applications for glass products, there are many product forms available. For purposes of this chapter, glass forms will be limited to continuous filament product forms. These forms fall into four major categories. They are continuous rovings, yarn for fabrics, or braiding, mats, and chopped strand. They are available with a variety of physical surface treatments and finishes. Most structural applications utilize fabric, roving, or rovings converted to unidirectional tapes. Perhaps the most versatile

Table 13.1 Types of glass fiber.

Type	General Indications
E	Compositions with electrical properties, strength, and durability appropriate for most applications
D	Compositions with a low dielectric constant
A	High alkali content compositions typically with lower electrical resistivity, strength, and durability than E-glass
C	Composition resistant to corrosion by most acids
S	Compositions in the system SiO_2-Al_2O_3-MgO with high tensile strength and high elastic modulus
R	Compositions in the system SiO_2-Al_2O_3-MgO-CaO with high elastic modulus
AR	Compositions resistant to corrosion by alkali
E-CR	E-glass compositions modified for improved resistance to corrosion by most acids

Table 13.2 Chemical compositions of glass fibers in wt%.

Type	SiO_2	Al_2O_3	TiO_2	B_2O_3	CaO	MgO	Na_2O	K_2O
E	55.0	14.0	0.2	7.0	22.0	1.0	0.5	0.3
C	64.6	4.1	—	5.0	13.4	3.3	9.6	0.5
S	65.0	25.0	—	—	—	10.0	—	—
A	67.5	3.5	—	1.5	6.5	4.5	13.5	3.0
D	74.0	—	—	22.5	—	—	1.5	2.0
R	60.0	24.0	—	—	9.0	6.0	0.5	0.1
E-CR	61.0	13.0	—	—	22.0	3.0	—	0.5

fiber type to produce glass product forms is "E" glass. "E" glass is identified as such for electrical applications. This type or grade of glass has eight or more standardized filament diameters available. These range from 3.5 to 13 μm. This facilitates very thin product forms. The "S" glasses are identified as such to signify high strength. Although woven rovings may be considered a fabric product form, they should be noted for their importance for military applications. Also, there are glass product forms which could be considered as complementary products for advanced structures. These would include milled fibers and chopped strand.

13.2.1.5 Advantages and Disadvantages

Regarding weight to strength, metals and other conventional materials are no match for glass composites. Despite the fast growth of carbon and aramid fibers, glass composites continue to dominate in several fields. This includes the cost per unit of weight or volume, specific weapons uses, chemical or galvanic corrosion resistance, electrical characteristics, and a wide range of product shapes. Carbon composites' thermal expansion coefficient and

Table 13.3 Physical and mechanical properties of glass fiber.

Type	Density (g/cm^3)	Tensile Strength (GPa)	Youngs Modulus (GPa)	Elongation (%)	Coefficient of Thermal Expansion ($10^{-7}/°C$)	Refractive Index	Dielectric Constant	Dielectric Loss
E	2.58	3.445	72.3	4.8	54	1.558	6.7	0.0017
C	2.52	3.310	68.9	4.8	63	1.533	—	—
S	2.46	4.890	86.9	5.7	16	1.521	5.4	0.0015
A	2.44	3.310	68.9	4.8	73	1.538	—	—
D	2.11–2.14	2.415	51.7	4.6	25	1.465	4.3	0.005
R	2.54	4.135	85.5	4.8	33	1.546	—	—
E-CR	2.72	3.445	80.3	4.8	59	1.579	—	—
AR	2.70	3.241	73.1	4.4	65	1.562	8.5	—

Table 13.4 Properties of glass fiber.

Mechanical strength	Glass fiber has a specific resistance greater than steel. So, it is used to make high performance.
Electrical characteristics	Glass fiber is a good electrical insulator even at low thickness.
Incombustibility	Since glass fiber is a mineral material, it is naturally incombustible. It does not propagate or support a flame. It does not emit smoke or toxic products when exposed to heat.
Dimensional stability	Glass fiber is not sensitive to variations in temperature and hygrometry. It has a low coefficient of linear expansion.
Compatibility with organic matrices	Glass fiber can have varying sizes and has the ability to combine with many synthetic resins and certain mineral matrices like cement.
Non-rotting	Glass fiber does not rot and remains unaffected by the action of rodents and insects.
Thermal conductivity	Glass fiber has low thermal conductivity making it highly useful in the building industry.
Dielectric permeability	This property of glass fiber makes it suitable for electromagnetic windows.

modulus are two common drawbacks. Glass has a lower tensile strength than aramid composites, but its higher ultimate compression, shear strength, and moisture absorption make up for it.

There are several uses for glass goods. Automobile and recreational vehicle structural components, filtration devices, thermal and electrical insulation, pressure and fluid containers, and insulation are a few examples. Military and aerospace goods also have a wide range of uses. Some examples include the replacement of asbestos, electronics, optical equipment, radomes, helicopter rotor blades, and ballistics. Structural applications are almost limitless in terms of manufacturing due to the wide range of product forms. Low thermal and electrical conductivity, as well as greater melt temperatures compared to carbon fibers, may be limitations compared to other fibers. Table 13.5 lists the specific mechanical properties of cured epoxy/glass.

13.2.1.6 Common Manufacturing Methods

Standard practice calls for mixing and premelting raw materials (and any additions) into marbles. As a raw material shape for automated feeding to the various melt furnaces, this form makes it easy to examine samples. Dried raw materials can also be fed directly into the batch cans via hopper utilizing this strategy. It doesn't matter what the natural form of the material is; it will go into a furnace and be melted down to about 1500 degrees Celsius. To draw individual filaments out of the molten material, small orifices in the bushings have microscopic holes in them. Sometimes the temperature of a bushing may be precisely regulated to within 0.6 degrees Celsius. The diameter of the filament is governed by the glass melt's viscosity and the extrusion's velocity. As it leaves the bushings as a filament, the glass rapidly cools or solidifies depending on the surrounding conditions. Water sprays and/or

Table 13.5 Typical cured epoxy/glass mechanical properties.

E glass, Woven 7781 Style	Standard Structural	Dual-Purpose Structural
Tensile strength (MPa)	430	330
Tensile modulus (GPa)	36	19
Compressive strengh (MPa)	410	340
Compressive modulus (GPa)	25	22
Flexural strength (MPa)	550	450
Flexural modulus (GPa)	26	23
Interlaminar shear (MPa)	18	26
Specific gravity (g/cm^3)	1.8	1.6
Cured resin content (% Wt)	33	48

binders are commonly used to cool down food and drink. They gather and wrap individual, untwisted filaments in "cakes" at high speeds on tubes. Adding finishes to the strands after they have been wrapped around the tubes and conditioned is not uncommon (dried). Instead of "S" filaments, the strands used in these items are "C" filaments (continuous). It is then necessary to peel the strands and unwind them to generate ends, or a large number of untwisted strands to create roving. This collection or mixing operation must be repeated multiple times (yards per pound). Weaving and braiding yarns are formed from twisted strands, whereas single yarns are made from single twisted strands. Two-strand construction refers to a single yarn made from two twisted strands. Plied yarns are created by twisting together several different yarns. Throwing is a common term for twisting and plying.

Variables in 'C' filament processing include the amount of repeated tensioning needed to make different product forms. Tensioning devices such as disc-type or "whirls," gate-type, tension bars or "S" bars, and compensating rollers are used in the delivery from the creels. Humidity is an additional element that may be controlled in the twiddling and plying and braiding and weaving industries. Maintaining relative humidity of 60–70% facilitates these processes. In glass manufacturing, it's essential to monitor the surface abrasion. Wear and tear on many devices need their replacement. Examples include guiding eyes, spacer bars, and rollers. Wear may also affect tensioning. Steel, chrome plating, and ceramics are some of the materials utilized to manufacture these contact devices. Table 13.6 lists the various strand fiber categories and their corresponding strand counts in detail.

13.2.1.7 Applications of Glass Fiber in Various Industries

In the case of industrial gaskets, high-temperature insulating materials provide a superior thermal barrier. Because of its durability, safety, and ability to offer good thermal insulation, glass fiber is a preferred choice for industrial gaskets. Not only do they provide better insulation, but they also help to preserve machines as well as save energy and keep the professionals on-site safe. This might explain why the following industries commonly use glass fiber:

Beverage industry: Beverage bottling lines and brewhouses frequently use glass fiber grating.

Table 13.6 Basic strand fiber designations and strand counts.

Filament Diameter		Strand Count (Number)		
SI (μm)	Letter	TEX (g/km)	100 Yd. (Cuts/Lb.)	Yds./Lb.
5	D	11	450	45 000
7	E	22	225	22 500
9	G	33	150	15 000
10	H	45	110	11 000
13	K	66	75	7500

Car washes: Glass fiber grating is now widely utilized for rust resistance and to add contrast color to formerly prohibited regions. It illuminates the interior of the car-wash tunnel, making the automobile appear cleaner than it was before.

Chemical industry: Glass fiber grating is utilized in this industry for the anti-slip safety characteristic of the embedded grit surface as well as the chemical resistance of various resin compounds. The chemicals utilized are compatible with the resins.

Cooling towers: Because cooling towers are continuously moist, they must be protected against rust, corrosion, and other hazards. Because of the outstanding qualities of glass fiber, it is utilized as screening in these structures to keep humans and animals away from hazardous zones.

Docks and marinas: The saline ocean corrodes, rusts, and damages the docks. As a result, glass fiber is employed for protection in this case.

Food processing: Glass fiber grating is used in chicken and cattle processing industries for slide resistance and to withstand acidic blood. Glass fiber is also used in most food preparation sectors since alternative grating materials are ineffective.

Fountains and aquariums: Glass fiber is used to support rocks in all sizes of fountains and aquariums to aid in circulation and filtering from beneath the rocks. Glass fiber grating is utilized in giant public fountains to protect spray heads and lighting from damage. This also helps to protect individuals from drowning in the fountains.

Manufacturing: The integrated grit surface of glass fiber grating ensures slip resistance in wet regions or areas with hydraulic fluids or oils.

Metals and mining: Glass fiber grating is utilized in chemically corrosive regions of electronic refining. Other grating materials are not permitted here.

Power generation: Glass fiber grating is utilized in chemically corrosive regions of electronic refining. Other grating materials are not permitted here.

Plating plants: Because of the anti-slip feature of the surface, this application employs glass fiber grating.

Pulp and paper industry: Glass fiber's ability to withstand chemical corrosion is beneficial in pulp and bleach mills. Glass fiber has recently been employed in various applications due to its corrosion resistance and anti-slip characteristics.

Automobile industry: Glass fiber is widely utilized in the automotive sector. Glass fiber components and body kits are found in almost every automobile.

Aerospace and defense: Glass fiber is utilized to make parts for military and commercial aircraft industries, such as test equipment, ducting, and enclosures.

13.2.2 Polymers

There are a variety of polymers, precursor materials, and/or combinations and formulations of polymers with various additives or chemically reactive components that are referred to collectively as resin. In composite material manufacture, production, and final quality, the resin and its chemical composition and physical properties all matter greatly. There are a variety of factors that can affect the processability and handling of resins. These factors can also affect the lamina/laminate properties of resins, as well as their performance in composite materials. For polymer matrix composites and adhesion, resin materials are employed. This section addresses resin chemistry, composition, impurities, contaminants, and their impact on resin processing features, as well as on resin and composite qualities.

13.2.2.1 Epoxy

Molecular building blocks that comprise epoxide groups are referred to as "epoxy" molecules. Oxirane structures are epoxide groups because they have three members: one oxygen, two carbon atoms. These resins have one or more epoxide groups and are polymerizable, thermosetting, and can be changed by reacting with amines or acids, amide or phenols, acid anhydrides or mercaptans. Liquid to solid viscosity ranges is available in polymer materials. Prepregs and structural adhesives commonly use epoxy resins as resins. Epoxies are strong and flexible, have low volatile levels, good adhesion, shrinking is limited, and they are chemically resistant. When exposed to moisture, their primary disadvantages include brittleness and a loss of properties. Polyester resins cure and process faster than epoxy resins. As compared to polyester, resin is a lot more costly. All of these types of processing, including autoclave molding and filament winding, are examples of processing. Surface curing takes place in a temperature range of ambient to around 180 degrees Celsius. Cure temperatures typically range from 120 to 180 degrees Celsius. As the cure temperature changes, so will the structure's operating temperatures. The better treatment is at higher temperatures, the more resistant it is to such temperatures. When people think about cure pressures, they typically think of low-pressure molding, which ranges from vacuum up to 700 kPa. Epoxy resin characteristics are shown in Table 13.7.

13.2.2.2 Polyester (Thermosetting)

It's orthophthalic polyester resin or isophthalic polyester resin when someone refers to "thermosetting polyester resin.". Polyester resins are popular because they are inexpensive and easy to process. Specialized fillers can be added to these materials to increase their breakdown resistance in electrical arc and track environments. Thermal stability, dimensional stability, and creep resistance are all improved using isophthalic polyester resins. Fiber-reinforced resin systems can benefit from polyester's low cost and speed of production.

Table 13.7 Characteristics of epoxy resins.

Class	Chemical Base	Viscosity at 23 °C and Shear Rate of 1 s^{-1}	Epoxy Equivalent	Organic Modifiers or Solvents	Density at 23 °C
1	Bisphenol A/glycidyl ethers	≤0.25	≤115	None	<1.1
2	Aromatic glycidyl ethers (or esters)	>0.25–1	116–150	Reactive agent	1.10–1.14
3	Aliphatic glycidyl ethers (or esters)	>1–5	151–175	Non-reactive agent	1.15–1.19
4	Cycloaliphatic glycidyl ethers (or esters)	>5 but fluid	176–210	Organic solvent	1.20–1.29
5	Epoxy-cycloolefins	Semi-solid	211–290	Reactive agent and organic solvent	1.30–1.39
6	Epoxy-novolac	Solid	291–525	Non-reactive agent and organic solvent	1.40–1.59
7	Halogenated epoxides	Thixotropic	536–1025	—	1.60–1.80
8	Other glycidic nitrogenous compounds	—	1026–2050	—	>1.80
9	Heterocyclic compounds	—	>2050	—	—

Polyester fiber-reinforced materials can be handled using a variety of ways. Methods such as pultrusion and autoclaving as well as matching metal mold wet setting up and pressing mold are frequently employed. The thermoset molding technique can speed up the cure time of polyesters while delaying the cure time of phenolics. When it comes to processing, polyester has a high-temperature sensitivity, whereas processing phenolics is more time/temperature-independent, for example. Polyesters can be treated at temperatures ranging from ambient to as high as 180 °C, depending on the formulation used. The patient will soon recover if the correct temperature is utilized. The resin/catalyst combination will stay plasticized if there is not enough heat applied during the reaction. When it comes to working with polyesters, they're easier since they're stiffer and more resistant to cracking than epoxies. On the other hand, phenolics are brittle and prone to breaking, but they can sustain greater service temperatures than polyesters.

13.2.2.3 Phenolic

Commercial production of phenol-formaldehyde resins and their direct precursors began in the early 1900s. For use at lower temperatures, urea-formaldehyde and melamine-formaldehyde were introduced in the 1920s and 1930s. Phenolics often heal by condensing and off-gassing water. There is less smoke and harmful breakdown products in the resulting matrix because of its chemical and thermal resilience.

In addition to phenolic resole and novolacs resins, phenolic polymers are also known as phenolic resole and novolacs resins, which are condensation polymers produced by the reaction of excess formaldehyde with an acidic catalyst and formaldehyde (resole) (novolac). Because novolacs lack methylol groups, an extension agent such as paraformaldehyde, hexamethylenetetraamine, or additional formaldehyde must be used as a curative instead of resoles. When compared to their original components, these additions have higher molecular weights and viscosities. Consequently, they're perfect for working with parts that have unusual forms or complex curves. The resins may be cured in a press or an autoclave and allow for free-standing positions at quite high temperatures.

13.2.3 Manufacturing Methods

Composite fibers that have previously been "pre-impregnated" with a matrix material, such as epoxy, are referred to as "prepreg." Fibers often have a weave pattern, and the matrix is used to bind the fibers and other components together during manufacturing. A portion of the matrix has been partially cured to facilitate handling; this is known as B-Stage material and must be kept cold to prevent complete curing. B-stage prepreg is always stored in a cool location since heat speeds up the polymerization process. As a result, curing in an oven or autoclave is nearly always required for prepreg-based composite structures. The B-Stage prepreg procedure offers several advantages over the hot injection method. For hot injection processes, prepreg makes it possible to soak the fibers and mold them into a useful form. Prepreg allows you to do this on a level working surface, or rather in an industrial process. Additionally, you may use Prepreg to saturate a huge volume of fiber and then cure it later in a cool area. Because it takes longer than the hot injection, the approach is less useful for prepreg preparation than the hot injection method. The prepreg procedure is shown in Figures 13.2–13.4.

If you want a particular kind, size, configuration, or quality out of your Fiber-Reinforced Plastic (FRP), you may make it in a variety of methods. The following section summarises and briefly introduces FRP composite material production processes.

Hand lay-up molding, also known as contacting molds or contacting moldings, was the first technology used to make composites from fiber-reinforced plastic (FRP). Or, to put it another way, the process involves injecting a polymer resin mold with a fiber fabric (matrix).

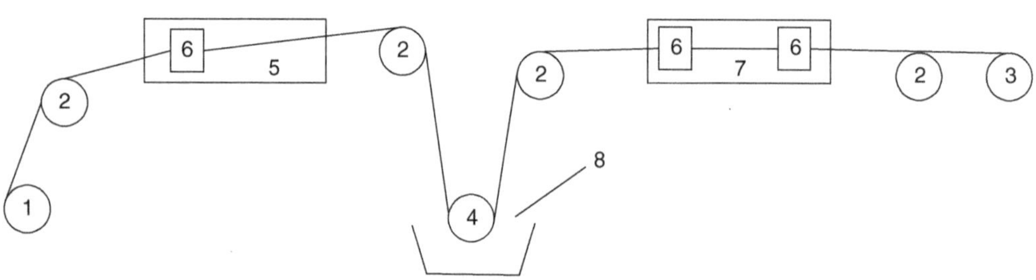

Figure 13.2 Horizontal glue-dipping machine. 1-cloth roll, 2-guide roll, 3-traction roll, 4-rubber roller, 5-heat treatment furnace, 6-air outlet, 7-drying box, 8-dip trough.

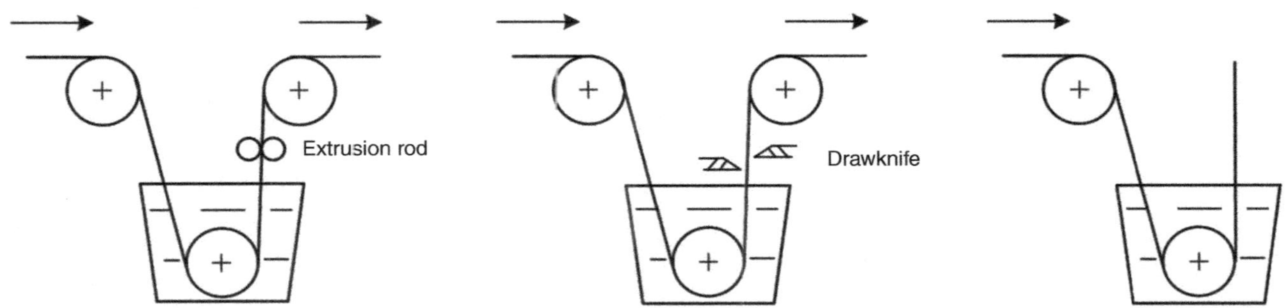

Figure 13.3 Three types of devices to control the amount of glue. 1-extrusion roller type, 2-scraper type, 3-glue type.

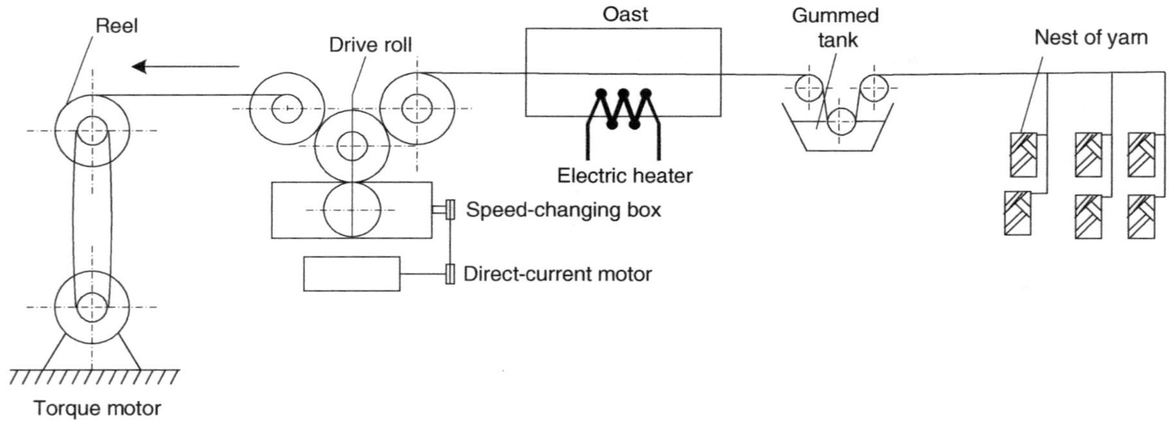

Figure 13.4 The glue tape processing.

Figure 13.5 Hand lay-up molding process.

The cloth is pushed with a roller or rubber to force the polymer resin into the fabric and eliminate as much air as possible. The process is repeated with new resin and fabric until the required thickness is achieved. It's easy and inexpensive to make a hand lay-up mold using only a few simple pieces of equipment. And it's not constrained by the object's dimensions or shape. However, it is inefficient, labor-intensive, and subpar in every way. As a result, avoid using this approach unless you have a high level of reliability on hand. The molding process is shown in Figure 13.5.

Chopper gun: With a hand-held gun, fiberglass is supplied in continuous strands, which are chopped and combined with catalyzed resin-like polyester. Chopped glass coated with resin is sprayed over the mold surface in a thickness determined by the design and by the human operator. This approach is suitable for large production runs at a cheap cost, but it creates geometric shapes that are weaker than other molding methods and have poor dimensional precision (Figure 13.6).

Compression molding: Compression molding is used to make up for the shortcomings of manual lay-up molding. Compression molding uses high pressure and high temperature to thoroughly integrate fiber fabrics and polymer resins instead of physically pressing them together. Compression molding is shown graphically in Figure 13.7.

Vacuum molding: When compared to compression molding, vacuum molding applies pressure to the fabric/resin combination. A vacuum pump removes the air under a plastic sheet that has been placed on top of the mold mixture. Because of this, any trapped air or gas in the resins is forced to the surface by atmospheric pressure. Figure 13.8 depicts the vacuum molding process.

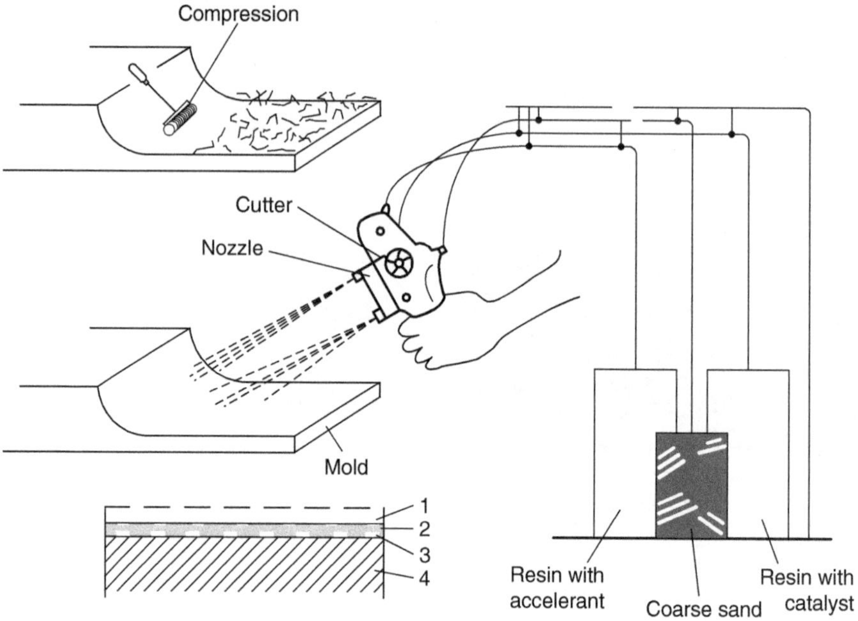

Figure 13.6 Spray-up technique. 1-laminated compound, 2-gel coat, 3-mold discharging agent, 4-mold.

Autoclave–vacuum bag: The vacuum-assisted resin transfer molding technique is built on vacuum molding. A plastic vacuum bag is used to encapsulate fiber fabrics in a mold. A vacuum pump attached to the mold drives the infusion of liquid polymer resin into the fabric. Air or disable gases are pushed out of the spaces between the fiber laminates as part of this process. Vacuum-assisted resin transfer molding is commonly used in the marine industry. A vacuum bag molding method is shown in Figure 13.9.

Filament winding: Filament winding is a manufacturing technique that entails wrapping polymer resin-impregnated fiber bundles around a spinning mandrel before spinning. The winding patterns, fiber angle, and quantity may all be automatically controlled by a computer. To demonstrate the filament winding molding procedure, see Figure 13.10.

Pultrusion: When fiber bundles are pulled successively between an impregnator and a forming die, they are pultruded. In the manufacture of tubular goods, bars/rods, square goods, and so on, it is widely used this technique. For example, in Figure 13.11, we see the pultrusion molding process in action, whereas, in Figure 13.12, we see it in dried form.

The final two processes listed above are the most often utilized to produce electrical components in applications to power transmission networks. Figure 13.13 depicts the hollow core insulator manufacturing process, for example.

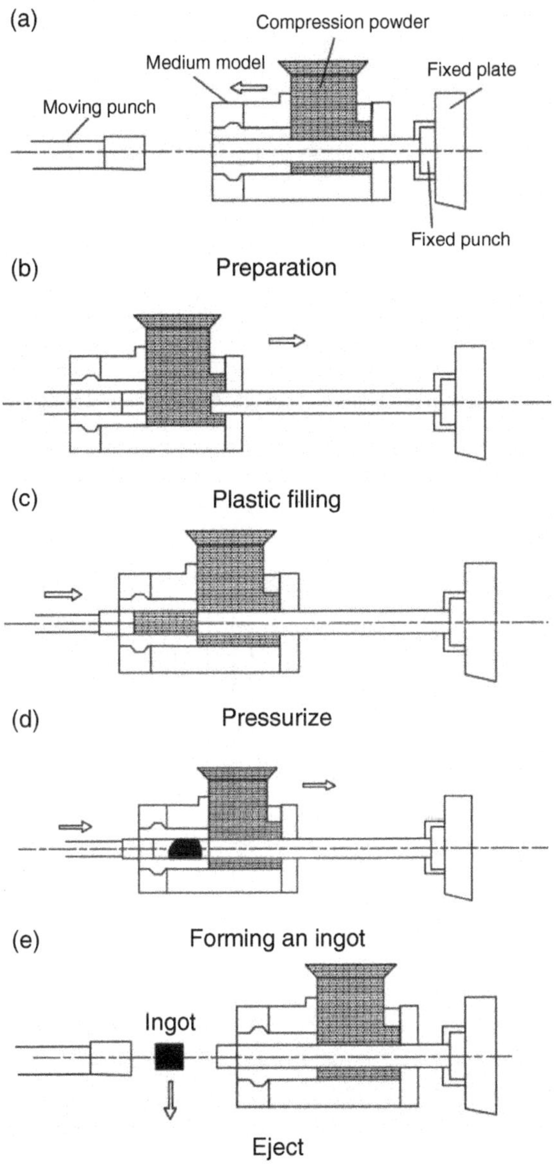

Figure 13.7 Compression molding process. (a) Preparation. (b) Plastic filling. (c) Pressurize. (d) Forming an ingot. (e) Eject.

13.2.4 Specifications of Several Kinds of GFRP Materials

13.2.4.1 Rigid Laminated Sheets

The types of industrial rigid laminated sheets based on epoxy resins are shown in Table 13.8, and the property requirements of industrial rigid EPGC and EPGM laminated sheets are shown in Table 13.9.

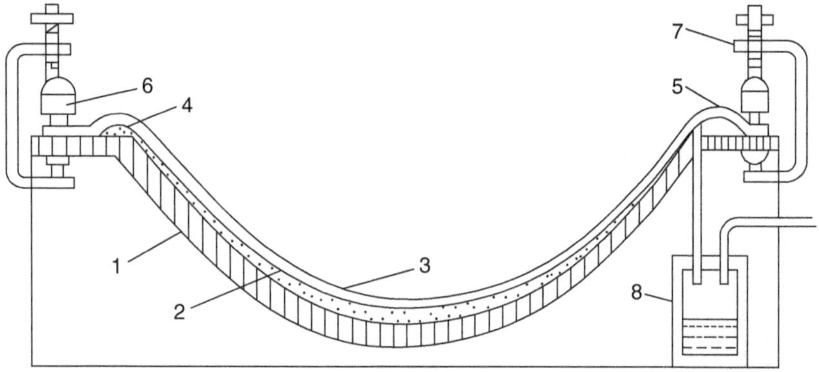

Figure 13.8 Forming technique of vacuum bag with die. 1-mold, 2-roughcast, 3-rubber blanket. 4-gas collector, 5-washer, 6-bolt, 7-bow clip, 8-resin collection tank.

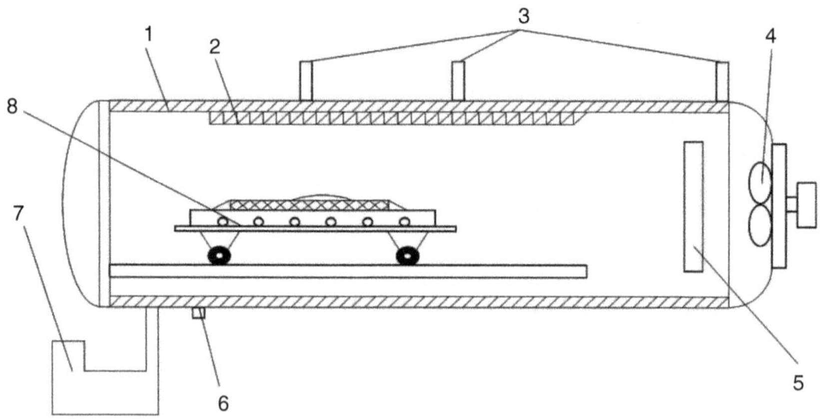

Figure 13.9 Autoclave structure diagram. 1-Tank, 2-Heating, 3-Pressurizing, 4-Blowing, 5-Cooling, 6-Vacuum, 7-Control, 8-Rack.

13.2.4.2 Industrial Rigid Round Laminated Rolled Tubes (Tables 13.10 and 13.11)

The typical types of industrial round rolled tubes are shown in Table 13.10, such as, EPGC 21, 22, 23 and PFGC 21, and their applications and distinguishing characteristics are listed. The property requirements for round rolled tubes are shown in Table 13.11, including flexural strength perpendicular to laminations, axial compressive strength, breakdown voltage, electric strength, thermal endurance etc.

13.2.4.3 Insulated Pipe (Table 13.12)

The properties of GFRP pipe for 1100 kV AC composite hollow insulators are shown in Table 13.12, including moisture absorption, flexural modulus, electrical strength at parallel to laminations, electrical strength at perpendicular to the lamination.

13.2.4.4 Insulated Pull Rod

The recommended specification of insulated pull rod for GIS below 550 kV is shown in Table 13.13 and the one above 550 kV is shown in Table 13.14.

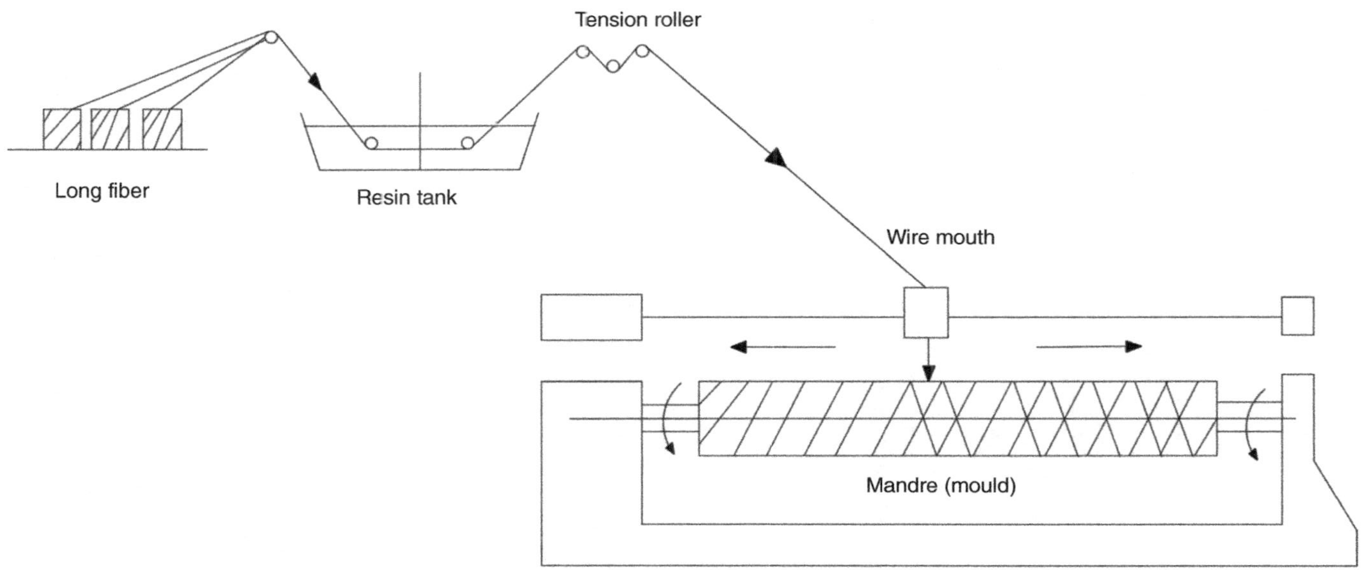

Figure 13.10 Filament winding molding process.

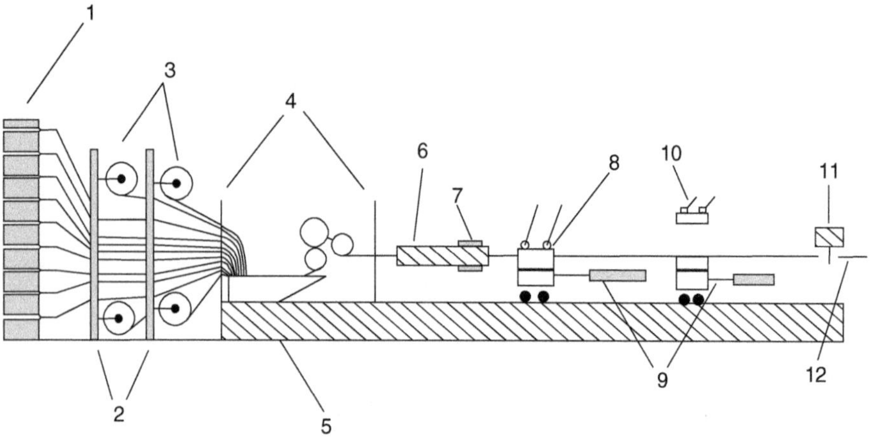

Figure 13.11 Schematic diagram of wet pultrusion. 1-fiber rack, 2-preforming guide, 3-fiber cloth support, 4-material guide, 5-resin tank, 6-pultrusion die, 7-heater, 8-pulling mechanism engagement, 9-hydraulic cam, 10-pull mechanism disengaged, 11-motion cutting saw, 12-finished product.

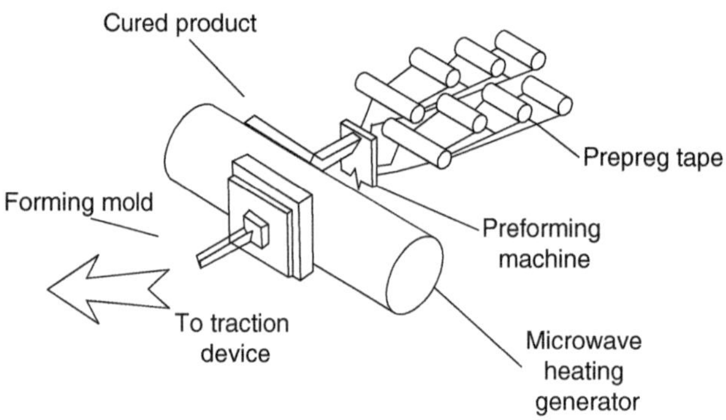

Figure 13.12 Schematic diagram of dry pultrusion.

13.3 Application of Glass Fiber-Reinforced Polymer Composites

13.3.1 Laminated Sheets

Glass cloth-reinforced epoxy natural is typically yellowish to light green in color. The most versatile all-around laminate grades are continuous glass-woven fabric impregnated with an epoxy resin binder. Epoxy resins are among the most versatile and widely used plastics in the electronics field, primarily because water absorption is virtually nil, rendering it an outstanding insulator. Beyond its electrical insulating properties, epoxy resins exhibit great dimensional stability (shrinkage is usually less than 1%) and superior adhesive properties. This type of laminated sheet has extremely high mechanical strength, good dielectric loss

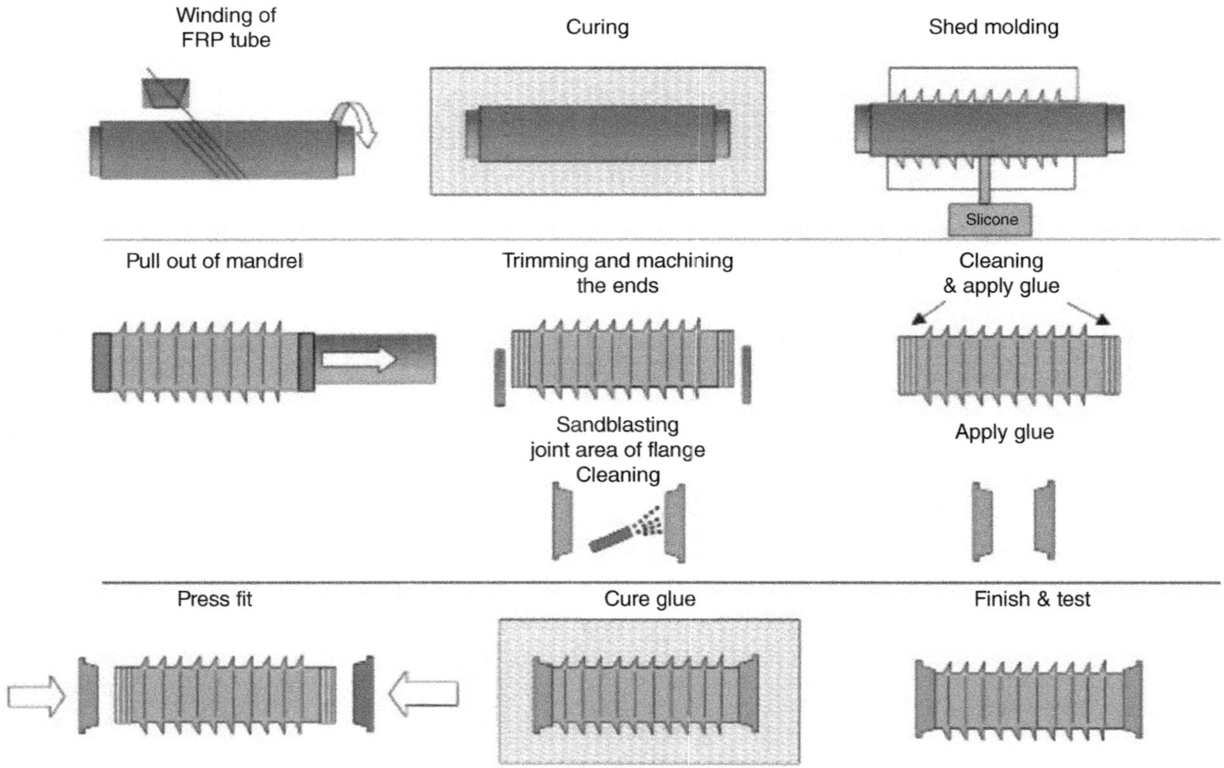

Figure 13.13 Hollow-core insulator-manufacturing process.

Table 13.8 Types of industrial rigid laminated sheets based on epoxy resins.

Laminate Type			
Resin	**Reinforcement**	**Serial Number**[a]	**Applications and Distinguishing Characteristics**
EP	GC	201	Mechanical, electrical, and electronic applications. Extremely high mechanical strength at moderate temperature. Very good stability of electrical properties in high humidity
		202	Similar to type EPGC201, low flammability
		203	Similar to type EPGC201, high mechanical strength at elevated temperature
		204	Similar to type EPGC203, low flammability
		205	Similar to type EPGC203, but with roving cloth
		306	Similar to type EPGC203, but with improved tracking indices
		307	Similar to type EPGC205, but with improved tracking indices
		308	Similar to type EPGC203, but with improved thermal endurance properties
		309	Similar to type EPGC201, but with defined mechanical strength at elevated temperature
		310	Similar to type EPGC202, but with halogen-free compound
		311	Similar to type EPGC204, but with halogen-free compound
	GM	201	
		202	Similar to type EPGM201, low flammability
		203	Similar to type EPGM201, high mechanical strength at elevated temperature
		204	Similar to type EPGM203, low flammability
		305	Similar to type EPGM203, but with improved thermal endurance properties
		306	Similar to type EPGM305, but with improved tracking indices

[a] This specification was originally based on ISO 1642, which is now obsolete.

properties, and good electric strength properties, both wet and dry. It is suitable for power equipment in marine applications, circuit breakers, arc barriers, arc shoots, motor wedges, insulator panel, and switchboard parts (Figure 13.14).

13.3.2 Composite Long Rod Insulators

The composite long rod insulators combine the highest levels of electrical insulation and mechanical tensile strength in a compact, lightweight one-piece housing design with two different sealing options depending on the customer requirements. The FRP rod with ECR glass (boron-free) has the brittle fracture-resistant performance (hydrolysis- and acid-resistant). It has the HTV silicone rubber with a hydrophobicity transfer mechanism for the highest pollution performance. The direct injection silicone-molding process prevents partial discharges (Figure 13.15).

Table 13.9 Property requirements of industrial rigid EPGC- and EPGM-laminated sheets.

Type	Flexural Strength/MPa	Impact Strength/kJ/m^2	Electrical Strength[a]/kV/mm	Breakdown Voltage[b]/kV	Insulation Resistance[c]/MΩ
EPGC201	340	42	11.8	45	5×10^4
EPGC202	340	42	11.8	45	5×10^4
EPGC203	340	50	11.8	45	5×10^4
EPGC204	340	50	11.8	45	5×10^4
EPGC205	340	70	—	45	1×10^4
EPGC306	340	50	11.8	45	5×10^4
EPGC307	340	70	—	45	1×10^4
EPGC308	340	50	11.8	45	5×10^4
EPGC309	340	42	11.8	45	5×10^4
EPGC310	340	42	11.8	45	5×10^4
EPGC311	340	50	11.8	45	5×10^4
EPGM201	340	50	10.5	35	5×10^3
EPGM202	320	50	10.5	35	5×10^3
EPGM203	320	50	10.5	35	5×10^3
EPGM204	320	50	10.5	35	5×10^3
EPGM305	320	50	10.5	35	5×10^3
EPGM306	320	50	10.5	35	5×10^3
Nominal thickness, mm	≥1.5	≥5	≤3	≥3	All

[a] Electrical strength at 90 °C in oil perpendicular to laminations with 2 mm thickness.
[b] Breakdown voltage at 90 °C in oil parallel to laminations, one minute proof test or 20 seconds step-by-step test.
[c] Insulation resistance after immersion in water.

Table 13.10 Types of industrial round rolled tubes.

Resin	Reinforcement	Serial Number	Applications and Distinguishing Characteristics
EP	GC	21	Mechanical, electrical, and electronic applications. Extremely high mechanical strength at moderate temperatures. Very good stability of electrical properties when exposed to high relative humidity.
		22	Similar to EPGC21, but with high mechanical strength at elevated temperature.
		23	Similar to EPGC21, but with improved flame resistance.
PF	GC	21	Mechanical and electrical applications. Very high mechanical strength at moderate temperatures.

Table 13.11 Property requirements for round rolled tubes.

Property	Unit	EPGC21	EPGC22	EPGC23	PFGC21
Flexural strength perpendicular to laminations	MPa	300	300	300	—
Axial compressive strength	MPa	175	175	175	—
Cohesion between layers	MPa	200	200	200	—
Breakdown voltage[a]	kV	40	40	40	—
Electric strength[b]	kV/mm	9.0	9.0	9.0	5.6
Insulation resistance	MΩ	1000	1000	1000	—
Dissipation factor	%	5	5	5	—
Permittivity	—	5.2	5.2	5.2	—
Thermal endurance	TI	130	155	130	130
Water absorption	mg/cm^2	1.5	1.5	1.5	—
Density	g/cm^3	1.70–1.90	1.70–1.90	1.70–1.90	1.70–1.90

[a] Breakdown voltage at 90 °C in oil parallel to laminations, one minutes proof test or 20 seconds step-by-step test.
[b] Electrical strength at 90 °C in oil perpendicular to laminations for round rolled tubes with 2 mm thickness.

Table 13.12 Property of GFRP pipe for 1100 kV AC composite hollow insulators.

Property	Unit	Hollow Insulator
Moisture absorption	%	≤0.05
Moisture absorption of inner-liner	%	≤0.02
Flexural modulus	GPa	≥15
Electrical strength at parallel to laminations	kV/mm	≥6
Electrical strength at perpendicular to the lamination	kV/mm	≥15

13.3.3 UHV-Insulated Pull Rod for GIS

At present, most manufacturers in China use glass fiber as reinforcement material, and basically have the manufacturing capacity of glass fiber insulation rods of all voltage levels except UHV. The infiltration of glass fiber and epoxy is better, and the process is easy to master. Compared with glass fiber, aramid fiber has lower density and higher tensile strength and electrical properties, but it is difficult to process and often needs to be used together with polyester fiber. However, in the field of UHV, the insulation rods for 1100 kV GIS circuit breakers include solid plate-type insulation rods produced by Toshiba, Hitachi, and Takaoka company of Japan, and hollow tubular insulation rods produced by TE company (formerly Axicom company of Europe). Solid plate-type insulation rod is mainly used

Table 13.13 Recommended specification of insulated pull rod for GIS below 550 kV.

Property	Unit	GFRP	AFRP
Fiber content	wt%	≥60	≥60
Density	g/cm^3	—	—
Moisture absorption	%	≤0.3	≤0.3
Interlayer adhesion strength	MPa/N	300	200
Electrical strength at parallel to laminations	kV/mm	≥8	≥10
Electrical strength at perpendicular to laminations	kV/mm	≥10	≥12
Volume resistivity	Ω·cm	$\geq 1.0 \times 10^{15}$	$\geq 1.0 \times 10^{15}$
Dielectric loss	%	≤0.7	≤0.5
Dielectric constant		3~5	3~5
Maximum permissible PD level	pC	<3	<3
Glass transition temperature	°C	≥100	≥100

Table 13.14 Recommended specification of insulated pull rod for GIS above 550 kV.

Property	Unit	GFRP	AFRP
Fiber content	wt%	—	—
Density	g/cm^3	—	—
Moisture absorption	%	≤0.1	≤0.2
Tension strength	MPa	≥300	≥430
Compression strength	MPa	≥200	≥100
Shear strength at parallel to laminations	MPa	≥30	≥25
Interlayer adhesion strength	MPa/N	200	200
Electrical strength at parallel to laminations	kV/mm	≥8	≥10
Electrical strength at perpendicular to laminations	kV/mm	≥12	≥13
Volume resistivity	Ω·cm	$\geq 1.0 \times 10^{15}$	$\geq 1.0 \times 10^{15}$
Dielectric loss	%	≤0.7	≤0.7
Dielectric constant		3~5	3~5
Maximum permissible PD level	pC	<3	<3
Glass transition temperature	°C	≥105 (<1100 kV), ≥115 @1100 kV	≥105 (<1100 kV), ≥115 @1100 kV

Figure 13.14 EPGC201-laminated sheets.

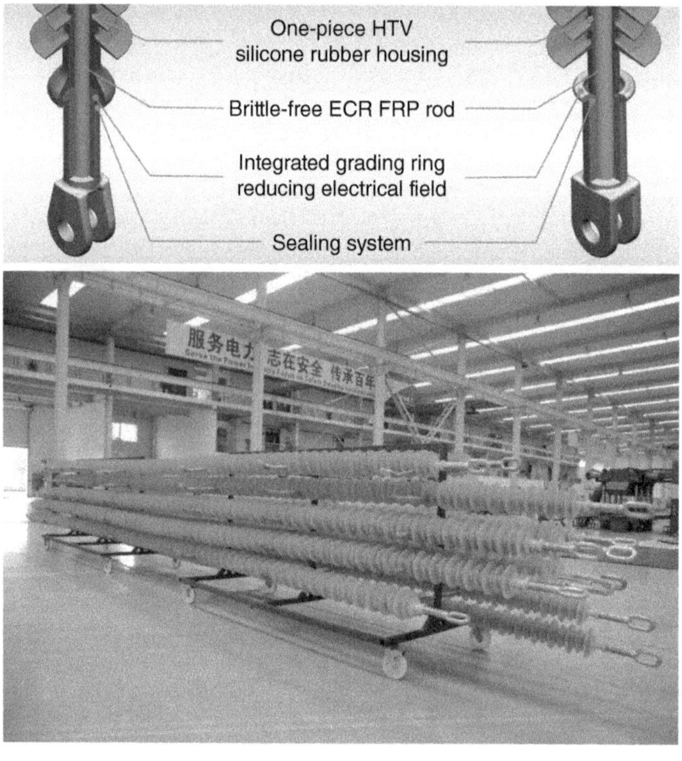

Figure 13.15 Composite long rod insulator.

in GIS (double fracture structure of circuit breaker) which adopts the Japanese technology route. The reinforced material of the insulation rod is glass fiber, which is rectangular after molding, large in size and heavy in weight. The hollow tubular insulation rod is mainly suitable for GIS (circuit breaker with four-fracture structure) which adopts European (ABB, Siemens, etc.) technology route. The reinforcement material of this kind of insulation rod is aramid fiber and polyester fiber. After molding, it is hollow tubular,

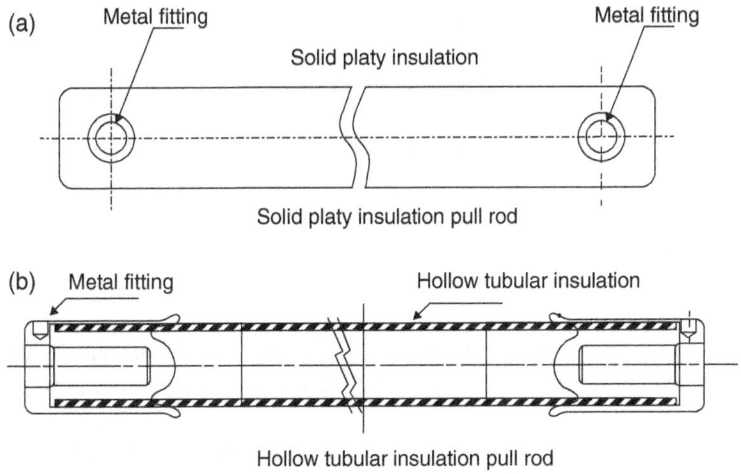

Figure 13.16 Two types of insulation pull rod for UHV GIS. (a) Solid platy insulation pull rod. (b) Hollow tubular insulation pull rod.

Figure 13.17 TE Axicom hollow tubular insulation pull rod.

thin, and long, with light weight, which can greatly reduce the operating power of the circuit breaker (Figures 13.16–13.18).

13.3.4 Composite Pole

Significant portions of the utility grid were installed decades ago. Aging structures endure constant attack from rot, corrosion, woodpeckers, and termites and are regularly challenged by ice storms, hurricanes, tornadoes, vandals, and even vehicular impact. New line

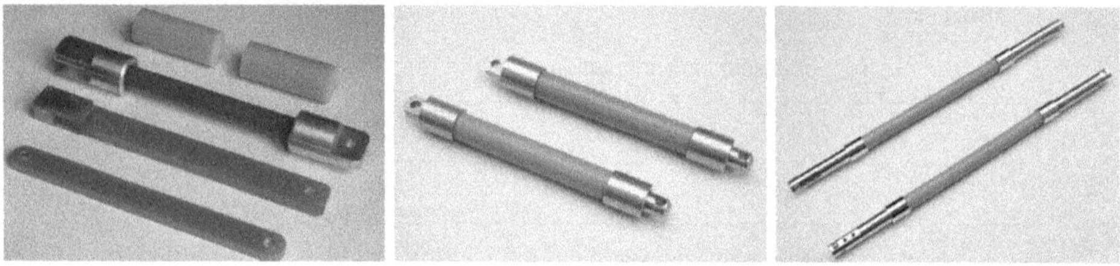

Figure 13.18 China XD tubular and solid platy insulation pull rod.

construction and pole replacement can be problematic with long lead times, challenging terrain, right-of-way issues, environmental assessments, disposal costs, power interruptions, and costly equipment requirements.

The composite utility pole is made of a glassfiber-reinforced thermoset epoxy tube with an outer thermoplastic shell. The innovative exterior protective layer allows for the use of traditional climbing and safety equipment. The surface layer also protects the load-bearing composite from UV radiation and mechanical wear. The pole is suitable for use in environmentally sensitive areas since there is no leakage of toxic chemicals. The pole is designed to meet both strength requirements and the stiffness requirements of the standard including combined load cases. Normal EBR kits can be used provided fully threaded screws are used and screw holes are predrilled.

Polyurethane-based composite poles are environmentally inert, contain no VOCs (volatile organic compounds) or HAPs (hazardous air pollutants), and use no styrenes during the manufacturing process. With composite poles, there are no concerns about toxic chemicals leaching into surrounding groundwater (as can occur with other pole materials) or contaminating the soil around the pole. This is becoming more significant at a time when environmental issues are an increasing concern for utilities and the public in general. RS Composite Utility Poles are constructed from combinations of up to eight standard-sized tubular modules to create poles with heights ranging from 30 ft. [9.1 m] to 155 ft. [47.2 m] that use standard industry hardware. RS poles deliver the following: lowest logistics costs with industry-best lead times, more efficient transportation, fast installations, and cost-effective inventory management. Lowest liability with a limited 41-year warranty, high dielectric strength providing improved safety for workers and the public, better storm and high wind resilience, faster response times in emergencies, and minimal environmental impact. Longest life with 80-year service life, integrated UV protection and immunity to rot, corrosion, woodpeckers, and termites (Figures 13.19 and 13.20).

13.3.5 Aluminum Conductor Composite Core in an Overhead Conductor

The ACCC/TW conductor was developed to improve several key performance metrics in comparison with conventional ACSR conductors. The primary design objectives included increasing the strength of the conductor (by replacing the stranded steel with a stronger material), increasing the rated ampacity, and improving the resistance to sag at high temperatures. To meet these objectives, a carbon/glass fiber, polymer matrix (CGPM) composite core was developed to replace the stranded steel strength member used in ACSR. The CGPM composite core offers several advantages over a stranded steel core, such as much

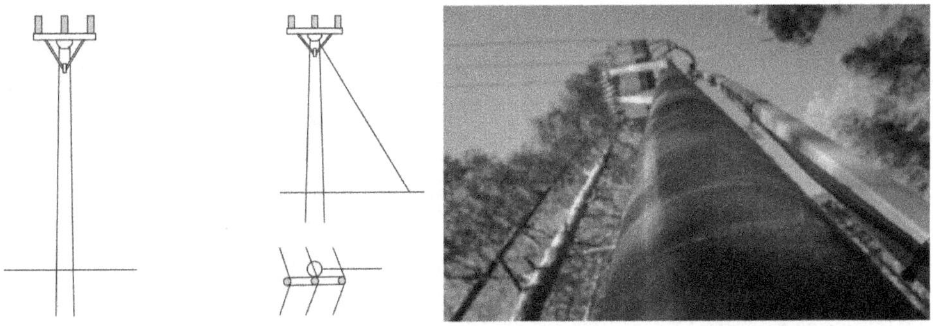

Figure 13.19 Hitachi ABB power grids' composite utility poles.

Figure 13.20 Typical composite utility pole installed in energized power transmission grid by R.S. Technologies.

lower density and a much lower coefficient of thermal expansion (CTE). The composite core was designed as a solid, one-piece rod with no interstices, unlike the steel core in ACSR, which is comprised of stranded round wires with interstitial volumes. This led to a core with a greater cross sectional area, but a smaller core diameter. The increased cross sectional area of the composite permits larger loading with reduced tensile stresses over that of the steel core subjected to the same loads. Figure 13.21 is a comparison of the cross sections of the ACCC/TW and the ACSR (Figure 13.21).

13.3.6 Composite Station Post Insulators

The load-bearing insulator tubes are made of epoxy resin reinforced with electrical grade fiberglass using a wet filament-winding technique. The whole process ensures high-grade insulator tubes with narrow manufacturing tolerances and excellent mechanical and electrical properties. The fiberglass tubes are fitted into high-strength and high-resistant aluminum alloy-casted flanges using a special process, which results in a combined shrink fit and adhesive bond. We use only high-quality, high-temperature vulcanized (HTV) silicone rubber to ensure the highest possible durability of sheds, as well as outstanding tracking and erosion resistance. The superior silicone material has an ATH filler level of more than 45% in weight. The fiberglass tubes are equipped with HTV silicone rubber weather sheds using a void-free helical extrusion process. All insulators are manufactured in one step, resulting in continuous silicone housings without molding lines (where dirt and salt can accumulate) (Figure 13.22).

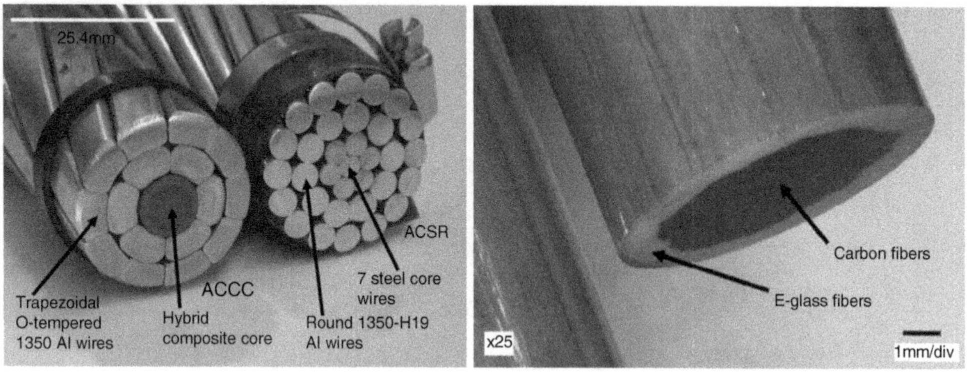

Figure 13.21 Comparison between the ACCC/TW (left) and the ACSR (right) conductors. The number of trapezoidal aluminum wires does not represent the real number of wires and is shown just for comparison reasons. A picture of the latest ACCC/TW cross section is not available. ACCC composite core showing the carbon composite in the center and the glass composite shell around it.

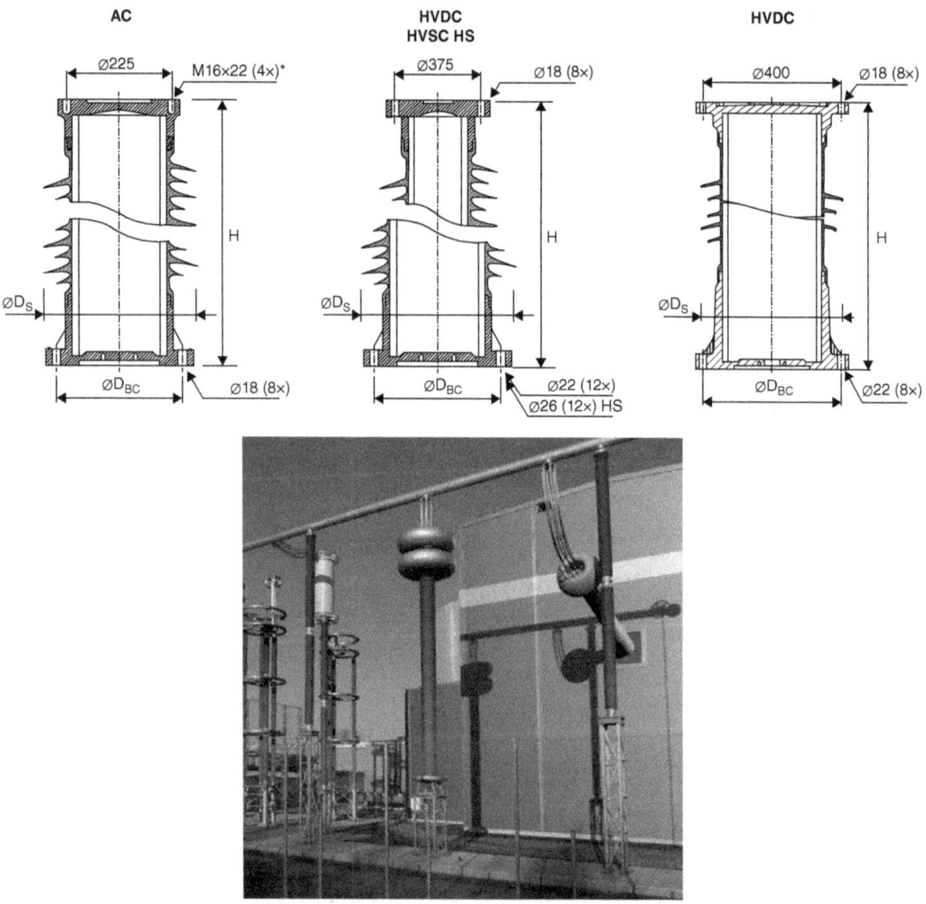

Figure 13.22 Composite station post insulators.

13.3.7 Composite Hollow Insulators

The composite hollow insulators for all high-voltage applications including high-voltage breakers, instrument transformers, surge arresters, cable terminations and bushings up to 1200 kV AC and 1100 kV DC. The load-bearing insulator tubes are made of epoxy resin reinforced with electrical grade fiberglass using a wet filament-winding technique. The whole process ensures high-grade insulator tubes with low manufacturing tolerances and superb mechanical and electrical properties. All insulator tubes should be made in-house and as one continuous piece. Some manufacturers can make the world's largest one-piece insulators in lengths up to over 15 m. The high-quality, high-temperature vulcanized (HTV) silicone rubber should be used to ensure the highest possible durability of sheds, as well as outstanding tracking and erosion resistance. It is well documented that HTV silicone rubber enriched with ATH filler outperforms low viscosity silicone rubbers, such as liquid silicone rubber, regarding long-term performance. The superior silicone material involves an ATH level of more than 45% in weight. The fiberglass tubes are equipped with HTV silicone rubber weather sheds using a void-free helical extrusion process. Even the longest insulators are manufactured in one step, resulting in continuous silicone housings without molding lines or other potential weak joints. Major development efforts have been carried out to achieve our optimized shed profiles. Both the bottom surface and top surface of the sheds are angled. This ensures optimal creepage distance and protected areas, resulting in the lowest possible leakage currents, even in severe environments (Figure 13.23).

13.3.8 Composite Crossarms

For new and current overhead transmission lines running at 132 kV or higher, a unique insulating crossarm (ICA) has been created. The ICA may be used to directly replace the steel cross-arms and insulators on existing lattice towers, enabling for the upgrading of existing infrastructure, or to build new towers with a smaller footprint than would otherwise be feasible. The crossarm is made up of four insulating components, end fittings, field-grading devices, and a nose connection to which the conductor is joined (Figures 13.24 and 13.25). (Figures 13.24 and 13.25). The compression insulators, the assembly's two primary structural elements, possess a unique non-cylindrical form that offers them superior mechanical characteristics to typical overhead line insulators. A compression insulator's profile, in particular, is based on a patented T-shaped column section that shows a considerably enlarged second moment of the area when compared to a cylindrical profile of the same cross-sectional area. This design decreases the weight of the compression insulator by 2.4 times while allowing the crossarm to be positioned in the same way as a normal steel lattice crossarm, i.e. at shallow raking angles and without the necessity of diagonal bracing.

Figure 13.26 shows a 400 kV voltage level steel lattice tower, a T-Pylon, and a fully composite pylon. Figure 13.26 depicts a composite pylon half the size of the lattice tower and with a more appealing look. In addition, suspension insulators, which are present in the composite crossarm design, such as the unibody crossarm, are absent from the completely composite pylon design. The unibody crossarm will be held together by conductor clamps. Insulation for unibody crossarms is being explored using shed profiles. It is seen

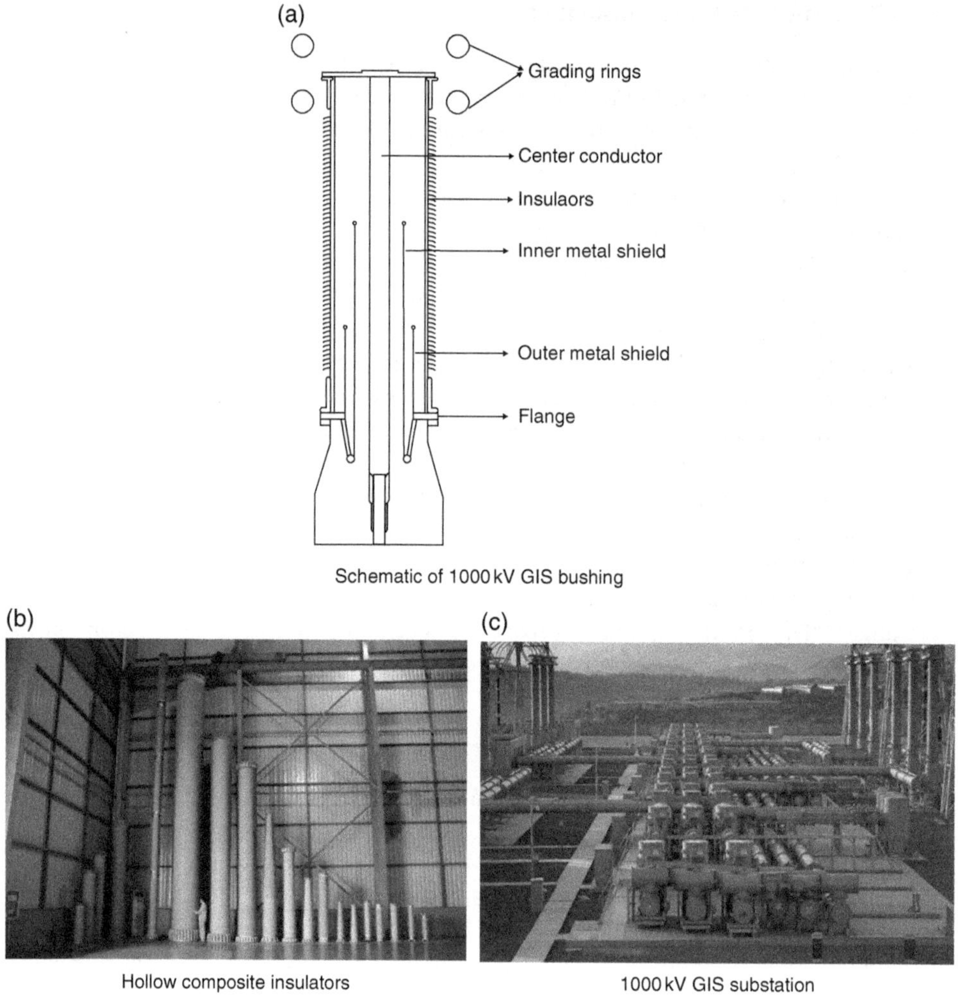

Figure 13.23 1000kV GIS layout of UHV AC power system from Zhebei to Fuzhou. (a) Schematic of 1000 kV GIS bushing, (b) Hollow composite insulators, (c) 1000 kV GIS substation.

in Figure 13.26 that the unibody crossarm is inclined by 30 degrees, indicating a particular aesthetic appeal. And to sum it all up, the all-composite pylon is remarkable in the following ways.

The pylon holds two circuits of 400 kV alternating current wires. The pylon includes a built-in crossarm with no insulator strings. In other words, the one-piece crossarm functions as an insulator, electrically isolating three-phase wires while simultaneously providing mechanical support. To manage the extraordinary moment generated by the one-piece crossarm, the crossarm insulator is made of lightweight composite materials. It's because of this that the pylon is now referred to as a composite transmission pylon. Polygonal towers can be constructed from a variety of materials, such as a concrete mix, or steel, or aluminum. The unibody crossarm is wired up by connecting all of the wires to the appropriate

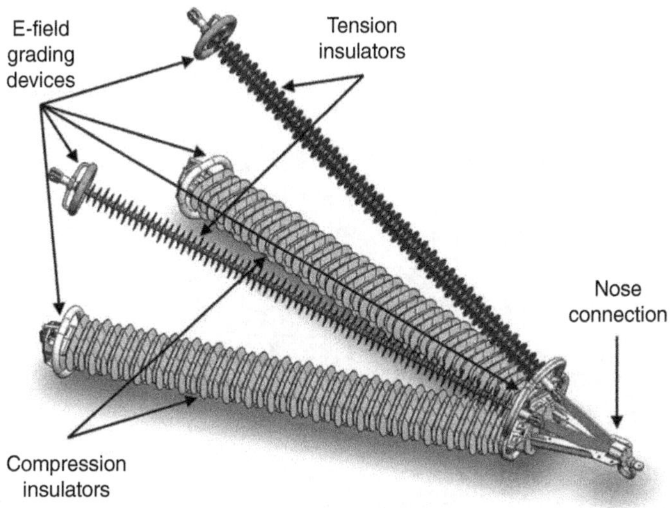

Figure 13.24 The insulating crossarm. It consists of four insulating members, end fittings, electric-field grading devices, and a nose connection for the attachment of the conductor.

conductor clamps. The arcing distances are the same as the pylon's stage distances, making them comparable. Two shield wires are held in place by clamps on the unibody crossarm's tips. The shielding angle of the pylon is therefore negative. Because of its peculiar and unique structure, the completely composite pylon stands at just 22.5 m tall, far lower than typical lattice pylons of the same voltage level (see Figure 13.26). A smaller transmission channel is needed for the pylon compared to its competitors. The pylon has a span length of 250 meters. The pylon's modest form allows it to blend into the environment and have a minimum apparent effect on residences nearby. As a result, it is likely to acquire popular support more quickly. The project is being carried out in conjunction with five well-established European partners, each of whom has different qualifications and years of knowledge. BYSTRUP, Valmont SM, Reinhausen Power Composites, Nexans, and Dervaux are among the partners.

The 750 kV composite tower with a height of 62 m is a single-circuit straight tower, as depicted in Figure 13.27a. It employs the local insulation system, which means that the tower's body is composed of steel, and the crossarm is primarily built of composite materials. The tapered composite crossarm is made up primarily of tension insulators (long and short), post insulators, intermediate flanges, and fittings. The bottom part of the crossarm, which acts as compression members, is primarily made up of two V-shaped post insulators that are horizontally positioned. An intermediate flange splits them into two sections. The two middle flanges are connected by an additional post insulator. The crossarm's upper half is made up of tension insulators and serves as a tension member. Connector flanges are used to join two different parts of the tower. Figure 13.27b depicts the insulators' chemical makeup. The post insulator comprises three layers. The interior layer is filled with polyurethane. A hollow insulating tube lines the inside of the sheath, while a silicone rubber sheath covers the outside. There are two layers to a tension insulator. The inner layer is made up of a core rod, while the outside layer is made up of silicone rubber

Figure 13.25 Assembled 132 kV ICA (left) and three ICAs installed on one circuit of a 132 kV Scottish and Southern Energy Power Distribution line (right).

(a)

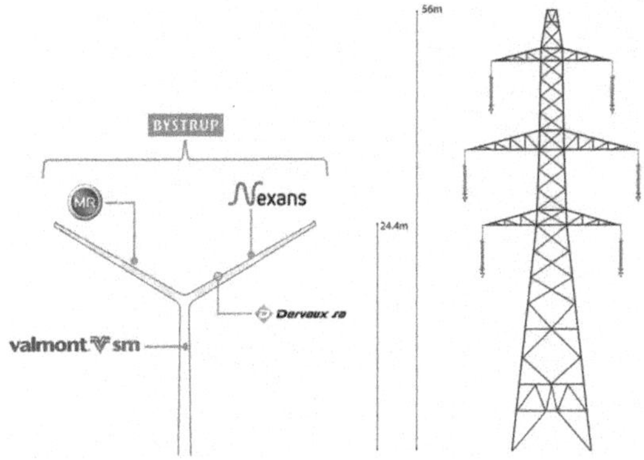

Comparison between T-Pylon and traditional tower

(b)

Landscape with T-Pylon

Figure 13.26 T-Pylon for 400 kV composite tower. (a) Comparison between T-Pylon and traditional tower, (b) Landscape with T-Pylon.

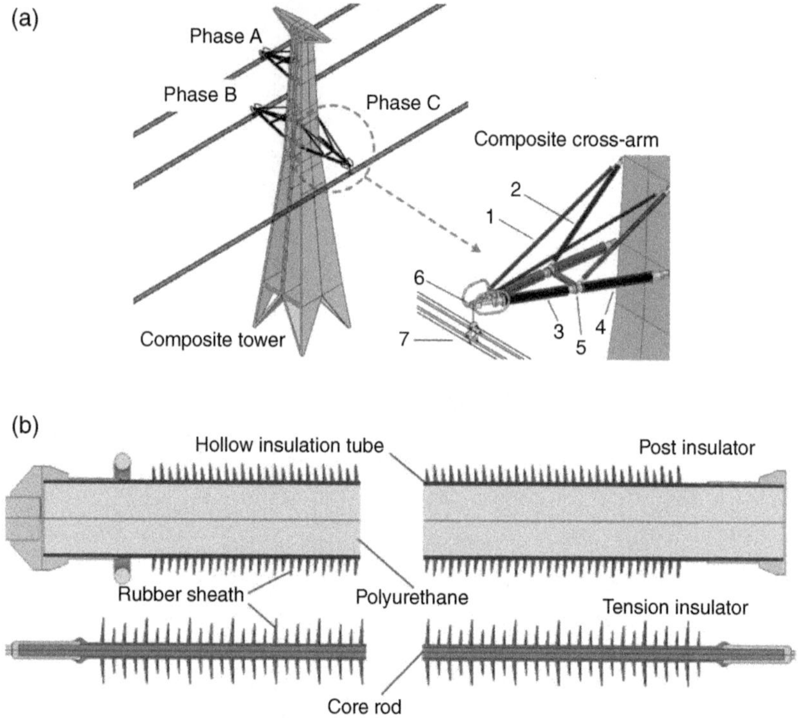

Figure 13.27 AC750 kV composite tower (a) 3D model of a 750 kV composite tower and its composite crossarm. 1-long-tension insulator, 2-short-tension insulator, 3-the 1st section of the post insulator, 4-the 2nd section of the post insulator, 5-intermediate flange, 6-end fittings, 7-yoke plate. (b) Constitution of a post insulator and a tension insulator.

Figure 13.28 Composite crossarm of 750 kV AC overhead transmission line.

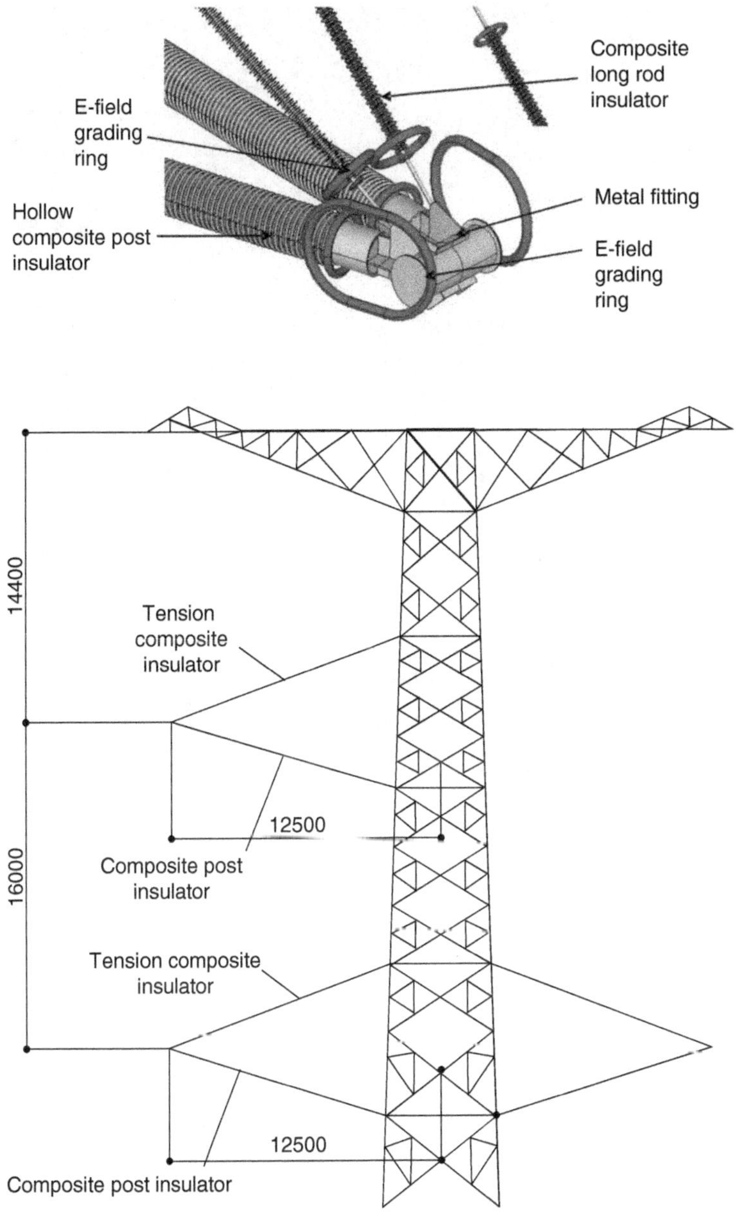

Figure 13.29 1000 kV composite tower and its composite crossarm.

sheath. The real picture of composite crossarm of 750 kV AC overhead transmission line is shown in Figure 13.28.

The tower of the UHV transmission line is high and the load is large. So, the tower needs to have enough stiffness. Although the composite material is light and high-strength, its elastic modulus is low. If the composite material is used in the whole tower of the self-supporting tower, the deformation of the tower will be too large to meet the requirements of the design specification. Therefore, for UHV transmission line tower, only the crossarm

Figure 13.30 Real picture of composite crossarm of 1000 kV AC overhead transmission line.

is made of composite materials, and the tower body is still made of steel, which can make full use of the advantages of composite materials and steel, and optimize the overall cost of the tower. When the composite material is used in transmission line engineering, it should have the following characteristics: good electrical insulation performance; high specific strength; good impact resistance; and moderate cost and specific modulus. According to the performance requirements of composite materials in power transmission line engineering, glass fiber-reinforced composite pipe is more suitable, and the resin matrix composite with glass fiber as reinforcement material is referred to as "glass fiber-reinforced plastic." The 1000 kV AC composite tower and its composite crossarm are shown in Figure 13.29, and its real picture is shown in Figure 13.30.

References

1 Rajak, D.K., Pagar, D.D., Menezes, P.L. et al. (2019). Fiber-reinforced polymer composites: manufacturing, properties, and applications. *Polymers* 11 (10): 1667.

2 Sathishkumar, T.P., Satheeshkumar, S., and Naveen, J. (2014). Glass fiber-reinforced polymer composites – a review. *Journal of Reinforced Plastics and Composites* 33 (13): 1258–1275.

3 Prashanth, S., Km, S., Nithin, K. et al. (2017). Fiber reinforced composites – a review. *Journal of Materials Science and Engineering* 6 (3): 1000341.
4 Ling, C., Fangfang, W., Heng, M.A. et al. (2016). Applications of FRP in electric power network. *Science & Technology Review* 34 (8): 77–83.
5 Composite Materials Handbook, Volume 1. Polymer Matrix Composites Guidelines for Characterization of Structural Matertials, MIL-HDBK-17-2F, Volume 1 of 5, 17 June 2002.
6 Composite Materials Handbook, Volume 2. Polymer Matrix Composites Materials Properties, MIL-HDBK-17-2F, Volume 2 of 5, 17 June 2002.
7 Composite Materials Handbook, Volume 3. Polymer Matrix Composites Materials Usage, Design, and Analysis, MIL-HDBK-17-3F, Volume 3 of 5, 17 June 2002.
8 ISO 2559-2011, Textile glass – mats (made from chopped or continuous strands) – designation and basis for specifications.
9 ISO 2797-2017, Textile glass – rovings – basis for a specification.
10 ISO 8516-2011, Textile glass – textured yarns – basis for a specification.
11 ISO 3598-2011, Textile glass – yarns – basis for a specification.
12 ISO 2113-1996, Reinforcement fibers – woven fabrics – basis for a specification.
13 ISO 2078 AMD 1-2015, Textile glass – yarns – designation; amendment 1.
14 Huazhen, W., Hengchun, L., and Yulong, Z. (2017). *The Technology of Composite Materials*. Beijing: Chemical Industry Press.
15 IEC 61212-3-1-2013 Insulating materials – industrial rigid round laminated tubes and rods based on thermosetting resins for electrical purposes – part 3: specifications for individual materials – sheet 1: round laminated rolled tubes (Edition 3.0).
16 IEC 61212-3-2-2013 Insulating materials – industrial rigid round laminated tubes and rods based on thermosetting resins for electrical purposes – part 3: specifications for individual materials – sheet 2: round laminated moulded tubes (Edition 3.0).
17 IEC 60893-1-2004, Insulating materials – industrial rigid laminated sheets based on thermosetting resins for electrical purposes – part 1: definitions, designations and general requirements.
18 IEC 60893-3-1-2012 Insulating materials – industrial rigid laminated sheets based on thermosetting resins for electrical purposes – part 3-1: specifications for individual materials – types of industrial rigid laminated sheets (Edition 3.0).
19 IEC 60893-3-2-2011 Insulating materials – industrial rigid laminated sheets based on thermosetting resins for electrical purposes – part 3-2: specifications for individual materials – requirements for rigid laminated sheets based on epoxy resins (Edition 2.1).
20 T/CEC xxx-2020, Specification for insulated tension pole used in 550 kV and above AC gas-insulated metal-enclosed switchgear.
21 T/CEC 188-2018, Insulated tension pole for 550 kV and lower gas insulated swichgear (GIS).
22 Bin, M.A. (2006). Implement and design of hollow composite insulators used in AC 1100 kV GIS filled with SF6. *Power System Technology* 30 (12): 98–101.
23 Yun, C., Pengfei, Z., Boyuan, C.U.I. et al. (2019). Application status of insulation pull rod for 1100 kV GIS circuit breaker. *High Voltage Engineering* 45 (9): 2699–2706.
24 Alawar, A., Bosze, E.J., and Nutt, S.R. (2005). A composite core conductor for low sag at high temperatures. *IEEE Transactions on Power Delivery* 20 (3): 2193–2199.

25 Zachariades, C., Rowland, S.M., Cotton, I. et al. (2016). Development of electric-field stress control devices for a 132 kV insulating crossarm using finite-element analysis. *IEEE Transactions on Power Delivery* 31 (5): 2105–2113.
26 Sarmento, M. and Lacoursiere, B.A. (2006). State of the art overview: composite utility poles for distribution and transmission applications. IEEE 2006 IEEE/PES Transmission & Distribution Conference and Exposition, Latin America – Caracas, Venezuela, 15 August 2006–18 August 2006.
27 Jahangiri, T., Wang, Q., da Silva, F.F., and Bak, C.L. (2020). *Electrical Design of a 400 kV Composite Tower*. Springer.
28 Wang, Q., Yang, X., Li, H. et al. (2016). Computation and analysis of the electric field distribution and voltage-sharing characteristics for 1000 kV composite tower. *2016 IEEE International Conference on Dielectrics (ICD)*, 1048–1051.
29 Jahangiri, T., Wang, Q., Bak, C.L. et al. (2017). Electric stress computations for designing a novel unibody composite crossarm using finite element method. *IEEE Transactions on Dielectrics and Electrical Insulation* 24 (6): 3567–3577.
30 Yuxi, Z., Gengxin, X., Qiang, S. et al. (2013). Research on the design of composite cross arm for 750 kV transmission line. *Power System and Clean Energy* 29 (1): 1–8.
31 Zheng, N., Zhu, A., Wang, Z. et al. (2013). Progress and prospect of application of composite external insulation in the 750 kV power transmission project. *Power System and Clean Energy* 29 (12): 1–10.
32 IEC 61462-2007, Composite hollow insulators – pressurized and unpressurized insulators for use in electrical equipment with rated voltage greater than 1000 V – definitions, test methods, acceptance criteria and design recommendations.
33 GB/T 24840-2018, Technical specification for bushings of 1000 kV AC system.
34 DL/T 1181-2012, Technical specification of 1000 kV AC composite long rod insulators.
35 DL/T 1277-2013, Technical specification of 1100 kV AC composite hollow insulators.
36 Q/GDW 11124.1-2013, Technical code for design of composite crossarm to 750 kV overhead transmission line part 1: design technology.
37 Q/GDW 11714.1-2017, Composite cross arm of 1000 kV AC overhead transmission line part 1: design technology.
38 Yang, X., Li, N., Peng, Z. et al. (2014). Potential distribution computation and structure optimization for composite crossarms in 750 kV AC transmission line. *IEEE Transactions on Dielectrics and Electrical Insulation* 21 (4): 1660–1669.
39 Yang, X., Wang, Q., Wang, H. et al. (2016). Transient electric field computation for composite crossarm in 750 kV AC transmission line under lightning impulse voltage. *IEEE Transactions on Dielectrics and Electrical Insulation* 23 (4): 1942–1950.
40 Zhang, S., Wu, H., Li, H. et al. (2015). Electric field calculation and shield optimization for 1000 kV GIS bushing. IEEE International Conference on Properties & Applications of Dielectric Materials.
41 Peng, L., Hong-Ling, Z., Shi-Ling, Z. et al. (2013). Optimization on the internal insulation system of gas-insulated bushing for 1100-kV GIS based on multiobjective optimization method. *IEEJ Transactions on Electrical and Electronic Engineering* 8 (S1): S20–S29.
42 Yuan, Z., Xinqiang, Y., Feng, L. et al. (2018). Key technology research of GFRP cross arm tower of AC UHV project on the transmission lines. *High Voltage Engineering* 44 (7): 2402–2409.

43 Xiao-lu, B., Yu-ze, L., Wen-xun, L. et al. (2020). Research on the application of composite cross arm for 1000 kV ultra high voltage AC transmission line. *Electric Power Servey & Design* 5 (1): 120–126.

44 Steve Dewdney, TE Connectivity'S Solutions for Harsh Environments-HV and EHV HTV Composite Hollow Core Insulators, technical brochure, TE Connectivity, 2017.

45 http://www.taiwanglassgroup.cn

46 https://www.rspoles.com

47 https://www.hitachiabb-powergrids.com

48 https://library.e.abb.com

49 https://www.te.com

50 https://www.powerpylons.com

Index

Polymer Composites for Electrical Engineering, First Edition. Edited by Xingyi Huang and Toshikatsu Tanaka.